Solid State Development and Processing of Pharmaceutical Molecules

Methods and Principles in Medicinal Chemistry

Bachhav, Y. (Ed.)

Innovative Dosage Forms
Designs and Development at Early Stage

2019
ISBN: 978-3-527-34396-6
Vol. 76

Gervasio, F. L., Spiwok, V. (Eds.)

Biomolecular Simulations in Structure-based Drug Discovery

2018
ISBN: 978-3-527-34265-5
Vol. 75

Sippl, W., Jung, M. (Eds.)

Epigenetic Drug Discovery

2018
ISBN: 978-3-527-34314-0
Vol. 74

Giordanetto, F. (Ed.)

Early Drug Development

2018
ISBN: 978-3-527-34149-8
Vol. 73

Handler, N., Buschmann, H. (Eds.)

Drug Selectivity

2017
ISBN: 978-3-527-33538-1
Vol. 72

Vaughan, T., Osbourn, J., Jalla, B. (Eds.)

Protein Therapeutics

2017
ISBN: 978-3-527-34086-6
Vol. 71

Ecker, G. F., Clausen, R. P., and Sitte, H. H. (Eds.)

Transporters as Drug Targets

2017
ISBN: 978-3-527-33384-4
Vol. 70

Martic-Kehl, M. I., Schubiger, P. A. (Eds.)

Animal Models for Human Cancer
Discovery and Development of Novel Therapeutics

2017
ISBN: 978-3-527-33997-6
Vol. 69

Solid State Development and Processing of Pharmaceutical Molecules

Salts, Cocrystals, and Polymorphism

Volume 79

Edited by
Michael Gruss

WILEY-VCH

Volume Editor

Dr. Michael Gruß
Solid State Concepts
Hermannstr. 8
52062 Aachen
Germany

Series Editors

Prof. Dr. Raimund Mannhold
Rosenweg 7
40489 Düsseldorf
Germany

Dr. Helmut Buschmann
Sperberweg 15
52076 Aachen
Germany

Dr. Jörg Holenz
GRT Therapeutics
Grunenthal Boston Pain Innovation
Hub1 Broadway
Cambridge MA 02142
United States

Cover Illustration: Gunther Schulz using iStock #17264342 / alwyncooper

Library of Congress Card No.:
applied for

British Library Cataloguing-in-Publication Data
A catalogue record for this book is available from the British Library.

Bibliographic information published by the Deutsche Nationalbibliothek
The Deutsche Nationalbibliothek lists this publication in the Deutsche Nationalbibliografie; detailed bibliographic data are available on the Internet at <http://dnb.d-nb.de>.

Print ISBN: 978-3-527-34635-6
ePDF ISBN: 978-3-527-82305-5
ePub ISBN: 978-3-527-82306-2
oBook ISBN: 978-3-527-82304-8

Cover Design SCHULZ Grafik-Design, Fußgönheim, Germany
Typesetting Straive, Chennai, India
Druck und bindung CPI Group (UK) Ltd, Croydon, CR0 4YY

Printed on acid-free paper

C094817_210821

Contents

Series Editors Preface

Michael Gruß works as an independent scientific consultant for the pharmaceutical, chemical, and nutrition industry, and he is known and well recognized as an expert in this field. During his industrial career, he conducted solid-state investigations on pharmaceutical drug substances and intermediates. He is author or co-author of more than 15 patent applications in the field of salts, cocrystals, and polymorphs, and some of the drugs are globally marketed.

Polymorphism presents opportunities as well as challenges. Investigation of the properties of different forms of a commercial drug can lead to new products with improved onset time, greater bioavailability, sustained release properties, or other therapeutic enhancements. New forms can bring improvements in manufacturing costs or API purity.

The potential for solid form variation does not end at API production. Solid form issues remain through formulation, manufacture, storage, and use of drug product. It is common to observe form transformation during standard manufacturing operations like wet granulation and milling. Excipient interactions and compaction can induce form changes. Changes can occur in the final dosage form over time.

Whenever there is a specification failure in drug product or drug substance, solid form changes should be considered in the search for causes. Particularly, symptomatic is failure to meet melting point or dissolution specifications. Changes in humidity, crystallization conditions, or crystallization solvent can produce unwanted forms.

Due to the importance of solid-phase characteristics in pharmaceutical industry, several brilliant review articles and books have been published.

The current volume entitled *Solid-State Development and Processing of Pharmaceutical Molecules* in our book series *Methods and Principles in Medicinal Chemistry* is focused on specific aspects of salts, cocrystals, and polymorphism in drug development. The book compiles general concepts and their latest applications along with well-selected case studies considering experiences and publications from the last 10 to 12 years. It considers new developments in science and knowledge that has been contributed by academic and industrial researchers during the last decade.

This book clearly goes beyond the classical description of solid-state properties. It encourages out-of-the-box thinking and allows a look beyond the science. In defining the scope of the book, one intention was to allow the identification of beneficial synergies for successful integration of solid-state form development workflows into general management strategies for R&D and production.

The editors would like to thank Michael and all contributing authors for such a successful compilation, which will stimulate the awareness of the impact of solid-state-related topics and enhance the importance of salts, cocrystals, and polymorphic forms throughout the pharmaceutical value chain.

Aachen, Düsseldorf and Boston
July 2021

Helmut Buschmann
Raimund Mannhold
Jörg Holenz

Preface

"[...] Cuius rei demonstrationem mirabilem sane detexi. Hanc marginis exiguitas non caperet." Pierre de Fermat wrote a note that states that no three positive integers a, b, and c satisfy the equation $a^n + b^n = c^n$ for any integer value of n greater than 2 in a copy of his book of the Arithmetica of Diophantos of Alexandria. The quote translates to "I have discovered a truly remarkable proof of this theorem which this margin is too small to contain."

The present edition of "Solid-State Development and Processing of Pharmaceutical molecules" may be considered as a margin that is likewise too small to cover all aspects of relevance and interest in this vast field addressed in the title. However, it is an attempt to collect and consider at least some aspects which are important from an industrial perspective on developing and processing crystalline forms in the pharmaceutical value chain.

Being a topic of some interest already since many years, solid-state investigation of crystalline pharmaceuticals gained more and more attention and commercial relevance in the last 20–25 years. Nevertheless, approaches to cover the challenges and obstacles coming along with the peculiarities of crystalline pharmaceutical molecules and their development and processing are manifold. Although sophisticated analytical devices are available on the market to investigate solid forms, not every project can be supported by all technical and personal intelligence. There are limitations framed by limited resources of financial funding and manpower. Eventually, no excuses count when safety and efficacy of medicines are of concern. However, if not all, but many routes lead to Rome.

Some of the routes are presented in this edition. Readers are called to wisely select the right means and measures for their particular projects. Besides the essential and unavoidable basics to reach the goal, there are add-ons and nice-to-haves that may shorten the route to process understanding and market entrance.

Starting with important general considerations in Chapter 1, the subsequent chapters lighten the road along solid-state development and processing. Chapter 2 describes entry points into a solid-state-related project. Investigation of the polymorphic landscape is exemplified in Chapter 3 and complemented by Chapter 4 which is divided into 10 subchapters describing various solid-state characterization techniques. Treatment of solid pharmaceutical drug substances upon scale-up is covered in Chapter 5, while Chapter 6 considers the challenges for Drug Product development and manufacturing. From my perspective often neglected but essential support processes addressing simplification of daily routines, best practices and documentation are described in Chapters 7 and 8. Chapter 9 covers ensuring the return of investment by the protection of intellectual property, while Chapter 10 addresses the regulatory aspects of relevance. Last but not least, the perspective of extension of the value chain of pharmaceutical molecules is given in Chapter 11.

Clearly, without all the authors who accepted to contribute, this edition would never have been possible. Thank you a lot. You made this book to a valuable collection of practical knowledge and expertise. It was my pleasure and honor to cooperate with you. Sincere thanks to Wiley-VCH, namely Dr. Frank Weinreich, Stefanie Volk, and N. Kiruthigadevi for answering my questions and acting behind the scenes to drive this publication, as well as Felix Bloeck who worked with the graphic designers on the book cover. Finally, I would especially like to thank Dr. Helmut Buschmann for his everlasting supportive initiative to motivate and develop people.

Aachen
July 2021

Michael Gruss

1

Aspects for Developing and Processing Solid Forms

Michael Gruss

Dr. Michael Gruß – Solid State Concepts, Hermannstr. 8, 5062 Aachen, Germany

1.1 Aspects for Developing and Processing Solid Forms

1.1.1 Introduction

Due to progressing through time and space, we constantly learn and forget. We lose things out of sight and focus on the ones that are of most importance and interest. Consequently, we cannot keep pace on every single field of technology and science. Career pathways are manifold. It is fairly natural that people in management positions who take over the responsibility to direct companies, departments, or groups cannot be experts in all the domains falling under their responsibility. Therefore, they have to educate themselves and rely on their staff, employees, team members, suppliers, contract organizations and consultants, and the assessments delivered. Decisions are made on such a base. Those decisions determine the commercial fate of the business, which in turn determines the well-being of those who gave input into the decisions. A circle of life? Will just the fittest survive?

1.1.2 Education and Personal Background

Some lack of knowledge and understanding of impact and relations may occur especially in fields that are typically not in the focus of a general chemical, medicinal, or pharmaceutical curriculum. Organic compounds constitute the majority of active pharmaceutical ingredients (API) very often in the form of crystalline solids. Nevertheless, solid-state-related topics for organic compounds are treated in introductory organic synthetic textbooks, like the *Organikum* [1], just on a few pages that do not go much into the details. Crystallization (including selection of the solvent, recrystallization, and crystallization from the melt) is explained in two pages, and structure analysis by means of X-ray is mentioned in another page.

Besides, still a predominant perception of solid-state characterization techniques, in particular X-ray powder diffraction (XRPD), is that the investigations are expensive. Admittedly, XRPD is not as widely distributed and readily accessible as spectroscopic and chromatographic techniques. Consequently, solid compounds

Solid State Development and Processing of Pharmaceutical Molecules: Salts, Cocrystals, and Polymorphism, First Edition. Edited by Michael Gruss.

and dosage forms are primarily characterized by the analytical techniques that are easily available. It is perfect to assess purity profiles and determine the solubility and dissolution profiles. Unfortunately, it is not sufficient to analyze the liquid state because that does not reveal much about the properties of the material in the solid state. Additional processing knowledge is also needed to design and modify solid-state properties.

Our current position, our educational background, as well as the social and technical environment that surrounds us determine the perception of threads and opportunities.

As long as during chemical development or production, solid-state-caused obstacles can be overcome by some, maybe magic and not really understood, measures that everything is fine or, more precisely, appears to be fine, at least for the moment. No further resources, time, and money are invested to understand the cause–effect relations. How long will this satisfaction for saving money last? Who is eventually paying the bill for lack of thoroughly understanding the processes and interdependencies? The advice is to implement solid-state experts into CMC or other development and processing teams. Taking the advantage and benefit from the different perspective, they can add to discussions and innovations.

In the past 20–30 years, solid-state development became more and more important in the pharmaceutical industry. Many treatises in print and online cover a broad variety of aspects that can be subsumed under the roof of solid-state development. This concerns not just molecules that are API but also many other compounds classified as fine chemicals, agrochemicals, explosives, or those having a relevance for nutrition products. The eye-catching word is mainly "polymorphism" that, along with similar terms like "polymorph" and "pseudo-polymorph" or terms often discussed in the context like "hydrate", "solvate", "salt", "cocrystal", "co-crystal", or "amorphous", was and still is worthwhile to cause attention.

The attention culminated to a hype that has today become a scientific and commercial important field of sound investigation to ensure proper development and subsequent successful marketability of products in general. However, in the world of pharmaceuticals, the interest is beyond commercial aspects related to ensuring safety and efficacy of the products dedicated to reduce suffering from diseases or curing patients.

Consequently, many stakeholders are involved in solid-state research, development, manufacturing, and commercialization. But it is also the other way round. Chemists and pharmacists who are active in the field of development and processing of solid compounds have lots of interfaces to other departments from which they get or to whom they deliver information and materials as exemplified for a crystallization laboratory in Figure 1.1.

Obviously, every single discipline can consider itself as the most important and thus justifies its position in the center of this representative arrangement. Actually, the center of Figure 1.1 only represents the point of view and maybe the self-conception. Solid-state development, particularly for pharmaceutical applications, is a complex and ever-changing setup involving and needing lots of disciplines for successfully mastering the development and manufacturing workflows.

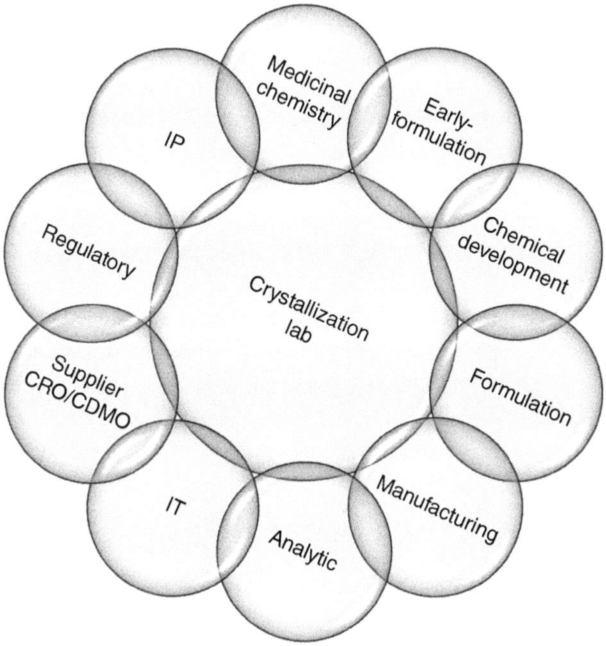

Figure 1.1 The crystallization laboratory – integrated in pharmaceutical development and manufacturing.

Looking behind the scenes, stepping somewhat back from the science but without neglecting the importance of a sound understanding of the basics and the implementation of solid-state-related processes is the intention of the following chapters constituting this edition. Every chapter was contributed by someone who is an expert in his or her particular field, someone who illuminates the aspects of his or her domain with the awareness of being a part of the whole. While reading the chapters, consider its topic as standing in the center of the arrangement (Figure 1.1). Development and processing of solid compounds and forms is apparently enlightened from different perspectives. Therefore, some aspects are covered not just by one author. When there is light, then there is shadow. Not every aspect can be treated, especially because the chapters were intentionally written from a subjective standpoint and perspective. Personal preferences, experiences, and peculiarities directed the content. Other perspectives and considerations may well exist.

Solid-state development and processing is nothing that can be handled as an isolated aspect of pharmaceutical R&D or manufacturing, although a lack of general understanding of the foundations, disregarding the essential impact the solid state has on process and eventually product quality, is surprisingly still around in the industries.

This is understandable from one perspective. Dedicated experts are typically required to address the many topics showing up in the course of pharmaceutical research and particular pharmaceutical development and manufacturing. All of them are well trained and have skills in their particular sciences or businesses.

Education at university is focusing on the formation of domain experts. As a drawback, less time is typically available to look to the left and to the right, forward and beyond of the own field of expertise.

As a consequence, all those experts eventually involved in versatile R&D and manufacturing teams have the duty to ensure that their colleagues (or "interfaces") also get an understanding of the impact of their particular field of expertise for the whole process and vice versa. All participants in R&D and manufacturing processes have the obligation to accentuate the need and significance of the topics addressed by themselves and their laboratories or departments. There is a necessity for people having the capability to act as coaches, teachers, and trainers; devoted to their field of expertise but not trapped therein; on the contrary, open minded and willing to share knowledge. People dedicated to draft, construct, and maintain their part of the development and production, as well as of business processes.

Getting the knowledge out of the heads. What is important to share? What should others know about your business? Besides sciences, what is also important to communicate to make industrial and commercially oriented environments work?

The intention is to cover the most essential topics in industrial, mainly pharmaceutical, environments that are related to the manifold of properties and characteristics solid compounds have.

1.1.3 Societal Impact – Fishing in Foreign Waters

The societal impact or the impact of society – human being are creating the society they live in and individuals are formed by the society they live in. Dealing with societal, economical, and historical impact on science and business is typically not a field harvested by a solid-state expert. Other businesses and sciences treating sociological relations, management, or national economy usually consider topics like this. Professionals in these fields have experiences, tools, and know the sources for research and how to investigate developments and interrelations. Therefore, the following is layman's reflection on societal aspects.

1.1.3.1 Motivation

Why trying to cover this topic? First, because it is interesting. Second, it has some relevance to start with this consideration now. As long as individuals are present who eyewitnessed or even designed the initial pharmaceutical solid-state development, or worked on the implementation of the current status, it is possible to get insights not just based on figures and statistics. They can talk and report about motivations, desires, challenges, dead ends, hopes, and obstacles. The dimension of personal experiences and observations is important and worth to be told.

A comparison might help. It is a bit as if a chemist, well educated and trained in organic chemistry, attempts to cover topics related to solid-state aspects, like designing a crystallization process or investigating a solid-state landscape. Why are not the experts doing the job? Answers are manifold: "Not a big task, let's just have a look", or there is no solid-state expert around, or the expert is packed with other projects, or nobody is aware that a solid-state expert would probably have a different perspective and appropriate education, better tools, and experiences to do the job.

Usually the attempt to "just let's do it" fails. It might lead to some results. Maybe that these results are good enough to decide that, probably at a first glance, no further investigation is necessary or meaningful. Time will show whether the project dies or whether it is interesting enough to follow up.

"Just let's do it" reflects the approach in this part of the section to consider mutual interaction between people or society and pharmaceutical solid-state chemistry. Unfortunately, these economical and societal aspects of pharmaceutical solid-state development and about the historical development-related interactions between society and science are not yet covered accordingly. There is a hope that in future someone will do so properly.

While reflecting this challenge with others, the response was like "who could be interested and would spent time on reading?" Maybe nobody is interested. However, there are always people acting for some years in their field of expertise and have observed and contributed to changes. Certainly, these individuals are curious to read a summary and a reflection about the topic and fields of expertise they spent years of their lives on.

The likelihood for realizing such a project is naturally larger for widespread general topics like computer technology or popular sciences like flights to the moon, exemplified by [2–6].

The hope is that someone picks up the idea and starts with economical, societal, and historical research on this small but important niche in pharmaceutical science with more adequate means, knowledge, and resources than attempted here.

1.1.3.2 The Personal Dimension

Any action consequently leads to a reaction. Probably, every student of natural science has heard in the first physics lectures about Isaac Newton who formulated this as his third law of motion "Law III: To every action there is always opposed an equal reaction: or the mutual actions of two bodies upon each other are always equal, and directed to contrary parts" [7].

This consideration is not limited to physical bodies moving in the three-dimensional physical world. There is also always an impact in a nonmaterial sense, an impact on ideas, thoughts, and desires. This impact is the essence of advancement in general and technological progress in particular. Realization of progress is only possible because the fourth dimension, time, is also affected. Progress is not possible without a change of states in time. The qualitative and quantitative identification of change is only possible if a prior state is compared to a later state of a system or society.

Being asked by one of the series editors if I am interested to edit a book on pharmaceutical solid-state development, it took me a while to make my decision. I wondered, what can I, what can we as authors, contribute to the society of solid-state experts or community of people who strive to learn about solid-state development and processing that has not yet been reported before? Moreover, it was not just me among the authors who has asked himself this question.

Before I decided to dive into this project, my understanding was that everything of importance for investigating the solid state was already told. In this very moment, while typing these lines and editing the contributions of the other authors, I am convinced that this is certainly not the case.

For sure, progress is always made and there will always be new developments and new inventions and discoveries, as in all sciences. In addition, there will be old, sometimes forgotten, stories told in a new fashion. Actually, this is not what I had in mind.

What in particular is untold is our subjective perception and perspective. Having the opportunity to write from and about someone's subjective position in this area is an argument that is convincing to start editing and writing. The burden and chance to communicate our very personal view on the "Solid-State development and processing of pharmaceutical molecules" is appreciated. We have the chance to communicate about those aspects and topics we personally consider of being of interest and value for the community. We have the chance, and responsibility, to make a personal selection of scientific themes, topics, references, and examples that are worthwhile to address in the context of industrial tasks. We have the chance to act, to influence, and impact. This will for sure lead to reactions: affirmative or contrary. Regardless of its nature, every reaction will itself be of further impact on personal thinking and therefore on thinking of the community.

This is already a societal impact of solid-state development. People, certainly not for the first time, agreed to spend a significant part of their time to restitute partly to the society what they received from others. There were many who taught and trained them to acquire and prosper particularly specialized and sometimes singular and rare skills. The motivation that drives each individually can be manifold. It might be the chance to spread the personal view, it might be the chance to summarize and share a particular sequence of business life, it might also be to face a particular personal challenge, and it might well be the option to publish and to promote personal career. It might be something else or a mixture of the preceding. These willing authors differentiated themselves from those who objected to contribute. Some of those who rejected concluded a personal calculation that taking the time to contribute is not what they like to do or that it would not pay off for them or their business. Some questioned if anyone might read the chapter, the book. Some were stopped by illnesses, by other duties, or by supervisors who preferred seeing them to work (in a more direct manner) for the company or protected them to not burn themselves. Solid-state development and processing as well as many of the other branches in the pharmaceutical value chain with all their facets have for sure a societal impact and are impacted by society.

1.1.3.3 Beyond the Impact on Individuals

The aforementioned addresses the personal dimension of societal impact. There is another dimension, an even larger one worth to be addressed. What is the impact that the investigation of the solid state has on organizations? How do these organizations act, develop, and influence science, regulations, businesses, and consequently societies?

It was Albert Einstein who realized that the experience of a chronological sequence, which we usually call history, depends on the perspective of the observer and whether events appear to be (i.e. are) simultaneous or one after the other. Apart from this relativistic point of view on events, the speed of experiencing and learning for human beings is usually chronological. "Standing on the shoulder of giants", often attributed to Isaac Newton, but going back at least to the twelfth century [8], summarizes how the accumulation of knowledge and progress establishes. Individuals learn from and build on the experiences of others. The essence of enterprises, often praised, sometimes forgotten (as reflected by the term human resources), is the sum of all the individuals working for these companies. One way to reflect on their collective achievements is that economical figures illustrate the financial value of a company. Besides, there is also a social value. There are examples where the social value of an organization correlates with the economic value [9].

1.1.3.4 Understanding the Market – Not an Easy Task

In particular, the pharmaceutical industry in Europe itself, represented by the European Federation of Pharmaceutical Industries and Associations (EFPIA), annually presents their economic and societal contribution [10, 11].

There are other potential sources for data, like statistical offices. Unfortunately, figures are usually not broken down to an extent where the impact of solid-state-related activities becomes visible. This same issue occurs with official statistical data acquired, for example, by the German statistical office (Statistisches Bundesamt, DESTATIS) as published in the GENESIS-online database (https://www-genesis.destatis.de) or the European Commission as published in the Eurostat database (https://ec.europa.eu/eurostat/web/science-technology-innovation/data/database).

Because the categories are very generally defined (see Table 1.1), no information can be derived and assigned particularly to the impact of such a niche like "solid-state activities" on turnover or number of employees.

Even harder it is to get corporate data broken down to the affect and effect of pharmaceutical solid-state activities. According to national laws, companies may have

Table 1.1 Categories in the German DESTATIS GENESIS-online database (as of 2019-08-14).

Code	Content (category)	Translation of content
WZ08-72	Forschung und Entwicklung	Research and Development
WZ08-721	Forschg.u.Entwicklg. in Natur-u.ä. Wissenschaften	Research and Development in natural and similar sciences
WZ08-7211	Forschung und Entwicklung in Biotechnologie	Research and Development in biotechnology
WZ08-7219	Sonstige Forschg.u.Entwicklg. von Naturwiss. u.Ä.	Other Research and Development in natural and similar sciences
WZ08-74	Sonst. freiberufl.,wissenschaftl. u. techn.Tätigk.	Other self-employed, scientific, and technical activities

the obligation to publish their financial results. In Germany, e.g. public companies (AG) listed on the stock exchange and companies with limited liability (GmbH) disclose their accounting documents on a yearly basis to the operator of the Federal Gazette (i.e. Bundesanzeiger) or deposit them in the business register (i.e. Handelsregister) [12]. The data can be accessed via the Internet [13].

Following this approach to get insights into the historical development of solid-state-related businesses, it has to be taken into account that the level of detail is quite coarse as it lists the cumulative financial results of the enterprises per year. Therefore, possibly none of the financial figures can be attributed to solid-state departments or even smaller working groups being just a part of a bigger chemical or pharmaceutical company.

Nevertheless, if the business of a company is mainly focused on solid-state-related services, the historical financial figures of this company can be considered, with some approximation, to reflect the development of solid-state services over the years. Maybe not as a representative *pars pro toto* but at least as an example.

On the German market, the company Solid-Chem GmbH, located in Bochum, is one of these focused companies. According to the company philosophy [14], they investigate the solid-state chemistry of products and offer consulting and scientific support with also covering aspects dealing with drug products (DPs). Data about the company are available in the Bundesanzeiger. The company may serve, at least for the period of data published, as an example that reflects, based on its financial results, the necessity and importance of solid-state investigations and services for the chemical and pharmaceutical industry. Unfortunately, data published in the Bundesanzeiger are not as comprehensive and telling as one might wish. In general, nothing is said about the number of employees and the wages paid nor is there information on the types of projects.

Another indication for the increased interest in pharmaceutical solid-state activities is the acquisition of companies formerly specialized in the investigations of solid-state topics and their implementation into larger contract development and manufacturing organizations (CDMO), exemplified by

- SSCI, originally founded in 1991, acquired in 2006 by Aptuit, and since 2015 a division of Albany Molecular Research (AMRI).
- Crystallics, originally founded in 2000, acquired by Ardena in 2016.
  In the years between, the solid-state facilities were part of several splits and mergers.
- Pharmorphix, originally founded in 2003, acquired by Johnson Mathey in 2015.
- Solid Forms Solutions, acquired by Avista Pharma early in 2018 subsequently acquired by Cambrex end of 2018.

Another observation is expansion of formerly focused expertise into service providers with an extended portfolio as well as formation of associations, as there are, for example,

- Solvias
- Crysforma, as part of ICIQ (Institut Catalá d'Investigació Quimica)

- APC
- CMAC

Although there are examples that the pharmaceutical industry divests solid-state expertise, several CDMO have established in-house solid-state expertise to support development and manufacturing but may or may not offer this expertise as a separate service for third parties.

1.1.3.5 Benefits of an Interdisciplinary Mindset

There may be other approaches to get telling data. Maybe professional data providers or chemical or pharmaceutical societies have a deeper insight into the figures of their individual or corporate members. Maybe newspapers or major business consulting companies have a database that might give more answers about the societal impact and impact on society if properly analyzed and publicly distributed. Business, literature, or patent databases may, and certainly do, serve as a source, and the contents might be crawled by data mining algorithms. Data analytical tools could eventually process the results with respect to

- publications, posters, conferences, whitepapers, webinars, and seminars
- business figures, financial data, merger & acquisitions
- applications, granted patents, application status (active, inactive)
- dates, to put everything in chronological order
- occurrence of terms in title, abstract, description, example, and claim
- terms like polymorphism, salt, cocrystal, modification, Form A … Form Z, Form I … Form XVIIIC, Form α … Form Ω, and so on
- various solid-state characterization techniques
- companies, assignees, inventors
- … and more …
- … and combinations thereof

As an example to demonstrate, the opportunities of combining several sources like patent and financial data may serve the valuation analysis for Norvasc® in which the attractiveness, i.e. sales threshold, to circumvent a secondary salt patent was estimated for products in the United States to be about $1.5 bn [15].

Such results combined with insights gained from sociological or historical research might generate new knowledge and understanding of relationships. This will certainly happen if data use rights and privacy laws allow and someone identifies a financial advantage or business case.

1.1.4 The Basis for Mutual Understanding

Terms used in science typically get a definition to provide a common base for conversation. Unfortunately, different actors think that different definitions are meaningful and correct to express the situation. Therefore, a range of definitions for terms evolve throughout time and space and a jargon is formed, while we are permanently exchanging thoughts to simplify and speed up communication and interaction.

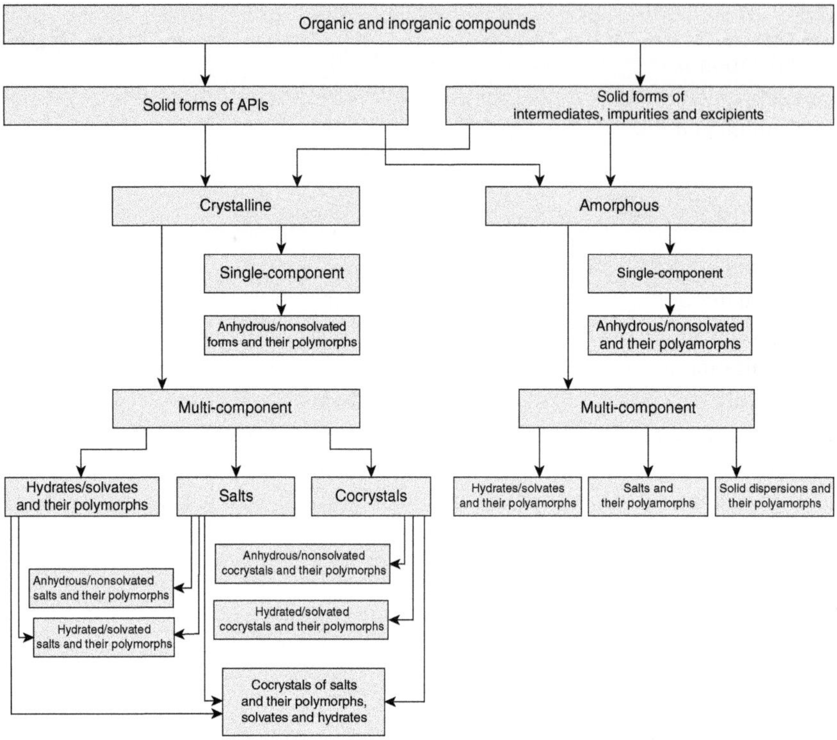

Figure 1.2 The complexity of solid compounds in pharmaceutical applications. Amended chart based on Aitipamula et al. [16].

Figure 1.2 illustrates an attempt to categorize the diversity of solid forms. Various potential combinations of solids can be formed by combining single components. In principle, the following categories apply:

- Nature of compound
- Degree of crystallinity (amorphous–crystalline; i.e. unordered–ordered)
- Number of components (single–multiple) and ratio
- Type of components (liquid–solid; aqueous–nonaqueous)
- Nature of bonding, type of interaction (ionic–π–π interaction–van der Waals)
- Number of solid forms having the same composition (polymorphism)
- Homogeneity (purity, ideality) of the solid arrangement (pure–mixed phases; solid solutions)

To make the image complete, combine elements from these categories.

A consideration of basic principles, terms, and aspects is provided and discussed throughout the years in details elsewhere, exemplified, without claiming completeness, by [17–35] and all the references mentioned therein.

The following sections and chapters emphasize some of the prerequisites and aspects to provide a base for a better understanding for solid-state aspects of small molecules and the enormous impact on chemical and pharmaceutical development and processing.

1.1.5 Crystallization is a Separation, Not a Separated Process

Crystallization is the separation process that leads to solid crystalline material and is as such, together with other separation techniques, responsible for 40–90% of the spent capital and operating costs of industrial production [36].

Why crystallization? Crystallization is an acknowledged and well-established purification step suitable to deplete impurities like by-products, unreacted reagents, catalysts, or residual solvents introduced by the preceding synthesis. Therefore, crystallization is a perfectly suited unit operation to separate valuable material from waste or material that is supposed to be used in other value chains.

Moreover, why crystalline solid material? Materials in crystalline solid state generally exhibit superior stability properties than in amorphous or liquid state. Alteration processes like decomposition and chemical reaction (e.g. oxidation) are reduced due to low mobility of the constituents compared with amorphous or liquid state.

In course of the organic synthesis and the subsequent crystallization procedure, the solid material is in the suspension still in contact with solvent and impurities from the reaction. At this stage, in the slurry, changes in the solid form may continue to happen. Thermodynamically, more stable forms can evolve from metastable modifications or solvate formation can happen since at lower temperature another region of stability might have been reached. Due to increased concentration unintended precipitation of byproducts or impurities can happen or addition compounds may form.

To further purify the desired material, typically filtration or centrifugation followed by washing of the filter cake is applied subsequently. With respect to the solid form, this should happen with some care because solvate formation or exchange of the solvate in solvated forms can take place as well as pre-drying with flow of air or inert gas may lead to desolvation of solvated forms.

Often the separation is followed by a drying routine to remove residual solvent and increase flowability of the solid material. Besides changes in particle size caused by agglomeration or breakage, also a change of the solid form may happen as described earlier. In addition, since generally thermal energy is introduced by raising the temperature level, polymorphic forms may interconvert. Reduced pressure is commonly applied, which may lead to unexpected behavior in contrast to atmospheric pressure conditions. Especially drying of solvates to a defined stage requires thorough knowledge and investigation of the phase diagram to avoid producing material that is not according to the desired outcome.

Transport and storage may expose the drug substance (DS) or other materials, like reactants, intermediates, or excipients, to conditions that can also affect the solid state. Consider temperature changes by moving drums from the production hall to a storage place that has a different temperature or exposes the material from time to time to sunlight. Besides condensation of moisture, additional drying due to temperature changes may have an effect on the polymorphic or solvate form. Furthermore, mechanical forces can result in changes of particle sizes and morphology.

Eventually a formulation process like tableting can affect the solid state of the DS as the DP is produced. The variety of formulation processes comprises multiple chances to impact the solid form of the DS. Thermal stress and mechanical forces may alter the physical form of the DS as well as bring it in contact with excipients being potential sources for moisture or water or even partners for chemical interaction. Upon storage and transport, the final DP should also not alter its properties.

This process chain described earlier [31] illustrates the embedding of the crystallization in a broader context. Although crystallization is the first step where solid properties can and should be addressed, subsequent process steps may alter and modify solid characteristics and must therefore be considered and understood as well.

Additional aspects complete the image of integrated solid-state development and processing. Figure 1.3 illustrates interrelating tasks and duties that have to be managed and handled during solid-state development and processing. It all starts with setting up the scene. Define the objective(s) that shall be reached. Organize and establish the infrastructure necessary to reach the goal. This comprises availability of suitable equipment and materials as well as experienced and well-educated staff or teams who are prepared and trained to perform the required tasks. Next is designing the experiments and/or processes. Think about what is required and necessary to reach the objective as fast and as cheap as possible with sufficient quality. Select the types of analytics suitable to monitor proceedings and confirm quality. Develop and establish analytical methods and sampling (drawing and handling) procedures. Confirm that initial thoughts were correct or adopt if necessary. Subsequently, solid formation processing sets in. Experiments to investigate and optimize or processes to reproducibly conduct crystallization and formulation as well as downstream procedures like isolation and drying have to be set up. Not to forget the installation of efficient delivery pathways to analytical operators, clients, or patients. As part of all these processes, knowledge generation happens. People and organizations learn permanently. Learning should be as effective as possible to reduce development efforts or enhance manufacturing capacity. Additional value can be generated by efficient communication between all stakeholders participating in those development and manufacturing processes. Digitalization of

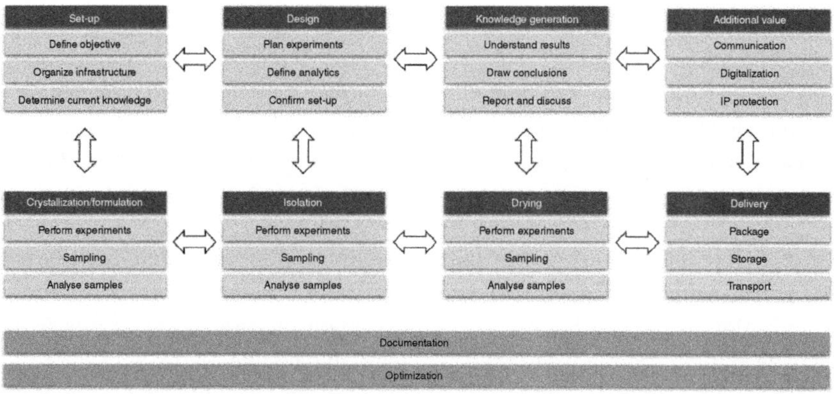

Figure 1.3 Tasks and duties for solid-state development and processing.

technical and business procedures is advised wherever meaningful and applicable to ensure effectiveness and reliability. Understanding mutual requirements and needs is important to reduce overall efforts, enhance performance, and reduce costs. Certainly, the efforts should pay back and value preserved by protecting the intellectual property (IP) generated. Along all these procedures and processes, documentation and optimization must take place. Otherwise, quality cannot be guaranteed and knowledge gets lost if not distributed within the organization while costs increase and performance stagnates.

Always keep in mind that tasks should never be considered from an isolated perspective. Interact, exchange, and attempt to resolve issues jointly. However, not necessarily at the stage where they pop-up or cause trouble. Strive to overview the complete processing chain. Try to get external input by consultancy, give internal teams the chance to reflect their achievements, and support further progress. Identify causes and effects and establish the solution where overall costs and burden are least. Probably, isolated budgets and task forces optimizing solely their particular field of responsibility are the worst approaches – at least from a scientific or an engineering point of view.

1.1.6 Some Early Information About Solid-state Properties

It may happen that synthetic chemists may not even be aware that they are describing important properties of the materials used as starting materials (reactants) for a reaction or as outcome thereof (products, by-products) in the synthesis procedure that is documented in the (electronic) laboratory notebook. This information may be very valuable initial information for a solid-state or formulation scientist when it comes to further investigation and development of the compound. Table 1.2 illustrates some types of solid-state-related information that could be extracted and questions that may arise from synthesis descriptions. As a consequence, the more detailed such a procedure is documented, the more information can be derived.

A basic training for synthetically working staff is recommended to teach them about what may be relevant information for subsequent solid-state investigations. For example, failing attempts with other solvents for extraction or some early solubility tests are worth to be documented and to be handed over when the synthesis is transferred to another lab, e.g. for resynthesis, further optimization, or scale-up. However, even more important is a clear, extensive, and unambiguous documentation that enables someone who wants to reproduce the results or wants to scale-up the reaction to do so.

1.1.7 Digitalization (Not Only) in the Laboratory

1.1.7.1 Prerequisites – Technology and People

Digitalization is more than digitization. While the latter means transferring analogue data into a digital version, i.e. scanning a piece of paper to create an electronic JPEG or PDF document, the former term "digitalization" is more comprehensive. The Gartner IT Glossary [37] defines digitalization as "the use

Table 1.2 Example for a synthesis procedure.

Description	Information on properties
A 250-ml three-necked flask was charged with 105 ml ethanol 1.30 g (56.5 mmol) and reactant **1**	Solubility of reactant **1** in ethanol is >1300 mg/105 ml, i.e. >12.4 mg/ml
After the reactant 1 was dissolved upon stirring, 4.70 g (49.9 mmol) compound 2 in 20 ml ethanol and 8.56 g (69.6 mmol) compound **3** were added.	Reaction time for dissolution (not specified) Was compound **2** dissolved or suspended?
The solution was boiled under reflux and moisture exclusion for 2 ½ hours.	Was it a solution from the beginning or dxfissolve the reactants slowly upon reaction? The reaction (or reaction product) is sensitive to hydrolysis.
The next day, the ethanol was distilled off and the residue was dissolved in 35 ml of 5% sodium hydroxide (NaOH) solution	Reaction product is stable at elevated temperature, stable under basic conditions and soluble in NaOH ($c = 5\%$) Was it a solution or suspension the other day?
The solution was extracted five times with 20 ml *tert*-butyl methyl ether (TBME) in a separating funnel	Product is soluble in TBME
Subsequently, the solvent was stripped off yielding some grams of a yellowish crystalline powder	Evaporation crystallization is possible The color may indicate that some impurities remained How has crystallinity been proved? by XRPD? Which crystal form resulted? Evaporation may lead to amorphous material or metastable polymorphs Is cooling crystallization also possible?

of digital technologies to change a business model and provide new revenue and value-producing opportunities; it is the process of moving to a digital business."

Referring the example of scanning the paper, digitalization addresses the causes and effects of doing the scan. "Scanning" could bring additional value by reducing shelf or storage room for archiving; applying object character recognition techniques to the electronic document enables searching for terms and implementing the data in more complex business process, e.g. for electronic implementation of the scanned data into an Electronic Lab Notebook (ELN).

However, looking into laboratories today, one might get the impression that digitalization is yet a challenging task, and digital transformation, as remarked by Gartner [38] "can refer to anything from IT modernization" and is indeed interpreted as anything related to IT. Initiatives by senior management [39] eventually force actions in the laboratories. However, this desirable situation finds its limitations in reality. Machine learning (ML), for example, is a topic of general interest and concern. Unfortunately, there is a lack of experiences and competences and if there are no resources or budget for data analysts or data scientist than it is advised to companies to educate employees from R&D (i.e. domain experts) in these technologies [40]. As long as this expertise has not yet entered laboratories, even implementation of basic digitalization solutions does not take place everywhere. Still, experiments are documented in paper lab notebooks and Excel sheets are (mis)used as databases and represent the foundation of the daily work in many enterprises. Basic principles of database practices (e.g. normalization) are not always applied. Information is filled into Excel sheets in a manner that would require additional, yet unnecessary workload to tidy the data for automated analysis by ML techniques. As an example, spreadsheet cells become united to look nicely or become filled with unformatted and inconsistent types of data. Hopefully unknowingly, but consequently increased efforts are necessary to extract the information automatically, yet even manually. Therefore, training in basic IT principles is essential to raise the chance to modernize, i.e. digitalize, laboratory workflows.

In the end, it is all about people, their attitude, habits, and "kingdoms." The latter refers to another important preventer of rapid and exhaustive implementation digitalization: the tendency of protecting and shielding domain areas, like synthesis, biological or analytical data, and reports from external (but in-house) inspections or accesses. Being afraid of nonexperts to misinterpret data or tear information out of the original context leads to establish protection routines, either physical or organizational barriers.

Consequently, advice for starting (or accelerating) digitalization is to identify domain experts who have an affinity and attitude toward IT topics and love to develop visions and who have the ability to transport these visions to their colleagues and encourage progress in digitalization and empower these talents to reorganize workflows, if necessary beyond borders of laboratories or departments. To prevent frustration, it is recommended to harvest the low-hanging fruits first, i.e. start small in an area, e.g. laboratory, that is under supervision of the domain expert and select primarily staff willing to follow suggestions for innovative solutions. Educate domain experts not just in technological topics, but also encourage continuous professional development of soft skills, like innovation management or leadership.

1.1.7.2 Connect Data and the Right Information from Synthesis and Analysis

Not only for solid-state experiments, there is a necessity to connect information from synthesis with analytical data. Especially in light of digitalized workflows, proper naming of samples is necessary to facilitate the interpretation of results. In industrial environments, often multiple experiments are performed in short time or even in parallel.

Awareness of which experiments shall be and were performed and at what stages samples should be or were taken as well as proper coding of the experiments and especially the samples and their characterization, i.e. the analytical results,

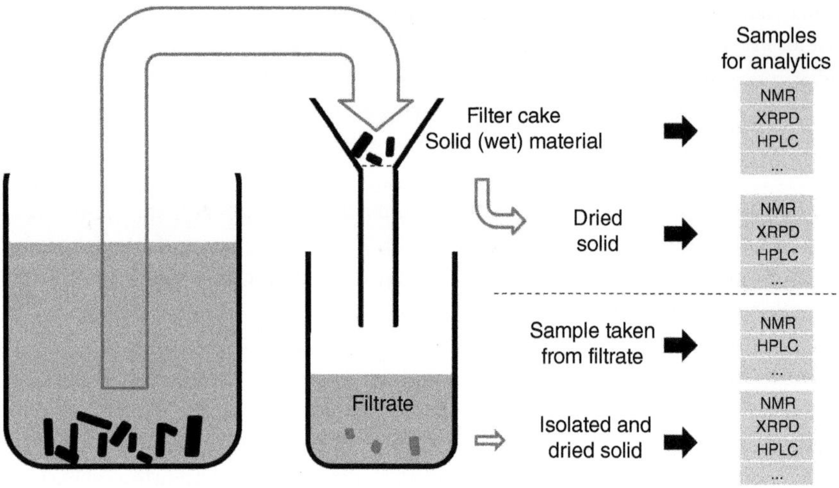

Figure 1.4 Sampling procedure for a filtration process.

is key for accurate and quick interpretation and consequently knowledge generation.

Figure 1.4 illustrates potential sampling steps for a filtration. During isolation of a solid from a suspension, various solid materials occur. First, there is the filter cake. Depending on the filtration procedure applied, it may be pre-dried applying an air or nitrogen flow going through the wet solid. This filter cake is typically transferred to a dryer (e.g. a vacuum oven), where it is usually dried until no further mass loss is observed. Below the filter, there is the filtrate, which is a liquid and it occurs that subsequent precipitation or crystallization happens out of the filtrate. This accidently generated solid material should subsequently also be isolated and analyzed in a wet or dried state. These considerations on awareness on sampling procedures apply not only to laboratory experiments but also to piloting and manufacturing of DSs and DPs.

To understand material properties and changes throughout the process, first the right information must be derived. For example, in case of the filtration experiment, the wet filter cake should not be exposed unconsciously to subsequent sample preparation when the solid form, maybe a solvate, shall be determined by XRPD. Sample preparation like drying or storing the sample in a fridge may alter the solid form, and consequently, false conclusions can be drawn from the analytical investigations. Even worse, when the sample shall be investigated by various analytical methods (probably in different laboratories) and each sample is treated differently on its way to the measuring device. The results coming out of these investigations may not reveal a clear picture, i.e. the results are inconsistent. For example, it maybe that thermal gravimetric analysis (TGA) reveals a mass loss indicating existence of a mono-solvate, but the XRPD shows the pattern of an ansolvate.

A prerequisite to be able to collate analytical results accordingly is meaningful naming or coding of experiments, materials, and samples. It sounds obvious but unfortunately, sometimes not enough care is spent on the coding of samples that enables

(a) correct assignment to the process step where and when the samples were taken
(b) automatic collation in a data management system of the results of various analytical methods.

An inefficient approach of sample naming is attempting to type "all" information about the synthesis conditions (e.g. temperature, solvent, catalyst used) or material properties (e.g. enantiomer, solid form) in the sample name. This may work for exploratory experiments where one single researcher tries to keep track of his (initial) efforts for a small project but is doomed to fail in a professional industrial workflow involving multiple users and laboratories.

The solution to keep track of synthesis conditions and analytical information efficiently is some kind of data management system. That may be a paper notebook, but the analogue format of paper lacks flexibility to replicate, analyze, visualize, aggregate, and report information. Today, an electronic format, a digitalized solution, appears to be mandatory. Worksheets may do the job but far more efficient, especially in complex industrial organizations, are dedicated systems like electronic laboratory notebooks (ELNs), laboratory information systems (LIMSs), databases for the various analytical datatypes, knowledge management platforms, or interconnected combinations of those systems. These systems can simplify ensurance of data integrity and should be properly selected to match the individual needs of the laboratories and users and their roles. Furthermore, it should not be forgotten that implementing, maintaining, extending, and developing such systems enforces overhead which typically is not appreciated by senior management. The business case considering efforts and returns must be made up by every organization itself.

1.1.7.3 Contributions and Choices

Sometimes it is hard to find the appropriate equipment, be it hard- or software, that fits to the needs of the laboratory or the staff working there. One reason might be that decisions on product features are often driven by manufacturers on technical reasons, causes to ensure compatibility to older systems, or by innovative minds prospecting and anticipating future needs. Quite intelligent is approaching the customers and listening to the "voice of the customers" [41] or even better identifying the needs unknown or not even expressed by the customer as propagated by innovation management. It is attributed, probably falsely, to Henry Ford that customers would rather wish "faster horses" instead of thinking in new, innovative dimensions like "cars" [42]. Fortunately, the latter approach is followed as witnessed by numerous questionnaires, conferences, or user group meetings organized or sponsored by manufacturers. Serving not only to raise awareness about latest product developments (i.e. promotion) but also to get a better understanding about the market's desires is a driving force. Eventually raising market share and turnover is the ultimate goal for the manufacturer. Nevertheless, reaching this objective is under control of the customers in the laboratory that claim their needs, decide for systems, and continuously raise their voice and actively contribute to device, system,

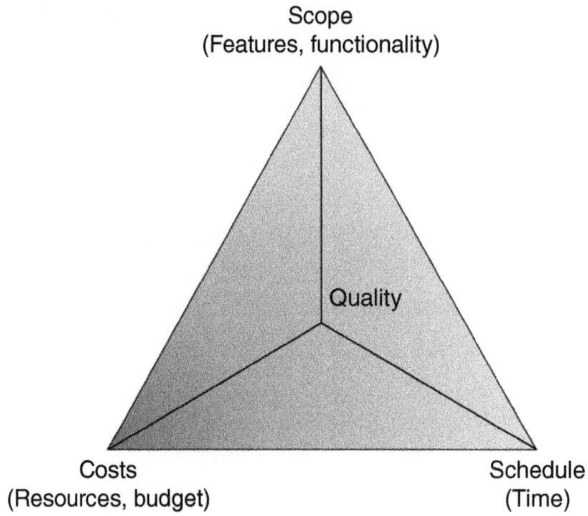

Figure 1.5 Triangle of project management.

and software development. It is worth to spend some resources (time and budget) on contributing to systems that are business critical for the respective laboratory. Giving feedback on bugs, naming desired changes or modifications for interfaces, or expressing functional requirements is essential to direct system suppliers in the right, the customer's, direction. If the supplier is not willing or able to listen and act, then customers have the choice to vote with their feet.

However, what is the right direction to go? Which device or system fulfills the needs of the various stakeholders best? One foundation is to get an overview on the market and understand pros and cons of the various solutions. Define your objective and requirements first, identify your current standpoint, and then identify the various paths to proceed and options you have. Evaluate appropriate solutions and suppliers. Make use of the various tools to get market transparency. Go to fairs, attend conferences, attend webinars and seminars, and search the Internet for product information. However, be careful with experiences posted anonymously. Find advice you can trust. Either by recommendation of colleagues or by other customers you might have identified on focused user group meetings. Weigh disadvantages and advantages. What is a disadvantage for the other one might be advantageous for you and vice versa. Be aware that good decisions require efforts on your side. The options you have reflect the triangle of project management (Figure 1.5). In the center, you get the maximum of quality with the most comprehensive scope instantaneously at no costs. Real live projects are always a trade-off. If you do not have persons with the right skills, you can spend money hiring them. If you do not have the money, you will get poorer quality or a less number of requirements sufficiently addressed or it will take longer time to finish the project. Eventually risk management tells at what levels the project is managed at best.

1.1.7.4 Application of Digitalization

Certainly, the main and basic starting point is digitalization of the general daily work. Historically, introduction of calculating machines and substitution of type- and telewriters can be considered as the starting point for the digital revolution reaching to the masses. Today, knowledge in basic office solutions, like text processors, spreadsheets, email programs, and ability to search the Internet, is a basic request in all job announcements.

In addition, more specific talents are ask when it comes to complex tasks and applications that are more and more part of industrial workflows, exemplified by the ability to insert basic information into information systems and knowledge of how to query databases. Often welcomed expertise are skills to create databases and write software applications, starting with macros or python scripts for data analysis but leveling up to implementation of automated data analysis by means of ML or artificial intelligence (AI) or more complex software solutions implemented in the laboratory workflow.

A part of these requirements is a necessity to apply and transfer general digitalization solutions to applications in the expert domain. Staff and management need to familiarize with domain-specific software. In the field of solid state, there are numerous analytical devices in use that create data files that need to be evaluated, interpreted, and understood. This process is often be accompanied by third-party software that, for example, calculates physicochemical properties, simulates XRPD pattern, visualizes, analyses, or predicts crystal structures. Last, but not least, it is important to have an idea, if not a vision and thorough experience, how individual available and potential solutions can be assembled and integrated to enhance comfort, efficiency, and reliability of daily tasks and build a seamless workflow. Figure 1.6 might give an impression on the various levels and complexity for digitalization solutions in a laboratory environment. However, it has to be considered that the complexity of implementation and use of software solutions always go hand in hand with

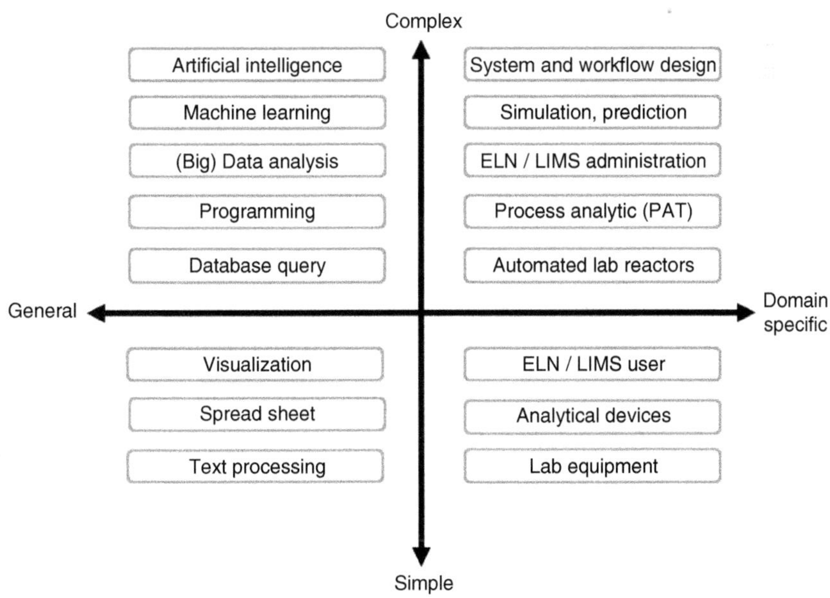

Figure 1.6 Level and complexity of digitalization.

the skills of the users. Furthermore, application of general tools for domain-specific purposes always requires domain experts that support design, implementation, and maintenance.

1.1.7.5 Fully Digitalized Infrastructure

Categorization of the various IT systems used within a laboratory in hierarchical order with respect to complexity and criticality for business aspects is meaningful. This supports definition of objectives and where to start (or continue) digitalization. A pyramid naming the types of systems visualizes for a single department or laboratory this hierarchy. From bottom to top, the number of devices typically decreases, whereas the complexity and criticality increase. First, interdependencies with this departmental pyramid should and must be considered. Remove ideological and technological barriers. Ensure seamless mutual interoperability of the systems as well as data integrity. Next, identify and implement critical interfaces to ensure effortless operation of the departmental systems. Focus on satisfying the needs of all stakeholders – within the pyramid and beyond. Start with simple and most useful tasks and progress with the more complex and less important ones. Never forget justification of actions by ensuring return of investment, competitive advantages, or fulfilling regulatory requirements.

Last, try to interconnect systems of various departments and external contributors wherever meaningful (Figure 1.7).

Exemplified for a solid-state laboratory and displayed on a more detailed IT and laboratory system and technology level, a fully integrated IT environment (Figure 1.8) comprises systems like

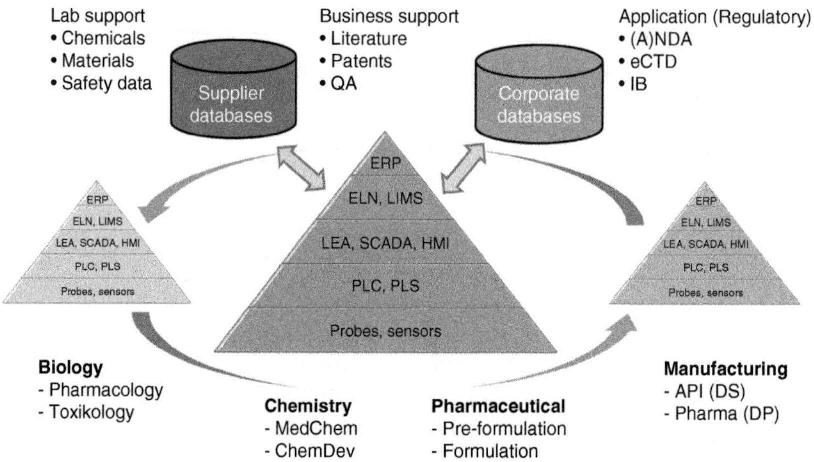

Figure 1.7 Hierarchy and Interoperability of departmental IT systems. ERP, enterprise resource planning; ELN, electronic laboratory notebook; LIMS, lab information system; LEA, lab execution and analysis system; SCADA, supervisory control and data acquisition; HMI, human–machine interface; PLC, programmable logic controller; PLS, Prozessleitsystem, i.e. distributed control systems (DCS); DS: drug substance; DP: drug product; QA: quality assurance; (A)NDA: abbreviated new drug application; eCTD, electronic Common Technical Document; IB, investigators brochure.

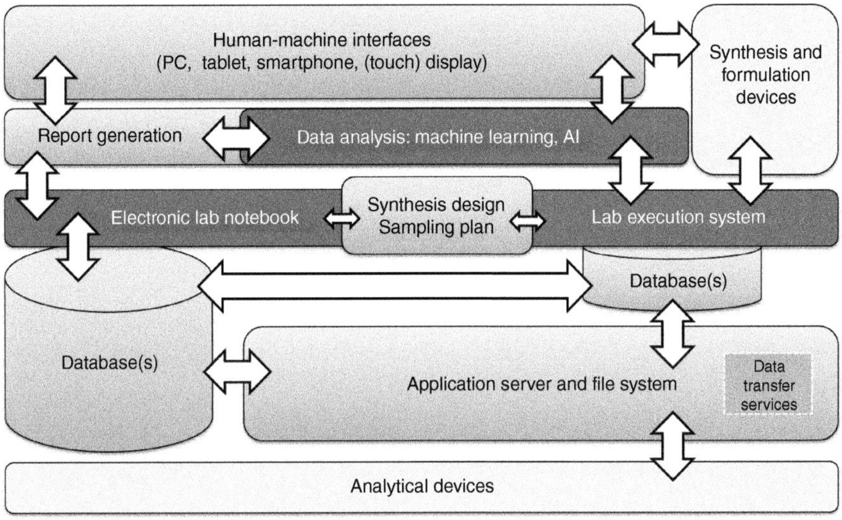

Figure 1.8 Fully digitalized IT and lab system landscape for a solid-state laboratory.

- (Automated) lab reactor systems (synthesis and formulation devices)
- analytical devices (X-ray, spectroscopy, imaging, chromatography, stability, dissolution, PAT probes)
- server and automation services (hardware, software)
- design and planning tools
- ELN and database systems
- data analysis and reporting tools
- human–machine interaction (HMI) devices (PC, tablets, smartphones, etc.)

Such fully digitalized infrastructures (Figure 1.8) have already been successfully implemented in the past. Many efforts are required to configure and adopt out-of-the box solutions and interfaces between the systems. However, attempts to harmonize data exchange standards may simplify implementation and maintenance. Yet, skilled staff and professional support is a prerequisite for implementation, optimization, and maintaining integrated high-end systems operable. One of the most important aspects, as for many projects and business processes, is communication. Communicate and discuss

- needs to keep stakeholders satisfied and
- intended changes to ensure seamless and continuous interoperability of systems.

1.1.8 Basic Terms and Concepts in the World of Solid State

1.1.8.1 Crystalline and Amorphous

Typically, a crystalline material is understood as having long-range order in all three dimensions of space (see Figure 1.9a).

Order means periodic arrangement of smaller units (atoms or molecules) that is mathematically described by symmetry operators. The International Tables of

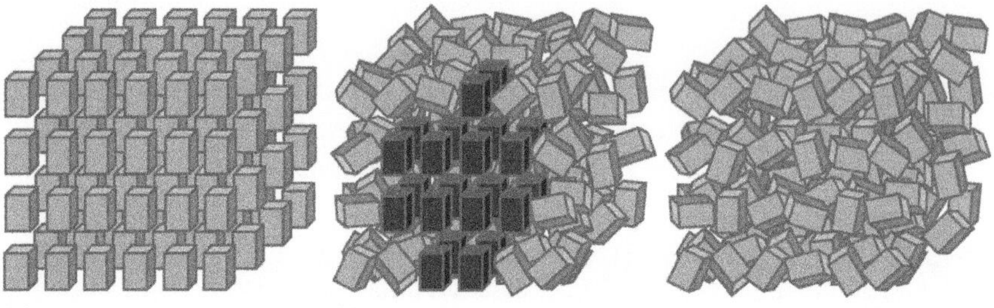

Figure 1.9 (a) Arrangement with long-range order, (b) mainly amorphous, arrangement without long-range order, to some extent local order exist, (c) completely amorphous.

Crystallography lists 230 space groups that represent the manifold of arrangements of potential symmetry elements that are used to describe periodicity in crystal structures [43].

Amorphous material lacks the periodicity over a longer range (see Figure 1.9b and c). Therefore, an X-ray powder investigation would not result in a pattern with narrow and distinct reflexes but with one or more broad halos. The borders may be fuzzy. Local order and its extent may be recognized with solid-state nuclear magnetic resonance (ssNMR) spectroscopy [19, 44] or pair distribution function (PDF) analysis applied to XRPD [45–48].

Various analytical methods can be used to determine qualitatively and with additional efforts quantitatively whether a sample is mainly or partially crystalline or amorphous. For example,

- Microscopy
  - Crystalline samples of pharmaceutical compounds often exhibit birefringence. The sample must be analyzed using polarized light using cross-polarized optical filters.
- X-ray diffraction
  - Crystalline samples diffract radiation in the wavelength of X-rays (e.g. copper radiation has a wavelength of c. 1.54×10^{-10} m). This results in either distinct diffraction spots in case of single-crystal diffractometry or reflexes in an X-ray powder diffractogram.
  - Even in case of no distinct Bragg reflexes present, a powder diffractogram can be used to determine the extent of short-range and long-range order, e.g. by applying the methodology of fast Fourier transform (FFT) analysis and PDF analysis.
  - Single-crystal X-ray diffraction (SCXRD) data are important to simulate the XRPD based on the structure to confirm phase analysis. Furthermore, SCXRD enables the determination of identity including the absolute stereo configuration of compounds.
- Thermal analysis
  - Differential scanning calorimetry (DSC) may reveal glass transition, melting, and phase transitions.
  - TGA may show mass loss (solvent) or decomposition.

1.1.8.2 Crystallization and Precipitation

Precipitation (i.e. rapid, uncontrolled solidification) and crystallization are not the same. To get a hand on the material, synthetic chemists often crash out (precipitate) the solid form and believe they have crystallized. However, common understanding of crystallization is that crystallization is a controlled process step, whereas precipitation is mostly uncontrolled and can lead to amorphous products, too (which are not crystalline).

Crystallization and precipitation processes both have their justification and application. Thorough investigation of the underlying process parameters and knowing the intended product profile is essential for successful development, implementation, and integration of such processes.

1.1.8.3 Understanding the Phase Diagram – Analytical Characterization of the Solid–Liquid and Solid–Solid Systems

Besides microscopy, X-ray diffraction, and thermal techniques, there is a variety of additional analytical techniques more or less suitable to investigate and characterize the solid state. There are ssNMR spectroscopy, vibrational spectroscopy like Raman, infrared spectroscopy (IR) and terahertz spectroscopy, dynamic vapor sorption (DVS) methods, electron microscopy (scanning and transmission techniques), and electron diffraction crystallography, to name the most commonly used. With a particular focus on industrial applications, other chapters of this edition describe in further detail some of the techniques mentioned along with examples for application.

In addition, investigating the liquid phase that is in contact, under equilibrium or nonequilibrium conditions, with the solid phase are relevant to foster understanding the complete picture leading to formation or stability of one or the other solid phase. Knowing the composition of the liquid phase as well, e.g. supersaturation, impurity profile, solvent composition, in particular water activity, is important. Appropriate means to determine concentration profiles are spectroscopic methods like Raman and IR, titration (e.g. Karl Fischer), and chromatography and coupled methods (high-performance liquid chromatography [HPLC]; liquid chromatography coupled with mass spectroscopy [LC–MS]).

Collecting and putting together information gathered about the liquid and solid phase under various conditions (like temperature, composition, pressure) constitutes a phase diagram that eventually indicates regions of stability and potential transformation pathways between solid phases. Phase diagrams are, for example, treated in detail in the chapter "Thermodynamics of Polymorphs and Solvates" written by Coquerel in [22], the construction of phase diagrams by means of DSC is described [49] and the utility of phase diagrams is discussed also in context with cocrystals [50–52]. The concept of phase diagram investigation is extendable to the exploration of polymorphic landscapes by computational methods. For example, to predict thermodynamic stability regions for crystal structures along with subsequent attempts to prepare those polymorphic forms, by application of high-pressure experiments [53].

1.1.8.4 Polymorphism

Polymorphic per se means multiple morphic forms. It stems from the ancient greek πολύς (polús, "many, much") [54] and μορφή (morphḗ, "form, shape") [55].

The morphology, in terms of shape and outer form, of a crystal corresponds not necessarily one-to-one to the crystal structure, meaning the construction of the inner matter as constituted by the internal arrangement of atoms or molecules.

Unfortunately, this can lead to confusion. Especially, the frequently used term "crystal form" does not inherently clarify if it addresses the inner form of a crystal. If not specified, crystal form can refer to the inner form, the crystal structure, or the outer form, the morphology or shape. For further clarification, the chapter "Form vs. habit" in [24] is recommended reading.

In the context of solid-state development of inorganic or organic compounds, the terms polymorphism and polymorphic forms are undoubtedly connected with the inner form of the matter.

Scientific literature extensively dealing with polymorphism as well as consideration on the regulatory treatment of polymorphism and solid-state-related topics are available [31]. Various scientific definitions for polymorphism can be understood as covering the spatial arrangement in a crystal of a single molecule or a single entity formed by atoms as well as that of a substance that consists of two or more molecules or other constituents. Sharma states "the term 'polymorphs' has in-fact all-encompassing through its application to different crystalline forms of an element or a compound with different atomic arrangements." [56]

The European regulatory perspective considers polymorphism as "the ability of a compound in the solid state to exist in different crystalline forms having the same chemical composition" and that it may be exhibited in the solid state by all types of compounds "single as well as multiple entities, such as salts, hydrates, cocrystals, etc." It is acknowledged that "different forms may possess different physico-chemical properties" [57].

Other terms in this context are

- *Allotropic forms*. "The phenomenon that a substance exists in various solid states, depending on the conditions (temperature, pressure), is found not only in sulfur, but also in many other substances [...]. This is called 'allotropy' in the case of elements, and 'polymorphism' in the case of compounds." Translated from: [58]. A discussion of the use of the terms allotropes and polymorphs is provided in [56] that concludes with the recognition that the terms have taken on the same meaning.
- *Modification*. Typically, the term modification is used synonymously with polymorph or polymorphic form. In English, the word "modification" has several meanings like change, alteration, limitation, deviation, and also deformation [59], which explains the synonymous use.

  However, a statement found in a reference from 1966 [60] indicates that there formerly might have been some differentiation in meaning of the terms. The reference states

> *"Wann liegt eine Modifikation vor? Der klassische Polymorphiebegriff und seine Definition der Modifikationen erscheint eindeutig, wenn man etwa an den Schwefel oder den Phosphor denkt. Er verliert jedoch an Klarheit, wenn nahe verwandte Strukturen vorliegen, zwischen denen noch Übergänge möglich sind."* which translates to
> *"When is there a modification? The classic concept of polymorphism and its definition of modifications seems clear when one thinks of sulfur or phosphorus. However, it [the concept] loses clarity when there are closely related structures between which transitions are still possible."*

In particular, "*concept of polymorphism and its definition of modifications*" suggest a difference in meaning. Unfortunately, no further description of what the authors meant could yet be found. One understanding could be that" a modification exist only under thermodynamic conditions" (R. Glaum [2019]. What is a modifiation? personal communication). This would mean that enantiotropic polymorphs are (both) also a modification because there are conditions under which either polymorph is thermodynamically stable. Whereas in the case of monotropically related polymorphs, only the thermodynamically stable one is a modification of the compound.

- *Morphic form*. The term "morphic form" is used, e.g. by Saal (see Chapter 10), to express the singular of polymorphic form. Saal understands the term as a synonym by stating "polymorphic forms – also called morphic forms". Therefore, the term in that chapter refers clearly to the inner structure.

Critical may be the use of "polymorphic" if a crystalline form is denominated as such if no different crystal structures, i.e. polymorphic forms, exist (or are known) from that compound. Since the term "poly" means more than just one, it is assumed that more than just crystalline form exists, which is contrary to the belief that only one crystalline form exists. This may justify the use of "morphic form". In reality simplicity wins, therefore using "polymorphic" may be acceptable in daily use for such cases, too. However, the term "morphology" is typically used to describe the outer shape of materials. Therefore, it is recommended to think twice what is expressed with the term "morphic form" when read or written.

1.1.8.5 Multi-component Compounds – Salt, Cocrystal, Solvate, and Hydrate

An overview about various definitions for multi-component compounds and alternatively used terms was collected by Stahly [61]. His broad definition of a cocrystal is "a crystalline structure with unique properties that is made up of two or more components. A component may be an atom, ionic compound, or molecule". By stating that the component "may be ionic," this cocrystal definition also comprises salts. Solvates and hydrates are included as a subset of cocrystals. In the case of solvates, the molecule is a compound that is also known or used as solvent. In case the solvent is water, the solvate is called hydrate.

The term clathrate describes multi-component compounds where one component is contained in spaces within the crystal structure of the second component.

Other terms may be synonymously used for cocrystal. So expressions like co-crystal, molecular complex, or multi-component molecular crystal can also be found in literature and may have a subtle different meaning.

Various discussions about definitions when a solvent is a solvent and the wording are published [30, 61, 62].

One definition about when a multi-component compound can be considered a salt and when a cocrystal is based on the difference in pKa values of the particular components shall contribute to the extent of the proton transfer between two components in the crystal structure. The situation has been discussed as the salt–cocrystal continuum [63, 64] with the conclusion that the crystal environment and other factors like temperature are decisive for the extent of proton transfer and not a definition (pK_a) based on equilibrium in aqueous environment. Sometimes the decision about the position of the proton cannot be easily made [65] or answering the question takes time [66]. Discussion of these definitions may seem to be of academic nature; the terms mentioned are used in regulatory documents. This means potential commercial impact. Therefore, meaning, understanding, and interpretation of words, data, and experiments can make an important and decisive difference. For example, the perspective of the European Medicines Agency (EMA) on the topic is that "solvates including hydrates can be considered as a subgroup of cocrystals. The solvent, or the water, acts as a co-former in the same way as other co-formers" [57]. The EMA acknowledges that there is no strict borderline between complete and no proton transfer at all. As a criterion of relevance, salts and cocrystals are considered to have defined stoichiometries. EMA considers the properties that determine the suitability for the intended objective and application as decisive: "Ultimately, the resulting material properties are the critical factors that determine the suitability of a developed solid-state form for the designated purpose, regardless of the molecular bonding involved" [57].

Consideration of regulatory implications may alter as time goes by. Nevertheless, the current perspective of regulatory aspects is important for registering and marketing DPs (Chapter 10, [31, 67], or [68]).

One important conclusion is that classification is desirable but "researchers do not always agree on what does or does not belong in a particular category or what the definition of each category is" [64].

1.1.8.6 Solvates, Hydrates, Non-solvated Forms, or Ansolvates

A crystalline compound that does not contain a solvate is called an ansolvate or non-solvated form. If a solid phase, e.g. a crystal, is in proximity of a liquid phase, e.g. a solvent, the solid can attract the solvent. If the liquid phase remains at the surface of the solid, the liquid is adsorbed. If it penetrates through the surface, it becomes absorbed. If the liquid is eventually incorporated into the crystal structure, a solvate or solvated form is formed. In case the liquid is water, the solvate is called a hydrate. In analogy, there are names for solvates formed by the various organic solvents (Table 1.3). If an ansolvate is formed from a previously solvated form that lost the liquid, the resulting polymorphic form may be called a desolvate or desolvated form. In the case of water, the form is called dehydrated form or anhydrate. There is

Table 1.3 Naming of common solvate forms.

Solvent	Name of the solvate
Water	Hydrate
Methanol	Methanolate
Ethanol	Ethanolate
Propanol	Propanolate
Alcohol	Alcoholate

a specific nomenclature indicating the amount of water per mol parent compound (see Table 1.4 and Figure 1.10).

Cave! In organic chemistry, some compounds are called hydrates (e.g. carbohydrates, diols, aldehyde hydrates), which actually do not contain any molecular water. The name stems from water added by a chemical addition reaction (see Figure 1.11). These types of compounds must not be mixed up with those containing water in the crystal lattice.

"Hygroscopicity" names the tendency of a compound to attract water. Deliquescence describes beginning dissolution of a compound in water that it has attracted from the surrounding atmosphere. Characterization of the degree of hygroscopicity (or attractiveness to water) of a compound as a function of humidity and temperature is possible by DVS experiments [69]. These kinds of investigation provide information on the kinetics of adsorption and desorption as well as the determination of threshold values for humidity levels where sorption and desorption happen. A rather simple setup is storing the compound in an exsiccator or glass

Table 1.4 Nomenclature for compounds with crystal water.

Amount of crystal water	Name of the hydrate
0	Anhydrate
0.5	Hemihydrate
1	Monohydrate
1.5	Sesquihydrate
2	Dihydrate
3	Trihydrate
4	Tetrahydrate
5	Pentahydrate
…	…
12	Dodecahydrate
Uneven amount	Non-stoichiometric hydrate
Variable amount	Variable hydrate

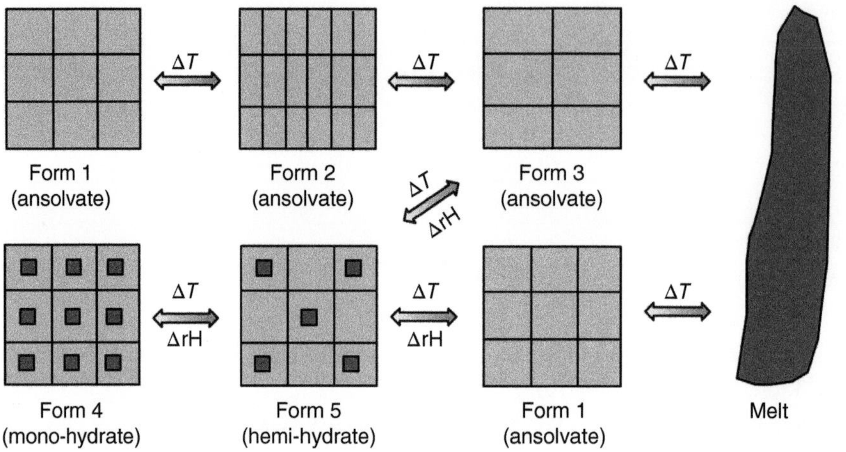

Figure 1.10 Schematic representation of potential interconversions between polymorphic and hydrated forms and melt upon changing temperature (Δ*T*) and/or relative humidity (ΔrH). Pathways depend on conditions.

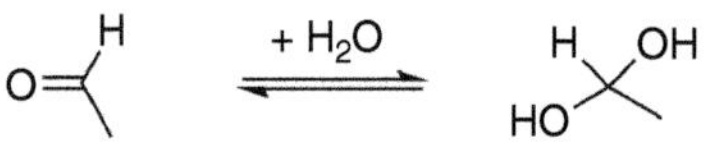

Figure 1.11 Acetaldehyde, a geminal diol – an organic compound "containing" water.

vial exposed to saturated salt solutions [70]. This setup enables quantification of water take-up (sorption) or loss (desorption) by gravimetry of larger (100 mg to g) quantities of material. Analytical characterization of the physico-chemical nature by solid-state analytical techniques such as XRPD, DSC, TGA, and the like as well as by chemical analytical methods such as HPLC prior and after storage is recommended. Hygroscopicity classification schemes are reported in [69]. The scheme of the European Pharmacopeia classifies percent (w/w) water uptake at 25 °C and 80% relative humidity (RH)

- 0–0.12% (w/w) as non-hygroscopic
- 0.2–2% (w/w) as slightly hygroscopic
- 2–15% (w/w) as moderately hygroscopic
- >15% (w/w) as very hygroscopic

Determination of stability information to be submitted in registration applications is documented in the ICH Q1A (R2) guideline "Stability Testing of New Drug Substances and Products" [71]. For chemical and pharmaceutical process purposes, the characterization of hydrates in suspensions of organic solvents with water (binary or ternary mixtures) with specified water activity is recommended [72, 73]. For further and detailed description of terms and relations, the chapter "Hygroscopicity and Hydrates in Pharmaceutical Solids" in [22] is recommended reading.

Eventually, the solid that incorporates a solvent may have various natures. It may be a neutral form, a salt, a cocrystal, or combinations thereof. Consequently, numerous variations of compounds may be formed. It appears to be a science on its own to classify and name such compounds accordingly [74].

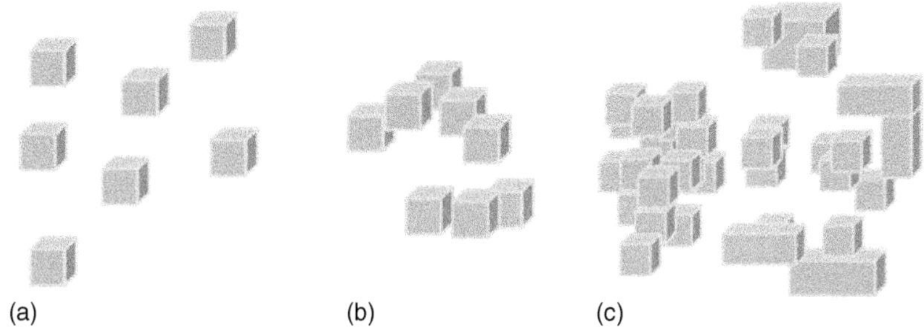

Figure 1.12 (a) Dispersed primary particles, (b) aggregates, and (c) agglomerates.

1.1.8.7 Dispersed Primary Particles, Aggregates, and Agglomerates

Crystals distributed as isolated primary particles in a suspension form a dispersion, i.e. a phase that is equally distributed in another phase. Due to attractive forces, the individual crystals may find each other and stick together and eventually form bigger particles constituted by the smaller ones. Depending on the attractive forces and the nature of bridges formed, aggregates and agglomerates may be distinguished as secondary particles (see Figure 1.12). While in aggregates (from latin aggregare, "group, attach"), multiple particles are connected by physical interaction or adhesion, agglomerates (from latin agglomerare, "mass together, join forces") are formed by multiple particles that are grown, sintered, or melted together, and thus form new or bigger particles that cannot be easily separated into the original constituents. However, in literature the two terms are often used interchangeably.

Aggregates may easily separate during post-crystallization procedures like filtration, washing, or application of tiny mechanical forces processing into primary particles, whereas agglomerates are more stable against size reduction. However, mechanical forces may cause formation of smaller particles by attrition or breakage of the agglomerates.

1.1.8.8 Particle Size and Particle Size Distribution (PSD)

Particle size considers individual particles with respect to length, width, and height or the volume as the product of these three measures as well as the morphology. Properties of these individual particles are homogeneity, stability against breakage, solubility, and dissolution rate.

Particle size distribution (PSD) considers a collective of particles, namely, their amount, distribution (with respect to size and mass), volume (e.g. tapped volume), and surface. Collective properties are separability, miscibility, tendency for agglomeration or aggregation, flowability, tapped density, and bulk density.

1.1.9 Investigating and Understanding the Polymorphic Landscape

Defining the objective at the beginning is one of the most important advices to follow in project management. Project goals may be formulated, e.g. according to the S.M.A.R.T. principles, i.e.

- specific
- measurable

- achievable
- relevant
- time bound

Easier said than done when it comes to the solid form. Assuming the intention is to administer the medication as a solid dosage form, e.g. as a tablet or as capsule, then probably the desired dose range can be estimated and the desired range for particle size can be specified, at least to a certain level. However, the number of isolatable polymorphic forms, potential solvates, types of salts, or cocrystals that can be achieved along with the particular properties relevant for successful drug development are unpredictable based only on the molecular structure. The timeframe for the project can certainly be set by management. Unfortunately, this may not be enough time to explore the polymorphic landscape with all its valleys, mountains, bright plains, and dark rivers. Every solid form project has its peculiarities. Every single compound behaves differently. Therefore, approaches to explore the landscape must be defined. Yet, the number of potential parameters to set for screening studies, like temperatures, pressures, solvents, methods for preparation, mixing, additives, and so forth, constitutes a really big experimental space. Reduction to practice necessarily decreases the number of experiments to an affordable and executable subset. However, the selection made depends on individual expertise and experiences of the operator(s) and the limits set by regulations and the institution the team works for.

These investigations yield, under the selected conditions, materials that have specific properties. This determines the first aspect of discovering the polymorphic landscape, the formation routes and their parameters lead to a smaller or bigger zoo of new compounds, i.e. polymorphs, solvates, salts, and the like. When determining the properties of compounds formed, a part of the investigation is aiming to identify the chemical and physical stability. The latter refers to potential transformation pathways and interrelationships, i.e. phase transitions, between polymorphic forms including solvation and desolvation.

A summary of all the findings or "polymorphic landscape" collected for a compound over time supports future development, manufacturing, and evaluation of next-generation products. It guides further optimization of synthesis, crystallization, and downstream processing, as well as formulation efforts with respect to operational space.

An example for a polymorphic landscape is given based on a paper on a methanol solvate of thiamine hydrochloride [75]. This may not reflect all possible polymorphic forms and solvates; however, the abstract of the paper mentions that thiamine hydrochloride (**THCl**) forms a monomethanolate (**MM**) upon exposure of crystalline thiamine phases (thiamine hydrochloride hemihydrate/**HH**, nonstoichiometric hydrate/**NSH**, and anhydrate/**AH)** to anhydrous methanol (solvent and vapor). Also, desolvation of **MM** at 50–80 °C resulted in the formation of a poorly crystalline intermediate which crystallized to **AH** at elevated temperatures (≥150 °C). When exposed to water vapor (11–75% RH, RT), **MM** transformed to **HH** (**NSH** was detected at ≤40% RH), while exposure to polar solvent vapor resulted in direct formation of **AH**. **MM** was stable in the presence of nonpolar (benzene and hexane) solvent vapor […] [75]. Hence, the abstract can be visualized as "polymorphic landscape" in a diagram (Figure 1.13).

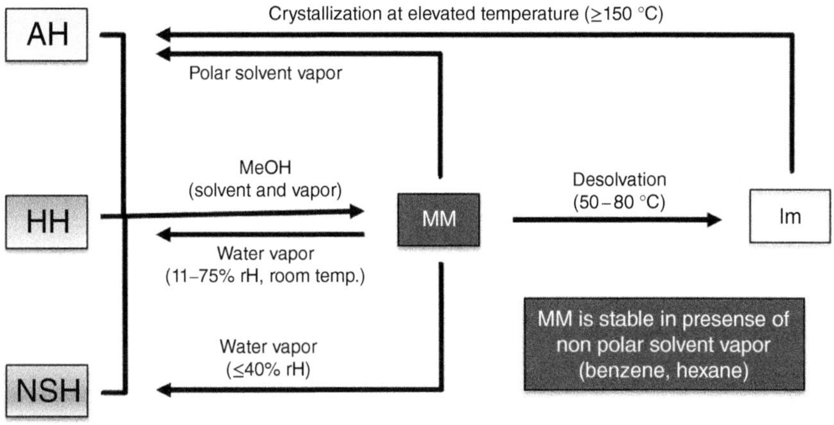

Figure 1.13 Polymorphic landscape derived from experimental findings. MM, mono-methanolate; AH, anhydrate; HH, hemi-hydrate; NSH, non-stoichiometric hydrate; Im, intermediate form; MeOH, methanol; rH, relative humidity.

1.1.10 Performing the Crystallization

The rational starting point for every crystallization (process) is the understanding of the solubility profile of the compound that shall be isolated [76–78]. In addition, it is advisable to know the accompanying impurities and their solubility profiles, too. It is worth to mention that the impurity profile can vary depending on the selected synthesis route [24].

In pharmaceutical industry, crystallization is predominantly conducted from solution. Reasons for this is that the organic synthesis is typically performed in solution and the reaction product can readily be crystallized from the system. Crystallization from the melt might be an option if the compound has sufficient thermal stability. Crucial for the selection of the crystallization method from solution (evaporative, cooling, anti-solvent, pH shift, or a combination thereof) is the solubility curve of the compound and the meta-stable zone width (MSZW). Once these data are determined and available for the selected compound–solvent system, the crystallization behavior can be investigated.

The MSZW determines the conditions under which the compound can remain in a supersaturated solution without spontaneous nucleation. The meta-stable zone is typically the range of conditions under which seed material (dry or in suspension) is added to initiate controlled crystallization. The seed has to be thoroughly characterized and selected with respect to polymorphic form, morphology, PSD, and amount. In general, it is recommended to perform crystallizations with seeding because this allows better reproducibility and control of PSD, yield, and crystalline form of the product.

As supersaturation is the driving force for crystallization, the process conditions have to be adjusted accordingly over time as by formation of the solid material, the supersaturation decreases. Typically, process conditions like mixing, temperature, anti-solvent addition, or vacuum are governed so that the growth of the particles is according to the desired crystalline form, morphology, PSD, and impurity profile.

Inclusion of impurities (residual solvent, by-products, reagents) into the product is usually not desired.

In addition to textbooks [32, 33, 35], further viable sources for an introduction into the basics of detailed investigation of solubility profiles and approaches for sophisticated industrial crystallization process optimization is presented on the websites of suppliers, like Technobis crystallization systems (www.crystallizationsystems.com) or Mettler Toledo Autochem (www.mt.com).

1.1.11 Objectives for the Optimization of Crystallization Processes and Solid-State Properties

There are various reasons to spent time and resources during development and manufacturing on the optimization of crystallization processes and solid-state properties.

Of particular importance is that the synthesis route and related conditions like process parameters and chemicals may change during the lifespan of a compound and its way of production. Consequently, any change may have an effect on the resulting solid form. This lesson is taught in reports about "Concomitant" [79], "Disappearing", "Reappearing" [25–27, 80], or "late appearing" [81] polymorphs. Joel Bernstein has summarized this insight with "...the polymorph obtained, or the polymorphic mixture obtained, depends on the synthetic route to the desired material. It is probably more correct to state that as usual, the polymorph or polymorphic mixture depends on the crystallization conditions, and these will clearly differ in the solvent/reagent/product compositions resulting from different synthetic conditions and routes" [24].

Synthesis routes typically change from early R&D, over chemical, process, and pharmaceutical development until DS and dosage form manufacturing, in general caused by optimization attempts. It is of utmost importance to understand that all these efforts have to take in count and consequently require surveillance and control of the resulting solid polymorphic form and its properties. Considerable impact may have all steps that define or deal with the solid form, including, but not limited to, crystallization, separation, drying, storing, formulating, transporting, and packaging along with parameters and conditions of those processes. In addition, transformation, like scale-up, technological transfer to or from Contract Research Organizations (CRO) and Contract Development and Manufacturing Organizations (CDMO), even in-house transfer to other production sites or manufacturing equipment, should be accompanied by solid-state expertise and responsible risk management.

1.1.12 Implementation of In Silico and Simulation Techniques

One appropriate mean to support risk management is consulting in silico techniques that simulate the physical or chemical behavior of a reaction system or process. Simulation is based on mathematical models and therefore the fields of applications, efficiency, and limitations depend on the information about the reaction or process that is available and can be drilled down to descriptive equations and numbers.

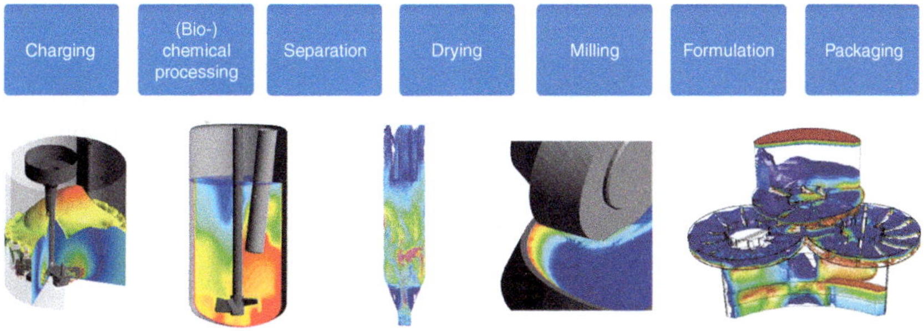

Figure 1.14 Various applications for simulation of chemical and pharmaceutical unit operations. Source: Courtesy of aixProcess GmbH, Aachen, Germany.

Knowledge about computational methods addressing pharmaceutically relevant topics for solid-state applications like crystal structure prediction, solubility prediction, and formulation design in industrial contexts was collected by Abramov in 2016 [82].

Fortunately, engineering aspects are well addressed by simulation, too. Simulation is applicable for many unit operations and in the world of chemical and pharmaceutical sciences (Figure 1.14). The basic principles as well as the evolving capabilities due to development in underlying theories, algorithms, and increasing availability of computational power are well reported [83–87].

It is worth noting that engineers and chemists typically have different educational backgrounds and probably as a consequence different perspectives to address and look at processes. Both may think in formulas. Yet the understanding and viewpoint are different. While an engineer primarily visualizes his process understanding in "mathematical formulas" and flow diagrams, the chemist illustrates processes with "molecular formulas" and chemical reactions thereof. The use of simulation packages to describe and understand processes is a helpful mean to bring the worlds together and illustrate engineering and chemical aspects to simplify mutual communication.

Theories about, e.g. fluid dynamics, mechanical, material, and thermal properties are readily available through databases and simulation software packages for various engineering tasks.

Consequently, basic information packages from laboratory scale experiments along with easily accessible geometrical information of equipment for scale-up or manufacturing can be fed into models. The subsequent insights enhance understanding of the underlying situation and potential challenges as well do they simplify communication according to the proverb "a picture says more than a thousand words" (see Figure 1.15).

Very often mixing is identified as critical process parameter (CPP) for crystallization processes. Affecting process parameters like dosing rates, locations for addition, order of addition, filling volumes, temperature profiles, propeller geometries, and related properties like power number and shear rates are quite often just estimated based on experiences or simple assumptions. On a higher level, crystallization

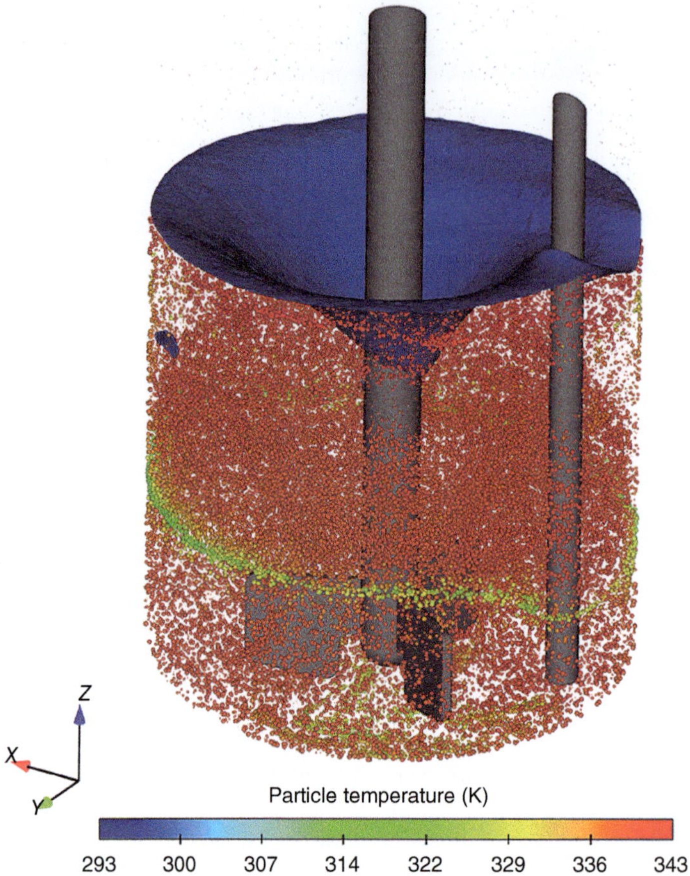

Figure 1.15 Calculated temperature distribution after seeding in a cooled mixed vessel in 100 l scale. Source: Courtesy of aixProcess GmbH, Aachen, Germany.

can be described by nucleation and growth kinetics and chemical reactions by reaction kinetics. Introducing such information into the simulation models require more efforts but could then lead to additional insights on the evolution of PSDs or by-product formation or decomposition.

Upon scale-up or site transfer, not all conditions and properties from the original (e.g. lab scale or CRO) equipment can be kept constant. Decisions, naturally supported by a risk assessment, must be taken.

As an example may serve the in silico investigation of a stirred vessel and a look at the calculated temperature distribution (compare Figure 1.15). Due to specific mixing properties of the equipment under consideration, stagnant zones evolve during mixing. Therefore, some regions are exposed somewhat longer to the cooled wall of the vessel. Consequently, this leads to cold spots (or zones) where unintended spontaneous nucleation of a supersaturated solution may happen well before the temperature probe that is located in another zone, that might probably better mixed, indicates that the temperature elaborated for seeding in lab experiments is reached.

In analogy, for chemical reactions this may be transferred to a heated vessel, where hot spots evolve inducing accelerated decomposition related with an undesired (and certainly unexpected) out-of-specification purity profile.

According to quality by design (QbD) principles, implementation of simulation approaches supports multidimensional exploration of the design space. This comprises rational process design and understanding of processes and equipment parameters. Identification of CPP combined with application of scale-up principles may serve to minimize, e.g. batch-to-batch variations and enhance overall process and product quality.

Simulation-related engineering efforts may be overcompensated by reduced costs for less chemicals, reagents, energy, re-working, disposal of failed attempts, as well as avoiding selection of unsuitable equipment and the like. In addition, mechanical forces can be derived from the calculations that may point to material stress or enhanced exposition to corrosion. This information may support preventive maintenance of manufacturing equipment. Many industries like automotive, aerospace, and chemical industry established the opportunity for more efficient and rapid development as well as optimization of manufacturing processes provided by simulation techniques. As part of a rational and properly coordinated development and manufacturing strategy, simulation techniques make sense to be implemented responsibly also into pharmaceutical R&D and manufacturing.

As such, surveillance of processes as well as continuous improvement and operational excellence is feasible. Besides data and information acquisition and interpretation by enhanced statistical interpretation through ML techniques or AI approaches, physical modeling by means of simulation brings additional efforts and insights.

As an entry point to overcome hesitations, often the application of simulation techniques as a "firefighting" tool is used in critical projects. This is an appropriate and suitable approach to learn about the opportunities before implementing simulation as part of strategy and daily business.

1.1.13 Saving the Investment – Addressing Intellectual Property Rights

Eventually, all measures can be drilled down to enhance profit or safety. However, approaching those final objectives, several intermediate targets can be accessed by investigating the crystallization process or the solid-state landscape.

As there are

- increasing yield
- reducing initial amounts of materials
- increase energy efficiency (e.g. heating, cooling, and drying routines)
- reduce potential threads for people and environment (e.g. avoid dust by larger particles)
- reduce reaction and overall processing times
- identification and control of CPPs for chemical and pharmaceutical operations
  - improve purity, reduce amount of by-products

  - understand the affect of synthesis routes, different impurity profiles, and the fate of impurities on resulting solid forms and morphology
  - understand the impact of morphology (flowability, filterability, compatibility)
  - optimize isolation steps (filtration, centrifugation), consider morphology
  - increase washing efficiency (e.g. reduce solvents or repetitions)
  - optimize drying (e.g. reduce time or energy, target desired polymorphic form)
  - understand impact of process conditions and routines
- enhance physical or chemical stability
  - understand impact of
    - light
    - humidity or moisture
    - temperature
    - mechanical forces (e.g. pressure, shear forces)
  - prevent alteration of solid form
  - optimize storage conditions and packaging
  - enhance shelf live

Besides understanding and gaining control of technical aspects of solid form properties and processing, the other important aspect to invest into solid-state activities is an additional chance to protect IP rights. Particularly, the pharmaceutical industry invests huge amounts of money into R&D and commercialization. However, there is high risk to not get a return on investment for many projects. Therefore, successful projects, i.e. those where eventually a medication reaches the patient, have to cover also the investments of the failed attempts. A system, worldwide established to protect IP rights, i.e. preventing others from exploiting efforts invested into R&D and commercialization, is the international patent system (PCT). It is recommended to educate (solid-state) scientist with the basics of the patent system. A starting point may be training on proper documentation and communication of experiments and results. Building close relationships and establishing cooperation in interdisciplinary teams, including scientists and patent attorneys, enhance mutual understanding for limitations, requirements, and challenges. Furthermore, it encourages innovation, which starts with an idea but takes a long way until market decides on failure or success of a product.

1.1.14 Concluding Remarks

As with all attempts to describe a complex matter, this chapter has been able to address only some of the topics that are relevant in the context of solid form development and processing, foremost in the pharmaceutical industry. It is of utmost importance to understand that properties and behavior of all types of solid materials, not just API, require attention along the development and manufacturing chains. Knowing and controlling the particularities of solid materials is an essential asset for all stakeholders, regardless of dealing with scientific, technical, or business aspects.

Risk-based management should include timely investment in solid-state activities and foster an appropriate work environment and infrastructure to avoid huge time lagging and money-intensive efforts to resolve problematic situations and ensure undisturbed marketization of DSs and DPs. Besides ensuring proper and timely technical development, the protection of IP rights and eventually the freedom to operate is a must and convincing aspect to invest into solid-state-related activities.

List of Abbreviations

(A)NDA	abbreviated new drug application
API	active pharmaceutical ingredient (used as synonym for DS)
bn	billion (10^9)
CRO	Contract Research Organizations
CDMO	Contract Development and Manufacturing Organizations
CPP	critical process parameter
DP	drug product
DS	drug substance
DSC	differential scanning calorimetry
DVS	dynamic vapor sorption
eCTD	electronic Common Technical Document
ELN	electronic laboratory notebook
ERP	enterprise resource planning
HMI	human–machine interface
HPLC	high-performance liquid chromatography
IB	investigators brochure
ICH	International Council for Harmonisation (of Technical Requirements for Pharmaceuticals for Human Use)
ICIQ	Institut Catalá d'Investigació Quimica
IP	intellectual property
IR	infrared spectroscopy
LC–MS	liquid chromatography coupled with mass spectroscopy
LEA	lab execution and analysis system
LIMS	lab information system
MSZW	meta-stable zone width
NMR	nuclear magnetic spectroscopy
PCT	patent cooperation treaty (international patent system)
PDF	pair distribution function
PLC	programmable logic controller
PLS	Prozessleitsystem, i.e. distributed control systems (DCS)
PSD	particle size distribution
QA	quality assurance
QbD	quality by design
R&D	research & development
rH	relative humidity
SCADA	supervisory control and data acquisition
SCXRD	single-crystal X-ray diffraction
ssNMR	solid-state nuclear magnetic resonance spectroscopy

T	temperature
TGA	thermogravimetric analysis
XRPD	X-ray powder diffraction

References

1 Beckert, R., Fanghänel, E., Habicher, W.D. et al. (2015). *Organikum: Organisch-chemisches Grundpraktikum*, 24e. Wiley-VCH Verlag: Weinheim.

2 Kranz, G. (2009). *Failure is Not an Option: Mission Control from Mercury to Apollo 13 and Beyond*. Thorndike Press.

3 Feist, B., Slater, S., Bennet, C. et al. (n.d.) Apollo 11 in real-time: a real-time journey through the first landing on the Moon. https://apolloinrealtime.org/11/ (accessed 27 January 2021).

4 Hansen, J.R. (2018). *Aufbruch zum Mond (engl. First man): Neil Armstrong - die autorisierte Biografie*. München: Wilhelm Heyne Verlag.

5 Heise, C., Heise, A., and Persson, C. (eds.) (2018). *c't Retro, Heise Medien GmbH & Co KG*. Germany: Hannover.

6 HNF – Heinz Nixdorf MuseumsForum (n.d.) The world's biggest computer museum in Paderborn. https://www.hnf.de/en/home.html (accessed 27 January 2021).

7 Wikipedia (2019). Newton's laws of motion. https://en.wikipedia.org/w/index.php?oldid=886801669 (accessed 27 January 2021).

8 Schultz, J. (2019) Standing on the shoulders of giants: reach great heights by building upon the achievements of others. https://www.citavi.com/en/planned-accidents/articles/standing-on-the-shoulders-of-giants (accessed 27 January 2021).

9 Bellostas, A.J., López-Arceiz, F.J., and Mateos, L. (2016). Social value and economic value in social enterprises: value creation model of Spanish sheltered workshops. *VOLUNTAS: International Journal of Voluntary and Nonprofit Organizations* 27 (1): 367–391.

10 EFPIA (2019). *The Pharmaceutical Industry in Figures – Key data*. EFPIA.

11 EFPIA (2019). The economic and societal footprint of the pharmaceutical industry in Europe: technical report.

12 Federal Office of Justice (n.d.). Offenlegungspflichten. https://www.bundesjustizamt.de/DE/Themen/Ordnungs_Bussgeld_Vollstreckung/Jahresabschluesse/Offenlegung/Offenlegungspflichten/Offenlegungspflichten_node.html (accessed 27 January 2021).

13 Bundesanzeiger Verlag (n.d.). Bundesanzeiger - search form. https://www.bundesanzeiger.de (accessed 27 January 2021).

14 SOLID-CHEM. Company Philosophy. https://solid-chem.de/en/philosophie/ (accessed 27 January 2021).

15 Stowasser, R. (2009). IP-Manager in research based pharmaceutical industry. In: *IP-Manager* (ed. A.J. Wurzer), 289–298. Cologne: Heymann.

16 Aitipamula, S., Banerjee, R., Bansal, A.K. et al. (2012). Polymorphs, salts, and cocrystals: what's in a name? *Crystal Growth & Design* 12 (5): 2147–2152.

17 Haleblian, J. and McCrone, W. (1969). Pharmaceutical applications of polymorphism. *Journal of Pharmaceutical Sciences* 58 (8): 911–929.

18 Brittain, H.G. (ed.) (1999). *Polymorphism in Pharmaceutical Solids*. New York, NY: Marcel Dekker.

19 Hilfiker, R. (ed.) (2006). *Polymorphism in the Pharmaceutical Industry*. Weinheim: Wiley-VCH.

20 Stahl, P.H. and Wermuth, C.G. (eds.) (2002). *Handbook of Pharmaceutical Salts: Properties, Selection, and Use*. Zürich/Weinheim: Verl. Helvetica Chimica Acta/Wiley-VCH.

21 Stahl, P.H. and Wermuth, C.G. (eds.) (2011). *Handbook of Pharmaceutical Salts: Properties, Selection, and Use*, 2e. Zürich/Weinheim: Verl. Helvetica Chimica Acta/Wiley-VCH.

22 Hilfiker, R. and von Raumer, M. (2018). *Polymorphism in the Pharmaceutical Industry: Solid Form and Drug Development*. Newark: Wiley Incorporated.

23 Byrn, S.R., Pfeiffer, R.R., and Stowell, J.G. (2003). *Solid-state Chemistry of Drugs*, 2e. West Lafayette, IN: Ssci Inc.

24 Bernstein, J. (2007). *Polymorphism in Molecular Crystals*. Oxford: Clarendon Press.

25 Dunitz, J.D. and Bernstein, J. (1995). Disappearing polymorphs. *Accounts of Chemical Research* 28 (4): 193–200.

26 Barsky, I., Bernstein, J., Stephens, P.W. et al. (2008). Disappearing and reappearing polymorphism in *p*-methylchalcone. *Crystal Growth & Design* 8 (1): 63–70.

27 Henck, J.-O., Bernstein, J., Ellern, A., and Boese, R. (2001). Disappearing and reappearing polymorphs. The benzocaine: picric acid system. *Journal of the American Chemical Society* 123 (9): 1834–1841.

28 Pindelska, E., Sokal, A., and Kolodziejski, W. (2017). Pharmaceutical cocrystals, salts and polymorphs: advanced characterization techniques. *Advanced Drug Delivery Reviews* 117: 111–146.

29 Brittain, H.G. (2013). Pharmaceutical cocrystals: the coming wave of new drug substances. *Journal of Pharmaceutical Sciences* 102 (2): 311–317.

30 Bond, A.D. (2007). What is a co-crystal? *CrystEngComm* 9 (9): 833.

31 Gruss, M. (2019). Regulatory aspects for formulation design – with focus on the solid state. In: *Innovative Dosage Forms: Design and Development at Early Stage*, 1e (eds. Y. Bachhav, R. Mannhold, H. Buschmann and J. Holenz), 155–208. Weinheim: Wiley-VCH.

32 Hofmann, G. (ed.) (2005). *Kristallisation in der industriellen Praxis*. Weinheim: Wiley-VCH.

33 Beckmann, W. (ed.) (2013). *Crystallization: Basic Concepts and Industrial Applications*. Weinheim: Wiley-VCH.

34 Tung, H.-H. (2009). *Crystallization of Organic Compounds: An Industrial Perspective*. Hoboken, NJ: Wiley.

35 Myerson, A.S. (ed.) (2002). *Handbook of Industrial Crystallization*, 2e. Butterworth-Heinemann: Boston, MA.

36 Wikipedia (2019). Trennen (Verfahrenstechnik). https://de.wikipedia.org/w/index.php?oldid=188114558 (accessed 27 January 2021).

37 Gartner (n.d.). Gartner glossary – Digitalization. https://www.gartner.com/en/information-technology/glossary/digitalization (accessed 27 January 2021).

38 Gartner (n.d.). Gartner Glossary Information Technology – Digital Transformation. https://www.gartner.com/en/information-technology/glossary/digital-transformation (accessed 27 January 2021).

39 Brudermüller, M. (2019). Auf dem Weg in das Labor der Zukunft, Grußwort, in Trendreport 2019 – Analysen-, Bio- und Labortechnik: Märkte, Entwicklungen, Potenziale (ed Spectaris), p. 3.

40 Seeberg, P. (2019). Machine learning in der industrie. *Selbst lernen* IX (11): 64–68.

41 Griffin, A. and Hauser, J.R. (1992). The voice of the customer. *Marketing Science* 12 (1) Winter 1993: 1–27.

42 Vlaskovits, P. (2011). Henry Ford, innovation, and that "Faster Horse" quote. https://hbr.org/2011/08/henry-ford-never-said-the-fast (accessed 27 January 2021).

43 Brock, C.P., Hahn, T., Wondratschek, H. et al. (2016). *International Tables for Crystallography Volume A: Space-group symmetry*. Chester, England: International Union of Crystallography.

44 Lubach, J.W. and Munson, E.J. (2006). Solid-state nuclear magnetic resonance of pharmaceutical formulations. *Encyclopedia of Analytical Chemistry* 2: 1–18. Meyers (Hg.).

45 Young, C.A. and Goodwin, A.L. (2011). Applications of pair distribution function methods to contemporary problems in materials chemistry. *Journal of Materials Chemistry* 21 (18): 6464–6476.

46 Prill, D., Juhás, P., Schmidt, M.U., and Billinge, S.J.L. (2015). Modelling pair distribution functions (PDFs) of organic compounds: describing both intra- and intermolecular correlation functions in calculated PDFs. *Journal of Applied Crystallography* 48 (1): 171–178.

47 Prill, D., Juhás, P., Billinge, S.J.L., and Schmidt, M.U. (2016). Towards solution and refinement of organic crystal structures by fitting to the atomic pair distribution function. *Acta Crystallographica Section A: Foundations and Advances* 72 (Pt 1): 62–72.

48 Thakralab, S., Terbanc, M.W., Thakralad, N.K., and Suryanarayanana, R. (2016). Recent advances in the characterization of amorphous pharmaceuticals by X-ray diffractometry. *Advanced Drug Delivery Reviews* 100: 183–193.

49 Threlfall, T.L. (2009). Turning DSC charts of polymorphs into phase diagrams: a tutorial paper. *Organic Process Research and Development* 13 (6): 1224–1230.

50 Chiarella, R.A., Davey, R.J., and Peterson, M.L. (2007). Making co-crystals – the utility of ternary phase diagrams. *Crystal Growth & Design* 7 (7): 1223–1226.

51 Nehm, S.J., Rodríguez-Spong, B., and Rodríguez-Hornedo, N. (2006). Phase solubility diagrams of cocrystals are explained by solubility product and solution complexation. *Crystal Growth & Design* 6 (2): 592–600.

52 Leyssens, T. (2018). Constructing and using cocrystal phase diagrams in the context of dissolution studies. https://www.crystallizationsystems.com/training-and-events/webinar-on-constructing-and-using-phase-diagrams (accessed 27 January 2021).

53 Neumann, M.A., van de Streek, J., Fabbiani, F.P.A. et al. (2015). Combined crystal structure prediction and high-pressure crystallization in rational pharmaceutical polymorph screening. *Nature Communications* 6: 7793.

54 Wiktionary.org poly (2020). https://en.wiktionary.org/wiki/poly- (accessed 27 January 2021).

55 Wiktionary.org –morphic (2017). https://en.wiktionary.org/wiki/-morphic (accessed 27 January 2021).

56 Sharma, B.D. (1987). Allotropes and polymorphs. *Journal of Chemical Education* 64 (5): 404.

57 EMA (2015). Reflection paper on the use of cocrystals of active substances in medicinal products. CHMP/CVMP/QWP/284008/2015.

58 Hollemann, A.F. and Wiberg, E. (1953). *Lehrbuch der Anorganischen Chemie*, 32e. Berlin: Walter de Gruyter & Co.

59 leo.org translation of "modification" to German and from the German words back to English (n.d.). https://dict.leo.org/englisch-deutsch/modification.

60 Schäfer, H., Gruehn, R., and Schulte, F. (1966). Die Modifikationen des Niobpentoxids. *Angewandte Chemie* 78 (1): 28–41.

61 Stahly, G.P. (2009). A survey of cocrystals reported prior to 2000. *Crystal Growth & Design* 9 (10): 4212–4229.

62 Thayer, A.M. (2007). War of the words – dispute over crystal structure nomenclature takes center stage. *Chemical and Engineering News* 85 (25): 28–29.

63 Gadade, D.D. and Pekamwar, S.S. (2016). Pharmaceutical cocrystals: regulatory and strategic aspects, design and development. *Advanced Pharmaceutical Bulletin* 6 (4): 479–494.

64 Childs, S.L., Stahly, G.P., and Park, A. (2007). The salt-cocrystal continuum: the influence of crystal structure on ionization state. *Molecular Pharmaceutics* 4 (3): 323–338.

65 Rossi, F., Cerreia Vioglio, P., Bordignon, S. et al. (2018). Unraveling the hydrogen bond network in a theophylline–pyridoxine salt cocrystal by a combined X-ray diffraction, solid-State NMR, and computational approach. *Crystal Growth & Design* 18 (4): 2225–2233.

66 Putra, O.D., Yoshida, T., Umeda, D. et al. (2016). Crystal structure determination of dimenhydrinate after more than 60 years: solving salt–cocrystal ambiguity via solid-state characterizations and solubility study. *Crystal Growth & Design* 16 (9): 5223–5229.

67 Ringle, S. (2018). Regulatory implications using API cocrystals for generic medicinal products within the EU and US. Friedrich-Wilhelms-Universität. Master thesis.

68 Schulze, B. (2016). *Different Salts of a Drug Substance –Comparison of Regulatory Pathways in the EU and USA*. Rheinischen Friedrich-Wilhelms-Universität Bonn.

69 Newman, A.W., Reutzel-Edens, S.M., and Zografi, G. (2008). Characterization of the "hygroscopic" properties of active pharmaceutical ingredients. *Journal of Pharmaceutical Sciences* 97 (3): 1047–1059.

70 Krzyzaniak, J.F., Williams, G.R., and Ni, N. (2007). Identification of phase boundaries in anhydrate/hydrate systems. *Journal of Pharmaceutical Sciences* 96 (5): 1270–1281.

71 ICH (2003). ICH guideline Q1A (R2) stability testing of new drug substances and products.

72 Variankaval, N., Lee, C., Xu, J. et al. (2007). Water activity-mediated control of crystalline phases of an active pharmaceutical ingredient. *Organic Process Research and Development* 11 (2): 229–236.

73 Li, Y., Chow, P.S., Tan, R.B.H., and Black, S.N. (2008). Effect of water activity on the transformation between hydrate and anhydrate of carbamazepine. *Organic Process Research and Development* 12 (2): 264–270.

74 Grothe, E., Meekes, H., Vlieg, E. et al. (2016). Solvates, salts, and cocrystals: a proposal for a feasible classification system. *Crystal Growth & Design* 16 (6): 3237–3243.

75 Chakravarty, P. and Suryanarayanan, R. (2010). Characterization and structure analysis of thiamine hydrochloride methanol solvate. *Crystal Growth & Design* 10 (10): 4414–4420.

76 Alsenz, J. and Kansy, M. (2007). High throughput solubility measurement in drug discovery and development. *Advanced Drug Delivery Reviews* 59 (7): 546–567.

77 Petereit, A. and Saal, C. (n.d.) What is the solubility of my compound? Assessing solubility for pharmaceutical research and development compounds. http://www.americanpharmaceuticalreview.com/Featured-Articles/36993-What-is-the-Solubility-of-My-Compound-Assessing-Solubility-for-Pharmaceutical-Research-and-Development-Compounds/ (accessed 27 January 2021).

78 Reus, M.A., van der Heijden, A.E.D.M., and ter Horst, J.H. (2015). Solubility determination from clear points upon solvent addition. *Organic Process Research and Development* 19 (8): 1004–1011.

79 Bernstein, J., Davey, R.J., and Henck, J.-O. (1999). Concomitant polymorphs. *Angewandte Chemie, International Edition* 38 (23): 3440–3461.

80 Bučar, D.-K., Lancaster, R.W., and Bernstein, J. (2015). Disappearing polymorphs revisited. *Angewandte Chemie, International Edition* 54 (24): 6972–6993.

81 Desikan, S., Parsons, R.L., Davis, W.P. et al. (2005). Process development challenges to accommodate a late-appearing stable polymorph: a case study on the polymorphism and crystallization of a fast-track drug development compound. *Organic Process Research and Development* 9 (6): 933–942.

82 Abramov, Y.A. (ed.) (2016). *Computational Pharmaceutical Solid State Chemistry*. Wiley.

83 Fujiwara, M., Nagy, Z.K., Chew, J.W., and Braatz, R.D. (2005). First-principles and direct design approaches for the control of pharmaceutical crystallization. *Journal of Process Control* 15 (5): 493–504.

84 Rantanen, J. and Khinast, J. (2015). The future of pharmaceutical manufacturing sciences. *Journal of Pharmaceutical Sciences* 104 (11): 3612–3638.

85 Gil Chaves, I.D., López, J.R.G., García Zapata, J.L. et al. (2016). *Process Analysis and Simulation in Chemical Engineering*, 1e. Cham: Springer International Publishing.

86 Boehling, P., Toschkoff, G., Just, S. et al. (2016). Simulation of a tablet coating process at different scales using DEM. *European Journal of Pharmaceutical Sciences: Official Journal of the European Federation for Pharmaceutical Sciences* 93: 74–83.

87 Böhling, P., Khinast, J.G., Jajcevic, D. et al. (2019). Computational fluid dynamics-discrete element method modeling of an industrial-scale Wurster coater. *Journal of Pharmaceutical Sciences* 108 (1): 538–550.

2

Determination of Current Knowledge

Andriy Kuzmov and Ronak Savla

Catalent Pharma Solutions, 14 Schoolhouse Road, Somerset, NJ 08873, USA

2.1 Why Is it Important to Search for Relevant Information Before Starting a Solid-State Project?

It is important to understand the current state of knowledge before embarking on a solid-state project for several reasons. Mainly, you do not want to spend valuable time, effort, and resources repeating the work someone has already completed or repeat their mistakes. There may be aspects of your project that have been previously studied, or there may be patents that limit what you can commercialize. Examples include the following:

(a) What is an acceptable upper limit of salt former in oral formulations? [1]
(b) What are the effects of the acidity of the counterion on the solubility and performance of pharmaceutical salts? [2]
(c) What are some of the challenges in characterizing salts and cocrystal forms? [3]
(d) What are the challenges in accurately measuring and interpreting the solubility of pharmaceutical salt forms in buffers and simulated intestinal fluids? [4]

The typical process before starting a research project is as follows:

1. Identify problem or challenge
2. Develop and execute a knowledge gathering process
3. Understand what is still left unaddressed or not fully addressed
4. Begin research

In this chapter, we will focus on the second point, the knowledge gathering process. The first step many researchers take is to reach out to colleagues who may have experience working on similar projects to help you create a development strategy for your project. If the compound has previously been within your institution, your colleague can help provide you with information about the challenges associated with the compound (e.g. solvent compatibility, solubility in various solvents, and manufacturing procedures). If working in an established laboratory, in-house protocols can be a great place to get the project off the ground, determine some of the internal

Solid State Development and Processing of Pharmaceutical Molecules: Salts, Cocrystals, and Polymorphism, First Edition. Edited by Michael Gruss.

expertise and capabilities, and identify parts of the project that may need external expertise for completion.

These are just some of the first places to begin the knowledge gathering process and they can always be supplemented with a robust search strategy to ensure the most up-to-date information is driving the development work. By performing a robust search, you can identify areas that present a challenge to solid-state development projects, gain insights into approaches others have taken to solve complicated issues, and determine whether your project has the potential to generate intellectual property rights. A robust search can also help you identify the experts in the field that you can contact if your team requires additional expertise.

If your project is going to be the targeting markets around the world, the team needs to understand the regulatory perspective on what is considered the "same" compound both from an innovator perspective as well as a generic product perspective. Looking at the regulations from the Food and Drug Administration (FDA) and European Medicines Agency (EMA), the regulations are generally the same with some nuanced differences. According to FDA guidance, "same as" means identical "active ingredient(s)" when compared to a reference listed drug (RLD). A different polymorphic form of a compound is not considered to be a different active; however, there is a need to show that the drug product has similar stability and bioavailability to the RLD [5]. According to a FDA guidance released in February 2018, cocrystals are considered to be a polymorph of the active pharmaceutical ingredient (API), meaning a cocrystal form is not considered to be a new API. This is in contrast to the current US regulations pertaining to different salt forms of the same active which are considered to be different APIs [6]. According to EMA regulations, oral administration of different salts, polymorphs, cocrystals, hydrates, or solvates of an API is not considered to be a new active substance, unless there is data to demonstrate that the new form differs from the RLD in terms of safety and efficacy [7]. The idea of "same as" seems simple enough; yet, the slightly different points of view on the solid state of the API by the FDA and EMA highlight the complexity and importance of getting this right. Table 2.1 summarizes the current viewpoints of the FDA and EMA when it comes to different solid-state forms. For example, a new polymorph of a compound can be marketed as a generic version for an RLD once the material patent has expired even though the currently approved drug product uses a polymorph which still has patent protection [8]. These considerations have regulatory implications; however, many of these solid-state nuances have intellectual property implications as well.

Table 2.1 Are different solid states considered to be the "same" by regulatory authorities?

	FDA	EMA
Cocrystals	Same	Same
Salt forms	Different	Same
Polymorphs	Same	Same
Hydrates and solvates	Same	Same

Source: © John Wiley & Sons, Inc.

When developing the solid-state form of a new chemical entity (NCE), it is important to understand the patent and intellectual property aspects as they relate to the material and solid-state patents of the API.

1. How to prepare a comprehensive patent application centered around the solid state of the API [9].
2. Precedent set by previous litigation involving material patents and solid-state patents and learn from successful and unsuccessful intellectual property challenges [10].
3. Strategies that competitors may use to circumvent solid-state patents such as filing a 505(b) [2] application instead of an ANDA [11]. A good understanding of the intellectual property landscape can help guide your early development work to ensure the appropriate data are collected to support claims that can improve the patent protection of your novel compound [10].
4. If you are a generic company trying to enter the market of a drug that has an expiring material patent but an active solid state or polymorph patent, it is important to understand the types of patent claims that have been challenged. As an example, Celgene and Dr. Reddy's have a legal dispute concerning whether an amorphous form of lenalidomide developed by Dr. Reddy's infringes Celgene's patents on different polymorphs of lenalidomide [12]. With the appropriate experience and knowledge, generic companies can identify patent applications that have weak or no data to support a patent's innovation claims and have a higher chance of getting approval of their generic product for an earlier entry into the generic market [10].

2.2 Where to Begin a Literature Search for a Solid-State Project?

The project should have a clear objective with specific actions and procedures to help track the progress of the work. The project and the search strategy need to begin with a clear objective and a thorough understanding of the steps and procedures that need to be completed. For example, you need to gather information about synthesis and manufacturing procedures, analytical techniques to identify polymorphs, polymorphic landscape in drug development, market data, formulation approaches for the desired delivery route, or a regulatory perspective of various solid states in the drug approval process. Each of these questions is related to any solid-state project in some way and needs to be addressed. Answering all these questions can be a complex process that one person may have difficulty completing on their own. Instead, a multifunctional team comprising of a mixture of chemists, analytical experts, product development scientists, marketing specialists, regulatory specialists, and intellectual property specialists is recommended to develop a comprehensive evaluation of the project's potential. Each can focus on answering the questions related to their individual area of expertise while learning from their colleagues the parallel considerations involved in completing a successful solid-state project.

To determine the current state of the art in a given field, a literature and patent search using several reputable databases can serve as a foundation of knowledge to get a project off the ground. To gather the necessary information, it is crucial to have a detailed search strategy in place which can include the following steps:

- Develop a research question
- Develop a search strategy
- Run the search
- Refine the search
- Analyze the hits and draw conclusions

Developing the question that is asked can have a large effect on the results that are obtained. The broader the question, the more likely it is to yield many results that are not relevant to the issue you are trying to overcome. If the question is too narrow, only a limited number of results will be returned which limits the ability to properly evaluate the scope of the project. The motivation of the patent and literature search (e.g. determine the current state of the art, landscape report, freedom to operate analysis, and data needed to support patent filing) can help guide the search strategy and determine the resources required for the search.

2.2.1 Literature Search

A thorough literature search can provide a foundation for the beginning of a solid-state project. Reading recently published papers is the best way to get information on the latest research and state of the art. A literature search can begin by using a database such as PubMed, Europe PMC, Google Scholar, Scopus, SciFinder, Web of Science, and others. There are many similarities between the literature databases, but they differ in terms of the journals that they cover and the method that they use to index publications. Simply having a greater number of titles covered does not mean that you will get better results. The focus should be on quality and if the database covers the relevant journals of interest.

Another option is to directly search the websites of publishers such as American Chemical Society, SpringerNature, Wiley Online Library, Elsevier (ScienceDirect®), and others. This is a good approach if there are prominent journals in the scientific field such as *Crystal Growth and Design or Organic Process Research and Development*. PubMed and Google Scholar search a wide range of journals and publishers and return results for both subscription-based and open access journals with the option to display only the open access content. Publishers like Elsevier (ScienceDirect), Springer (SpringerLink®), and many others provide subscription-based access to journals as well as individual article purchases. Several open access journals and publishers provide access to a subset of their content for free. PubMed Central (https://www.ncbi.nlm.nih.gov/pmc) is a search database within PubMed that focuses literature searches to journals and articles that are open access. Open access journals cover many of the same topics and subject areas as those covered in the subscription-based databases; however, they often have less articles in their databases. For a comprehensive search, it is important to collect data from a variety of journals and access content from both subscription and open access journals.

2.2.1.1 Focusing Your Literature Search

When performing a search using these literature databases, how the search is performed is just as important as the question or search terms used during the search. Most databases allow for advanced searches that apply Boolean operators to help customize a search to return the most relevant content. The three Boolean operators used by most search engines are AND, OR, and NOT. AND limits the search to include results that contain all the terms used to run the search (Figure 2.1a). The operator OR increases the number of results by returning text that containing either one, or all the search terms connected with OR (Figure 2.1b). NOT excludes search results that contain the specified phrase or term (Figure 2.1c) [13]. Furthermore, using parentheses in combination with the Boolean operators can help tailor the search to the needs of a given project. Many search databases also allow you to search specific fields such as the title, author, journal, date of publication, affiliation, and others. This can be useful to help build a custom search that will only find the documents of interest to you.

After the search results are returned, the databases allow for the filtering of the data. For example, PubMed allows to filter the results based on the article type (clinical trial, review, systematic review, etc.), availability (abstract, full text, free full text), publication date, and species. It may be beneficial to order search results by date of publication. The default setting on PubMed is to display results with the chronologically most recent publications. The potential downside is that more relevant publications may get buried within the results. It is recommended to order the results both chronologically and by relevance. At the start of a search reviewing the state of the art, it may be useful to search for recent review articles that can provide a snapshot of the field. The reference list in review articles is an excellent source to find related articles. As you have a better understanding of the field, you can begin to focus the search and evaluate primary literature that is related to the specific question you are trying to answer (research articles and clinical trials).

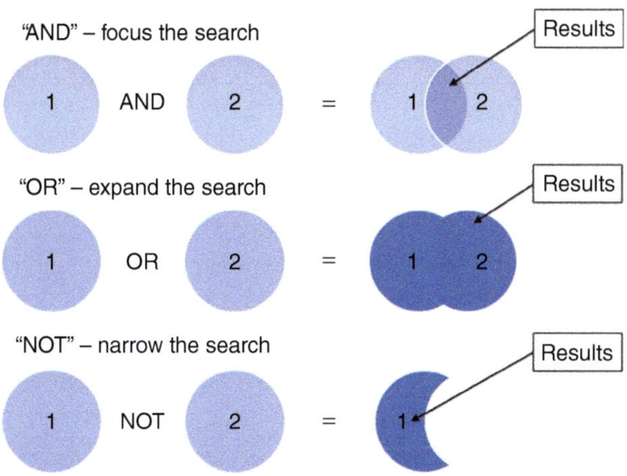

Figure 2.1 Visual depiction of the search results that will be returned when using "AND," "OR," and "NOT" to build a search string.

Example: "Ibuprofen Solid-State Polymorphism" Literature Search As an example, let us say you wanted to consider starting a solid-state project looking at ibuprofen polymorphism. You might begin by searching "ibuprofen," "solid state," and "polymorphism" in the PubMed Database. Table 2.2 shows a number of search results that are retrieved when different search strategies were used during the preparation of this chapter (week of 8 July 2019). Based on how the search is entered in the database using parentheses and the Boolean operators, vastly different numbers of results are obtained.

A simple search of "ibuprofen solid-state polymorphism" leaves us with only seven results which may or not contain the information you need to determine whether to pursue this project. To broaden the search, entering "ibuprofen polymorphism" into the database returns a plentiful but manageable 70 results. Upon examination of the results, 43 results address the effects of genetic polymorphism, especially the cytochrome P 450 (CYP450) family of metabolizing enzymes, on the pharmacokinetics and utility of ibuprofen [14–19]. The search needs to be refined to exclude text that refers to genetic polymorphism which was included in the search results due to the use of similar terminology in a related field. To help focus the search toward the original topic, careful use of parentheses and the use of "NOT" to exclude "CYP450" and "Genetic" yield a manageable number of results. Searches #4 and #5 show how to expand the search to incorporate ibuprofen solid state into the search strategy without growing the results into the tens of thousands. Searches #7–9 show how a sub-optimal search strategy can lead to an unreasonably high number of results that likely have little to do with the initial question, the goals of the project, or the issues you are trying to address with you project.

Table 2.2 PubMed search results when entering "ibuprofen solid state polymorphism" and related searches into the PubMed database using various search strategies during the week of 8 July 2019.

Search	Query	Items found
#1	Ibuprofen solid-state polymorphism	7
#2	Ibuprofen polymorphism	70
#3	((Ibuprofen polymorphism) NOT CYP450) NOT genetic	27
#4	(Ibuprofen) AND (solid state OR polymorphism)	240
#5	((Ibuprofen) AND (solid state OR polymorphism) NOT CYP450) NOT genetic	196
#6	(Ibuprofen OR solid state) AND polymorphism	1625
#7	((Ibuprofen) OR solid state) OR polymorphism	503 072
#8	(Ibuprofen and polymorphism) OR solid state	116 956
#9	Solid state	116 893
#10	Ibuprofen crystal structure	133
#11	Ibuprofen crystallization	505

Searches #10 and #11 show how to optimize the search strategy by using different but related terms such as "ibuprofen crystal structure" [20–24] or "ibuprofen crystallization" [25–29] (some of the returned references for each search are cited as an example of the returned results). This is just an example of the variables to consider when conducting your search strategy.

When keeping a record of the search strategy, it is important to record the database, the date of the search, the search terms used, how the search was input into the data base, and any filters applied to the search results. This example highlights the importance of keeping a record of the search strategy. Some benefits include:

1. Avoid repeating searches and work
2. Allowing someone to reproduce your search
3. Being able to pick up where you left off
4. Ensure the search was comprehensive
5. Help you track innovation and identify newly published research

2.2.2 Staying on Top of the Latest Publications

Keeping up with various journals and the latest publications can easily become overwhelming. One strategy is to set aside dedicated time to search databases and journals. However, this is a time-consuming process and likely to spend time looking at articles that are not of relevance. A better strategy would be to set alerts for new publications in journals or for searches. For PubMed, one can easily set up alerts for new publications by clicking the "Create alert" button under the search bar. Many journals and publishers also email alerts for new articles, articles in Press, and the table of contents of new issues. Embracing social media such as twitter is a useful way to find latest articles as most publishers promote new publications via social media. Social media also provides a platform to connect with peers. There are research specific social media sites such as Academia.edu and ResearchGate designed for researchers to share publications. However, these websites have faced litigation from publishers for inappropriate sharing of articles that violate copyright agreements.

2.3 Patent Search

A literature search is only the beginning of a comprehensive search strategy with a thorough patent search helping complete the evaluation of the state of the art. Patents and patent applications are an excellent source of information that has not been published in other forums with one study from the mid 1980s finding that 80% of information found in patents cannot be found anywhere else [30]. A thorough patent search alongside a literature search therefore provides a more complete picture of the state of art. A patent search also allows for the determination of the probability your invention can be patented, identification of potential roadblocks before you proceed with your product development, or provision of prior art to help you challenge an existing patent.

2.3.1 Types of Patent Reports

When evaluating patent literature, it is important to consider whether to evaluate patent applications, granted patents, or both. Patent applications are published by the patent office 18 months after their earliest filing date. The corresponding patents, if granted, are published with the granted claim set. Patent applications are a good source of information to reveal the direction of a specific technical field as they may represent the cutting-edge advances in a given field. Granted patents provide information regarding the type of technical advances considered to be truly innovative and worthy of patent protection. A patent search can be conducted to support the development of a patent landscape report (or a technology landscape report, together with a literature search), a patent map, patent watch reports, a freedom-to-operate report, and a prior art or validity report [31].

A patent landscape report is generated as an overview of the patenting activity and trends in a given technical field. Generally, the report is generated to present complex data in a simple manner to key decision makers. For industry applications, these reports can drive strategic investment decisions, research and development directions, or competitor evaluations. Governments may use patent landscape reports to help drive the creation of legislation and policy as it applies to a given technological field. A patent landscape report can include different types of information depending on who is requesting and executing the report and can be expanded to include literature searches, market analysis, patent review, and other data [31].

The patent map, often part of a patent landscape report, is a graphical representation of patent data to help improve the interpretation of the data. A patent watch or alert is a systematic process developed to monitor newly published patent applications, track changes in the status of patent applications, or monitor granted patents to stay up to date on the most recent innovations in a field. The systematic patent watch can help you track the activity of key competitors and such information can be used to help the leadership team plan the next steps for your company [31].

Freedom to operate searches can be of great importance for a company before or during a development project to ensure that their products do not infringe on any valid granted patents. The search is focused on identifying active granted patents related to the project at hand [31]. For solid-state projects, this could relate to the current patents centered around the API that you want to use for the development of a new product. It is critical to identify the status of the patent for the compound (i.e. API) you want to develop. If the compound patent is still active, it will be nearly impossible to launch a new product with the same compound until the patent has expired as compound patents are difficult to challenge [8]. Once the status of the compound patent has been determined, it is important to understand any secondary solid-state, drug product, and process patents that have been granted to the innovator or any third-party companies trying to enter the field once the compound patent expires.

As an example, a Google Patents search for granted US patents for "crystal forms of atorvastatin" returns 2700+ results with 10+ different assignees during the week of 8

July 2019. As with the literature search, it is important to refine your patent search to ensure patents that contain relevant information are returned. Scanning the results shows that some of the patents only reference atorvastatin in the abstract or body of the patent while discussing similar compounds or detailing novel applications. All the results are not going to be relevant to the question that needs to be answered which highlights the importance of refining the search strategy to obtain relevant results.

2.3.2 Understanding the Elements of Patents

Patents reports that review hundreds, thousands, or more patents are constructed to allow for the analysis of large sets of data into simple graphical or statistical representations. The metadata used in the creation of such reports is often collected from the front-page bibliographic data of a published patent or application. The metadata may include the following: applicant/assignee, inventor, dates, priority data, classifications, and citations.

The applicant or assignee "is the entity or person which or who presents an application for the grant of an industrial property right in an industrial property office, or in whose name an agent files such an application" [31]. The applicant can be the inventor, or the employer to whom the inventor assigned the rights of the invention. The applicant/assignee is the owner of the patent and controls the rights of the invention described in the patent. The name of the applicant/assignee on a given patent can change if the original applicant decides to sell the ownership rights to their patent.

The inventor is someone "who is the author of an invention" and "has the right to be mentioned as such in the patent" [31]. Unlike assignees, the inventor(s) listed in a patent or a patent application are comprehensive and usually do not change with time. Analysis of the inventors in a given field can provide a snapshot of the experts in the field, identify collaborations among different groups, and identify the companies or universities where the inventors work.

The dates that are important to consider and understand when evaluating patent literature are the priority, filing, and publication dates. The priority date is the earliest date to which a patent application can claim priority and largely determines what materials are prior art to the patent application. The filing date is determined by the patenting authority when certain requirements are met by the applicant. In certain cases, the priority date and the filing date can be the same date. The publication date is the date the patent application is made public. As stated before, this is most often 18 months after the priority date of the patent application.

The priority data are a set of three key pieces of information published on the first page of a patent document and includes the application number for the priority application, the priority filing date, and the country or organization where the earlier application was filed. The priority data can be used to identify a patent family or a group of patents for the same invention with filing in multiple countries/jurisdictions [31].

2.3.3 Patent Classification

Patents and patent applications are classified using a unique system that allows for categorizing patents into multiple groups based on their technical field. The Cooperative Patent Classification (CPC) and the International Patent Classification (IPC) are two examples of widely used patent classification systems, while there are other systems used by the various patent offices around the world [32, 33]. The IPC system is used by most jurisdictions around the world, including the World Intellectual Property Organization (WIPO). The IPC is regularly updated to include new classifications and sub-divisions as new technologies emerge [32]. Although the IPC is used around the world, there have been a few other classification systems that have gained popularity over the years. In 2010, the US Patent and Trademark Office (USPTO) and the European Patent Office (EPO) announced a joint agreement to work together to create a patent classification system shared by both offices and can be applied by any other patent office around the world. From this initiative, the CPC system was created. The CPC is designed to be comparable to the IPC classification system and a guide for the conversion of CPC codes to IPC codes can be found on the USPTO website (https://www.uspto.gov/web/patents/classification). The CPC codes are composed of a code of letters and numbers. More details about the structure of the CPC along with examples are provided below (Tables 2.3–2.5).

The section, class, subclass, main group, and sub-group of a patent can be easily determined by dissecting a single classification code [33]. The major sections of the CPC are listed in Table 2.3 [34]. For example, all classification codes that begin with the letter "A" describe patent that are in some way related to human necessities. Most applications pertaining to pharmaceuticals and the solid state of organic molecules will be classified under CPC section C (chemistry), class C07(organic chemistry) with the structure of the compound further described by the various subclasses, groups, and subgroups.

Table 2.3 CPC section codes.

CPC sections	Meaning
A	Human necessities
B	Performing operations; transporting
C	Chemistry; metallurgy
D	Textiles; paper
E	Fixed constructions
F	Mechanical engineering; lighting; heating; weapons; blasting
G	Physics
H	Electricity
Y	General tagging of new technological developments

Source: Adapted from European Patent Office [13].

A sampling of the common subclassifications that may be used to categorize patents that contain solid-state information is listed in Table 2.4. Patents classified in subclass A61K relate to a "preparation for medical, dental, or toilet purposes," as it relates to pharmaceutical products, this classification is often used for patents that describe the formulation used in creating the dosage form of a compound. Patents with this subclass classification may contain information relating to the polymorphic form used for the development of various tablets, capsules, solutions, and other dosage forms. It is worth noting that the CPC has a designated section under subclass C07B (general methods in organic chemistry) to classify patents that describe the physical–chemical properties of organic compounds.

Table 2.5 lists certain physical–chemical properties and the corresponding CPC codes. The CPC code C07B 2200/13 classifies patents describing the crystal form

Table 2.4 The most common CPC subclasses that may be used to categorize pharmaceutical patents containing solid-state information.

CPC subclass	Meaning
A61K	Preparations for medical, dental, or toilet purposes
A61P	Specific therapeutic activity of chemical compounds or medical preparations
C07B	General methods in organic chemistry
C07C	Acyclic or carboxylic compounds
C07D	Heterocyclic compounds
C07F	Acyclic, carboxylic, or heterocyclic compounds containing elements other than C, H, Halogens, O, N, S, Se, Te
C07G	Compounds of unknown constitution
C07H	Sugars, nucleosides, nucleotides, nucleic acids
C07J	Steroids
C07K	Peptides

Table 2.5 Physical–chemical properties indexed by the CPC system.

CPC codes for indexed properties	Meaning
C07B 2200/00	Indexing scheme relating to specific properties of organic compounds
C07B 2200/01	Charge–transfer complexes
C07B 2200/03	Free radicals
C07B 2200/05	Isotopically modified compounds
C07B 2200/07	Optical isomers
C07B 2200/09	Geometrical isomers
C07B 2200/011	Compounds covalently bound to a solid support
C07B 2200/013	Crystalline forms (polymorphs)

(e.g. polymorph) of a compound. However, every patent that has information about the crystal form of a molecule will not necessarily be classified under CPC code C07B2200/13. Some patents describe a systematic approach to polymorph screening, the manufacturing processes used to generate the API, the creation of the dosage form, or other related processes. In all the cases, the patents may claim the specific polymorphic forms that were observed, discovered, or used in the applications claimed by the authors. A complete listing of the CPC code classifications can be found on the USPTO website under the patent classification section [35].

A single patent may have only one classification or 30+ classifications depending on the number and nature of the patent claims. When conducting a patent landscape report, classification codes can be used to help narrow the search to return only patents related to a specific technology or application. Alternatively, if there are related classifications, they can be used to expand your search to include more patent data. US patent 7488750B2 for crystal forms of atorvastatin has two classifications: C07D207/34, A61K31/40 [36, 37]. The code C07D207/34 can be broken down to provide information about the section, class, subclass, group, and subgroup relating to the nature of the patent. The section for this classification is C for chemistry. The class C07 indicates organic chemistry and subclass C07D for heterocyclic compounds. The additional numbers 207/34 assign a main group (207) and sub-group [34]. The entire CPC code describes the chemistry of compounds in this classification as follows: "Heterocyclic compounds containing five-membered rings not condensed with other rings, with one nitrogen atom as the only ring hetero atom with only hydrogen or carbon atoms directly attached to the ring nitrogen atom having two double bonds between ring members or between ring members and non-ring members with heteroatoms or with carbon atoms having three bonds to hetero atoms, with at the most one bond to halogen, e.g. ester or nitrile radicals, directly attached to ring carbon atoms." The code C07D207/34 is not limited strictly to atorvastatin but can describe any number of compounds with a variety of applications, both pharmaceutical and non-pharmaceutical, that meet the chemical description provided by the code.

2.3.4 Patent Databases

The following is a list of some of the countries that provide a searchable patent database to the public: Australia, Canada, China, Denmark, Finland, Germany, Great Britain, India, Israel, Japan, Netherlands, Norway, Sweden, Switzerland, Taiwan, United States, and others. To increase access of the patents to a wider audience, many of the databases provide machine generated translations for the patents in their library. Searching the website of individual countries, patent offices may be too time consuming and miss valuable information from other patent offices. Several free patent databases that search patent documents from multiple countries are available to the public. These include Google Patents, Espacenet® (EPO database), Patentscope® (WIPO database), and Lens.

Along with the free databases, several subscription-based patent databases and search firms are available to help conduct a patent search and produce patent reports that meet the needs of the client, respectfully. The paid databases include Derwent World Patents Index™ by Clarivate Analytics, Orbit Intelligence by Questel, TotalPatent® by LexisNexis®, PatBase, and others. Numerous organizations offer intellectual property services to clients. If the help of outside search firms is required to complete your prior art or patent search, it is best to use a firm that has experience in your technical field to help you craft and implement an appropriate search strategy.

2.3.4.1 Free Patent Databases

To help gather large segments of patent data into one easily accessible platform, several groups have developed open access databases for searching the patent data from several patent authorities. Each of these databases contains a slightly different number of patents, use different search tools, and have unique user interfaces. This chapter will cover the following free patent databases: Google Patents, Espacenet, Patentscope, and Lens [38]. There are other patent databases and tools; however, these are easy to use, open to users across the globe, provide computer generated patent translations for multiple languages, and provide user interfaces in multiple languages. Although these tools draw their data from similar pools of patent data, each provides users with unique features.

Google Patents The Google Patents search engine references more than 120 million patents and applications with indexed full text documents from 22 different patent offices around the world. US patent documents date back to 1790 while the WIPO and European patent documents go back to 1978. If trying to conduct a comprehensive search, Google Patents allows the inclusion of non-patent literature in the search results by searching your search terms in the Google Scholar database. This can be useful when trying to gather a quick overview of the patent and scholarly landscape. If the patents were filed in a different language, the database provides machine generated translations of the patents, a feature seen with many patent offices around the world [39].

When using Google Patents, the search can be conducted using the patent, publication or application number, CPC codes, free text, or metadata such as the inventor, assignee, date, status, language, and country. The search engine supports the use of Boolean operators to help narrow the search strategy. Sometimes, when searching for a specific molecule, the search engine will return results for patents that claim broad applications for many compounds, including your compound of interest.

During the week of 8 August 2019, searching "efavirenz form" using Google Patents returned 5340 results. One of the results was for European patent EP1632479B1 titled "Pharmaceutical compositions comprising CCR5 antagonizing piperazine derivatives." The patent describes the structure and utility of CCR5 antagonists in combination with other approved anti-retroviral products for the treatment of HIV [40]. The patent contains both search terms, form and efavirenz, within its text, but it has little relevance to the polymorphic forms of efavirenz.

To better focus the search results, you can refine the search terms, or limit the inclusion of patents that contain your search term or terms in a specific section of

Table 2.6 Different numbers of results obtained when using varying search strategies using google patents during the week 8 July 2019.

Search	Query	Items found
#1	efavirenz form	5340
#2	(TI = efavirenz) (form)	63
#3	TI = efavirenz TI = form	9
#4	TI = efavirenz AB = form	29
#5	TI = efavirenz CL = form	46

the patent such as the title, abstract, or claims. Table 2.6 shows the number of results obtained using alternate search strategies. It is important to ensure the appropriate syntax is used when entering the terms of interest into the search bar. Google Patents also allows for the use of a proximity function (how close two terms are within the text) to improve the ranking order of the returned search results.

Espacenet Espacenet.com is an EPO website which enables a user to search a collection of patents from over 90 countries. The EPO maintains the site in the three official languages: English, French, and German. Many other countries have adopted Espacenet as the patent search engine for their county and provide access in the language of their country (change the country on the home page to access this option). Espacenet has three basic search types: smart, advanced, and classification [34].

The smart search is like the search bar found in Google Patents where it has a set of default search parameters and allows for searching different patent field identifiers. The Espacenet pocket guide provides a quick overview of the field identifiers that can be used with the smart search feature. These include inventor (IN), applicant (PA), title (TI), abstract (AB), priority number (PR), publication number (PN), application number (AP), CPC, claims (CLAIMS), and others [34]. To use these with the smart search, simply type the identifier, equal sign, followed by the search term (as seen with Google Patents). Smart search is a good place to begin when looking for a specific patent or beginning the patent search with the use of identifiers helping focus the search [13, 34].

Advanced search is set up to allow for the search of numerous patent fields as seen with the field identifiers that can be used in smart search; however, the fields have their own individual search bars. This can help in the construction of a specific search without the need to use the pocket guide to remember all the field identifiers. The Espacenet advanced search tool allows for the selection of one of four different databases: Worldwide – collection of published applications from 100+ countries (default); Worldwide EN – collection of published applications in English; Worldwide FR – collection of published applications in French; and Worldwide DE – collection of published applications in German. Outside of the different languages in which the patents can be searched, the worldwide collection from 100+ countries does not have the capacity to allow the user to search the full text for keywords as this function is limited to the language-specific databases. Advanced search looks at patents by the following criteria: key words in the tile, abstract, or full text (depending on the chosen database), numbers (publication, application, or priority), the publication date, applicant or assignee, the inventor, and CPC or IPC codes [34]. By applying multiple search parameters, a detailed search that meets your needs can be created. This can be especially useful if trying to evaluate the patent activity of a specific competitor over a specified time frame.

The last search tool on Espacenet is the classification search tool which allows the user to search all the patents by their assigned classification code where a specific section, class, subclass, main group, and sub-group can be searched. It is presented as the scheme of the entire CPC classification system and allows the user to check off different classifications to include in the search. This can be used to review the recent patent publications in a specific technical field or area. Once the search results are returned using the classification search tool, clicking "refine search" will take you to the advanced search with the selected CPC codes populating the CPC search bar [34]. This can help to limit the search by any of the previously mentioned criteria for the advanced search. For a complete guide on the utility and capacity of the Espacenet to build custom searches, please reference the resource book published by the EPO [34].

Patentscope Patentscope is the patent database maintained by WIPO and allows for users to search nearly 76 million [41] patent files from patent offices across the globe [42]. The database allows for the user to search patents in the various filing languages contained in the database while providing a search interface in 10 languages (Arabic, Chinese, English, French, German, Japanese, Korean, Portuguese, Russian, and Spanish). Patentscope provides the user with five different search tools: simple search, advanced search, field combination, cross lingual expansion, and chemical compounds search.

The simple search is the default tool presented to the user when navigating to the search tool. Here, one search bar is presented that allows the user to enter any number of different search criteria to retrieve related patent documents. The simple search allows the user to select eight different fields from a drop down manual where the search term will be queried: front page, any field, full text, English text, ID/number, IPC, names, and dates. As with previous databases, Patentscope supports the use of Boolean operators, parentheses, and a few other features. Patentscope also allows for the use of field codes with the simple search tool to build a custom strategy. The field codes relate to applicants, dates, language, names, and others. To use one of the field codes with the simple search tool, use the following format: "code : term" (e.g. title- "TI : polymorphism"). A complete list of the variety and the specific field codes can be found in the Patentscope user guide [42].

The advanced search is like simple search where it allows for the use of Boolean operators and field codes to build custom searches, but it also allows the user to select the language, the office where the patent was filed (if looking for a region of the world), expansion of the search using related terms, and the ability to stem the search terms. The stem option takes your original search term and searches for conjugated forms of the search term (e.g. "solid" will also include "solids" in the results). The field combination tool, a modification of the advanced search, allows the user a visual method to use the various field identifiers available where multiple search bars with adjustable field identifiers are presented to the user to help build a search strategy. The cross-lingual expansion has a similar interface as the advanced search and allows the user to include patent documents that were originally filed in a language different from the language used to run the search [42].

The final search tool available through Patentscope is the chemical compounds search which provides three methods to enter the compound of interest: compound name (commercial name, IUPAC, SMILES, etc.), structure editor (draw the structure of interest), or upload structure (MOL file, or image in PNG, GIF, TIFF, or JPEG). The tool can search for patents with the exact structure entered with the "exact structure search". The "substructure search" uses your in-put as a scaffold and identifies compounds containing the structure but with different functional groups and substitutions. The substructure search can identify the research direction of chemically related compounds [42, 43].

Lens.org Lens.org is an open source database that strives to provide a public tool for the discovery of cutting edge scientific and technological innovations (www.lens.org). The website provides access to literature search and analysis tools, patent search and analysis tools, PatCite, PatSeq, and QutIn4M influence tool. The literature database (200+ million works referenced) currently draws its data from PubMed, Crossref®, Microsoft Academic, CORE (https://core.ac.uk),[1] and PubMed Central and is likely to expand in the future. The patent data (117+ million patents) are sourced from the following: EPO bibliographic data from 1907 to present, EPO granted patents from 1980 to present, USPTO applications from 2001 to present, USPTO granted patents 1976 to present, USPTO Assignments, WIPO patent applications from 1978 to present, and full text from IP Australia. PatCite tool allows users to identify a relationship between specific literature and patent applications. The QutIn4M influence tool allows for the identification of institutions that contribute to innovations as determined by scholarly works cited in patents [44]. PatSeq is a search tool for biologics-based researchers that helps users search for the unique biological sequences found in patent literature. Along with the search tools, users can create a free account to access several additional features: search history, collections, notes, tags, and can save searches. These features can help the user keep track of their work.

The literature search interface allows for a structured search and a query text editor. The text editor simply allows the user to type free text connected via Boolean operators. The structured search is the tool that can be used to build out targeted searches by allowing the user to build the search by defining a field, publication date, ORCID® ID,[2] identifier type, flags, and publication type. Multiple fields

1 Collection of the world's open access research papers. https://core.ac.uk
2 Unique identifier for individuals engaged in research, scholarship, and innovation activities.

can be searched at the same time. The categories that can be searched include source, general, authors, external identifiers, subject matter, institution, funding, conferences, clinical trials, and open access with multiple fields in each category. After running a search, the tool provides data describing works in set (# returned results for the search), works cited by patents (all the results that have been cited by at least one patent), citing patents (patents that have cited at least one of the papers returned in the search), and works cited by scholarly articles (papers cited by other literature). The literature search tool also provides statistics describing the number of papers from the top publishing institutions, number of publications over time, leading authors, leading journals, and others. The data describing the number of scholarly citations and patent citations are available for individual articles as well. Alternatively, PatCite can be used to search for patents that cite-specific articles, or to search the literature cited by patents. When using PatCite, however, the inputs need to be the ID numbers or patent numbers of the individual texts that need to be analyzed.

The patent search tool has a similar interface to the literature search tool with a possibility for a structured or query text editor. The search can be built by searching various patent fields, dates, search language, ORCID ID, classification, jurisdiction, document type (application, granted patent, etc.), and others. The patent field categories include general (title, abstract, claims, etc.), family (patent families), classifications, citations, and sequences. After running a search, the tool shows the returned results along with data describing the publication year, applicants, document type, inventor, and others. The patent search tool connects to the literature by providing a listing of the cited scholarly works, on an entire search and an individual patent level.

2.4 Other Useful Resources for Solid-State Projects

There are numerous databases and resources maintained by organizations that focus on the properties of materials including Cambridge Structural Database, International Centre for Diffraction Data (www.icdd.com), Crystallography Open Database (http://www.crystallography.net/cod), National Institute of Standards and Technology (www.nist.gov), Protein Data Bank (www.rcsb.org), Nucleic Acid Database (ndbserver.rutgers.edu) [45], and others. Of these, the Cambridge Structural Database and Crystallography Open Database would be important for scientists working in solid-state chemistry.

2.4.1 Cambridge Structural Database

Cambridge Structural Database (CSD, https://www.ccdc.cam.ac.uk/solutions/csd-system/components/csd) is a collection of published chemical and crystal structure data. The original data goes back to 1965, and it contains over one million structures in the database. A Google Scholar search for "Cambridge Structural Database" returns over 46 000 papers, which speaks to the widespread use of this database. The database has several features: CSD materials, CSD discovery, access structures, and complete access to the CSD. CSD materials can help analyze intra and intermolecular interactions within a material to better understand its behavior and properties. CSD discovery is a tool designed for chemist discovering new molecules and allows for the modeling of protein–ligand interactions, interaction analysis, virtual screening, and geometric validation of the bound ligand. Access structures is a free search tool that allows users to search and retrieve a limited number of structures from the Cambridge Structural Database. The user can search the name or structure of a compound and obtain information about the compound's various crystal structures and, if available, links to the papers that reported the structure. Access to the full CSD allows users to retrieve more structures as well as providing access to other features to support the development of new compounds.

2.4.2 Crystallography Open Database

The Crystallography Open Database (http://www.crystallography.net/cod) in an open access collection of crystal structures founded in 2003. It covers organic, inorganic, metal–organic compounds, and minerals, but not biopolymers. As of August 2019, it contains more than 410 000 entries. Data can be browsed by journal of publication and year of publication. A detailed search for text, journal, year, volume, issue, DOI, Z, Z', chemical formula, unit cell volume, and elements is possible. Structure search is enabled via SMILES code and by drawing a structure or structural fragment. A search returns a table holding a link to a Crystallographic Information File (CIF), a link to HKL data (if available), chemical formula, space group, cell parameters, cell volume, and bibliographic data.

List of Abbreviations

API	active pharmaceutical ingredient
CIF	crystallographic information file
CPC	cooperative patent classification
CYP450	cytochrome P 450
EMA	European Medicines Agency
EPO	European Patent Office
FDA	Food and Drug Administration
IPC	International Patent Classification
IUPAC	International Union of Pure and Applied Chemistry
NCE	New Chemical Entity
RLD	Reference Listed Drug
SMILES	simplified molecular-input line-entry system
USPTO	United States Patent and Trademark Office
WIPO	World Intellectual Property Organization

References

1 Saal, C. and Becker, A. (2013). Pharmaceutical salts: a summary on doses of salt formers from the Orange Book. *European Journal of Pharmaceutical Sciences* 49 (4): 614–623.

2 He, Y., Orton, E., and Yang, D. (2018). The selection of a pharmaceutical salt-the effect of the acidity of the counterion on its solubility and potential biopharmaceutical performance. *Journal of Pharmaceutical Sciences* 107 (1): 419–425.

3 Cerreia Vioglio, P., Chierotti, M.R., and Gobetto, R. (2017). Pharmaceutical aspects of salt and cocrystal forms of APIs and characterization challenges. *Advanced Drug Delivery Reviews* 117: 86–110.

4 He, Y., Ho, C., Yang, D. et al. (2017). Measurement and accurate interpretation of the solubility of Pharmaceutical Salts. *Journal of Pharmaceutical Sciences* 106 (5): 1190–1196.

5 U.S. FDA, Center for Drug Evaluation and Research (2007). ANDAs: pharmaceutical solid polymorphism, guidance for industry. https://www.fda.gov/downloads/Drugs/GuidanceComplianceRegulatoryInformation/Guidances/UCM072866.pdf (accessed 1 April 2019).

6 U.S. FDA, Center for Drug Evaluation and Research (2018). Regulatory classification of pharmaceutical co-crystals, guidance for industry. https://www.fda.gov/downloads/Drugs/GuidanceComplianceRegulatoryInformation/Guidances/UCM281764.pdf (accesssed 1 April 2019).

7 European Medicines Agency, Committee for Medical Products for Human Use (2015). Reflection paper of the use of cocrystals of active substances in medicinal products. https://www.ema.europa.eu/en/documents/scientific-guideline/reflection-paper-use-cocrystals-active-substances-medicinal-products_en.pdf (accessed 1 April 2019).

8 Newman, A. and Wenslow, R. (2016). Solid form changes during drug development: good, bad, and ugly case studies. *AAPS Open* 2 (1): 1–11.

9 VanVliet, D. (2016). Patentability of pharmaceutical polymorphs in Europe. *Pharmaceutical Law and Industry Report* 14: 1–4.

10 Sodikoff MSMI, B.J. and Cole, D. (2015). Garrett. What form is it? Lessons from 13 polymorph pharmaceutical cases. *Pharmaceutical Law and Industry Report* 13: 1–9.

11 Gaby Longsworth, A.W. and Varughese, D. (2018). The 505(b)(2) drug approval pathway: a potential solution for the distressed generic pharmaceutical industry in an increasingly diluted ANDA marketplace? *Pharmaceutical Law and Industry Report* 16.

12 Celgene Corporation v. Dr. Reddy's Laboratories, Inc., No. 2:16-cv-07704-SDW-LDW (New Jersey District Court 2020).

13 European Patent Office (2019). Espacenet-pocket guide. http://documents.epo.org/projects/babylon/eponet.nsf/0/8C12F50E07515DBEC12581B00050BFDA/$File/espacenet-pocket-guide_en.pdf (accessed 05 February 2021).

14 Durrmeyer, X., Hovhannisyan, S., Medard, Y. et al. (2010). Are cytochrome P450 CYP2C8 and CYP2C9 polymorphisms associated with ibuprofen response in very preterm infants? *PLoS One* 5 (8): e12329.

15 Ochoa, D., Prieto-Perez, R., Roman, M. et al. (2015). Effect of gender and CYP2C9 and CYP2C8 polymorphisms on the pharmacokinetics of ibuprofen enantiomers. *Pharmacogenomics* 16 (9): 939–948.

16 Martinez, C., Garcia-Martin, E., Blanco, G. et al. (2005). The effect of the cytochrome P450 CYP2C8 polymorphism on the disposition of (R)-ibuprofen enantiomer in healthy subjects. *British Journal of Clinical Pharmacology* 59 (1): 62–69.

17 Glowka, F. and Karazniewicz, M. (2007). Enantioselective CE method for pharmacokinetic studies on ibuprofen and its chiral metabolites with reference to genetic polymorphism. *Electrophoresis* 28 (15): 2726–2737.

18 Krasniqi, V., Dimovski, A., Domjanovic, I.K. et al. (2016). How polymorphisms of the cytochrome P450 genes affect ibuprofen and diclofenac metabolism and toxicity. *Arhiv za Higijenu Rada i Toksikologiju* 67 (1): 1–8.

19 Yu, L., Shi, D., Ma, L. et al. (2013). Influence of CYP2C8 polymorphisms on the hydroxylation metabolism of paclitaxel, repaglinide and ibuprofen enantiomers in vitro. *Biopharmaceutics and Drug Disposition* 34 (5): 278–287.

20 Lou, B. (2015). Crystal structure of 1,3-dihydroxy-2-(hydroxy-methyl)propan-2-aminium 2-(4-iso-butyl-phenyl)propanoate: a simple organic salt of racemic ibuprofen. *Acta Crystallographica Section E, Crystallographic Communications* 71 (Pt 8): 923–925.

21 Rasenack, N. and Muller, B.W. (2002). Ibuprofen crystals with optimized properties. *International Journal of Pharmaceutics* 245 (1–2): 9–24.

22 Abioye, A.O., Chi, G.T., Simone, E., and Nagy, Z. (2016). Real-time monitoring of the mechanism of ibuprofen-cationic dextran crystanule formation using crystallization process informatics system (CryPRINS). *International Journal of Pharmaceutics* 509 (1–2): 264–278.

23 Ziaee, A., Albadarin, A.B., Padrela, L. et al. (2017). Spray drying ternary amorphous solid dispersions of ibuprofen – an investigation into critical formulation and processing parameters. *European Journal of Pharmaceutics and Biopharmaceutics* 120: 43–51.

24 Kamarudin, N.H., Jalil, A.A., Triwahyono, S. et al. (2014). Variation of the crystal growth of mesoporous silica nanoparticles and the evaluation to ibuprofen loading and release. *Journal of Colloid and Interface Science* 421: 6–13.

25 Hooper, D., Clarke, F.C., Docherty, R. et al. (2017). Effects of crystal habit on the sticking propensity of ibuprofen-a case study. *International Journal of Pharmaceutics* 531 (1): 266–275.

26 Soares, F.L. and Carneiro, R.L. (2014). Evaluation of analytical tools and multivariate methods for quantification of co-former crystals in ibuprofen-nicotinamide co-crystals. *Journal of Pharmaceutical and Biomedical Analysis* 89: 166–175.

27 Melzig, S., Finke, J.H., Schilde, C., and Kwade, A. (2018). Formation of long-term stable amorphous ibuprofen nanoparticles via antisolvent melt precipitation (AMP). *European Journal of Pharmaceutics and Biopharmaceutics* 131: 224–231.

28 Afrose, A., White, E.T., Howes, T. et al. (2018). Preparation of ibuprofen microparticles by antisolvent precipitation crystallization technique: characterization, formulation, and in vitro performance. *Journal of Pharmaceutical Sciences* 107 (12): 3060–3069.

29 Skorupska, E., Kazmierski, S., and Potrzebowski, M.J. (2017). Solid state NMR characterization of ibuprofen:nicotinamide cocrystals and new idea for controlling release of drugs embedded into mesoporous silica particles. *Molecular Pharmaceutics* 14 (5): 1800–1810.

30 Levine, L.O. (1986). Patent search strategies and techniques: Concept paper, Pacific Northwest Lab., Richland, WA (USA), . Report No.: PNL-SA-14428; Other: ON: DE87002470 United States Other: ON: DE87002470 NTIS, PC A03/MF A01. *PNNL English.*

31 World Intellectual Property Organization (2015). Guidelines for preparing patent landscape reports. https://www.wipo.int/edocs/pubdocs/en/wipo_pub_946.pdf (accessed 19 August 2019).

32 World Intellectual Property Organization (2019). Guide to the international patent classification, version 2019. https://www.wipo.int/edocs/pubdocs/en/wipo_guide_ipc_2019.pdf (accessed 05 February 2021).

33 USPTO and EPO (2017). Guide to the CPC (cooperative patent classification). Version 1, Revision 4. https://www.cooperativepatentclassification.org/wcm/connect/cpc/212f75e9-e9d4-4446-ad7f-b8e943588d1b/Guide+to+the+CPC.pdf?MOD=AJPERES&CVID= (accessed 1 May 2019).

34 European Patent Office (2019). Espacenet Resource Book. http://documents.epo.org/projects/babylon/eponet.nsf/0/c82219a43c976dbcc12581cd0052925d/$FILE/espacenet_resourcebook_v3.0_en.pdf (accessed 19 August 2019).

35 USPTO(2019). Classification Resources. https://www.uspto.gov/web/patents/classification/cpc/html/cpc.html (accessed 19 August 2019).

36 Tessler, L., Aronhime, J., Lifshitz-Liron, R., Maidan-Hanoch, D., Hasson, N., and inventors; Teva Pharmaceutical Industries Ltd, assignee (2009). Crystal forms of atorvastatin hemi-calcium and processes for their preparation as well as novel processes for preparing other forms. United States patent US 7,501,450, 10 March 2009.

37 Aronhime Judith, L-H.R., Valerie, N., Revital, L., Shlomit, W., and inventor; Teva Pharmaceutical Industries Ltd., assignee (2009). Crystal forms of atorvastatin hemi-calcium and processes for their preparation as well as novel procesess for preparing other forms. United States.

38 Paul Oldham, I.K. (2016). World intellectual property organization manual on open source patent analytics. https://wipo-analytics.github.io/introduction.html (accessed 22 August 2019).

39 Google (2019). About google patents: coverage 2019. https://support.google.com/faqs/answer/7049585?hl=en&ref_topic=6390989 (accessed 22 August 2019).

40 Bahige, M. Baroudy, J.W.C., Gilbert, E., Josien, H. B., Labroli, M. A., Laughlin, M. A., Mccombie, S. W., Mckittrick, B. A., Miller, M. W., Neustadt, B. R., Palani, A., Smith, E. M., Steensma, R., Tagat, J. R., Vice, S. F. (2000). Pharmaceutical compositions comprising CCR5 antagonizing piperazine derivatives, E.P. Office, Editor, Merck Sharp and Dohme Corp.

41 Organization WIP (2019). National collections – data coverage 2019. https://patentscope.wipo.int/search/en/help/data_coverage.jsf (accessed 22 August 2019).

42 World Intellectual Property Organization (2019). Patentscope: the user's guide. https://patentscope.wipo.int/search/help/ar/users_guide.pdf (accessed 05 February 2021).

43 World Intellectual Property Organization (2019). Chemical search.

44 Jefferson, O.A., Jaffe, A., Ashton, D. et al. (2018). Mapping the global influence of published research on industry and innovation. *Nature Biotechnology* 36 (1): 31–39.

45 Coimbatore Narayanan, B., Westbrook, J., Ghosh, S. et al. (2014). The nucleic acid database: new features and capabilities. *Nucleic Acids Research* 42 (Database issue): D114–D122.

3

Systematic Screening and Investigation of Solid-State Landscapes

Ulrike Werthmann

Boehringer Ingelheim Pharma GmbH & Co. KG, Drug Discovery Sciences, Birkendorfer Straße 65, 88397 Biberach/Riss, Germany

3.1 Introduction

The solid-state behavior of a drug compound could translate into measureable differences in properties of pharmaceutical importance. Therefore, the solid-state landscape of an Active Pharmaceutical Ingredient (API) has to be well investigated and known. The selection of the right solid form for development has a great impact on the timeline for admission of a drug product.

The tendency of pharmaceutical solids to crystallize in multiple crystal forms and the significance of this phenomenon (polymorphism) have been demonstrated [1]. Since polymorphism can affect the chemical, biological, and pharmaceutical properties of the drug, it is very important to detect and analyze polymorphic, solvated/hydrated, or amorphous forms of an API at early time points. The definition of solid forms of a New Chemical Entity (NCE) is a key step for candidate nomination at the Research (RES) and Development (DEV) interface. Therefore, it is very important to have a thorough understanding of the drug substance in terms of its solid form, physicochemical properties, and manufacturability for drug product design and DEV.

The strategies to investigate the solid form landscapes of an API differ considerably among the pharmaceutical companies. Their common goal is fast market approval of an NCE. Each team follows their own strategy based on long experiences with respect to invested time, used resources (personnel and financial), and success. Therefore, no standard procedure exists which is used by all pharmaceutical industries.

This chapter is intended to summarize important aspects of the strategy for the solid-state evaluation of small molecules, which is followed by an RES team working in a RES-based pharmaceutical company. Focus will be set on a pragmatic approach for investigations on solid forms and crystallization experiments in the late stage RES phase. There is no intention to raise any claim of describing the best approach on solid form selection. There are many ways to be successful. Besides a large number

Solid State Development and Processing of Pharmaceutical Molecules: Salts, Cocrystals, and Polymorphism, First Edition. Edited by Michael Gruss.

of publications describe general procedures [2], this book chapter will concentrate on daily routine in the NCE environment of a pharmaceutical business.

One example of daily project work with respect to solid form characterization will illustrate the procedure.

3.2 General Aspects of Solid-State Investigations in Early Drug Discovery Phase

The early Drug Discovery Process covers different phases of API characterization (Figure 3.1). In the very early RES period, called pre-lead optimization phase (pre-LO phase), usually only a few solubility data points are used as a guidance for optimization. The solubility data of these compounds can be collected by a high-throughput solubility screening method (shake flask method from dimethylsulfoxid (DMSO) stock solution). At the beginning of the lead optimization phase (LO-phase), the partition coefficient should be measured for some representative compounds. A first check regarding chemical stability in solution has to be performed as well. In case of liabilities, these properties should be improved to identify suitable lead structures. As a consequence, the first investigations in terms of DEV aspects are initiated with a more promising set of compounds. At that time, the amount synthesized for a particular compound increases and the first solid material is available and can be used for physico-chemical investigations. This includes a crystallinity check via X-ray powder diffraction as well as particle size and shape analysis by optical microscopy. Later in

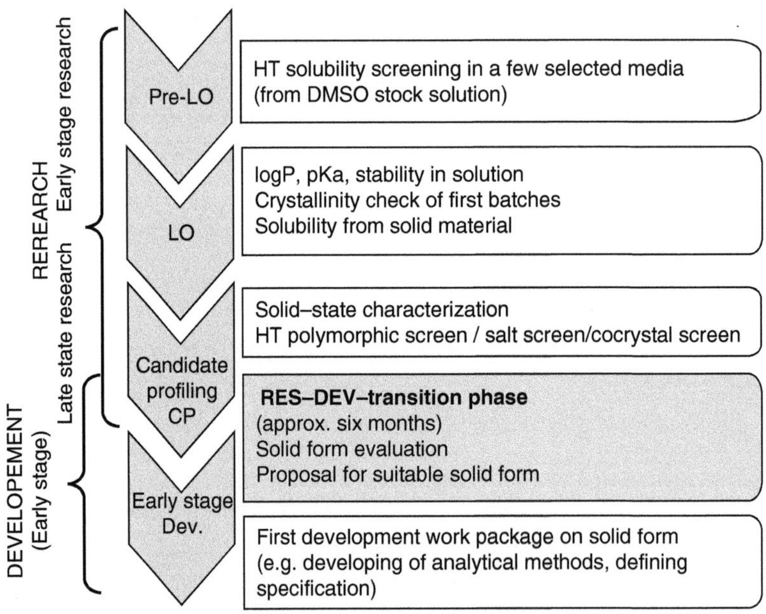

Figure 3.1 Different phases of API-characterization within Research.

the project timeline, interest in the crystallinity status (degree of crystallinity), particle shape, and particle size of the first powder material increase steadily, because this information is used to optimize API processing and formulation [3]. The influence of these material properties (e.g. crystallinity, amorphous, or micronized) on early in vivo experiments is already significant. The knowledge of the quality of the solid material used for first solubility tests and standard pharmacokinetic (PK) studies are very helpful for the interpretation of the data and misinterpretation of the results can be minimized. At that project step, no intensive optimization efforts on the solid form can be started due to lack of availability of sufficient amount of compound. This will be done at later stage.

With the beginning of the candidate profiling phase (CP phase), the most promising molecules are identified and selected for deeper evaluation. At this time, medicinal chemistry delivers larger batches of solid material. These batches are used for relevant biological studies. Next to basic physico-chemistry data like crystallinity or solubility, the formulation DEV uses these solid materials to support the non-clinical animal studies (e.g. PK study studies and safety pharmacological studies). These outcomes are of great interest for the further consideration of the compounds for DEV. A well-characterized formulation, with respect to the solid material within the formulation, even in this early stage is becoming increasingly significant.

Approximately six months before an NCE project enters the DEV stage, the project team identifies one compound to move forward, i.e. select a candidate for DEV. Crystallization of a drug, manufacturing of the bulk material, defining analytical protocols, setting up reasonable stability programs, generating product formulation, and so on are important aspects for DEV and delivery of pharmaceutical products. All these activities are defined as Chemistry, Manufacturing, and Control (CMC), which plays an important role during late RES phase and the complete DEV stage.

The focus from a CMC perspective is now to start crystallization experiments in order to find the most stable solid form with acceptable physico-chemical properties. This can be a particular polymorph, a hydrate, a salt, or a cocrystal. These screens are focused on finding at least one suitable solid form to enable a progression of the project to the next stage. Later in DEV, after proof of clinical concept, more material and resources are available, and comprehensive screens can be initiated to evaluate the polymorphic landscape in more detail.

3.3 Transition Phase from Late Stage Research to Early Stage Development

Within the late stage of RES, the solid form characteristics of a selected candidate are of great importance for many disciplines in RES and DEV.

Therefore, a transition team consisting of members of both RES and DEV is helpful to ensure high quality of DEV compounds and to provide a seamless transfer from RES to DEV.

Upon entering this phase, the indication of the API is well defined and most of the important biological studies, which show therapeutic efficacy, are completed.

Additional important preclinical studies like safety pharmacology investigations of the candidate to identify potential undesirable biological effects have to be set up. For these studies, an optimized and well-characterized formulation has to be provided because the required plasma levels for detecting potential side effects have to be achieved.

Due to limited solubility of many drug molecules, it is not always possible to provide a solution as an ideal formulation. The doses for safety pharmacological studies are normally 15–30 times higher than the expected human estimated doses and the concentration of the API within the formulation is rather high.

Most frequently, APIs are formulated as a suspension (insoluble particles dispersed in liquid medium). The degree of crystallinity of these insoluble particles has an impact on the solubility and/or dissolution rate of the API within the body at least for the biologically relevant data. Because of the impressive DEV of home lab X-ray equipment, powder patterns of formulations containing low amount of solid material (for low concentration suspension) can be easily measured in a very short time with sufficient resolution.

3.4 Solid-State Characteristics in Preclinical Formulations

The powder material, which will be used for in vivo studies, has to be analyzed with respect to solid-state characteristics regarding crystallinity and morphology. Additional solubility data in aqueous media, physiological relevant media like Fasted State Simulated Intestinal Fluid (FaSSiF) and Fed State Simulated Intestinal Fluid (FeSSiF), and/or relevant excipients are helpful for the formulation setup. The scope of these solubility tests is regulated by the quantity of compound, which is related to the complexity of the synthesis.

For highly soluble drugs, a clear solution can obtained. For low soluble drugs, a heterogeneous mixture containing solid particles is formed (suspension). Water-soluble methylcellulose derivates like Methocel or Natrosol are very common ingredients used in pharmaceutical formulations. For initial PK studies, a Natrosol suspension is a common formulation for oral applications. Figure 3.2 shows a microscopy image of a standard Natrosol suspension.

For relevant in vivo studies, the solid form of the API within the formulation has to be checked prior to the application. X-ray powder diffraction can be used for monitoring the crystallinity/amorphous state of a small molecule in a formulation. Prior to X-ray powder diffraction (XRPD) measurement, the formulation suspension (see Figure 3.2) is centrifuged to concentrate the amount of solid material. After discarding the supernatant liquid, the pasty solid sample is transferred to a sample holder, e.g. a silicon wafer, and spread as even as possible. Sufficient amounts of sample material are very important in order to achieve high-quality measurements. Hence, 15–20 mg of the pasty solid material is required. The solid form of the API can change to another polymorphic form (see Figure 3.3), potentially also resulting in a significant reduction of the degree of crystallinity. The material could also

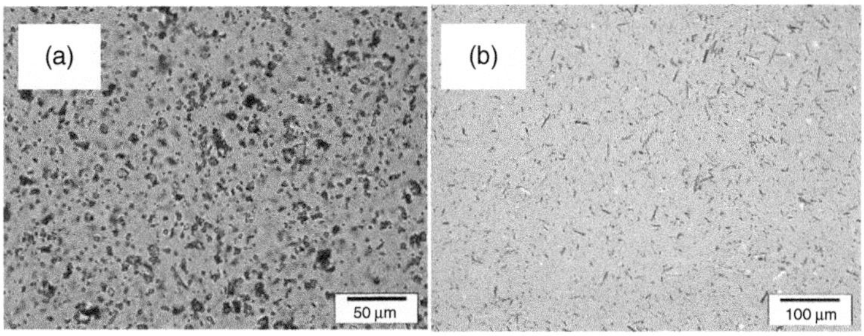

Figure 3.2 Microscopic picture of natrosol suspensions (heterogeneous mixture – API-particles not dissolved) (a) amorphous particles within natrosol; (b) crystalline particles within natrosol. Source: Dr. Ulrike Werthmann.

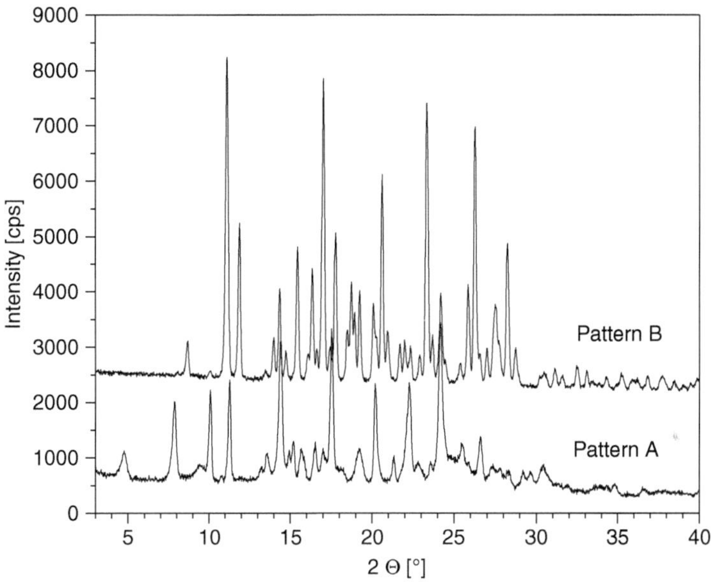

Figure 3.3 Polymorph transformation within natrosol suspension. Pattern A: XRPD of solid material prior to formulation (Form I). Pattern B: XRPD of solid material within natrosol formulation after 24 hours (Form II).

become completely amorphous (see Figure 3.4). Likewise, an amorphous starting material can crystallize into a known or new form within the formulation. Consequently, there is a risk that the formulation used for in vivo studies contains a different solid form of the API than initially planned. This can have significant impact on the outcome of the study. A short-term (up to three days) physical stability test is recommended and helpful for proper handling of the formulations before application.

In addition to crystallinity, the shape and the size, i.e. the morphology, of API particles play an important role within a formulation. The relationship between solubility and particle size is well investigated and known [4]. Microscopic images of the suspensions depict the shape, and a rough impression of the particle size distribution of the suspended matter can be obtained (see Figure 3.2). Modern techniques allow a

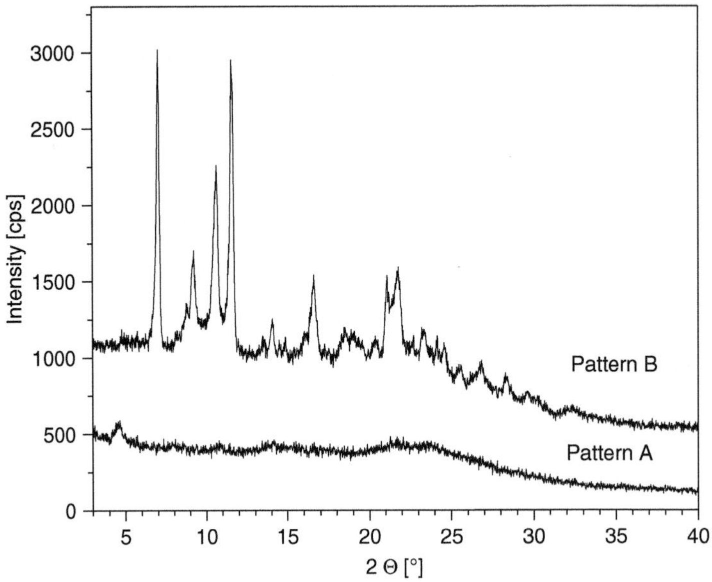

Figure 3.4 Crystallization of amorphous material within natrosol suspension. Pattern A: XRPD of solid material prior to formulation (amorphous). Pattern B: XRPD of solid material within natrosol formulation after 2 hours (crystalline Form I).

more precise measurement of particle size distribution (e.g. nanometer scale) even using a small amount of sample [5].

For specific drugs, decreasing particles down to nanosize improves solubility and thus achieves better biological parameters [6]. For the milling (i.e. micronization) process, the starting material has to be crystalline. Otherwise, no sufficient size reduction can be achieved via milling. Hence, the crystallinity has to be checked before and after micronization. After the micronization procedure, the material should still exhibit the same crystal form. For some drug substances, a mixture of amorphous and crystalline material is obtained [7]. The degree of amorphousness will not be determined at this early project stage.

It should be noted that the nanosized formulation approach is a very helpful tool in RES and early stages of DEV.

Another technology to improve the dissolution rate of a compound is the conversion of crystalline material to the amorphous state. Different technologies are currently available, like milling, spray drying, or lyophilisation [8]. A stabilized suspension of amorphous drug (e.g. solid dispersion) can lead to much better results in the in vivo studies. X-ray powder diffraction measurements should prove the stability against recrystallization of the amorphous state of the API within this formulation. Figure 3.5 shows representative patterns to display the recrystallization of amorphous substance as a function of time. For this compound, the solid dispersion formulation has to be used directly after preparation for getting reliable results.

Table 3.1 summarizes the important aspects of solid from control for formulations via XRPD.

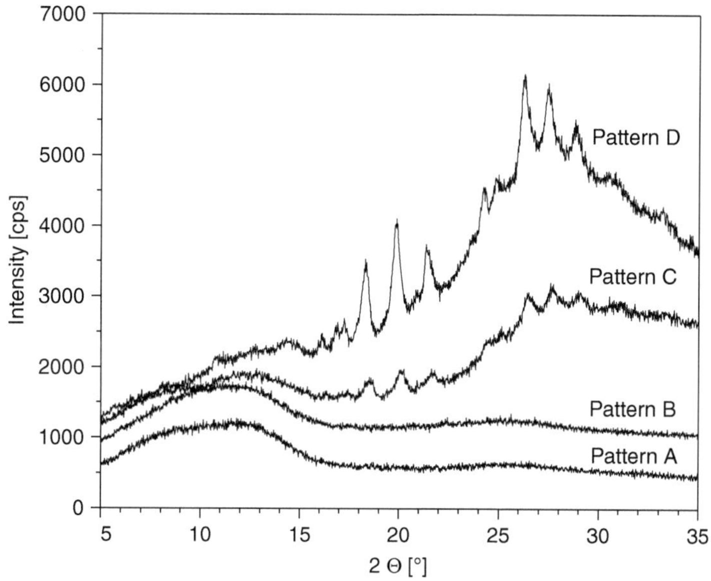

Figure 3.5 Recrystallization of amorphous material within a solid dispersion formulation. Pattern A: XRPD of freshly prepared solid dispersion (API = amorphous). Pattern B: XRPD of solid dispersion after 1 hour (API = amorphous). Pattern C: XRPD of solid dispersion after 3 hours (recrystallization of API started). Pattern D: XRPD of solid dispersion after 24 hours (recrystallization of API progressed).

Table 3.1 Important aspects of solid form control for formulation.

Formulation principle	XRPD measurements of	Result
Oral (natrosol suspension)	API + formulation	Detect change in degree of crystallinity up to amorphous state detect change in polymorphic form
Parenteral (solution)	API	Gather knowledge about solubility behavior of used batch (crystalline or amorphous)
Oral: nano-suspension	API + formulation	stability of solid form within formulation
Oral: solid dispersion (amorphous API included in gastroresistant polymers)	API (amorphous) + formulation	stability of solid dispersion (API in amorphous state)

3.5 API-crystallization Strategy in Candidate Profiling Phase

The first step in defining the solid form is to determine if chemical variations of the free form are possible at all, such as formation of salts, hydrates, solvates, and cocrystals. Thereafter, solid form screening protocols have to be set up to identify the scope of possible solid forms, i.e. the solid-state or polymorphic landscape. These include crystalline polymorphs of the free form and amorphous form. In case that multiple solid forms are found, one out of these has to be selected as a suitable form to be used for DEV [9].

For compounds exhibiting polymorphism, it is important that there is a clear scientific rationale for the chosen solid form which typically should be the thermodynamically most stable form [10]. The solid form also plays a role in achieving drug substance purity [11] and isolation efficiency of a drug substance.

Although significant advances have been made in the field of ab initio crystal structure prediction (CSP), robust methods for assessing polymorphic behavior of pharmaceutical compounds remain a sizable challenge. Organic CSP methods focus on searching for the most thermodynamically stable three-dimensional crystal structures. In cases where the existence of a known crystalline form is predicted, estimation of its stability relative to other crystalline packing arrangements is possible. However, even with current technology, the accuracy of relative ranking of the most thermodynamically stable predicted structures requires further DEV. Polymorphic calculations, with CSP and other programs such as Hydrogen Bond Propensity (HBP), may be used in risk assessment as well as in crystal structure identification [12].

Additionally, predictive tools, such as HBP, Molecular Complementarity (MC), Molecular Electrostatic Potential (MEP), are important in cocrystal design and serve as a good starting point for selecting solvents or cocrystal formers. DEV of automated scripts has enabled the application of these methods to become practical in routine use. In addition, a program for prediction of the propensity of a compound to form a solvate has been developed which provides insight into solvent selection in crystallization trials [13].

Currently, solid form discovery relies primarily upon experimental studies, where manual screening methods are employed to explore form diversity of a compound. Computational methods serve as a complimentary tool to advance and accelerate this process. Further DEV and enhancement of predictive technologies offers the opportunity to accurately guide investigations of crystal form identification and characterization and ultimately reduce the number of laboratory experiments [14].

In the last 20 years, the high-throughput (HT) crystallization method was well established as a rapid screening tool for evaluating the crystallization behavior of organic molecules and is used in many companies. Modern HT experiments, mostly carried out in 96-well plate systems in μl scale, can implement various salt formations and crystallization conditions. The main interest in following this approach is to identify key solvents and solvent combinations, counterions, and conditions for a successful crystallization of a specific molecule in a short period of time. HT analytical techniques (e.g. HT X-ray powder diffraction) allow a quick check regarding crystallinity. A high number of solvents and solvents mixtures can be tested as well as different acids, bases, and cocrystal formers. A huge number of experiments have been analyzed to evaluate potential anions/cations and solvents for a successful crystallization of an API. There are a number of excellent reviews on these screening techniques [15]. Full physico-chemical characterization of the hits (crystalline forms) has to be performed on larger quantities which are obtained by manual crystallization experiments following the recipes for the hits from HT screen. All these screening programs are designed to obtain a large range of solid forms (polymorphs, salts, and cocrystals) of a drug molecule.

Recently (about 2014), there has been a trend to reduce the number of crystallization experiments in the candidate-profiling phase with the focus of finding at least one suitable form. How extensively a polymorph screening can be conducted during late RES phase or early DEV phase is controversially discussed and depends on strategy within the company.

A reduction of the number of crystallization plates within the HT screen approach (e.g. using only one 96-well plate applying one crystallization procedure) is one way to safe material and to speed up the process. Additional lab-scale experiments using a low number of specific selected solvents, acids, or bases are getting more common. Manual crystallization methods (low-throughput approach) were applied successfully for many years in solid form screening. A typical scale for manual experiments can range from ~20 to 50 mg. Manual screenings offer the employment of a wider variety of crystallization conditions and a detailed monitoring of all the different manufacturing steps, e.g. observation of precipitation or change in morphology.

The availability of a drug substance (depending on the complexity of its synthesis routes) defines the scope of the initial solid form screening which can vary between HT solid form screen (96 or more experiments) and some selected single crystallization experiments. The amount of material for initial investigations typically varies roughly between 300 mg and 5 g. Table 3.2 summarizes the different screening types.

Key information about a compound has to be collected before a reasonable crystallization strategy can develop for a specific molecule:

(a) availability of the compound and purity of batches
(b) dissociation constants (pK_a and pK_b values)
(c) solubility data
(d) estimated human dose
(e) application route

The quality of the batch is of high importance. Purity testing (High performance liquid chromatography – ultra-violet detector [HPLC-UV] or high performance liquid chromatography – mass spectrometer [HPLC–MS]) is mandatory and the purity for batches used for crystallization experiments should be greater than 99%.

The knowledge of the pK_a/pK_b values, which define the acid and basic properties, allows the selection of acids/bases for a salt screen. If the API is a neutral molecule, only polymorph and cocrystal experiments can (and should) be conducted.

The safety assessment of a pharmaceutical salt regulates the selection of the acids/bases. Not all counterions are ideal because they may have an impact on toxicity. Well-known acids/bases (hydrochloric acid or sodium hydroxide), which are widely used on the market, should be tested first [16]. The introduction of novel salt forms can significantly increase the workload to evaluate the toxicological profile of an NCE.

Table 3.2 Key aspects of different screening types.

Compound requirement	Duration	Screen type	Outcome	Analytics
~300 mg	1 wk	**Exploratory experiments** 3–5 crystallization experiments using diff. solvents	Providing information regarding crystallization behavior	XRPD
1 g	1 wk	**Discovery screen** limited solid form screening on mg-scale using HT crystallization platform, e.g. 11 counterions in 8 solvents	Providing hit list for upscaling to 50–100 mg scale for further characterization; potential salt or cocrystal forms incl. polymorphic forms	HT XRPD
1 g	1–2 wk	**Form evaluation screen** crystallization of 50–100 mg of individual salt forms and first characterization	Providing hit list of promising salt forms for in depth characterization	XRPD, DSC, TG, DVS, microscopy
2–5 g	2–3 wk	**Process crystallization screen** crystallization of gram – quantities of most promising salt forms for full characterization	Providing one suitable form for development	XRPD, DSC, TG, DVS, microscopy, solid state stability (kinetic and chemical), solubility, intrinsic dissolution rate

Selection of appropriate solvents for crystallization is sometimes challenging. Solubility data of the API in organic solvents including water are helpful to get an idea which solvents will be tested in the beginning. Solvents that are known to cause unacceptable toxicities should be avoided in the first crystallization screen and procedure. From DEV perspective, a limited number of solvents are preferred regarding the ICH guideline [17] (class III and class II solvents) and should be tested first.

The term tolerable daily intake (TDI) describes exposure limits of potential toxic chemicals. The maximum acceptable limits of residuals (e.g. solvents, metals) are listed und regulated by a collection of guidelines [18]. Based on the estimated human dose of each API, the selection of solvents can be adapted. For low doses compounds, the number of solvents could increase.

Based on all these information, the crystallization strategy of the drug substance will be defined.

Because solid forms can have a significant influence on many physical properties, different analytical techniques to characterize solid forms need to identify and describe the different crystal forms (salts, solvates, hydrates, cocrystals). Characterization of solid forms by using multiple analytical techniques like microscopy, XRPD, differential scanning calorimetry (DSC), thermogravimetry (TG), dynamical vapor sorption (DVS), and spectroscopy (Infra-Red [IR], Raman, solid-state nuclear magnetic resonance [SSNMR]) is well established and described in literature [19]. For inorganic crystallized salts, ionic chromatography (IC) determines the ratio between API:acid/base, while for organic salts or cocrystals ^{1}H NMR is an appropriate mean. For DEV of a solid form, solubility behavior is essential and should be collected in different media like aqueous media, organic media, and bio relevant media like FaSSIF and FeSSiF [20]. The solubility can increase by modifying the API in a chemical manner (salt or cocrystal formation) or in a physical way (amorphization, selection of another polymorph).

3.6 Selection Criteria of a Suitable Solid Form

After collecting all important analytical data (see Table 3.2) of the most promising forms, it has to be discussed which form can be proposed for DEV. Only in rare cases, ideal physico-chemical properties are observed. The choice of the selected form for DEV is often a compromise and the risks and opportunities of the solid form have to be carefully evaluated.

Selection criteria for each parameter point out possible liabilities based on physico-chemical parameters, which is helpful to judge the different forms regarding DEV aspects. It is important that there is a clear rationale for the chosen solid form.

Four different categories can be specified to get an overview for an early pharmaceutical developability assessment. Solid-state physico-chemical properties can be divided into ideal, aacceptable, undesirable, and no go.

Table 3.3 defines the categories including their definition and the consequences for DEV. Additionally, the table lists exemplary the selection criteria for the melting point of a crystalline drug substance.

For each solid-state property, specific criteria have to be assessed for the next milestone (i.e. start of DEV). The criteria are intended to provide transparency on the success of identifying an ideal solid form. Each property will be ranked according to the four categories. The defined criteria of, e.g. ideal melting point or ideal sorption behavior are on the one hand based on the strategy of the company (risk assessment) and on the other hand on the deep experience of the solid-state experts within the company.

Table 3.3 Definition of different categories for physico-chemical properties.

Category	Melting point	Definition	Consequence
Ideal	>120 °C	Uncritical	Requires standard efforts Level of difficulty is low
Acceptable	80–120 °C	Challenging property identified, but unlikely to lead to discontinuation	Requires most likely increased capacities Level of difficulty is medium
Undesirable	<80 °C	Critical property identified, can lead to discontinuation	Requires distinct increased capacities and induce time delay Level of difficulty is high
No go	Not defined	Not acceptable property	Requires immediate discontinuation

The following solid-state properties are of relevance:

- Crystallinity
- Melting point
- Weight loss
- Hygroscopicity
- Mechanical stability
- Solubility in different media

In general, the criteria are defined for the DEV of oral dosage forms. For specific application routes, the criteria have to be adapted based on recommended dosage form. As an example for dry powder inhalation, the particle size must be suitable for inhalation (< 5.8 μm) [21], and the stability of the particle size must be maintained during storage of micronized drug substance. These values have to be defined within the criteria list for the inhalation application route.

By completion of the solid form screening and evaluation of the important physico-chemical properties of at least one solid form of the compound, all data have been evaluated according to the "criteria" for DEV, as described earlier. At the end, a head-to-head comparison of most attractive solid forms facilitates the decision for the selected solid from.

A well-understood API preparation, crystallization, and isolation process must be designed that will ensure that the API crystallizes in the desired solid form and that the subsequent processes maintain the physical form or alter it by design. The final isolation process must also ensure that the desired compound yield and purity levels are met.

A CMC team, consisting of members from RES and DEV, defines the final or at least one suitable solid form for DEV, based on all data.

3.7 Knowledge Management

All new data should be continuously distributed during the project as knowledge is gained. Regular team meetings have to be organized for an exchange of information and detailed scientific discussion. Additionally, the collected data have to be reported into a database, in which data are organized especially for rapid search and retrieval. All team members must have access to the database. It should be noted that all solid-state parameters of all compounds and their associated batches are added to the database. This means that the solid form parameters (e.g. melting point, degree of crystallinity of every batch) can be checked up by a defined search routine. Additionally, all physicochemical data of a compound and the respective batch can be linked with other parameters (e.g. in vivo or in vitro results). This allows a more or less complete overview on the properties of a batch and the specific usage of theses batches for other studies. A complete "One Compound Report" can be generated to get a fast and quick overview. After defining the solid form for DEV, a scientific report will be generated containing all important information regarding the selected solid from.

3.8 Control of Solid Form Properties in Development

Ideally, the solid form should not change throughout DEV. During DEV, modifications to the API synthetic scheme, scale, processing conditions, or equipment may impact the isolation with respect to the final solid form. In cases where the compound properties are poor with respect to filtration properties, size reduction (difficulty in milling), or use in drug product manufacturing (e.g. sticking), the last crystallization step in chemical DEV may involve some additional solid form considerations.

The disposability of compound increases during the DEV phase and additional studies can and should be conducted. This includes, but is not limited to, developing phase diagrams for enantiotropically related polymorphs, as well as thermodynamic boundaries for hydrates and solvates that can be formed. Based on this knowledge, the crystallization process can be designed properly and optimized to ensure consistent formation of the target solid form.

Additional work should be conducted on the drying and milling operations to understand any risks to the form that can occur during isolation and size reduction processes.

Comprehensive solid form screening activities should be initiated prior to Phase III to ensure sufficient knowledge to de-risk solid form control for commercial DEV. A Case Study – Solid State Activities in CP phase for a standard project.

Six months prior to the transfer to DEV, the solid form evaluation of a NCE is conducted. The nature of the compound cannot be disclosed, thus, it is just named *compound A*. The substance is intended for oral administration as a tablet. Based on early calculations, the estimated human dose for a once daily application is 50 mg.

Table 3.4 Synthesized batches from medicinal chemistry available at that time point.

Batch no	XRPD-result	Batch used for
1	Amorphous	• In vitro studies • Pharmacodynamics and pharmacokinetics studies
2	Amorphous	• In vitro studies • Pharmacodynamics and pharmacokinetics studies
3	Low crystallinity – Form I	• Pharmacodynamics and pharmacokinetics studies • First basic CMC analytics (log P and pK_a-values, solubility data in aqueous media, chemical stability in solution)
4	Low crystallinity – Form I	• Pharmacokinetics studies – 600 mg for some exploratory crystallization experiments • 500 mg for formulation development to support safety studies

Profiling efforts are undertaken to gain an initial understanding of the polymorph landscape of the new API and to find at least one solid form suitable for DEV.

Table 3.4 shows the required information of *compound A* that was available at the start of solid form evaluation.

Table 3.4 lists all batches of *compound A* that were synthesized before systematic solid-state characterization work begun. One crystalline form, identified only by X-ray powder diffractometry, was observed. Batch No. 4 was used for first solid form evaluation tests (see Table 3.2, "exploratory screen-type").

The analytical data of batch 4 are shown in Table 3.5.

The purity of batch No. 4 (>99%) is sufficient for crystallization studies. The aqueous solubility is rather low. The compound shows a strongly pH-dependent solubility with good solubility in acidic media. Solubility drops down to values lower than 1 μg ml^{-1} at pH values >4. The stability in solution over a wide pH range is given, even under sunlight. Therefore, a solution can be handled in the lab without any degradation protection equipment. *Compound A* exhibits an ionisable basic functional group in the structure which allows an ionic interaction between drug and acid. From that perspective, salt formation would be possible. The measured pK_a value of 3.8 allows only the use of strong acids for salt formation (e.g. hydrochloric acid is recommended).

3.9 Exploratory Crystallization Experiments

Due to limited amount of material, only 600 mg of batch No. 4 was available for these investigations at this point in time, 50 mg scale experiments were performed. The free form of *Compound A* was characterized using eight different solvents,

Table 3.5 Analytical characterization of batch no. 4.

Parameter	Result	
pK_a values	$pK_a = 3.8$ (basic)	
HPLC purity	99%	
Stability in solution (3 d@ 40 °C)	Media	Sum of degradation products
	0.1 N HCL	<0.5%
	Buffer pH 2.2	<0.5%
	Buffer pH 7.4	<0.5%
	Buffer pH 10	<0.5%
	24 h Suntester	<0.5%
Solubility data	Media	Concentration (mg ml^{-1}) after 2 h
	Water (pH 5.3)	<0.001
	0.1 N HCL	>1
	Buffer pH 2.2	0.3
	Buffer pH 4.5	<0.001
	Buffer pH 7.4	<0.001
	FaSSiF	0.008
	FeSSiF	0.05

select for this specific compound, in order to evaluate its crystallization behavior. The compound was dispensed into vials, and solvents were added subsequently. After two hours stirring at 50 °C, the material was filtered and dried in a vacuum oven (40 °C, 1 mbar). The solid materials were analyzed by XRPD (Table 3.6).

Evaluation of XRPD data of the exploratory crystallization experiments indicates six different patterns for *compound A*, including Form I crystallized before in Medicinal Chemistry. Due to confidentiality of *compound A*, the following figures show exemplary six XRPD diagrams of another compound (*compound B*) to illustrate six different crystalline materials (Figure 3.6).

In the meantime, two additional batches with sizes of approx. 30 g. A few grams of these batches were assigned for additional crystallization experiments. The compound supply now allows for a broader screening (see Table 3.2, type "Discovery screen") on solid forms. A high-throughput solid form evaluation was conducted using the automatic robotic setup to explore experiments in parallel. This screening covers also salt-, cocrystal, and solvate formation as well potential polymorphic forms. Due to the low pK_a value, strong acids were selected within this study. Amongst them, hydrochloric acid, hydrobromic acid, sulfuric acid, and phosphoric acid were used. Additionally, a few weaker acids were also implemented to the screen, which could act as cocrystal former. No different crystallization setups were applied for salt- and cocrystal formation. The screening of *Compound A* in μl scale consists of the slurry crystallization procedure. Slurry means that nondissolved

Table 3.6 Results of first crystallization experiments (exploratory screen type).

Used solvent	Result XRPD
	Low crystalline – Form I[a]
Ethanol	Crystalline – Form II
Ethylether	Crystalline – Form III
Methyl-ethyl-ketone (2-butanone)	Crystalline – Form III
Tetrahydrofurane	Crystalline – Form III
Cyclohexane	Crystalline – Form III
2-Propanol	Crystalline – Form IV
Acetone	Crystalline – Form V
Methylisobutylketone water (1 : 1 vol/vol%)	Crystalline – Form VI

a) Low crystalline batch no. 3 delivered by Medicinal Chemistry.

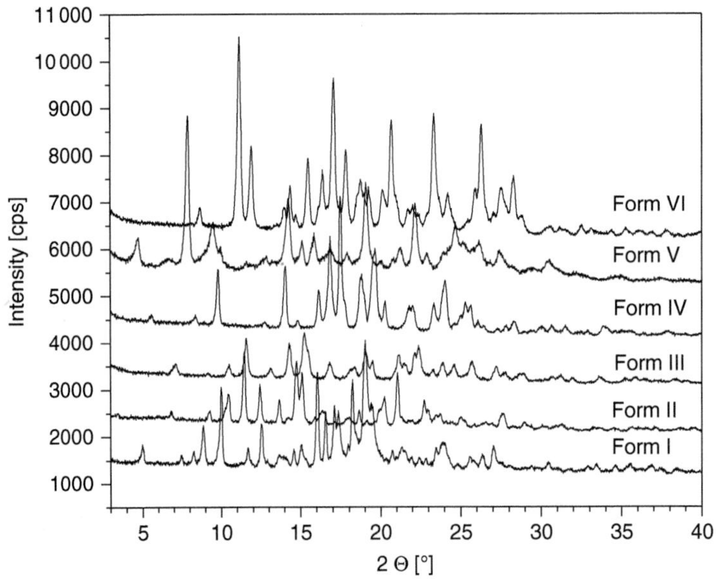

Figure 3.6 Exemplary XRPD-diagrams of compound B to illustrate six different crystalline materials.

solid material is suspended in the solution, which most often result in formation of a thermodynamically more stable, if not the most stable form.

Compound A was dissolved in ethanol and dispensed to the 96 wells into a crystallizer assembly (manufactured by Unchained Labs). Solutions of the acids in suitable solvents (methanol, ethanol, tetrahydrofurane, or water) were dispensed such that acids were delivered to each well in the stoichiometry specified in the experiment

Table 3.7 Experimental details of the HT screening setup.

Number of experiments	96
Amount of compound used	Approximately 1 g
Number of acids used	11
Ratio API : acid (mol/mol)	1 : 1.1
Temperature for crystallization	20–50–20 °C
Crystallization procedure	Crystallization by slurry

design (ratio API : acid = 1 : 1.1). After a period of two hours, the solvent was removed by evaporation and approx. 5–8 mg solid material remained in each well (Table 3.7).

Defined crystallization solvents were dispensed to each well, and the solutions were covered and subsequently heated up to 50 °C, stirred for 24 hours and equilibrated. By completion of the crystallization, the assemblies were opened and the residual solvents removed by evaporation using a vacuum oven (40 °C, 1 mbar).

Solid materials were formed on purpose-built kapton substrates. Kapton, a polyimide film material, is necessary if the XRPD analysis is performed on an instrument built up in transmission configuration. No reflection peaks appear from kapton in the 2-theta-range of 3–40 ° 2Θ. The substrate can be removed from the crystallizer assembly and can readily be used to obtain HT powder X-ray diffraction patterns without any manual manipulation of the samples.

Figure 3.7 shows the HT crystallization setup for *Compound A* on a 96-well plate. On behalf of the acids full name, a symbol is stated (CL: chloride, BR: bromide, TA: tartaric acid, CI: citric acid, FU: fumaric acid, MA: malic acid, SC: succinic acid, BZ: benzoic acid, LML: (L)-malic acid, PH: phosphoric acid, SU: sulfuric acid).

Solid materials (crystalline and amorphous) and transparent gels were obtained. Figure 3.8a shows an image of the final kapton crystallization plate, Figure 3.8b shows the results of the HT X-ray investigations. Within this figure, the four different blue-colored fields indicate four different crystalline XX forms of *compound A*,

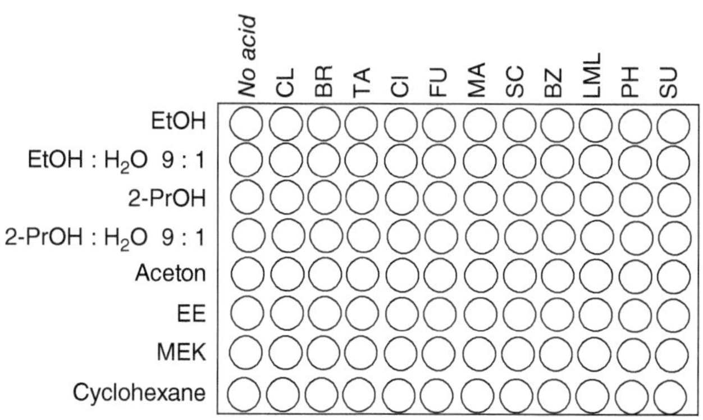

Figure 3.7 HT crystallization setup for compound A on a 96-well plate.

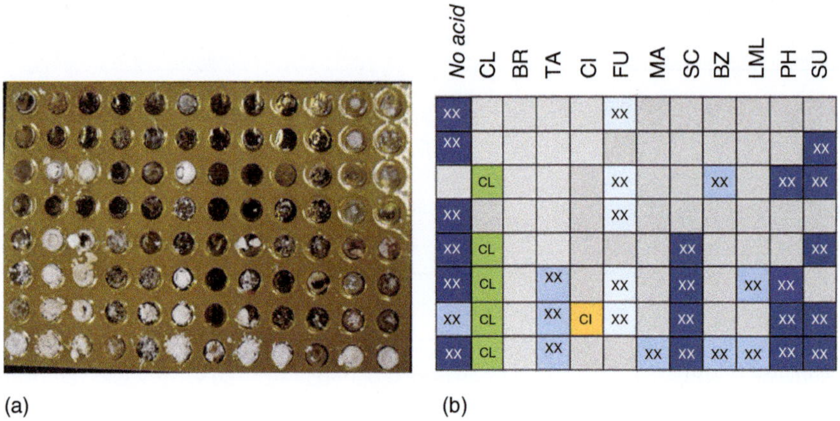

Figure 3.8 Image of the final kapton crystallization plate (a); results of the HT X-ray investigations (b). Source: Dr. Ulrike Werthmann.

which are already known from the first crystallization experiments. Salt formation was only successful by using hydrochloride acid (green field) and most likely a cocrystal was formed using citric acid (yellow field). In all other wells (grey), only amorphous material was formed.

In total, six different polymorphic forms of *Compound A* as well as two crystalline salt/cocrystal could be obtained by initial crystallization experiments. Due to time pressure, the focus was set on these eight forms and further evaluation of the forms was indicated. Four hundred milligrams-scale experiments (Table 3.2, type "form evaluation screen") were conducted to get sufficient material for additional analytics. For these experiments, the solvents that resulted in the best crystallinity within the HT screen were selected. All forms were successfully recrystallized, except Form V. Besides standard XRPD, the samples were also analyzed regarding melting behavior by using DSC. TG coupled with infrared spectroscopically identification of the vapor phase (TG-IR) provides knowledge about solvate and/or hydrate formation. These investigations allow a first interpretation regarding the most stable form and hydrate or solvate formation.

Table 3.8 summarizes the thermal analysis results. Three hydrates, one solvate, and two anhydrous solid forms of *compound A* were obtained.

Based on the DSC and TG data, the focus of further investigation was set on the two anhydrous solid forms (Form III and Form VI) and the form of the citric acid cocrystal. About 2 g of each form were manufactured in high crystallinity with high chemical purity.

Comparing the two melting points of the two anhydrous solid forms (Form III, mp at 171 °C, Form VI, mp at 186 °C), there is a clear indication that Form VI will be the more stable polymorph because of a 15 °C higher melting event. Long-term slurry experiments (3 days, 7 days, and 14 days) provide more confidence about the most stable form identified so far. These experiments were done using a 1 : 1 physical mixture of crystalline Form III and Form VI suspended in water and buffer at pH 7.4.

Table 3.8 Results of 400 mg crystallization experiments (screen type "Form Evaluation Screen").

	XRPD crystallinity	DSC melting point	TG weight loss and identification of residual solvents
Solid Form I hydrate	High	Broad endotherm at 100 °C Endotherm at 163 °C **Mp at 173 °C (relates to Form III)**	4.1% up to 100 °C (water)
Solid Form II hydrate	Medium	2 endotherms at 40 °C and 70 °C **Sharp melting point at 171 °C (relates to Form III)**	5.7% up to 80 °C (water)
Solid Form III anhydrous	High	**Sharp melting point at 171 °C**	0.6% up to 210 °C
Form IV hydrate	Medium	Broad endotherm at 88 °C Broad exotherm at 119 °C Broad endotherm at approximately 165 °C **Mp at 180 °C**	4.4% up to 200 °C (water)
Form V solvate	High	Broad exotherm at 131°**C** **Sharp mp at 171 °C (relates to Form III)**	NA due to less material
Form VI anhydrous	High	**Sharp mp at 186 °C**	0.4% up to 210 °C
Chloride compound form	Very low	Very complex diagram with several peaks	Not applicable
Citric acid compound form anhydrous	High	**Sharp mp at 163 °C**	0.8% up to 140 °C

All X-ray pattern collected on the materials coming from these slurry experiments showed only reflection peaks of Form VI, which could be explained by a conversion of the metastable Form III to the more stable Form VI.

Subsequently, the CMC data set of Form VI had to be completed by sorption isotherms, mechanical and chemical stability experiments, solubility, and intrinsic dissolution rate.

The crystalline citric form, most likely a cocrystal (not confirmed), which was first crystallized within the HT screen and then reproduced in 2 g scale, may offer a chance for improvement in solubility. A complete characterization including an extensive solubility screen of the citric acid compound form was conducted for a better comparison to the free Form VI.

For *Compound A,* the citric acid compound (CI Form) did not show a clear solubility advantage for DEV activities. In addition, slurry experiments of the CI Form in water hinted at the disproportionation of the crystalline material in free compound Form VI and citric acid after 30 minutes, which was seen in the XRPD pattern of the filtrated slurry material. Therefore, the chemical stability of the CI Form in water was not given and this may cause major difficulties within the manufacturing and formulation DEV.

The most critical issue of the solid materials of *Compound A* is the low solubility, which may require higher formulation efforts to support animal studies and clinical trials. As a consequence, this can have an impact on timelines and costs. Therefore, the calculation of a preclinical dose number pDo is helpful for a better assessment of the pharmaceutical DEV strategy [22]. The dose number is defined as

$$\mathrm{Do} = \mathrm{dose}/[\mathrm{Cs} * 250\,\mathrm{ml}] \tag{3.1}$$

with dose being the estimated human dose and Cs being the saturation concentration and can be calculated according Eq. (3.1) for any solubility data point. A dose number ≤ 1 indicates excellent solubility related to high drug exposure, high dose numbers are associated with poor solubility and low bioavailabilities which require additional formulation efforts [23]. The dose number provides a more robust assessment of the most appropriate pharmaceutical DEV strategy. Especially of interest are dose numbers in bio-relevant media. The estimated human dose (EHD) for *compound A* is 50 mg once daily. Low to moderate solubility is detected in simulated intestinal fluids FaSSiF and FeSSiF media, with corresponding dose numbers of Do (FaSSiF) = 25 and Do (FeSSiF) = 4, respectively. For *Compound A* with dose numbers of 25 and 4, a pharmaceutical DEV strategy with moderate complexity can be expected.

The results of all relevant solid-state properties of solid Form VI and the CI Form are summarized in Table 3.9 for a head-to-head comparison, and each property is classified in one of the different categories defined in Table 3.3.

Due to tight timelines, no additional crystallization experiments were planned because a suitable solid form for DEV had to be delivered in a short time.

Before the final decision regarding solid form can be made, a bridging PK study using Form VI had to be done in order to show bioequivalence to the data obtained using the first batches (amorphous and/or low crystalline Form I). This experiment (rat-PK, po-administration, low dose) turned out to be successful for Form VI.

After detailed discussion within the solid-state expert group, a separate CMC-team meeting had to be set up to present all the data to the colleagues from the DEV Department (Chemical-, Pharmaceutical-, and Analytical DEV). The goal of this meeting was to achieve a consensus between RES and DEV regarding solid form selection for DEV.

Table 3.9 Head-to-head comparison of solid form properties of Form VI and citric acid compound form.

Physicochemical properties	Form VI		Citric acid compound form	
Crystallinity	Ideal		Ideal	
thermal behavior (MP)	Ideal		Ideal	
Weight loss (TG)	Ideal		Ideal	
Hygroscopicity (0–90% r.h.)	Ideal		Ideal	
Polymorphism	2 anhydrates known so far		No other form known so far	
Solubility	pH 1–3 acceptable	pH 4.5–10 undesirable	pH 1–3 acceptable	pH 4.5–10 undesirable
	FaSSiF undesirable	FeSSiF acceptable	FaSSiF acceptable	FeSSiF acceptable
Dose number (EHD = 50 mg)	Acceptable	Acceptable	Acceptable	Acceptable
Intrinsic dissolution	pH 1 ideal	pH 3–pH 7.4 undesirable	pH 1 ideal	pH 3–pH 7.4 undesirable
Solid-state chemical stability	Ideal		Ideal	
Polymorphic stability slurry in water and pH 7.4	Ideal		Undesirable (disproportionation in the free compound Form VI and citric acid)	
Solid-state mechanical stress	Ideal		Ideal	
Feasibility in chemical development	Ideal – Form VI can be reproducibly manufactured		Undesirable problems with isolation	

Eventually, all CMC data including the basic properties of *compound A* were reported in a comprehensive RES document that summarizes all aspects of the compound (e.g. pharmacological profile, pharmacokinetic data, and nonclinical safety properties).

List of Abbreviations

API	Active Pharmaceutical Ingredient
BZ	benzoic acid
CI	citric acid
CMC	Chemistry, Manufacturing, and Control

CP	candidate profiling
CSP	crystal structure prediction
DEV	development
DMSO	dimethylsulfoxide
DSC	differential scanning calorimetry
DVS	dynamical vapor sorption
EHD	estimated human dose
FaSSiF	fasted state simulated intestinal fluid
FeSSiF	fed state simulated intestinal fluid
FU	fumaric acid
HBP	hydrogen bond propensity
HBr	hydrobromic acid
HCL	hydrochloric acid
HPLC–MS	high-performance liquid chromatography–mass spectrometer
HPLC–UV	high-performance liquid chromatography–ultra-violet detector
HT	high throughput
IC	ionic chromatography
IR-spectrometer	infra-red spectrometer
LML	(L)-malic acid
LO	lead optimization
MA	malic acid
MC	molecular complementarity
mp	melting point
NCE	new chemical entity
pDo	preclinical dose number
PH	phosphoric acid
PK	pharmacokinetics
po	per-os (i.e. oral administration)
RES	research
SC	succinic acid
SSNMR	solid-state nuclear magnetic resonance
SU	sulfuric acid
TA	tartaric acid
TDI	tolerable daily intake
TG	thermal gravimetry
XRPD	X-ray powder diffraction

References

1 Byrn, S.R. (1982). *Solid-State Chemistry of Drugs*. Academic Press.

2 Lee, E. (2014). A practical guide to pharmaceutical polymorph screening and selection. *Asian Journal of Pharmaceutical Sciences* 9: 163–175.

3 Croker, D.M., Kelly, D., Horgan, D. et al. (2015). Demonstrating the influence of solvent choice and crystallization conditions on phenacetin crystal habit

and particle size distribution. *Organic Process Research and Development*: 1826–1836.
4 Beezer, G.B. (1992). *The relationship between particle size and solubility. International Journal of Pharmaceutics* 82 (3): R7–R10.
5 Dhamoon, R., Popli, H., Aggarwal, G., and Gupta, M. (2018). Particle size characterization – techniques, factors and quality. *International Journal of Drug Delivery* 10 (1): 1–11.
6 Raida Al-Kassas, M.B. (2017). *Nanosizing techniques for improving bioavailability of drugs. Journal of Controlled Release* 260: 202–212.
7 Byard, S.L., Jackson, S.L., Smail, A. et al. (2005). *Studies on the crystallinity of a pharmaceutical development drug substance. Journal of Pharmaceutical Sciences* 94 (6): 1321–1335.
8 Newman, A. (2015). *Pharmaceutical Amorphous Solid Dispersions*. Wiley.
9 Kiang, Y., Bercot, E.H., Wu, Q. et al. (2015). Selection of a suitable physical form and development of a crystallization process for a PDE10A inhibitor exhibiting enantiotropic polymorphism. *Organic Process Research and Development* 19 (12): 1849–1858.
10 ICH (1999). ICH Q6A specifications: test procedures and acceptance criteria for new drug substances and new drug products: chemical substances.
11 Dikundwar, A., Pal, S., Chodon, P. et al. (2019). Solid state behavior of impurities during "in-process" phase purity analysis of an API. *Organic Process Research & Development* 23 (2): 269–273.
12 Bhardwaj RM, J. A. McMahon, J. Nyman, L. S. Price, S. Konar, I. D. H. Oswald, C. R. Pulham, S. L. Price, and S. M. Reutzel-Edens (2019). A prolific solvate former, galunisertib, under the pressure of crystal structure prediction, produces ten diverse polymorphs. *Journal of the American Chemical Society*, 141 (35):13887-13897.
13 Xin, D., Gonella, N.H., and Horspool, K. (2019). Solvate prediction for pharmaceutical organic molecules with machine learning. *Crystal Growth and Design* 19 (3): 1903–1911.
14 Oganov (2018). Crystal structure prediction: reflections on present status and challenges. *Faraday Discussion* 211: 643–660.
15 Morissette, S., Almarsson, O., Peterson, M. et al. (2004). High-throughput crystallization: polymorphs, salts, co-crystals and solvates of pharmaceutical solids. *Advanced Drug Delivery Reviews* 56 (3): 275–300.
16 Stahl, P.a. (2011). *Handbook of Pharmaceutical Salts: Properties, Selection, and Use*. Zürich: Wiley-VCH.
17 ICH (2016). ICH Q3C (R6) impurities: guideline for residual solvents.
18 ICH (2019). Quality guidelines. https://www.ich.org/products/guidelines/quality/article/quality-guidelines.html (accessed 27 October 2019).
19 Storey, R.A. and Ymén, I. (2011). *Solid State Characterization of Pharmaceuticals*. Wiley-Blackwell ISBN: 978-1-405-13494-1.
20 Jantratid, E., Janssen, N., Reppas, C., and Dressman, J.B. (2008). *Dissolution media simulating conditions in the proximal human gastrointestinal tract: an update. Pharmaceutical Research* 25: 1663–1676.

21 Jabbal, S.P. (2017). *Does size really matter?: relationship of particle size to lung deposition and exhaled fraction. The Journal of allergy and clinical immunology* 139 (6): 2013–2014.e1.

22 Oh, D.-M.C. (1993). *Estimating the fraction dose absorbed from suspensions of poorly soluble compounds in humans: a mathematical model. Pharmaceutical Research* 10 (2): 264–270.

23 Wuelfing, W., Daublain, P., Kesisoglou, F. et al. (2015). Preclinical dose number and its application in understanding drug absorption risk and formulation design for preclinical species. *Molecular Pharmaceutics* 12 (4): 1031–1039.

4.1

Solid-State Characterization Techniques: Microscopy

Luis Almeida e Sousa and Constança Cacela

R&D Analytical Development, Hovione Farmaciencia SA, Campus do Lumiar – Edificio S – Estrada do Paço do Lumiar, 1649-038 Lisboa, Portugal

4.1.1 Microscopy

Microscopy is widely used in the field of industrial and physical pharmacy as an analytical tool with applications in different stages of product development. Its use extends from early pre-formulation and research to formulation and manufacturing stages. Microscopy techniques are used in the pharmaceutical industry for four main purposes: solid-state characterization, size and shape analyses, and identification of contaminants. Other applications have been explored in recent years, e.g. the investigation of particle interactions with atomic force microscopy (AFM). However, the abovementioned applications are more established in the area and routinely used in development.

Microscopy techniques can be grouped into three main categories depending on the imaging principles: optical, electron, and scanning probe microscopy. Optical microscopy relies on the detection of refracted, diffracted, and reflected light produced after interaction of the electromagnetic radiation with the sample. In electron microscopy, an electron beam interacts with the sample with a variety of different mechanisms and produces a complex set of signals which are used to obtain a magnified image. On the other hand, scanning probe microscopy involves interaction/contact between a scanning probe and the surface of the sample. The principles and applications of these three microscopy techniques are described in more detail in the following sections.

4.1.1.1 Optical Microscopy

Optical or light microscopy uses visible light to irradiate samples. When light interacts with the sample, it can be reflected or transmitted and, in the process, may lose some of its energy to the sample. The complexity of these interactions allows

Solid State Development and Processing of Pharmaceutical Molecules: Salts, Cocrystals, and Polymorphism, First Edition. Edited by Michael Gruss.

for the versatility and power of the microscope. Two types of light microscopes are commonly used: simple microscopes that use a single lens for magnification and compound microscopes that use several lens. The majority of equipment are compound microscopes which combine two lenses, the objective lens and the eyepiece (or ocular), that work together to produce a magnified image of an object (typically up to 1000×). The magnification is dependent upon the shape and size of the lens, the distance from the object to the lens, the distance of the image from the lens, and, of course, the size of the object. Other important components of the microscope are the condenser lens, the lamp, filters, polarizers, retarders, and the microscope stage and stand with coarse and fine focus dials.

The main use of optical microscopes is solid-state analysis. Pharmaceutical applications range from simple images of drug substance or drug product to investigate particle size and shape to optical crystallography. Pharmaceutical applications depend on the illumination technique used. The following section will focus on the application of traditional techniques – bright-field, dark-field, and polarized light microscopy (PLM) – as well as others that may be relevant in the context of pharmaceutical sciences [1].

4.1.1.1.1 Bright-Field Microscopy

Bright-field illumination (dark sample in a bright background) can be used to observe the color of drugs as a result of white light absorption. That can be seen in both reflected and transmitted light. Despite color being an important attribute for identifying compounds, it is important to point out that the observation of color is dependent on the light source used and whether we are looking in transmission or reflection mode. In the case of crystalline drugs, observation under the microscope may also give information on the crystal shape and habit which may be useful to identify different polymorphs.

There is an important difference between habit and tracht [2]. Both together describe the external morphology of a crystal. Habit corresponds to the characteristic of the external crystal form (e.g. isometrical, planar, or prismatic) and the ratio of the size of the faces that leads to the formation of rod-, flake-, or needle-like crystals, while tracht refers to the number and occurrence of crystal faces. Agglomeration/aggregation of primary particles can also be observed with a light microscope. The tendency of particles to stick together and form larger entities is important in formulation development because these can impact flow properties and the downstream process. These observations also help the process engineer in the final crystallization process. Twinning, the formation of crystals that share a crystallographic face, is not uncommon in pharmaceutical crystals and can also be detected with a microscope. However, the presence of twinned crystals rarely has an impact on bulk powder properties. On the other hand, the observation of cleavage planes is important if the crystals show preferential breakage along a particular face. For example, a study with paracetamol showed that Form I cannot be used in direct compression due to the way it cleaves apart, while Form II can [3]. Surface texture, which is often related to solvent loss during crystallization or polymorphism, can also be observed with the light microscope. In the case of polymorphism, small crystals are often observed growing on the surface of the sample, e.g. during stability tests, and this indicates conversion of a less stable to a thermodynamically more stable form. Evaluation of sample transparency can also be helpful in development since the majority of crystalline drug substances is transparent. Sample opacity is often attributed to inclusion of gas or liquid in the lattice during the crystallization process which, in turn, is dependent on the type of solvent system used, stirring speed, or the presence/lack of seeding. Therefore, evaluation of sample transparency can add valuable information to optimize the crystallization process.

4.1.1.1.2 Dark-Field Microscopy

Dark-field conditions are obtained by illuminating the specimen at an oblique angle such that direct, nondiffracted radiations are not collected by the objective lens. As a result, a dark background is observed and bright objects are viewed. This illumination technique is useful to examine parenterals when looking for foreign particles because these are easier to observe in that type of lighting. The main disadvantage of dark-field microscopy is that it requires strong illumination which can damage the sample.

4.1.1.1.3 Polarized Light Microscopy

PLM is a contrast-enhancing technique that provides improved quality images for birefringent compounds when compared to bright-field and dark-field illumination techniques. Polarization of light refers to limitations on the direction of wave oscillation imposed by a filter (polarizer) that only allows light oscillating in one direction to pass. The polarized light microscope is used to observe compounds that are visible primarily due to their anisotropic character, i.e. due to their optical properties being dependent on the direction of light. To accomplish that, the microscope must be equipped with two polarizers that are placed in the optical pathway, before (polarizer) and after (analyzer) the sample. The contrast-enhanced image results from the interaction of polarized light with a birefringent, or double-refracting, sample. From that interaction, two individual waves are formed: the ordinary and extraordinary waves. The velocity of these waves is different and varies with the propagation direction through the sample. They are ultimately recombined upon entering the second polarizer, the analyzer, and the resulting interference, constructive or destructive, is responsible for the final image.

Some of the applications of PLM in the pharmaceutical field are similar to those of bright-field illumination. For example, PLM can be used to provide information on absorption color and optical path boundaries between crystals with different refractive indices. The greatest advantage of PLM is that it can also distinguish between isotropic and anisotropic materials and explore the optical properties of the latter to elucidate structure and composition of the materials.

To better understand this specific application of PLM, let us first introduce some important concepts. Materials can be classified into isotropic or anisotropic depending on the optical properties that they present when light passes through them. Isotropic materials have equivalent optical axes that interact with light in a similar manner. Light that enters the isotropic material is refracted at a constant angle and passes through it at a single velocity. Examples of such compounds are cubic crystals and amorphous compounds. On the other hand, anisotropic materials have optical properties that vary with the orientation of the incident light with the crystallographic axes. They show a range of refractive indices that depend on the propagation direction of light through the sample and on the vibrational plane coordinates. Anisotropic crystals are further subdivided into uniaxial – the refractive index (RI) of one axis is different from the other two – or biaxial – having three refractive indices corresponding to three principal optical axes of the crystal. Tetragonal and hexagonal crystals are uniaxial, whereas orthorhombic, monoclinic, and triclinic crystals are biaxial.

Identification and structure elucidation of these materials can thus be done by exploring and determining the different refractive indices of samples. For materials with only one RI, cubic crystals, and amorphous materials, a very simple and straightforward test is used (Becke line test) [4]. The sample is mounted in different reference liquids of known RI, and the contrast between particles and liquid is evaluated. When the refractive indices of particle and liquid are the same, it is not possible to distinguish between them since there is no refraction of light at their interface. The RI is thus determined by comparison. For uniaxial (two refractive indices) and biaxial crystals (three refractive indices), determination also requires comparison with reference liquids of known RI. However, these are significantly more complex because the different indices must be measured at the extinction positions, where the vibration direction of the crystal is aligned with the direction of the crossed polarizers.

Determination of the RI is important not only for identification purposes and investigation of polymorphism but also for the optimization of laser diffraction particle size methods that use the Mie theory. This theory requires knowledge of the average RI of compounds for accurate particle size measurements, as demonstrated in previous studies with Lisinopril and Enalapril Maleate particles [5]. In these studies, the RI of particles was determined with Saveyn's method [6], validated with the Becke line test, and used with Mie theory to determine the size distribution of particles. The size distributions obtained with the RI determined with both methods were very similar. However, when size distribution was obtained with Fraunhofer approximation, which does not include RI in calculations, the fine fraction was greatly over-reported. In addition to RI and birefringence determination, images obtained with PLM also provide information on interference colors, extinction position, and angles, among other properties, which are distinct features of anisotropic crystals.

Evaluation of birefringence and interference colors is particularly relevant in the context of stability testing of amorphous pharmaceuticals since these materials can crystallize during storage, due to their inherent metastability. PLM can thus be used to detect onsets of crystallization by monitoring the appearance of spots in the sample that show birefringence and interference colors [1, 7, 8]. Figure 4.1.1 compares the typical optical characteristics of amorphous and crystalline materials observed with the polarized light microscope.

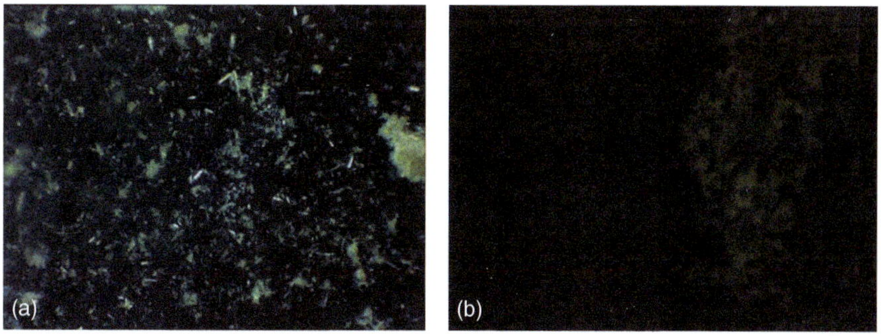

Figure 4.1.1 Polarized light microscope images of (a) crystalline drug particles (b) amorphous material.

4.1.1.1.4 Other Optical Microscopy Variants

In addition to the microscopy techniques described above, several variants of light microscopy are developed for specialized purposes.

Thermal Microscopy (or Hot-Stage Microscopy) Thermal microscopy refers to a variant of microscopy where a heating and/or cooling stage is adapted to an optical or electron microscope. Evaluation of the effect of temperature on the physical and crystallographic properties of samples can be very useful to understand the nature of hydrates, polymorphs, and solvates. Although thermal microscopy is not the only way to investigate the thermal behavior of compounds, its greatest advantage over the others is that phase transitions are observed directly. These direct observations help interpreting the phase transitions detected by other techniques such as differential scanning calorimetry, thermal gravimetric analysis, and other thermal techniques. The most relevant applications of thermal microscopy are determination of melting behavior and temperatures [9], solid–liquid and solid–liquid–solid phase transitions [10], dehydration, and desolvation [11].

Confocal Microscopy Confocal microscopy improves the depth of field of the microscope, without compromising resolution, by means of using spatial pinholes, in front of the objective and condenser, to block stray light and isolate only light that comes from the feature of interest. The laser scanning confocal microscope is able to capture two-dimensional images at different depths in a sample, which allows re-construction of three-dimensional structures within an object. The technique is typically used with fluorescent or fluorescence-stained materials, although that is not an absolute requirement. Since most drugs are not naturally fluorescent or easily stained, its application in pharmaceutical development is limited. Nevertheless, confocal laser scanning microscopy has been used, e.g. to determine the film thickness and coating quality of microparticles [12].

Fluorescence Microscopy The fluorescence microscope is a light microscope that measures the emission of fluorescence by samples that naturally fluoresce or are stained with a fluorescence probe. Fluorescence microscopy has been used to evaluate the distribution of the drug substance in tablets in cases where the drug naturally fluoresces, but it is more commonly used for topical drug products [1]. The reason for that is that there are better imaging techniques for tablets and capsules and because the cellulose excipients generally also fluoresce in the wavelength of interest and interfere with the drug substance. Fluorescence imaging has also been used in the analysis of spatial distribution of drug and carrier and in the characterization of cellular barriers to drug penetration [13].

Freeze-Dry Microscopy A freeze-dry microscope basically combines a cold stage, designed to work in vacuum, with an optical microscope. This microscope is mainly used to understand and determine critical parameters in the lyophilization process, such as the cake collapse temperature and ice morphology during the sublimation cycle [14].

4.1.1.2 Electron Microscopy

In electron microscopy, instead of electromagnetic radiation, an electron beam is used to interact with the sample and produce the magnified image. The most common types of electron microscopy, in the pharmaceutical field, are scanning electron microscopy (SEM), energy-dispersive X-ray spectroscopy (EDS) coupled to SEM and transmission electron microscopy (TEM).

4.1.1.2.1 Scanning Electron Microscopy

The SEM is widely used in pharmaceutical development for both drug substance and drug product. The major application is the examination of size and shape of both individual particles of drug substances or individual components of drug products. The technique may be considered quantitative in the sense that size and shape of particles are measured using image analysis, although most applications are qualitative.

In SEM, electrons are accelerated toward the sample and interact with the specimen with a variety of different mechanisms, producing a complex set of signals (e.g. backscattered electrons (BSE), secondary electrons (SE), Auger electrons, X-rays, etc.) which are captured by different detectors. SEM images are obtained by collecting the signals that result from interaction of the sample with the electron beam that scans its surface. Magnification is dependent on the sample area scanned by the beam and the size of the monitor. Although higher magnifications are obtained when the beam scans a small specimen area, the damaging effects of the electron beam are also greater due to an increase in overall energy per cm [2]. With respect to resolution, depending on the type of microscope, SEM can achieve resolutions of a few nanometers only. High resolutions are achieved with small electron beam diameters, high current, and low convergence angle [1, 15].

In order to manipulate the incident beam of electrons and minimize unwanted interactions, the column that directs the electrons to the sample should be at high vacuum. Low-vacuum SEMs are also available and, in fact, are very useful in pharmaceutical development for a number of reasons. Contrary to high-vacuum SEMs, low-vacuum microscopes can be used to image wet samples, electrically nonconductive compounds, and do not require additional coating of samples. The absence of coating is particularly advantageous in elemental analysis of samples using SEM–EDS.

As mentioned above, the interaction of incident electrons with the sample produces several signals. Those with more application to pharmaceutical development are BSE, SE, and X-ray emission. The latter will be discussed in the next section. BSE are electrons that are reflected back to the pole piece of the SEM after interacting with the sample. The yield of BSE increases for higher atomic number elements and with lower beam energy. Therefore, the images obtained with these electrons are based on atomic number differences. Because of that, BSE are preferred when doing X-ray mapping with SEM–EDS or metal contaminant identification. For example, BSE signals are very useful for studying the distribution of different tablet components in whole tablets. The downside of BSE signals is that they are much less sensitive to sample topology. Although image resolution is also lower with these electrons, the impact is not significant since the majority of drug or excipient particles are on the micrometer range.

SE emission results from interaction of the electron beam with valence electrons in the outer shell of the sample atoms. These electrons are knocked out of the atom, travel through the sample, and escape the surface. The yield of SE emission is not very sensitive to composition and atomic number. However, SE yield increases with lower beam accelerating voltage. This relationship is very useful for analyzing nonconductive materials, such as most organic compounds, that otherwise would have to be coated, e.g. with gold, to avoid charging problems. By reducing the accelerating voltage, it is possible to reduce the charging effect. Another advantage of SE emission is that it is very sensitive to irregularities in the sample surface and, therefore, produces high-resolution images [1, 15].

As previously mentioned, SEM is widely used in pharmaceutical sciences for evaluation of morphology, size, and shape of particles. Figures 4.1.2 and 4.1.3 show two examples where SEM was used to investigate the morphology of individual particles in formulations developed for dry powder inhalers (DPI): composite particles (Figure 4.1.2) and lactose carrier-based formulations (Figure 4.1.3). The SEM image of Cu, Zn-superoxide dismutase composite particles (Figure 4.1.2) was acquired to evaluate the shape of the individual particles and confirm that their size is within an acceptable range for efficient lung deposition (lower than 5 μm) [16]. On the other hand, Figure 4.1.3 shows an SEM image obtained for a lactose carrier-based formulation. In this image, it is possible to observe the presence of fine particles of lactose and drug adhered to the surface of the coarse lactose particles which act as a carrier.

SEM and microscopy techniques, in general, can also be used as auxiliary tools in quality control since they can provide information on batch homogeneity and batch to batch variability. Figure 4.1.4 shows a great example of such application. The SEM images obtained for different doxycycline monohydrate batches supported

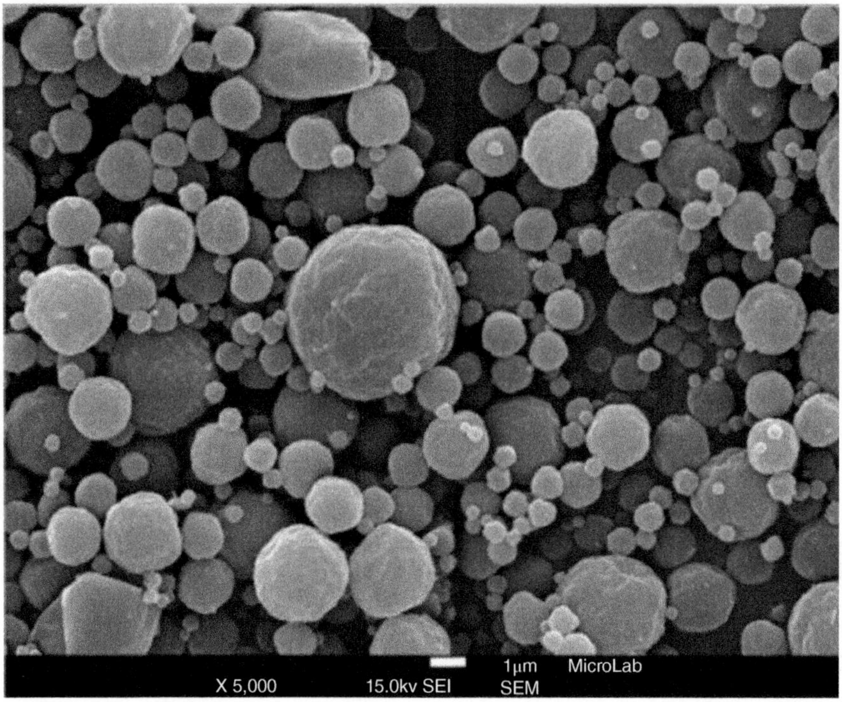

Figure 4.1.2 SEM image of Cu, Zn-superoxide dismutase composite particles containing trehalose and leucine.

the particle size results obtained by laser diffraction, which showed significant variability between the two batches.

Energy-Dispersive X-ray Spectrometry Coupled to SEM EDS is used with SEM to provide chemical information on the samples, in addition to imaging. EDS is based on the physical principles of electron interaction with specimen to produce X-rays of two types: Bremsstrahlung X-rays, also known as continuum X-rays, that result from deceleration of the incident electrons upon interaction with the atomic nucleus (background signal) and characteristic X-rays that result from relaxation of excited atom electrons to the ground state after removal of inner electrons that were knocked out of the atomic orbital by incident electrons. These characteristic X-rays are the basis of EDS analysis. Since different atoms have different distribution of electrons in their orbitals, the possibility of transitions to the ground state is also different. As a result, the intensity and energy of the X-rays generated are characteristics of each atom and so is the X-ray spectrum collected. Unfortunately, the resolution of SEM–EDS is low; therefore, it is inevitable that X-ray peaks of different atoms will overlap. EDS can only detect elements with atomic number between Beryllium (4) and Uranium (92). Light elements are particularly difficult to detect because they are not easily ionized and emission of X-rays is less likely to occur. Furthermore, the low wavelengths generated by light elements are easily absorbed by the sample [17].

Figure 4.1.3 SEM image of particles in a carrier-based formulation for DPI. The formulation contains fluticasone propionate, fine and coarse lactose monohydrate particles.

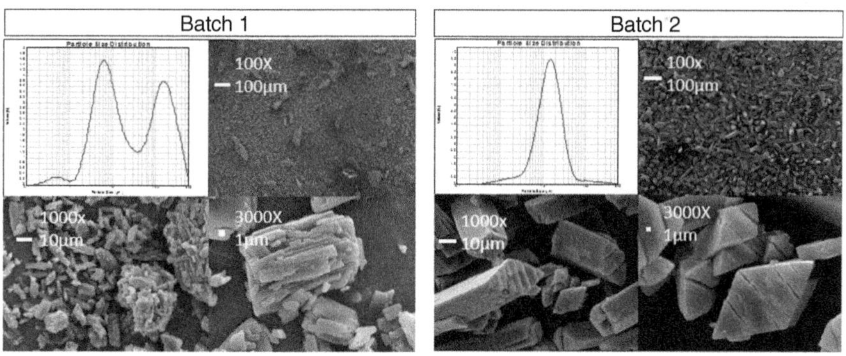

Figure 4.1.4 Comparison of particle size distribution of two different batches of doxycycline monohydrate using SEM and a laser diffraction method.

SEM–EDS is extensively used for qualitative elemental identification and distribution in a drug or drug substance. Spot mode should be used for elemental analysis to allow focusing the electron beam in one spot without rastering. Because the intensity of the electron beam is very high using spot mode, sample integrity should be monitored and sample damage should be minimized. On the other hand, the accelerating beam voltage should be sufficient to excite the high energy orbitals of all elements present. Therefore, an adequate balance is required. Sample preparation is simple, and analysis is relatively easy to do with current available software that assign elements automatically to the X-ray peaks in the spectrum [1].

The main application of EDS in the qualitative mode is the identification of contaminants and/or foreign particles [18, 19]. If the particle/spot under investigation is present in a tablet, the whole tablet should be mounted on the sample holder. If the particle is dispersed in a liquid, as in the case of pharmaceutical solutions, the solution should be filtered and the sediment is mounted onto the SEM holder. In the pharmaceutical industry, it is very useful to use SEM–EDS when unwanted spots are present in tablets, in cases where the powder or granule formulation contains foreign particles or, e.g. when the feed solution used in a spray-drying process contains undissolved material. In such cases, it is advisable to do an elemental analysis of potential sources of contamination for comparison purposes and to help assign a root cause of contamination. Equipment parts, such as rubber o-rings and metal components, starting raw materials or solvents from the reservoir are often considered for analysis in cases where foreign particles are detected.

Although the SEM and EDS were not designed with the purpose of quantification, quantitative analysis is also possible. That is because the X-ray counts are directly related to the concentration of each element in the sample. Unfortunately, quantification of elements cannot be done simply by determining ratios of peak areas and intensities. To improve accuracy of quantification, elemental standards must be analyzed under strictly controlled operating conditions, followed by sample analysis under the same conditions. Standardless analysis can also be done, but the errors associated may be significant. The American Society for Testing and Materials (ASTM) provides a "Standard Guide for Quantitative Analysis by Energy-Dispersive Spectroscopy" that covers recommended procedures for elemental quantification, including the methods that use standards and standardless methods [20].

SEM–EDS can also be done for elemental mapping of samples. In this application, the location of each element in the sample is represented by a different color on the image [21]. This mapping technique is usually superimposed onto the original SE or BSE image.

4.1.1.2.2 Transmission Electron Microscopy

In TEM, a beam of electrons is transmitted through the sample for imaging. After passing through the sample, the electrons are detected directly in a fluorescence screen, a layer of photographic film or a sensor attached to a charge-coupled device. Samples must be thin, in the order of 100–200 nm, and must be stable in an intense electron beam. Sample preparation is very demanding and may induce changes in the specimen. Because of these requirements and issues, application in pharmaceutical development is limited. TEM is commonly used in microbiology and for the study of cells morphology and chemistry (using EDS). In physical pharmacy, the most important application is size and morphology evaluation of nanoparticles [22].

4.1.1.3 Atomic Force Microscopy

AFM is the principal member of the scanning probe microscope family. AFM is a high-resolution imaging technique that uses a small tip prefabricated onto a cantilever to probe the surface of samples and provide topographical information. Since a small tip is used to scan the surface of samples, the requirement to focus light or electrons, as with light and electron microscopy, is eliminated. This overcomes the Rayleigh criterion resolution limit, allowing imaging and resolution in the nanometer, and sometimes atomic range [23]. The instrument operates in three modes: contact mode, tapping mode, and noncontact mode.

In contact mode, the probe (cantilever and tip) is brought into contact with the sample, until a small deflection of the cantilever, corresponding to a repulsive force (the set point), is detected via a displacement on the photodetector. The cantilever moves along the surface of the sample maintaining that set point force. If too much or too little force is detected, the height of the scanner is adjusted through a feedback loop. That avoids damage to the sample and/or tip. Tapping mode is as widely used as contact mode. In tapping mode, the cantilever is oscillated at its resonant frequency during scanning to minimize potentially damaging lateral forces being exerted on the sample. Measurements are done only at the moment of contact with the sample, and when the sample surface is scanned, the tip is safely retracted. Tapping mode is often used to image delicate biological samples.

Different signals can be acquired at the same time as topography images. For example, in contact mode, the error signal and the friction response may be obtained. In tapping mode, in addition to the error signal, the phase image and amplitude image are obtained. These modes often provide more visual contrast than the topographic images collected simultaneously. In addition to imaging, force data can be obtained by pushing the tip into or pulling the tip away from the sample surface. Force measurements, when combined with the scanning capability, produce local physical property maps that provide information on indentation, Young's modulus [24], adhesion, and friction.

A wide variety of probes are available depending on the application. These vary in the type of material that they are made of, shape of the cantilever, and tip. Small entities, such as cells or polymer spheres can also be adhered to a tipless cantilever to provide biological functionality or defined contact geometries. It is also possible to attach ligands on the tip and map receptor-binding sites on cell/substrate surfaces, whilst acquiring simultaneously topographic images. Probes can also be coated on the top side to increase laser reflection, on the underside to allow chemical functionalization and both sides. Florin et al. first used functionalized AFM tips and surfaces to measure forces between biotin-avidin ligand–receptor pairs [25].

Most applications in pharmaceutical sciences are related to structural biological studies, e.g. mapping of ligand–receptor binding on cell surfaces, investigation of protein–protein interactions or even research on cancer or other diseases. However, some applications to solid-state analysis and physical pharmacy are also found [23].

AFM has been used to investigate several aspects of tablet coating, as well as topographic features that impact dissolution of tablets. Studies conducted on tablets proved to be useful to obtain topographic information, quantitative surface roughness, and surface area measurements [26]. Data on porosity and distribution of components in tablets may also be obtained. In addition, it can be used to examine the quality of film coatings and film uniformity on the tablet surface [27]. Regarding dissolution of pure drugs, AFM was used to investigate the dissolution process, particularly the mechanisms involved in surface morphology changes and etching patterns [28]. Correlations between dissolution rate and topographic features have also been found with the help of AFM for aspirin crystals [29].

Crystal growth and polymorphism can also be investigated with AFM. The technique has been used to monitor the step growth and two-dimensional nucleation rates of lysozyme crystals. The impact of supersaturation on step growth was also investigated [30]. AFM was used to study the kinetics of crystallization from amorphous drugs (nifedipine) with and without polymer [31]. With respect to polymorphism investigation, several studies have been reported in the literature, such as [32].

AFM is excellent for visualizing particles in the nanometer range allowing quantitative size measurements. Since the equipment allows measuring distances in all three axis, parameters such as diameter, volume, and surface area can be calculated. Because surface properties and particle dimensions are easily obtained, AFM is suitable to evaluate nanoparticle stability with time [33].

In inhalation sciences, AFM has applications that range from the investigation of adhesive and cohesive forces in DPI systems [34] to the evaluation of capsule roughness (Figure 4.1.5) and its impact on the emitted dose from the capsule.

The list of applications is extensive, but the examples provided above demonstrate well the invaluable contributions that AFM can provide to pharmaceutical development.

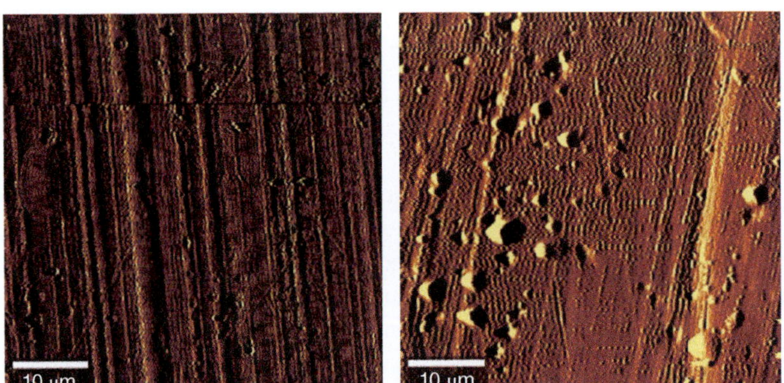

Figure 4.1.5 AFM image of the surface of clear (on the left) and colored (on the right) capsules typically used in dry powder inhaler (DPI) formulations.

4.1.1.4 Microscopy in Regulatory Documents

In an effort to standardize the nomenclature, vocabulary, and procedures used, regulatory authorities created a series of guidelines to help harmonization within the microscopy community. Of particular relevance, in the pharmaceutical field, are the European Pharmacopeia (Ph. Eur. 2.9.37) [35], the United States Pharmacopeia (USP <776>) [36], and the Japanese Pharmacopeia (JP 3.04) [37] that dedicate separate chapters or sections to optical microscopy and its application to particle characterization. The pharmacopeias are harmonized, and so the recommendations and guideline texts are similar in all three. In addition to instrumentation and recommended operation procedures, the guidelines provide standard terminology and recommendations for particle size, crystallinity, and shape characterization. The documents also describe a limit test of particle size by microscopy. Although the guidelines apply specifically to optical microscopy, they can clearly apply to SEM, as well.

In addition to the optical microscopy chapter, the USP has a separate chapter on SEM (USP <1181>), which provides an overview of the common microscopy technologies and techniques [38]. This chapter includes a section on X-ray generation and elemental compositional analysis.

Other regulatory documents that refer the use of microscopy in a pharmaceutical quality control context are ICH (International Conference on Harmonization) guidelines Q5A(R1) [39] on and ICH Q6A [40]. The first is a guideline on "Viral safety evaluation of biotechnology products derived from cell lines of human or animal origin" which recommends using electron microscopy in retroviruses testing for cell line qualification. Regarding ICH guideline Q6A on "Specifications: test procedures and acceptance criteria for new drug substances and new drug products: chemical substances," the text recommends different techniques, including optical microscopy and hot-stage microscopy, for the investigation of polymorphism in drug substances.

List of Abbreviations

AFM	atomic force microscopy
ASTM	American society for testing and materials
BSE	backscattered electrons
DPI	dry-powder inhalers
EDS	energy-dispersive X-ray spectroscopy
ICH	International Council for Harmonization
PLM	polarized light microscopy
RI	refractive index
SE	secondary electrons
SEM	scanning electron microscopy
TEM	transmission electron microscopy
USP	United States pharmacopeia

References

1 Allen Carlton, R. (2011). *Pharmaceutical Microscopy*, 321. New York: Springer-Verlag.

2 Gaedeke, R., Wolf, F., and Bernhardt, G. (1980). On the influence on tracht and habit of urea crystals (II). *Kristall und Technik* 15 (5): 557–564.

3 Nichols, G. and Frampton, C.S. (1998). Physicochemical characterization of the orthorhombic polymorph of paracetamol crystallized from solution. *Journal of Pharmaceutical Sciences* 87 (6): 684–693.

4 Choi, J.H., Eichele, C., Lin, Y.C. et al. (2008). Determination of effective refractive index of molecular diamondoids by Becke line method. *Scripta Materialia* 58 (5): 413–416.

5 Malvern Instruments Ltd. (2007). Rapid refractive index determination for pharmaceutical actives. Mastersizer 2000 application note: MRK529-03.

6 Saveyn, H., Mermuys, D., Thas, O., and van der Meeren, P. (2002). Determination of the refractive index of water-dispersible granules for use in laser diffraction experiments. *Particle & Particle Systems Characterization* 19 (6): 426–432.

7 Alonzo, D.E., Zhang, G.G., Zhou, D. et al. (2010). Understanding the behavior of amorphous pharmaceutical systems during dissolution. *Pharmaceutical Research* 27 (4): 608–618.

8 Bhugra, C., Rambhatla, S., Bakri, A. et al. (2007). Prediction of the onset of crystallization of amorphous sucrose below the calorimetric glass transition temperature from correlations with mobility. *Journal of Pharmaceutical Sciences* 96 (5): 1258–1269.

9 Li, J., Bourne, S.A., de Villiers, M.M. et al. (2011). Polymorphism of the antitubercular isoxyl. *Crystal Growth & Design* 11 (11): 4950–4957.

10 Leitão, M.L.P., Castro, R.A.E., Costa, F.S., and Redinha, J.S. (2001). Phase transitions of 1,2-cyclohexanediol isomers studied by polarised light microscopy and differential thermal analysis. *Thermochimica Acta* 378 (1): 117–124.

11 Wu, H. and Jones, A.G. (2012). Crystallization, dehydration, and phase transformations of diastereomeric salts: L-ephedrine and D-tartaric acid. *Chemical Engineering & Technology* 35 (6): 1031–1038.

12 Depypere, F., Van Oostveldt, P., Pieters, J.G., and Dewettinck, K. (2009). Quantification of microparticle coating quality by confocal laser scanning microscopy (CLSM). *European Journal of Pharmaceutics and Biopharmaceutics* 73 (1): 179–186.

13 Abhinav, A., Pushpendra, K.T., Shalini, T., and Narendra, K.J. (2008). Fluorescence imaging: applications in drug delivery research. *Current Drug Targets* 9 (10): 895–898.

14 Liu, J. (2006). Physical characterization of pharmaceutical formulations in frozen and freeze-dried solid states: techniques and applications in freeze-drying development. *Pharmaceutical Development and Technology* 11 (1): 3–28.

15 Egerton, R.F. (2005). The scanning electron microscope. In: *Physical Principles of Electron Microscopy: An Introduction to TEM, SEM, and AEM* (ed. R.F. Egerton), 125–153. Boston, MA: Springer International Publishing.

16 Jabbal, S., Poli, G., and Lipworth, B. (2017). Does size really matter?: relationship of particle size to lung deposition and exhaled fraction. *Journal of Allergy and Clinical Immunology* 139 (6): 2013–2014.e1.

17 AZoM (2013) Thermo Fisher Scientific – Materials & structural analysis. Using SEM/EDS for the quantitative analysis of light elements. https://www.azom.com/article.aspx?ArticleID=11975 (accessed 2 September 2019).

18 Sarecka-Hujar, B., Balwierz, R., Ostrozka-Cieslik, A. et al. (2017). Scanning electron microscopy and X-ray energy dispersive spectroscopy – useful tools in the analysis of pharmaceutical products. *Journal of Physics: Conference Series* 931: 012008.

19 Pajander, J., Haugshoj, K.B., Bjorneboe, K. et al. (2013). Foreign matter identification from solid dosage forms. *Journal of Pharmaceutical and Biomedical Analysis* 80: 116–125.

20 ASTM E1508-12a (2012). *Standard Guide for Quantitative Analysis by Energy-Dispersive Spectroscopy*. West Conshohocken, PA: ASTM International www.astm.org.

21 Scoutaris, N., Vithani, K., Slipper, I. et al. (2014). SEM/EDX and confocal Raman microscopy as complementary tools for the characterization of pharmaceutical tablets. *International Journal of Pharmaceutics* 470 (1): 88–98.

22 Malatesta, M. (2016). Transmission electron microscopy for nanomedicine: novel applications for long-established techniques. *European Journal of Histochemistry* 60 (4): 2751.

23 Lamprou, D. and Smith, J.R. (2014). Applications of AFM in pharmaceutical sciences. In: *Analytical Techniques in the Pharmaceutical Sciences* (eds. A. Müllertz, Y. Perrie and T. Rades). New York, NY: Springer.

24 Roa, J.J., Oncins, G., Diaz, J. et al. (2011). Calculation of Young's modulus value by means of AFM. *Recent Patents on Nanotechnology* 5 (1): 27–36.

25 Florin, E.L., Moy, V.T., and Gaub, H.E. (1994). Adhesion forces between individual ligand-receptor pairs. *Science (New York, N.Y.)* 264 (5157): 415–417.

26 Seitavuopio, P., Rantanen, J., and Yliruusi, J. (2003). Tablet surface characterisation by various imaging techniques. *International Journal of Pharmaceutics* 254 (2): 281–286.

27 Csontos, I., Rónaszegi, K., Szabó, A. et al. (2006). Controlled technology for forming a nanostructured polymer coating for solid pharmaceuticals. *Polymers for Advanced Technologies* 17 (11–12): 884–888.

28 Li, T., Morris, K.R., and Park, K. (2000). Influence of solvent and crystalline supramolecular structure on the formation of etching patterns on acetaminophen single crystals: a study with atomic force microscopy and computer simulation. *The Journal of Physical Chemistry B* 104 (9): 2019–2032.

29 Danesh, A., Connell, S.D., Davies, M.C. et al. (2001). An in situ dissolution study of aspirin crystal planes (100) and (001) by atomic force microscopy. *Pharmaceutical Research* 18 (3): 299–303.

30 Durbin, S.D. and Carlson, W.E. (1992). Lysozyme crystal growth studied by atomic force microscopy. *Journal of Crystal Growth* 122 (1): 71–79.

31 Miyazaki, T., Aso, Y., and Kawanishi, T. (2011). Feasibility of atomic force microscopy for determining crystal growth rates of nifedipine at the surface of amorphous solids with and without polymers. *Journal of Pharmaceutical Sciences* 100 (10): 4413–4420.

32 Danesh, A., Chen, X., Davies, M.C. et al. (2000). Polymorphic discrimination using atomic force microscopy: distinguishing between two polymorphs of the drug cimetidine. *Langmuir* 16 (2): 866–870.

33 Grama, C.N., Venkatpurwar, V.P., Lamprou, D.A., and Ravi Kumar, M.N. (2013). Towards scale-up and regulatory shelf-stability testing of curcumin encapsulated polyester nanoparticles. *Drug Delivery and Translational Research* 3 (3): 286–293.

34 Begat, P., Morton, D.A., Staniforth, J.N., and Price, R. (2004). The cohesive-adhesive balances in dry powder inhaler formulations I: direct quantification by atomic force microscopy. *Pharmaceutical Research* 21 (9): 1591–1597.

35 Council of Europe, E. P. C., 2.9.37 (2007). Optical microscopy. In: *European Pharmacopoeia*, 6e. Strasbourg: Council Of Europe.

36 United States Pharmacopeial Convention (2016). <776> Optical microscopy. In: *United States Pharmacopeial Convention, U.S. Pharmacopeia National Formulary 2017: USP 40 NF 35*. United States Pharmacopeial Convention.

37 The Ministry of Health, Labour and Welfare (2016). 3.04 Particle size determination. Method 1. Optical microscopy. In: *The Japanese Pharmacopoeia (JP XVII)*. Japan: Pharmaceuticals and Medical Devices Agency.

38 United States Pharmacopeial Convention (2016). <1181> Scanning electron microscopy. In: *United States Pharmacopeial Convention, U.S. Pharmacopeia National Formulary 2017: USP 40 NF 35*. United States Pharmacopeial Convention.

39 Q5A(R1) (1997). I. Viral safety evaluation of biotechnology products derived from cell lines of human or animal origin. In: *ICH Harmonised Tripartite Guideline*. European Medicines Agency.

40 Q6A (2000). I. Specifications: test procedures and acceptance criteria for new drug substances and new drug products: chemical substances. In: *ICH Harmonised Tripartite Guideline*. European Medicines Agency.

4.2

Standards and Trends in Analytical Characterization – X-ray Diffraction (XRD)

Clemens Kühn

Merck KGaA, Frankfurter Str. 250, 64293 Darmstadt, Germany

4.2.1 X-ray Diffraction

4.2.1.1 Introduction

The principle of diffraction is known to many analytical chemists from the diffraction gratings in optical spectrometers used to discern between wavelengths of light between the infrared and ultra-violet. In these, reflection of the incoming light takes place on a regular pattern of structures which size is in the same order of magnitude as the wavelengths. Interference of the reflected (diffracted) beams result in a diffraction pattern, where certain wavelengths can just be observed at certain angles. The effect, therefore, can be used to disperse the wavelengths of "white" radiation. Other commonly cited examples are the surface patterns of compact discs and digital versatile disk (DVDs) leading to iridescent reflection of white light.

The principle can be extended into three dimensions, which, for visible light, is visible in the mineral Opal. The gemstones are composed of a close packing of regular spheres of amorphous silica, whose size is in the order of magnitude as visible light – a few 100 nm.

The same effect works on the regular packing of crystals, which is in the order of magnitude as X-ray radiation which is hence called X-ray diffraction (XRD). The effect has been first postulated by Max von Laue 1912 and awarded Nobel Prize in physics in 1914. Already in 1913 Sir William H. Bragg used the effect to solve crystal structures, which he was awarded Nobel Prize for physics in 1915 together with his son.

The determination of crystal structures for the pharmaceutical industry serves basically two purposes: the determination and proof of a molecule's structure and the determination and examination of packing patterns of polymorphs, salts, cocrystals, hydrates, and solvates thereof.

Solid State Development and Processing of Pharmaceutical Molecules: Salts, Cocrystals, and Polymorphism, First Edition. Edited by Michael Gruss.

4.2.1.2 Measurement Principles

It is not within the scope of this chapter to give a full-fledged overview about XRD and its physical and mathematical background. But to understand use, applicability, and limits of the technology, some terms shall be introduced here. The interested reader should feel invited to consult the sources cited at the end of this excerpt for deeper insight.

4.2.1.2.1 The Crystal Lattice

A crystal structure is described by the positions of atoms in a unit cell, which is the smallest pattern, being repeated in all three dimensions to make up the crystal lattice. Imagine millions of similar shoes stacked in their corresponding boxes in a container. The form of the box or the container does not tell you much about what is inside. It does, however, tell you the largest possible size, the shoes might be (as they have to fit). And there are certain labels attached, which will tell you about certain parameters like the color. The boxes contain two shoes of the same kind, which are not identical as they are intended for the left and the right foot, they are chiral. The box itself, however, is not chiral as it contains both "enantiomers" of the shoe.

In crystallographic terms, the form of the "box" is given by the so called "lattice constants": The length of the three edges a, b, c, and the angles between them (α, β, γ). Typically, you will also see a label telling you about the symmetry of the unit cell: the crystal class (e.g. cubic, orthorhombic, or monoclinic) and the space group. Generally, the crystallographers will not solve and report all the molecules in the unit cell but only those, which cannot be derived from the symmetry elements. In crystallographic terms, they report the contents of the asymmetric unit, all other molecules of the crystal structure are symmetrically equivalent.

4.2.1.2.2 The Space Group Symmetry

Most crystal lattices also feature a certain level of symmetry. To come back to the example of a shoe box, the "unit cell" would comprise an inversion center between the two shoes. As symmetry further reduces entropy in the lattice, there has to be a physical reason behind symmetry formation. For the molecule crystals in our scope, this would for example be a favorable packing of dipole moments. Both entropy and enthalpy of the packing obviously determine the thermodynamics of crystal formation and stability; therefore, it is worth to keep an eye on the spacegroup, e.g. when comparing polymorphs.

The symmetry is described by certain symmetry elements, namely inversion centers, mirror planes, rotation, and screw axes. Only a limited set of combinations of these elements are possible in a 3D repeating lattice. These are known to the crystallographer as the 230 spacegroups [1].

The space group is of considerable interest for the crystallographer, but for the solid-state analyst, it is more or less a set of operations applied in the background to build up the 3D molecular arrangement. However, there are some bits of information hidden here, which should be noted. First of all, there are chiral and achiral spacegroups. A chiral molecule cannot crystallize in an achiral spacegroup (containing symmetry elements that require mirror or inversion operations). Achiral molecules can crystallize in chiral spacegroups, though. This is not as uncommon, as one would expect based on the entropy loss, the system experiences. A good and common example would be right and left-handed quartz crystals.

On a molecular level, it means that the achiral molecule adopts a chiral conformation. This is usually the case as the functional groups in the crystal lattice cannot rotate freely as opposed to a solution. Typically, in a racemate, both enantiomers of the chiral conformations are packed in a way that they will be connected via mirror planes or such. But intermolecular interactions (e.g. hydrogen bond patterns) may lead to 3D stacking of the same enantio-conformation to be more favorable. This may also point to the possibility of polymorphs featuring achiral packing patterns.

4.2.1.2.3 What Determines a Diffraction Peak

First of all, it is important to understand that the diffraction of electromagnetic radiation takes place on electron density rather than atom cores[1]. The diffraction pattern resembles a Fourier transform of the electron density distribution. Thus, the X-rays reveal the electron density distribution rather than the atom positions.

As has been mentioned, the inner three-dimensional structure gives rise to interference of the diffracted X-rays. For a given wavelength (λ) and distance of diffraction centers (d), constructive interference can just occur at certain defined angles. The relation between the distance and the diffraction angle (θ) has been formulated by Bragg [2] and is widely known as Bragg's law:

$$n \cdot \lambda = 2d \sin \theta \qquad (4.2.1)$$

or alternatively:

$$d = n \frac{\lambda}{2 \sin \theta} \qquad (4.2.2)$$

where n, an integer number represents the diffraction order. This basic version of Bragg's law can be used to calculate the "d-spacing" from powder X-ray patterns.

For crystallographic applications, the version of Bragg's law cited above has to be extended into the three dimensions, thus instead of a diffraction order on a single one-dimensional lattice, we observe diffraction an different sets of lattice plains. These are defined using the so-called Miller Indices h, k, and l describing orientation of the respective lattice plain in the three dimensions as well as the respective diffraction order. With these and the set of lattice constants, the diffraction angle of every possible peak can be calculated.

The combination of the reflection positions determined by the lattice constants and their relative intensities determined by electron density distribution is called the diffraction pattern.

In case of certain symmetry elements, the repetition of electron density patterns at regular intervals within the unit cell causes destructive interference and therefore can give rise for systematic absences of reflections (no intensity observed at

---

1 In contrast, neutron radiation is scattered on atom cores and can therefore be used to determine hydrogen positions with great accuracy. However, due to availability in industrial laboratories and technical difficulties, it is not addressed in the scope of this chapter.

theoretically possible positions). These systematic absences can in turn be used to narrow down or determine the symmetry and therefore the space group of the unit cell.

4.2.1.2.4 X-ray Scattering Technics

For the sake of completeness, it should be noted that there are scattering technics, which are not dependent on the crystalline state. While the vast majority of applications rely on the regular 3D arrangement, the X-ray beam itself does not. Even in an amorphous state, certain distances will reoccur. Considering quartz and silica glass, both materials feature the same principal arrangement of tip-connected tetrahedra. They might be more distorted in the case of glass, but the same distances between atoms will reoccur. Therefore, from the scattering pattern of amorphous material, one can derive a statistic of reoccurrence of certain distances between atoms (or more precisely: electron density maxima). This is called the atom pair distribution function or short PDF. Analysis of the same and comparison with known crystal forms can reveal important information about the amorphous state. For example, differences in amorphous arrangements (often dubbed polyamorphism) or beginning crystallization.

Also, if one measures a bulk of (sufficiently small) particles, these will lead to small-angle X-ray scattering. This is exactly the same effect of light scattering, which is used to determine particle size distributions of μm sized particles. Small-angle X-ray scattering using a shorter wavelength can be used to determine the size distribution of nanoparticles. As X-rays penetrate the material, the effect also occurs and thus can be used to characterize nanoscaled voids.

4.2.2 Technics

4.2.2.1 Single Crystal X-ray Diffraction

Single Crystal XRD serves the purpose of determining the arrangement of atoms in a material, also called "solving" the crystal structure. The method allows a direct determination of chemical identity including the absolute structure of molecules. It also yields the packing pattern and identity and arrangement of other entities in the structure-like counterions, water molecules, or co-crystal partners.

As the name suggests, single crystal XRD is the measurement of the diffraction pattern of one single crystal. Depending on material and instrument properties, a single crystal of suitable quality and size is necessary. The crystal has to be rotated in virtually all three dimensions and reflections recorded at all possible angles. The result is a set of thousands of individual reflections with their corresponding relative intensities.

During structure determination, a model of atoms with their element specific electron density distribution is build and refined against the experimentally determined data set. The fit between model and experimental data is reflected in various quality parameters. Of which the most frequently cited ones are the internal R-value R_{int} (reflecting quality of crystal and data set), $R1$, and the weighted wR^2. The widely accepted criteria for successful structure solutions are $R1 \leq 10\%$ (and close to R_{int}) and $wR^2 \leq 15\%$. The absolute structure can be derived as well based on the intensity differences of reflections [3], which should be equivalent, which are condensed into parameters like the Flack x [4] or Hooft y [5] values. However, depending on crystal size and quality and on how well any disorder may be modeled, values slightly above this limit may be acceptable; therefore, all these values will be assessed by the crystallographer.

For organic, light atomic structures of small molecules and lab source copper radiation, typically the crystal has to be between 100 and 300 μm in size. Recent developments in area detectors and X-ray sources gave rise to a significant reduction of these figures, but growth of a suitable single crystal still represents the most important, resource-consuming, and tedious step.

In a pharmaceutical setting, the single crystal structure solution serves as a proof of absolute structure and a means for qualification of a reference for powder X-ray diffraction (PXRD) patterns (see below). As a match of the PXRD pattern calculated based on the crystal structure and the experimental pattern of a batch proves (polymorph and chemical) identity of the batch to correspond to the crystal structure. This is important, as the single crystal manually and subjectively chosen for measurement may correspond to a minor component of the batch.

4.2.2.2 Powder X-ray Diffraction

P-XRD or – as it is more commonly (but not quite correctly, as we are not diffracting powders) called – X-ray powder diffraction (XRPD) is the strongest method to characterize the crystallinity of a bulk of material. The powder typically comprises millions of crystals in virtually all possible arrangements. Therefore, the Bragg equation for all possible diffraction planes in the crystal is fulfilled for at least part of the sample at all times. This eliminates having to rotate the sample through all possible orientations and allows much simpler technical set-up of the instrument and faster measurements. Geometrically, due to the arbitrary orientation, the individual reflections of the single crystals get combined into diffraction cones [6] observed as rings on a 2D detector. As explained later, most diffractometers feature 1D detectors (or integrate the rings of a 2D image into one dimension).

The details of sample preparation depend on the system used. But for all systems in common, ideally the sample should be a fine powder with grain sizes below 10 μm. Larger grain sizes often give rise to preferable orientation, especially in reflection mode. Therefore, larger particles have to be ground or at least disintegrated, taking care, not to induce phase transitions due to the energy introduction.

4.2.2.2.1 Alternative Methods for Structure Determination

In 2013, Prof. M. Fujita published a method [7] using a crystalline sponge as a means to avoid the necessity of crystallization. For this method, a suitable single crystal of a metal–organic framework (MOF) is used to take up ("soaking") and align the analyte molecules. Single crystal XRD of the soaked crystal reveals the absolute structure of the molecules as in conventional structure determination.

The crystalline sponge method has since been explored by many companies as a substitute for conventional crystallization, especially as it offers to work with substance amounts as low as few nanograms [8, 9]. Another important fact is that the method can be applied to substances, which at ambient conditions are liquid or gaseous as well [10, 11].

Another possibility to reduce the crystal size necessary would be electron diffraction (ED, often dubbed cryo-ED, as the sample has to be kept at very low temperature to avoid overheating). However, ED also is dependent on single crystals. While it can – or rather has to – work on nanoscaled crystals, these still have to be single crystals. Sample preparation and data evaluation are critical and require high degree of expert knowledge. Thus, its application will be bound to those cases, where material is available in larger scale, but single crystal growth is the limiting factor. It should also be noted that data sets from ED intrinsically feature significantly lower quality and completeness and thus their evidential value (in terms of a proof of structure) is significantly lower if not questionable [12].

4.2.3 Instrumentation

4.2.3.1 X-ray Sources

The X-rays used in laboratory XRD equipment typically is produced by a sealed tube. These make use of the effect of characteristic X-ray fluorescence triggered by the Bremsstrahlung, which is emitted by the electrons of a cathode ray hitting a metal anode. Thus, the spectrum emitted by a sealed tube consists of the "white" X-ray Bremsstrahlung of the cathode beam superimposed with sharp characteristic fluorescence peaks of the anode material. Obviously, the characteristic radiation emitted is a spectrum by itself with three major peaks labeled $K_{\alpha 1}$, $K_{\alpha 2}$, and K_{β}. The K_{α}-lines are rather close in energy and often are superimposed in practical application. K_{β} features a distinctively higher energy and about 20% of intensity of the α-lines. The most widely used anode material used for PXRD in the laboratory is copper, having a wavelength Cu $K_{\alpha 1} = 1.54056$ Å. Typically, the same material is now used in single crystal X-ray diffractometers especially for organic molecules. Some instruments, especially older ones, with smaller detectors or those tailored for inorganic materials often use molybdenum as anode material. As single crystal instruments typically are not equipped with monochromators, the resulting radiation is the combination of the two K_{α} lines with averaged wavelength Cu $K_{\alpha(\text{av})} = 1.54184$ Å and Mo $K_{\alpha(\text{av})} = 0.71073$ Å, respectively.

Radiation is emitted in all directions but will leave the X-ray tube through windows, typically made of beryllium (to keep the vacuum in the cathode ray tube) in divergent beams.

Unless a monochromator is used, the Bremsstrahlung will give rise to increased background of the pattern. Also, unless a monochromator is used, the different characteristic peaks of the anode material will lead to a pattern of diffraction peaks corresponding to the respective wavelengths for every calculated position of the pattern.

As this renders the pattern very difficult to interpret, there are various ways to avoid the additional peaks or minimize their impact.

The best way is to use a monochromator in the primary beam, which can serve as a focusing mirror at the same time and results in highly monochomatic radiation and therefore high fidelity diffraction patterns. The set-up only works in transmission geometries, though. In reflection mode, typically, monochromators in the secondary (reflected) beam are used. These typically are less efficient though and will not be able to discern closely adjacent $K_{\alpha 1}$ and $K_{\alpha 2}$ lines.

K_{β} may be reduced using a metal foil with corresponding X-ray absorbance.

As the relative intensities of the peaks are known, software of reflection diffractometers typically offers an option to programatically remove the additional lines from the diffraction pattern.

4.2.3.2 Diffractometer Geometries

There are a number of different geometries available, but they can be divided into two major groups: reflection and transmission geometries. Common features are an X-ray source, typically a sealed tube. Common for all geometries, the powder pattern as such is generally depicted as diffraction intensity vs. diffraction angle 2θ, which is the angle between the primary and diffracted beams.

4.2.3.2.1 Reflection Geometry

Most diffractometers feature a reflection setup, as this is easier to produce and offers high versatility. A divergent X-ray beam illuminates a powdered sample, which typically is horizontally arranged on the system. The reflected rays converge onto a circle, on which the detector counts the reflected photons (see Figure 4.2.1a).

Typically, polychromatic radiation is used, which serves for very high primary beam intensity. Also, the horizontal arrangement can accommodate a large variety of samples including wet to liquid samples. Sample preparation is straightforward, if the data quality is not a concern (e.g. for phase identification in pharmaceuticals).

Typically, one of two forms of sample holders is used: The standard sample holders are open cups being filled with a defined volume of powder. The size, namely the diameter, depends on the instrument manufacturer. But most require 1–2 mm

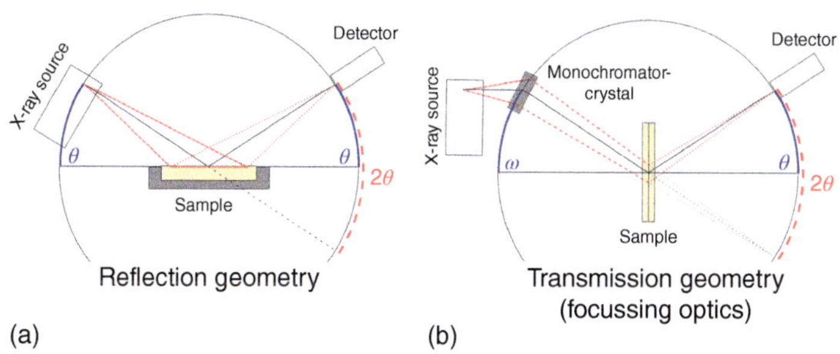

Figure 4.2.1 Schematics of the two basic geometries used in powder X-ray diffraction.

sample depth to fill completely, which can pose a problem, if sample amounts are scarce. In such cases, zero-background holder may offer an alternative. These consist of a flat plate of a nonreflecting material (typically specially oriented single crystalline quartz or silicon). These zero-background holders can ease the sample preparation significantly and require only minute amounts of material.

The reflection geometry is prone to sample preparation effects. Slight variations in the sample height (underfilling, overfilling, surface roughness, misalignment of the system, transparency of the sample packing) will lead to the X-ray's being no longer reflected on the center of the diffractometer. The effects are peak shifts and peak broadening due to geometry due to loss of convergence at the detector position. Due to peak broadening, signal-to-noise ratio will also decrease.

Some of these effects (e.g. due to the transparency of the packing) may be reduced using zero-background holders. The low amounts of sample obviously mean lower intensity as well.

Additionally, if polychromatic radiation is used, reflections are further broadened and additional reflections might occur due to the spectrum of the X-ray source. Use of a secondary monochromator can reduce these effects but reduce the signal intensity significantly.

The sample preparation for reflection geometry in the majority of cases leads to preferred orientation of the crystals. This is a very important effect, which will be further discussed later in this section.

4.2.3.2.2 Transmission Geometry

In diffractometers working in transmission mode, the sample is illuminated from the opposite side of the detector (see Figure 4.2.1b). The sample thus is a thin sheet of packed powder between X-ray source and detector. The sample therefore is typically prepared in a thin layer between amorphous films of polymer. Transmission diffractometers can only use nondiverging X-rays and therefore feature some kind of optics, ideally including a monochromator (a single crystal, e.g. of Silicon). These optics absorb a pronounced amount of primary beam intensity. This disadvantage, however, is partly made up by the fact that no secondary monochromator is needed. Also as the reflections do not have to converge onto the detector's circle, larger detectors may be used without significant loss of fidelity.

As the beam has to pass the sample packing, heavy elements (e.g. iodine) can significantly reduce transparency of the sample down to complete absorption. Due to the higher scattering power, such samples can typically be measured though using a lower amount of powder. For highly absorbing samples, this might result in somewhat delicate sample preparation and the necessity to check transmittance of the prepared sample before measurement.

On the upside, in transmission mode, many sample preparation effects do not occur in transmission or are less pronounced than in reflection measurements (e.g. surface roughness or the height error due to transparency). Therefore, in tendency, transmission measurements are more reproducible than reflection measurements and easier to handle in a quality controlled (QC) environment.

4.2.3.2.3 Benchtop Diffractometers

Normal powder X-ray diffractometers are bulky equipment with about half a ton in weight requiring high-voltage connection and an external chiller to cool the X-ray source. Recently, manufacturers developed systems with much smaller footprint comparable to gas chromatographs or other analytical devices. Some of these systems do not even require an external chiller.

Such a system is less versatile and does not feature the same level of X-ray power and resolution as a full-fledged diffractometer but also come at a significantly lower price. These systems mostly run in reflection mode (there is a very interesting transmission diffractometer available from Olympus, which is even portable and has been send to Mars with NASA's Curiosity probe [13]) with all the ups and downs of reflection geometry.

Given the performance and set-up of benchtop systems, these can be used for quick in-process control or identity testing. The data quality typically does not allow for more sophisticated data evaluation like quantification or refinements.

4.2.3.3 Detectors

There are various detector technics around but state of the art and thus de-facto standard nowadays are silicon strip detectors. These work very efficiently and are available in form of 2D detectors as well. In PXRD, typically 1D position-sensitive detectors are used offering a certain detection window and recording an angular range of the diffraction pattern at once. Most prominent of these are the detectors build by the Dectris Ltd. based in Switzerland, but many manufacturers offer their own silicon strip detectors as part of their systems.

2D detectors, originally developed for single crystal diffractometry or medical imaging, can significantly reduce measurement time and may be fitted in benchtop diffractometers, where they can also serve the purpose of reducing the number of movable parts.

Usually a system is delivered with just one detector technic, so there is no need to dwell on the specifics of the detection. More important is the relative performance of the whole system from X-ray source via optics and sample to the detector. Due to the various hardware options, in order to compare measurements from different systems, there is no way around actual measurements of the same sample.

4.2.3.4 Peak Asymmetry

There is one detector-contributed effect though, which can significantly affect the diffraction patterns peak shape. All detectors, at least to some extent, detect X-rays in a certain vertical window as well as horizontally. Even a 1D detector is actually sensitive in a thin band. As our diffraction peaks geometrically speaking are actually cones or – in the detector plane – rings, the detectors actually detect segments of circles. The diameters of these circles are defined by the diffraction angle and distance of the detector from the sample, the segment is defined by the vertical window.

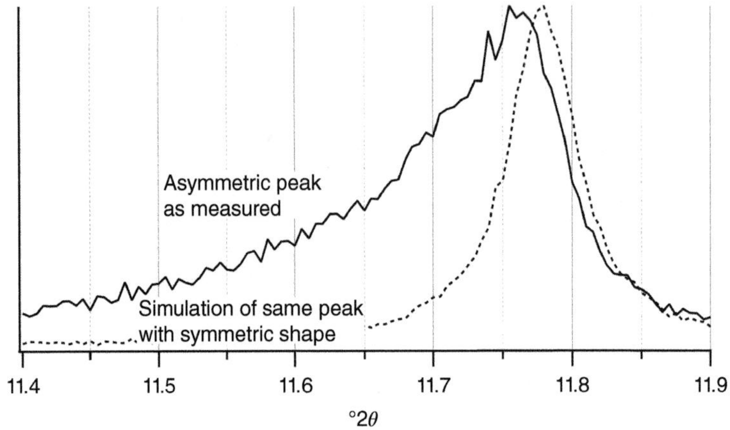

Figure 4.2.2 Example for the effect of asymmetry due to "umbrella effect" on the position of the peak maximum.

The effect called "umbrella effect" is that especially low-angle diffraction peaks look asymmetrical with a "fronting" toward the primary beam (zero degrees). The effect may be minimized adjusting sample to detector distance or using a mask to minimize the vertical detection window. However, this comes at a price in terms of intensity and fidelity of the pattern. As it can typically be neglected for pharmaceutical applications, it can normally just be ignored. It is mentioned here as different instruments may feature umbrella effect to a different extent.

The true peak position of an asymmetric reflection is slightly above the angular position of the peak maximum (see Figure 4.2.2). Dependent on the instrument set-up and the angle of the respective peak, the error may be in the first or second digit in two theta and may have to be taken into account, if comparing between peak positions of measured patterns and calculated ones or between different instruments.

4.2.3.5 Reproducibility of Diffraction Patterns: The Texture Effect (Preferred Orientation)

A major difficulty in PXRD, especially in regulated environments, is the reproducibility of the patterns between different batches and sample preparations. The effects will mainly influence peak positions and relative intensities. Errors in peak positions can be minimized by strict adherence to sample preparation protocols and of course a well-aligned instrument.

In a perfect preparation, the crystals are oriented randomly, with equal distribution of all possible orientations. This is assumed when calculating a pattern from a known crystal structure. Alas, such a situation is virtually impossible to achieve. Typically, crystals are not spherical and therefore will align more or less in a packing close to one another.

The first effect is the most prominent sample preparation effect and it is the most difficult to control. It is named "preferred orientation" or "texture effect," as it is due

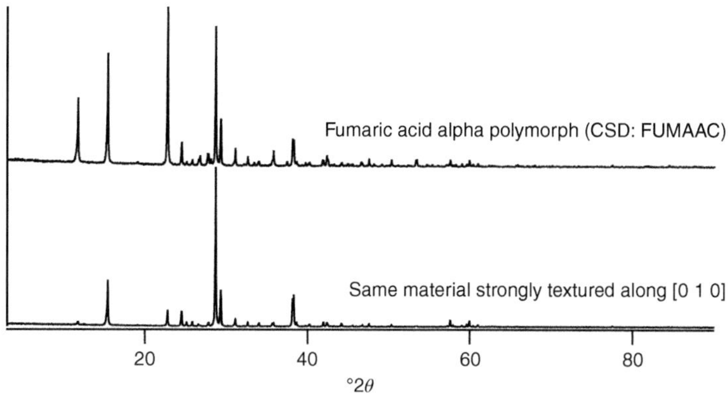

Figure 4.2.3 Effect of preferred orientation on relative intensities (simulation).

to crystal (or particle) texture, i.e. external morphology. Imagine needles or platelets in a dense powder packing. However well you prepare them, due to the texture of the crystals, they will tend to align in a parallel fashion. Therefore, certain crystal orientations are more probable than others and their respective diffraction planes more often fulfill the diffraction condition defined by Bragg's law. Thus, these diffraction intensities are more intense than they would be in a completely random arrangement. Vice versa, the intensities of the other peaks are lower than expected for a random orientation.

Pronounced preferred orientation may lead to peaks to show up, which would normally be below noise level or a number of strong peaks to seemingly disappear (see Figure 4.2.3). The effect can be so pronounced that patterns may result in a false-negative result. In regulated environments, this might trigger the question of acceptance criteria for a matching pattern: when is the pattern regarded as in or out of specification?

Texture effect has to be taken into account comparing R&D batches, which may differ in crystal size and shape (and therefore texture). And special care has to be taken, comparing patterns from different instruments, especially if they differ in diffraction geometry. The texture effect affects transmission and reflection patterns in a reciprocal way: reflections of reduced intensity in transmission are enhanced in reflection mode and vice versa. Also, the sample preparations for different instruments might trigger preferred orientation to a significantly different degree.

Therefore, in a regulated environment, it is advisable to measure the reference under the same conditions as the actual sample.

To minimize the effect of preferred orientation:

- Ensure small crystal sizes, e.g. by grinding (taking care not to induce phase transitions, i.e. change in polymorphism)
- Avoid compression of the powder
- Avoid thin preparations (e.g. zero background holder in reflection, very low sample amount in transmission mode)
- Define and strictly adhere to sample preparation protocols

In difficult cases, it may help to mix the sample with an inert amorphous material like aerosil, which will keep the crystals of the sample apart and therefore counteract preferred orientation. This, however, also introduces another phase (i.e. solid material), will reduce intensity, and will always show an amorphous fraction, so render it difficult to observe amorphous fractions in the original sample.

4.2.3.6 Databases of Known Diffraction Patterns

Unfortunately, besides the free American mineralogist crystal structure database (AMCSD) [14] database focused on minerals, there is only a rather expensive database. It is called Powder Diffraction File and provided by the International Center for Diffraction Data [15]. Different versions are available including one with the whole Cambridge structural database (CSD) (see below) content included. In pharmaceutical R&D, these databases come in handy, if dealing with unknown patterns, which could represent inorganic salts or organic excipients.

Besides databases of powder patterns, there are databases of crystal structures available, namely the Cambridge Structural Database (CSD) hosted by the Cambridge Crystallographic Data Centre [16], and the Inorganic Crystal Structure Database (ICSD) maintained by the Fachinformationszentrum (FIZ) Karlsruhe [17]. By definition, any compound containing a carbon–hydrogen bond will be entered into the CSD, all others in the ICSD. The crystal structures stored in these databases may be used to calculate theoretical powder diffraction patterns to be used as references in identity control by P-XRD. Another source for crystal structure data may be the crystallography open database (COD [18]).

4.2.4 Measurement

4.2.4.1 Instrument Calibration

The instrument is typically calibrated using a suitable standard (e.g. National Institute of Standards and Technology [NIST] SRM 640 Silicon). It is advisable to use not only the first, strongest reflection but also the whole pattern of the standard up to the maximum angle used in measurements of samples. If the systems set-up gives rise to umbrella effect (see Section 3.4), the calibration should be based on the refined peak positions using a suitable model for reflection asymmetry. If this is not possible, the error should at least be determined and assessed compared to acceptance criteria applied to phase analysis.

In pharmaceutical applications, one should also consider that most active pharmaceutical ingredient (API) and excipients feature diffraction peaks well below, i.e. having lower 2θ values than, the first reflections of the typical XRD standards (which are inorganic materials). So it is advisable to calibrate using a low-angle standard as well (e.g. NIST SRM 675 Mica).

Obviously, instrument calibration should not only focus on reflection position but also the line profile, e.g. full widths at half maximum (FWHM) and maximum intensity should be monitored. An increase of FWHM will reduce signal-to-noise ratio and is an indication of some geometric misalignment of the diffractometer or the monochromator. A decrease of intensity alone may indicate end of lifetime of the X-ray source (or the monochromator being out of alignment).

4.2.4.2 Sample Preparation

Depending on system set-up, the powder is then prepared in a way ensuring best and reproducible positioning on the diffractometer with as even a distribution as possible to avoid artifacts and misalignments. Most set-ups allow for rotation of the sample, which significantly enhances reproducibility of the intensity statistics.

Tablets – consisting of compacted powders – may be measured as well, provided form and size do fit to the respective instrumental setup. Very small and thin tablets may even be used in transmission mode.

Care has to be taken to ensure small enough and even distribution of crystal sizes. Macroscopically visible crystals (sparkling powder) must be avoided as well as pronouncedly different crystal sizes (beware of phase mixtures). This is usually accomplished with a simple mortar and pistil, although particular care is advised to avoid pressure or temperature-induced phase transitions. With pharmaceutical materials, the energy introduced in using a mortar may be enough to trigger phase transitions between polymorphic forms, especially in wet conditions. In these cases, the method for milling and homogenization has to be carefully verified (and probably validated for good manufacturing practice [GMP] compliance). Thus, if crystal sizes of the original sample allow, measurement should be made using the pristine sample.

Best crystal sizes are about few μm. NIST SRM 640 Silicon features a D_{50} of 5 μm. Smaller crystals will result in reflection broadening, and larger crystal sizes may result in spikes and bad statistics (unreliable relative intensities) in the diffractogram. This is especially important for low amounts of substance, e.g. with zero background holders in reflection mode or combinatorial attachments (well plate sample holders). As statistical errors can easily lead to a false negative identity check.

Therefore, it is vitally important in routine analysis to keep the prepared sample until the diffraction pattern has been evaluated. In case of a negative result, it will be worthwhile to check sample preparation and reprepare or remeasure where appropriate.

4.2.5 Data Evaluation

4.2.5.1 Qualitative Phase Analysis

The PXRD pattern of a pure crystalline substance basically represents a fingerprint for its crystal structure. As such, it can be used as a means to identify not only substances but also the respective crystalline modification, or polymorphic form. Especially for identification of polymorphs, XRD is the most important method, as it directly reflects the changes in crystal lattice and molecular arrangement (see Figure 4.2.4).

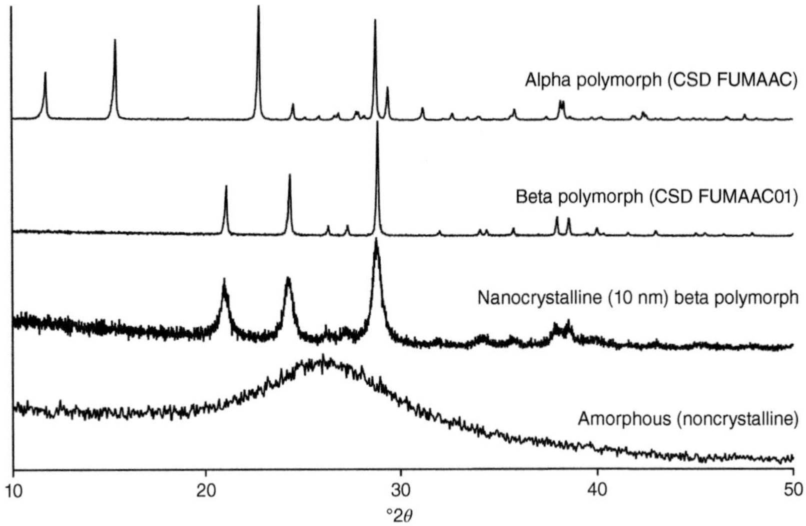

Figure 4.2.4 Powder X-ray diffraction patterns (simulations) of fumaric acid in different solid-state forms relative intensities, the maximum intensities of amorphous and nano-crystalline materials are much lower than those of crystalline materials.

The International Council for Harmonization (ICH) decision tree #4 [19] of the quality guideline Q6A on polymorphism in pharmaceutical APIs asks to specify the polymorphic identity if different polymorphs exist, which feature different properties. In pharmaceutical applications, therefore, PXRD is routinely used as a method for identification and control in a manner similar to infrared (IR) or Raman spectroscopy. To verify phase identity, the measured pattern (typically a fraction of the whole pattern) is overlaid or otherwise visually compared with a qualified reference. This can, for example, be a pattern calculated from a single crystal structure elucidation as part of a proof of structure study or a pattern of a batch of known identity. The European and US pharmacopeias contain a substantial monograph on the application of PXRD to test polymorph identity.[2]

Mixtures of crystalline phases will lead to their respective patterns to be linearly combined. As long as interference between the patterns is not too large, the components can safely be identified by the respective reference patterns. Reference patterns may also be taken or calculated from a database (see below). Such qualitative phase analysis is typically provided as base functionality within the instrument vendor's software.

4.2.5.1.1 Phase Identification or Identity Check

As has been discussed above, phase identification basically is a comparison of the measured pattern against a reference pattern. This may sound as straightforward as for spectroscopy or spectrometric data, but XRD pattern is way less reproducible

2 USP-NF <941>, Ph.Eur. Chapter 2.9.33.

than the spectrums of other methods, as many effects strongly depend on the individual sample preparation (due to transparency, distribution, height error, surface roughness). Dependent on sample and instrument properties, these namely affect the background, a peak offset, and most prominently the peak intensities due to texture effect.

Automated analyses (which are generally preferred in a regulated environment) therefore remove the background, determine peak positions, and use these for comparison, neglecting relative intensities (or applying a much lower weight to them) and allowing for a window of typically 0.1° 2θ for Cu-radiation. In visual comparison, however, the trained eyes of the crystallographers can easily verify the identity of phases regardless of background and texture effect and mostly regardless of the presence of other phases.

For semi-automated phase identification, the peak list is compared against similar lists of known phases, which may have been provided in user-defined or purchased databases. Due to the problems with reproducibility, the software will suggest a sorted list of candidate phases and leave it to the crystallographer to decide on the relevance and correct assignment.

If unknown phases are present (phases not in the database), the semi-automated identification may easily lead to false assignments. It is always a good idea to critically review the phase assignment for new R&D samples. Such a review should also take into account the results of other analytical technics like element analysis, mass spectrometry, or nuclear magnetic resonance (NMR) spectroscopy. A sample identified as a complex mixture may in fact be just a new unknown polymorph or salt.

It should also be noted that with pharmaceuticals at times, channel structures of solvates or hydrates are observed. These are often nonstoichiometric and may feature arbitrary amounts of solvent or water. In many cases, this will result in a notable swelling or shrinking of the crystal lattice, resulting in significant peak shifts, which may make the comparison with a certain reference with a specific composition difficult.

4.2.5.1.2 Amorphous Content

In contrast to the sharp, well-defined diffraction peaks of crystalline phases, amorphous phases give rise only to broad features, named halos. The area under the halo would roughly correspond to the area under the diffraction peaks of a crystalline phase of the same material. Therefore, amorphous fractions are only detectable, if they represent a major fraction of the sample, typically more than 20 mass% (see Figure 4.2.5).

Therefore, other technics such as IR or Raman spectroscopy take preference over XRD for determination and quantification of amorphous fractions in crystalline material. On the other hand, XRD is very sensitive determining crystalline fractions in amorphous material. Amorphous content may be quantified though by the addition of a known quantity of certified crystalline standards and quantification of the apparent content in the sample (see below). Amorphous fractions in the sample

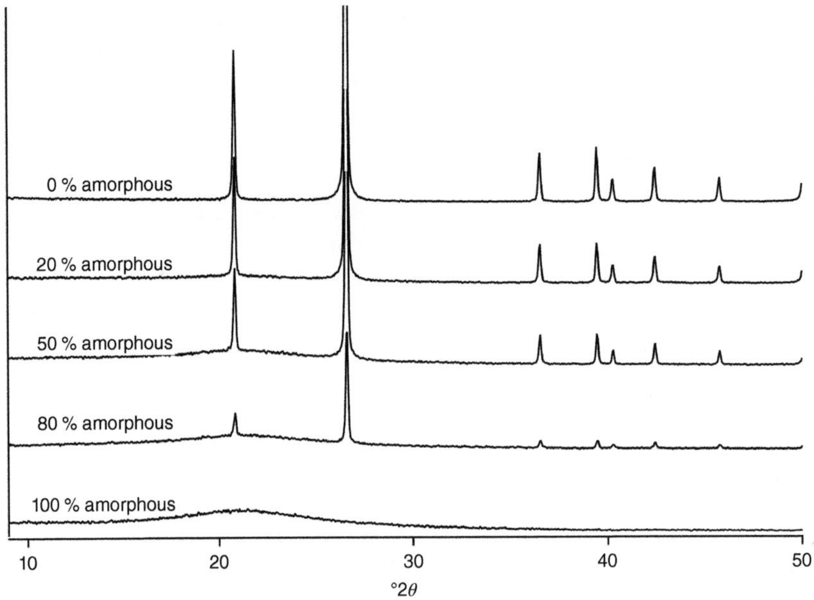

Figure 4.2.5 Diffractograms depicting visibility of amorphous fractions (powdered quartz/glass mixtures), low content of amorphous fractions only lead to a drop in absolute intensities. Clearly visible are amorphous fractions larger than 50%.

will seemingly increase the fraction of added standard. Therefore, the amorphous content (as a mass fraction) may be calculated as

$$c_{\text{amorphous}} = \left(1 - \frac{c_{\text{Std, added}}}{c_{\text{Std, determined}}}\right) / \frac{m_{\text{Sample}}}{m_{\text{total}}} \tag{4.2.3}$$

with: m_{Sample}, m_{Standard}: the respective masses
$m_{\text{total}} = m_{\text{Sample}} + m_{\text{Standard}}$; $c_{\text{Std, added}} = m_{\text{Standard}}/m_{\text{total}}$
$c_{\text{Std, determined}}$ = mass fraction of standard as per experimental quantification.

Note, however, that this method assumes the diffraction power to be exactly the same for the amorphous as for the crystalline materials. This holds true for amorphous fractions in the crystalline form of a molecule. It will not be applicable, if the amorphous material is significantly different in chemical nature (e.g. amorphous excipient in crystalline salt of an API).

4.2.5.2 Quantification

Having the constituent phases identified, the question of their relative content will follow. It has been said that the PXRD pattern resembles a linear combination of the patterns of the respective pure components. Thus, the relative composition can be derived rather easily from the patterns. However, due to the different factors impacting relative intensities in XRD (namely the texture effect described earlier), the deconvolution often becomes a rather difficult mathematical problem compared to quantifications with other technics such as IR spectroscopy or elemental analysis.

4.2.5.2.1 Based on Calibration Curve

It is often proposed to base quantification on the measurement of mixtures of known composition over the intended range of concentrations. This method requires both components to be available in pure form and having a method to produce "perfect" mixtures without inducing phase transitions. This can prove rather difficult, e.g. with pharmaceutical polymorphs. Even more difficult is the fact that the two components have to exhibit exactly the same properties in terms of crystallinity (crystal size, shape, amorphous content, etc.) as the components in the actual mixture. This is very difficult to ensure and especially to prove in a regulated environment. Such a procedure will also have to be revalidated for every change in the production process which may influence crystallinity of one or more components.

Dependent on the requirements for the method, it may still be a viable approach. In these cases, the intensities (after background subtraction) of selected reflections of the constituents are plotted in a calibration curve against the relative composition and the measurement evaluated accordingly against this calibration as one would do for spectroscopic or chromatographic measurements.

To minimize the influence of the texture effect, more than one nonoverlaid reflection of each component should be used spanning the range of orientations.

4.2.5.2.2 Based on Internal Standard Addition

The amount of a certain fraction may also be determined against an addition of a suitable internal standard. In principle, it would also be based on a calibration curve. But the standard has the advantage of being easily available in exactly the same quality and crystallinity as in the later measurement and will not undergo phase transitions. Typical standards would be Corundum or Silicon. The latter, as a cubic phase, also would be virtually free of texture effect. They also feature a limited number of strong reflections, which means, for pharmaceutical applications, these phases show low probability of overlap and thus will not interfere strongly with the reflections of the analyte components. The content of the respective phase can be calculated by definition of a "response factor," the relative intensity of a strong reflection of the phase in question compared to a suitable (close in 2θ and intensity) reflection of the standard. The response factor should be determined using a material of known high crystallinity. This method is strongly dependent on material crystallinity and texture though. At least peak areas should be used instead of maximum intensities to minimize the effect.

4.2.5.2.3 Based on Rietveld Refinement

The phase composition can also be calculated based on a simulation of the full pattern normally based on the known crystal structures of the constituents. This may also be combined with a standard addition.

The most widely accepted full-pattern analysis has been developed by Hugo Rietveld [20]. Many vendors offer Software add-ons to execute such refinements.

A free software, which is widely accepted in the community is the software general structure analysis system (GSAS) [21], which has for example been used by the NIST for refinement of their certified reference materials. In a Rietveld refinement, the full pattern is simulated using models for the background and the reflection profiles as well as preferred orientation. In principle, structure parameters such as atom positions may also be refined, but this is out of scope and should be avoided for quantification purposes. The relative phase composition is a by-product of such a refinement.

However, Rietveld refinements can be rather sophisticated. The sheer number of parameters, which need to be refined or held constant, can make them difficult to set-up and maintain in a regulated environment. There are software packages available trying to minimize the user's influence on the parameters and try to run the refinement in a black box mode. To the author's knowledge, however, none of these would be usable in regulated conditions (requiring audit trail and 21 CFR Part 11 compliance).

4.2.5.3 Advanced Phase Analysis

Section 4.2.5.2.3 introduced the concept of Rietveld full pattern refinement as a means to quantify phases. The refinement can yield much more though.

First of all, the refinement will yield the lattice constants of the phases. These may vary significantly for nonstoichiometric structures, and inspection of the lattice constant variations may reveal important facts of the respective batch.

The Rietveld refinement also needs a model of the reflection profile of the individual phases. This contains information about the crystallite size, the size of the truly single crystalline domains of the particles, as well as microstrain, induced by disorder, defects, or element substitutions. Both effects can be very important for pharmaceuticals as it affect the dissolution behavior.

To extract this information, the respective sample derived profile has to be deconvoluted from the effects caused by the instrument itself (e.g. X-ray monochromation, detector resolution, air scattering). Luckily the instrument contributions follow different (Gaussian shaped) statistics than the (Lorentzian) sample contributions.

The key for successful performance of such a peak shape analysis is the correct description of the "instrument function": the reference peak shape for a perfectly crystalline phase, where sample effects can be neglected. This should be derived from a well-prepared sample of highly crystalline material. Samples of choice for this task are the NIST SRM 640 Silicon and 660 LaB_6 offered as line profile standards for exactly this purpose.

Still, such a refinement and the deconvolution of reflection profile contributions is a delicate task and should only be performed by experienced crystallographers in order to avoid mis- or overinterpretation [22, 23].

Finally, the structure itself (the atoms element identity and atoms positions) may be refined. This is typically not in scope for pharmaceuticals, though, except in structure elucidations from powder diffraction data [24].

List of Abbreviations

1D	one dimensional
2D	two dimensional
3D	three dimensional
AMCSD	American mineralogist crystal structure database
API	active pharmaceutical ingredient
CFR	code of federal regulations
COD	crystallography open database
CSD	Cambridge structural database
D_{50}	median of particle size
DVD	digital versatile disk
ED	electron diffraction
FIZ	Leibniz Institute for Information Infrastructure (formerly Fachinformationszentrum)
FWHM	full widths at half maximum
GMP	good manufacturing practice
ICH	International Council for Harmonization
ICSD	inorganic crystal structure database
IR	infrared
MOF	metal organic framework
NIST	National Institute of Standards and Technology
NMR	nuclear magnetic resonance
PDF	pair distribution function
Ph. Eur.	European Pharmacopeia
P-XRD	powder X-ray diffraction
QC	quality control
R&D	research and development
SRM	standard reference material
US	United States
USP-NF	United States Pharmacopeia and the National Formulary
XRD	X-ray diffraction
XRPD	X-ray powder diffraction

References

1 Aroyo, M.I. (2016). *International Tables for Crystallography Volume A: Space-Group Symmetry*. Wiley, ISBN: 978-0-470-97423-0.

2 Bragg, W.L. (2014). The diffraction of X-rays by crystals. *Zeitschrift fuer Physikalische Chemie (Muenchen, Germany)* 228 (10–12): 957–968.

3 Thompson, A.L. and Watkin, D.J. (2009). X-ray crystallography and chirality: understanding the limitations. *Tetrahedron: Asymmetry* 20: 712–717.

4 Flack, H.D. (2003). Chiral and achiral crystal structures. *Helvetica Chimica Acta* 86: 905–921.

5 Hooft, R.W.W., Straver, L.H., and Spek, A.L. (2008). Determination of absolute structure using Bayesian statistics on Bijvoet differences. *Journal of Applied Crystallography* 41: 96–103.

6 For a nice illustration see https://en.wikipedia.org/wiki/Powder_diffraction (accessed 16 September 2019).

7 Inokuma, Y., Yoshioka, S., Ariyoshi, J. et al. (2013). X-ray analysis on the nanogram to microgram scale using porous complexes. *Nature* 495: 461–466.

8 Inokuma, Y., Ukegawa, T., Hoshino, M., and Fujita, M. (2016). Structure determination of microbial metabolites by the crystalline sponge method. *Chemical Science* 7: 3910–3913.

9 Sakurai, F., Khutia, A., Kikuchi, T., and Fujita, M. (2017). X-ray structure analysis of N-containing nucleophilic compounds by the crystalline sponge method. *Chemistry* 23: 15035–15040.

10 Ramadhar, T.R., Zheng, S.-L., Chen, Y.-S., and Clardy, J. (2017). The crystalline sponge method: a solvent-based strategy to facilitate noncovalent ordered trapping of solid and liquid organic compounds. *CrystEngComm* 19: 4528–4534.

11 Zigon, N., Kikuchi, T., Ariyoshi, J. et al. (2017). Structural elucidation of trace amounts of volatile compounds using the crystalline sponge method. *Chemistry – An Asian Journal* 12: 1057–1061.

12 van Genderen, E., Clabbers, M.T.B., Nederlof, I. et al. (2016). Ab initio structure determination of nanocrystals of organic pharmaceutical compounds by electron diffraction at room temperature using a Timepix quantum area direct electron detector. *Acta Crystallographica Section A Foundations and Advances* 72: 236–242.

13 https://www.olympus-ims.com/de/insight/olympus-xrd-on-mars (accessed 11 September 2019).

14 http://rruff.geo.arizona.edu/AMS/amcsd.php (accessed 11 November 2019).

15 http://www.icdd.com (accessed 11 November 2019).

16 https://www.ccdc.cam.ac.uk (accessed 11 November 2019).

17 https://icsd.products.fiz-karlsruhe.de (accessed 11 November 2019).

18 http://www.crystallography.net/cod (accessed 11 November 2019).

19 https://www.ich.org/fileadmin/Public_Web_Site/ICH_Products/Guidelines/Quality/Q6A/Step4/Decision_Trees.pdf

20 Rietveld, H.M. (1969). A profile refinement method for nuclear and magnetic structures. *Journal of Applied Crystallography* 2: 65–71.

21 Toby, B.H. and Von Dreele, R.B. (2013). GSAS-II: the genesis of a modern open-source all purpose crystallography software package. *Journal of Applied Crystallography* 46 (2): 544–549.

22 Shekunov, B.Y., Feeley, J.C., Chow, A.H.L. et al. (2002). Physical properties of supercritically-processed and micronised powders for respiratory drug delivery. *Kona: Powder Science and Technology in Japan* 20: 178–187.

23 Scardi, P., Ermrich, M., Fitch, A. et al. (2018). Size-strain separation in diffraction line profile analysis. *Journal of Applied Crystallography* 51: 831–843.

24 David, W.I.F., Shankland, K., van de Streek, J. et al. (2006). DASH: a program for crystal structure determination from powder diffraction data. *Journal of Applied Crystallography* 39: 910–915.

Further Reading

Pharmacopoias: USP-NF <941>, Ph.Eur. Chapter 2.9.33.

Etter, M. and Dinnebier, R.E. (2014). A century of powder diffraction: a brief history. *Journal of Inorganic and General Chemistry* 640 (15): 3015–3028.

Rissanen, K. (2012). *Advanced X-Ray Crystallography*. Springer, ISBN-10: 364244038X.

4.3

Standards and Trends in Solid-State Characterization Techniques – Thermal Analysis

Juergen Thun and Nikolaus Martin

F. Hoffmann-La Roche Ltd. Department for Solid State Sciences, Pharma Technical Development Synthetic Molecules Grenzacherstrasse 124, 4070 Basel, Switzerland

4.3.1 Introduction

Thermal analysis is a key technique to characterize materials with respect to their intrinsic properties and phase transitions between solid forms. The International Confederation for Thermal Analysis and Calorimetry defines thermal analysis as "the study of the relationship between a sample property and its temperature as the sample is heated or cooled in a controlled manner" [1]. Historically, the use of Kofler hot stages in the 1930s was a first milestone, allowing to observe temperature-dependent phase changes in a controlled environment [2–6]. In addition to thermo-optical analyses, a variety of techniques exist, including dielectric thermal analysis [7–9], dynamic mechanical analysis [10–13], thermomechanical analysis [13–15], differential thermal analysis [16–19], thermally stimulated current [20–22], and thermoluminescence. The most widely used techniques in the context of thermal analysis of organic materials, such as pharmaceuticals, are differential scanning calorimetry and thermogravimetric analysis. Therefore, this chapter focuses on these two techniques, while acknowledging however that all other techniques mentioned provide valuable information on the properties of substances.

In the pharmaceutical industry, thermal analysis is used in the development of small synthetic and large biological molecules. In connection with biological molecules, the main objectives are to identify folding structures in proteins, test thermal stability [23, 24], or understand and develop freeze-drying processes [25]. However, this chapter concentrates on thermal analysis in the context of small-molecule development.

Solid State Development and Processing of Pharmaceutical Molecules: Salts, Cocrystals, and Polymorphism, First Edition. Edited by Michael Gruss.

4.3.2 Thermal Analysis in Drug Development

4.3.2.1 Solid form Landscape

Selection of the solid form of an active pharmaceutical ingredient represents an essential milestone during drug development. Identification of the best suited solid form of the active pharmaceutical ingredients (API) relies on sound understanding of the solid form landscape, including the identification of the solid form that is thermodynamically stable at ambient conditions. In addition, hydrates and process-relevant solvates have to be identified. Ultimately, selection of the solid form for further development hinges on the sound understanding of transition pathways and transition temperatures between solid forms [26].

All solid forms identified are routinely characterized by a variety of orthogonal analytical techniques such as thermal analysis, X-ray diffraction, vibrational spectroscopy, and dynamic vapor sorption. The solid form landscape is established after characterization and careful interpretation of the analytical data.

Knowledge of the solid form landscape and possible transition pathways is also essential for the development of a holistic solid form control strategy to ensure the integrity of the selected solid form during manufacture and shelf-life determination of the final drug product.

Thus, thermal analysis contributes markedly to characterizing the solid form landscape of an API. For example, differential scanning calorimetry (DSC) determines transition and/or melting enthalpies, amorphous contents, glass transitions, and crystallization events, while thermal gravimetric analysis (TGA) provides information on weight loss, stability of hydrates/solvates, and, by applying coupling techniques, the nature/identity of volatile chemical species released upon heating.

4.3.2.2 Compatibility Studies

Apart from its use in the characterization of pure materials, thermal analysis is also applied in compatibility studies of drug substances and polymers [27–29]. Amorphous APIs can be stabilized by embedding the API molecules in a polymer matrix. Only APIs that are completely miscible with the polymer matrix provide the necessary stability of such solid dispersions. DSC is used to determine the miscibility of the two materials by analyzing glass transition temperatures. In miscible systems, solid dispersions exhibit a single glass transition temperature that lies between the glass transition temperatures of the pure phases. This fast method is suitable for determining long-term stability of both fresh solid dispersions and stressed samples (e.g. samples exposed to high relative humidity and/or high temperatures).

4.3.2.3 Other Applications

Thermal analysis is also used to study the sublimation behavior of a compound at different temperatures. Furthermore, thermal analysis is applied in process optimization with respect to technical applications. For example, bake-on siliconization of cartridges for medical devices is studied using thermogravimetric analysis [30, 31].

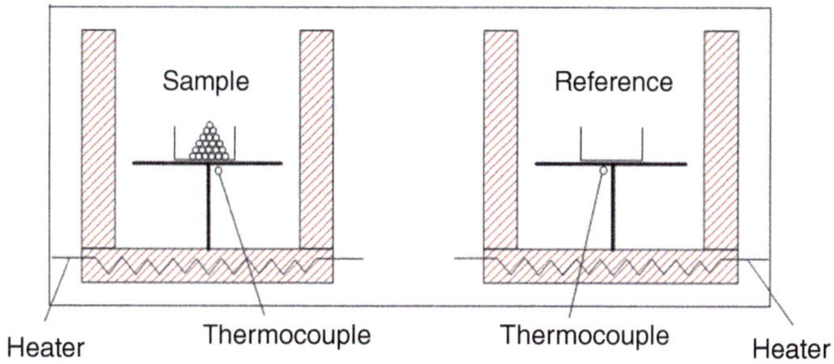

Figure 4.3.1 Principle of power compensation differential scanning calorimeter.

In principle, thermal analysis can be used to identify all types of phase transitions or, more generally, any changes in inter- or intramolecular interactions.

4.3.3 Methods

4.3.3.1 Differential Scanning Calorimetry

4.3.3.1.1 Techniques

In DSC, the difference between the heat flux to a sample and the reference substance is recorded as a function of temperature or time. Two subtypes of this technique are used, i.e. *power compensation DSC* and *heat flux DSC*. The two techniques differ with respect to the construction of the instrument and principle of measurement. Power compensation DSC can generally heat or cool faster and provides better resolution of sharp events. Heat flux DSC is usually associated with a straighter and more reproducible baseline. However, after decades of technique refinements, the differences have become marginal.

The measuring principle of a power compensation differential scanning calorimeter (see Figure 4.3.1) is based on two independent heating devices, i.e., one for the sample and one for the reference substance. During measurements, the individual heating power of the sample and the reference substance is controlled such that the temperatures of the sample and the reference substance are kept equal. In case of an endothermic or exothermic event in the sample, the difference in the heating power between the sample and the reference substance is recorded and is displayed as a function of temperature or time.

Heat flux DSC (see Figure 4.3.2) is based on a single heating device for both the sample and the reference substance. The temperatures of the sample and the reference substance are measured individually. In case of an endothermic or exothermic event, a difference between sample temperature and reference temperature is observed. This temperature difference is directly proportional to the heat flow.

4.3.3.1.2 Sample Preparation and Measuring Parameters

Depending on the type of sample measured and objective of measurement, the following parameters have to be considered:

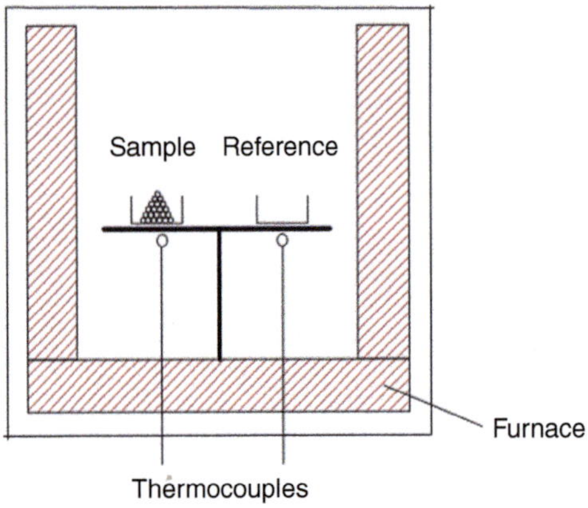

Figure 4.3.2 Principle of heat flux DSC.

Sample Pan A wide range of pan types for DSC measurements are provided by the instrument vendors. For typical pharmaceutical samples (e.g., small organic molecules, excipients, or polymers) in the temperature range up to about 500 °C, aluminum pans are most commonly used. However, if a chemical reaction between the sample and aluminum pan has to be anticipated, alternative pans made of gold, platinum, or aluminum oxide may be used. For measurements performed under internal pressure in a closed pan (e.g. for the study of hydrates/solvates, samples containing solvents, and drug product mixtures), intermediate-pressure or high-pressure pans are available.

Sample Weight Typical sample weights for DSC measurements range from 2 to 10 mg. Resolution improves with lower sample weights, but the measuring sensitivity decreases. To improve thermal conductivity within the sample, it might be helpful to gently compact the sample during sample preparation.

Heating Rate Typical heating and cooling rates range from 1 to 20 K min^{-1}. Deviations from typical heating rates may sometimes be necessary to separate peaks, prevent or enable phase conversions, and detect low-enthalpy events. It should be borne in mind that lower heating rates increase the resolution but lower the sensitivity, while the opposite is true for higher heating rates. Optimum settings need to be established during method development.

Gas Atmosphere DSC measurements are performed under a constant gas flow, usually nitrogen or air. For special purposes, other gases might be selected, e.g., helium, argon, or oxygen.

4.3.3.1.3 Evaluation

The main principle of DSC is the measurement of energy into and/or out of a sample as a function of temperature. This is described in the basic heat flow equation:

$$\frac{dQ}{dt} = C_p \cdot \frac{dT}{dt}$$

$\frac{dQ}{dt}$ is the heat flow, C_p is the heat capacity of sample, specific heat capacity of material × sample weight, and $\frac{dT}{dt}$ is the heating rate.

The product of the heating/cooling rate and heat capacity of the sample results in the measured heat flow. However, this simplified equation must be corrected for a time-dependent and hence scan rate-independent component (e.g. chemical reactions or curing phenomenon of resins):

$$\frac{dQ}{dt} = C_p \cdot \frac{dT}{dt} + f(t, T)$$

$f(t; T)$ is the time-dependent component of heat flow (kinetic component).

The DSC equipment generates an electrical potential obtained from either the power compensation or the difference in temperature, depending on the type of system used. The resulting DSC curve represents energy vs. time/temperature. Generally, power (e.g., mW or W/g) vs. temperature (e.g. °C or K) is displayed in line with the heat flow equation. We emphasize that clear indication of the direction of the endo- or exothermic events on the y-axis is essential (see Figure 4.3.3).

From the DSC diagram, the relevant temperatures are determined using *onset* and *peak* temperatures, and the corresponding enthalpy is determined by integrating the area under the peak. Changes in the specific heat capacity, e.g., glass transition of amorphous substances, can also be evaluated (see Figure 4.3.3). The onset temperature does not depend on measuring conditions, whereas the peak temperature depends on the sample mass, heating rate, and thermal resistance of both the sample and the measuring system.

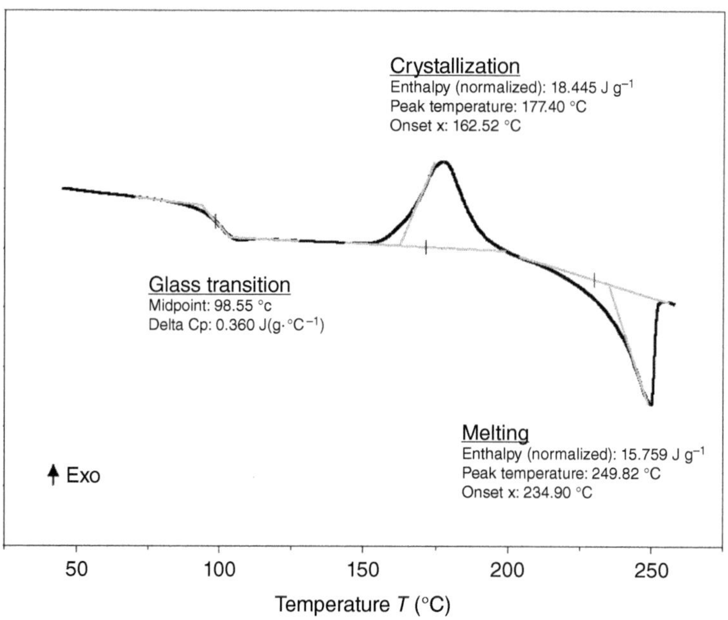

Figure 4.3.3 Standard DSC thermogram showing glass transition, crystallization, and melting of a sample.

In special cases, first and second derivatives of the DSC curve can also help to identify glass transitions (i.e., a step in the DSC trace converts into a peak in the first derivative) or the purity of a peak (i.e., shoulders in a peak convert to a peak in the second derivative).

4.3.3.1.4 Special Applications

Besides the typical DSC equipment described in Section 4.3.3.1.1, some more sophisticated systems have been implemented during the last decades. These include fast scanning calorimetry, modulated DSC, and cyclic measurements. Details of these systems are given below.

Fast scanning calorimetry: While typical DSC measurements employ scanning rates ranging from 1 to 20 K min^{-1} (see Section 4.3.3.1.2), higher heating rates up to 500 K min^{-1} may be achieved with certain instruments. The availability of fast scanning calorimeters has made heating rates as high as 3 000 000 K min^{-1} feasible [32]. Such extreme heating rates are used to identify the true melting point by separation of melting and thermal decomposition events (see example in Section 4.3.4.2). In addition, these heating rates may suppress solid–solid conversions into crystal forms with higher melting points (allowing to determine heat of fusion and melting point of this form that is metastable at that particular temperature, see example in Section 4.3.4.2) or suppress crystallization of amorphous materials, thus allowing to determine the glass transition temperature.

Modulated DSC: Here, a sinusoidal heating rate is applied to the sample. This technique allows the separation of thermodynamic and kinetic effects with improved sensitivity and resolution. The advantage is high resolution at low heating rates, eliminating the influence of a crooked baseline and separating overlaying thermal effects (see corrected heat flow equation and example in Section 4.3.4.3).

Cyclic measurements: These are used to study the reversibility of polymorph conversions or in the preparation and subsequent characterization of amorphous materials (see example in Section 4.3.4.1).

4.3.3.1.5 Detection Limits

When using standard DSC equipment and measuring conditions, the detection limit of amorphous content in samples of crystalline drug substance is usually below 5% but may be below 2% in certain cases. The detection limit for different polymorphs depends markedly on the nature and absolute enthalpy of the thermal event evaluated, e.g., solid–solid conversions or melting. High heating rates increase the sensitivity and, hence, lower the limit of detection for low-enthalpy events.

4.3.3.2 Thermogravimetric Analysis

4.3.3.2.1 Technique

TGA assesses the mass or change in mass depending on the temperature or time. Three arrangements of the measuring system are possible, i.e. top loading (Figure 4.3.4a), bottom loading (Figure 4.3.4b), and side loading (Figure 4.3.4c) (see Figure 4.3.4).

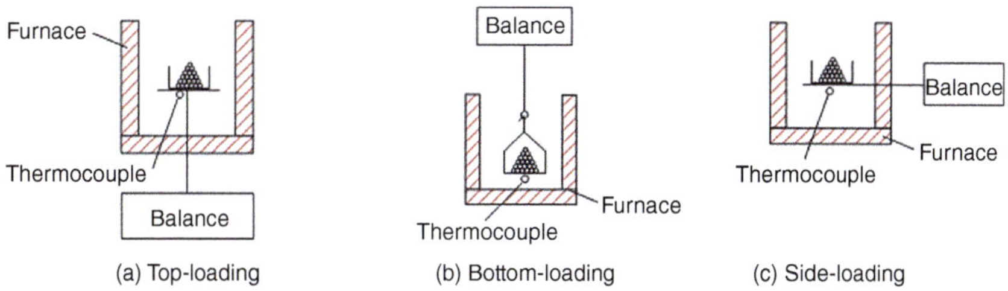

Figure 4.3.4 Arrangements of (a) top-loading, (b) bottom-loading, and (c) side-loading measuring systems for thermogravimetry.

In TGA, the sample is typically heated at a constant heating rate, but in certain cases, isothermal measurements or combinations of dynamic and isothermal segments may be applied. Most thermobalances allow to open the closed sample pan immediately before the measurement, for example, by piercing the lid. In this way, mass loss during the lag phase between sample preparation and measurement can be avoided. Undesired mass loss is a critical issue when using an automatic sampling system with open pans. In addition, hygroscopic materials may gain weight by the uptake of moisture during storage in an open sample pan.

4.3.3.2.2 Sample Preparation and Measuring Parameters

Depending on the type of sample measured and objective of measurement, the following parameters have to be considered:

Sample Pan All applications, restrictions, and recommendations for TGA are the same as those for DSC (see Section 4.3.3.1.2). The only exception is that TCA employs only open or pierced pans.

Sample Weight The sample weight for TGA measurements usually ranges from a few milligrams to several grams, depending on the sensitivity of the balance used in the instrument. The option to analyze heavy samples is advantageous (e.g. for heterogeneous materials, tablets, or samples of drug product).

Heating Rate Typical heating rates range from 1 to 20 K min^{-1}. Depending on the type of furnace installed in the thermobalance, heating rates up to 500 K min^{-1} can be achieved.

Gas Atmosphere TGA measurements are performed under a constant gas flow, usually nitrogen or air. For special purposes, other gases might be selected, e.g. helium, argon, or oxygen.

4.3.3.2.3 Evaluation

Thermobalances measure the change in mass (recorded as an electrical potential) relative to the temperature or time. Thus, the TGA curve usually represents the absolute or relative (compared to the starting mass) weight change vs. temperature and/or time.

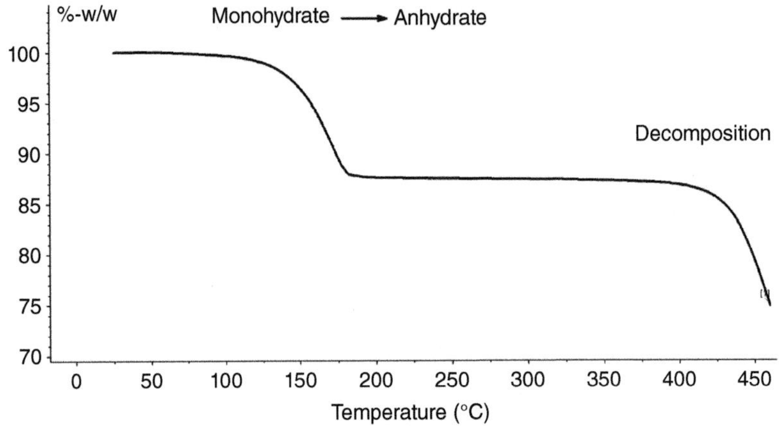

Figure 4.3.5 Standard TGA thermogram of calcium oxalate monohydrate.

From the diagram, the relevant steps in weight loss are determined and correlated with the respective temperatures (see Figure 4.3.5).

4.3.3.2.4 Special Applications

A widely used application of TGA is evolved gas analysis, where the evolved gas is transferred to an instrument that detects and identifies the gas. The evolved gas correlates with the weight loss. Usually, the thermobalance is coupled with Fourier-transformed infrared (FTIR) or mass spectrometry (see example in Section 4.3.4.4).

In another application, the heating rate is adapted simultaneously to the observed weight loss (high-resolution TGA). By applying this method, the heating rate can be set to very low (or even to isothermal conditions), if significant weight loss occurs at a certain temperature. Otherwise, rather high heating rates are applied, as no resolution of overlapping events is needed, resulting in accelerated measurements and saving measurement time.

4.3.4 Case Studies

4.3.4.1 Understanding Polymorphic Transitions

As pointed out in Section 4.3.2, understanding the transition pathways of solid forms is essential to define the solid form landscape of APIs. Cyclic DSC measurements help to understand solid–solid conversions of polymorphs. The nature of the conversion (i.e., endothermic or exothermic) provides additional information on the enantiotropic or monotropic relationship [33, 34].

The selected example of a development compound shows an extensive solid form landscape involving four known polymorphs. The DSC experiments started with either polymorph Form A or Form B. A low-temperature polymorph was detected upon cooling Form A to approx. −150 °C. Upon heating, an endothermic event at

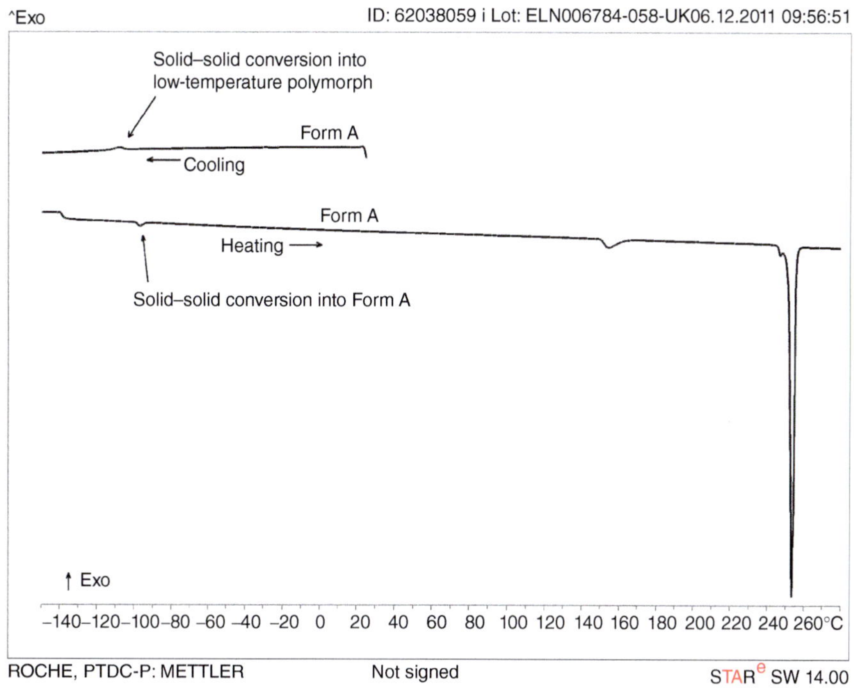

Figure 4.3.6 Low-temperature DSC experiment.

approximately −100 °C indicated the enantiotropic relationship between Form A and the low-temperature polymorph (see Figure 4.3.6). Knowledge of the presence of a low-temperature polymorph is essential as many single-crystal X-ray diffraction experiments are performed at low temperatures to minimize thermal motion of the atoms.

Figure 4.3.7 shows the significant differences between the simulated X-ray powder diffraction (XRPD) pattern of the low-temperature polymorph and the experimental XRPD pattern of the related polymorph at ambient conditions. Alternatively, low-temperature XRPD analysis could have been performed to identify the nature of this mismatch. However, a single DSC experiment determined the transition temperature and enantiotropic relationship within less than 1 hour.

A second DSC experiment also involved Form A, showing endothermic transition into Form D upon heating, thus illustrating the enantiotropic relationship of Form A and Form D with an onset temperature of about 150 °C (see Figure 4.3.8). Upon cooling of the melt, Form C crystallized (T_{onset} at about 185 °C) and showed exothermic transition into Form D (T_{onset} at about 190 °C) upon subsequent heating, suggesting a monotropic relationship between Form C and Form D.

Finally, Form B was heated, and endothermic transition (T_{onset} at about 195 °C) into Form D was observed, again indicating an enantiotropic relationship between Form B and Form D (see Figure 4.3.9). However, upon cooling of Form D, an endothermic transition (T_{onset} at about 140 °C) to Form A (the more stable

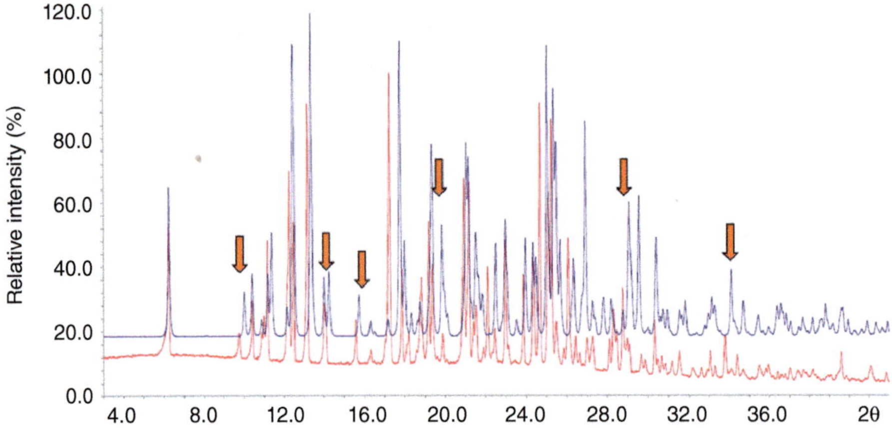

Figure 4.3.7 Simulated XRPD pattern from single-crystal structure solution of a low-temperature polymorph (top) vs. measured XRPD pattern of the enantiotropically related stable polymorph at ambient conditions (bottom).

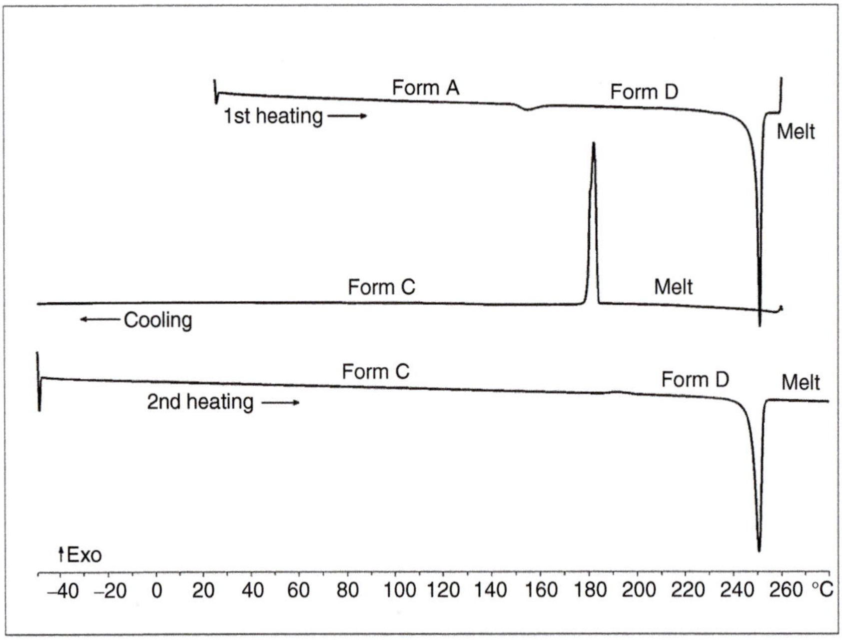

Figure 4.3.8 Cyclic DSC experiment starting with Form A.

polymorph at this temperature range) was observed. Hence, the reversible conversion of Form D into Form B was not confirmed by this cyclic DSC experiment. Instead, an endothermic transition (T_{onset} at about 150 °C) of Form A to Form D was observed upon heating.

In conclusion, using cyclic DSC experiments in combination with XRPD allowed to identify several polymorphs and provided information on the enantiotropic or

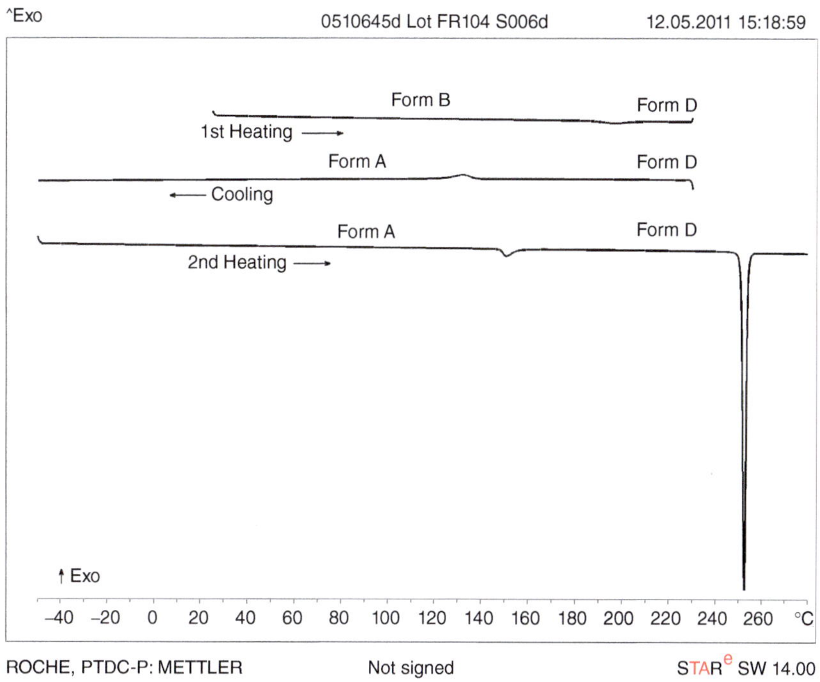

Figure 4.3.9 Cyclic DSC experiment starting with Form B.

Table 4.3.1 Overview of thermal data obtained by standard DSC.

Polymorph	Melting temperature (°C)	Transition temperature in Form D (°C)	Heat of fusion (J g^{-1})
A	—	154	—
B	—	196	—
C	230	190	93.5
D	251	—	103.5

monotropic relationship between the various polymorphs. Thus, by varying the heating/cooling rates and the type of solid form, a number of experiments to enlighten the solid form landscape of the substance in question can be performed.

4.3.4.2 The Power of Ultra-fast Heating Rates

In the example detailed in Section 4.3.4.1, the relationship between Form A and Form B was not clarified by applying standard DSC measurements and the heat of transition rule [33].[1] Another option is to apply the heat of fusion rule [33, 34].[2] However, this approach requires previous knowledge of the melting points and heats of fusion of all polymorphs (see Table 4.3.1).

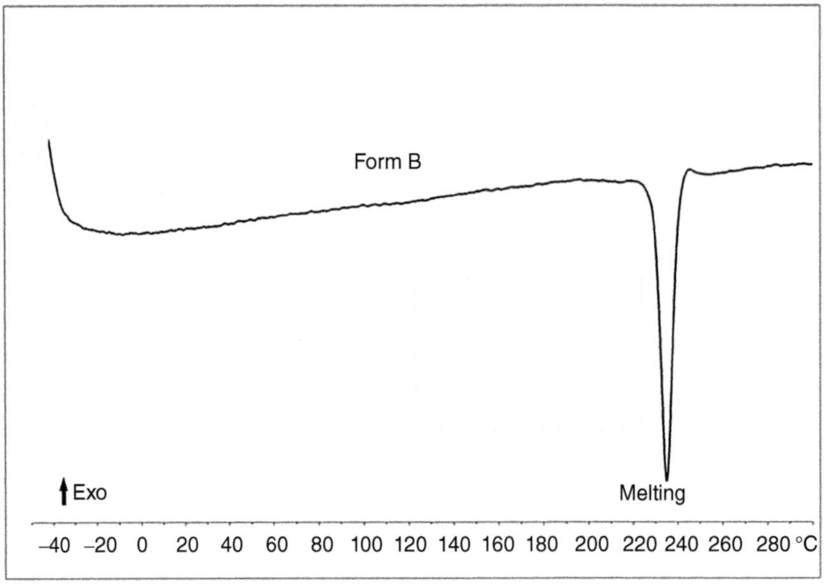

Figure 4.3.10 DSC of Form B using a heating rate of $1000\,K\,s^{-1}$.

Table 4.3.2 Overview of thermal data obtained by standard DSC and by applying ultra-fast heating rates.

Polymorph	Melting temperature (°C)	Transition temperature in Form D (°C)	Heat of fusion ($J\,g^{-1}$)
A	249[a)]	154	130[a)]
B	231[a)]	196	117[a)]
C	230	190	93.5
D	251	—	103.5

a) Data retrieved from measurements applying ultra-fast heating rates.

Ultra-fast heating rates ($1000\,K\,s^{-1}$) suppressed the transitions of Form A and Form B into Form D allowing to analyze the melting points and heats of fusion (see example of Form B in Figure 4.3.10) of the solid forms of interest.

Data analysis using the heat of fusion rule confirmed the monotropic relationship of Form C with Form A and Form B. Form C had a lower melting point and lower heat of fusion compared to Form A and Form B (see Table 4.3.2).

The relationship between Form A and Form B was more complex. Using ultra-fast heating rates limits the sample mass to only a few nanograms of material (i.e., a single crystal). To obtain a statistically significant result, the measurement was repeated multiple times. Because the sample is placed on the sensor of a fast scanning calorimeter without sample pan, the sample mass cannot be determined

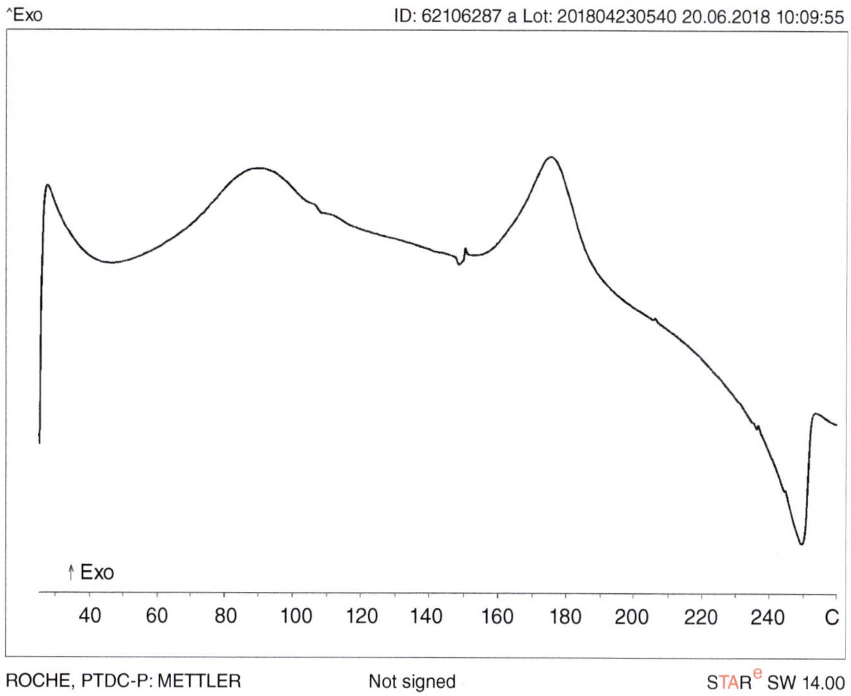

Figure 4.3.11 Standard DSC experiment of an amorphous material.

directly. Therefore, the approach described by Chua et al. [35] was used, allowing estimation of the sample mass and hence of the heat of fusion in J/g. Since the error of this estimation is >10% and heats of fusion for Form A and Form B were similar, the true nature of their relationship could not be identified using this fast method. Thus, competitive suspension equilibration experiments were needed. In other cases, this approach might be more successful.

4.3.4.3 Understanding Amorphous Phases

The glass transition temperature in a standard DSC experiment is sometimes difficult to determine because it may be overlaid in the thermogram by broad endothermic events, e.g., the release of water and/or residual solvents (see Figure 4.3.11).

Applying more sophisticated techniques, such as cyclic DSC or modulated DSC, helps to elucidate the glass transition temperature. Measurements are performed in an open or closed pan.

During a cyclic DSC experiment employing an open pan, residual solvent is released, and glass transition of the dry amorphous material can be determined (see Figure 4.3.12). However, previous heating of the sample above the glass transition temperature during the first cycle may trigger changes in the amorphous sample, which has to be taken into account in the experimental design.

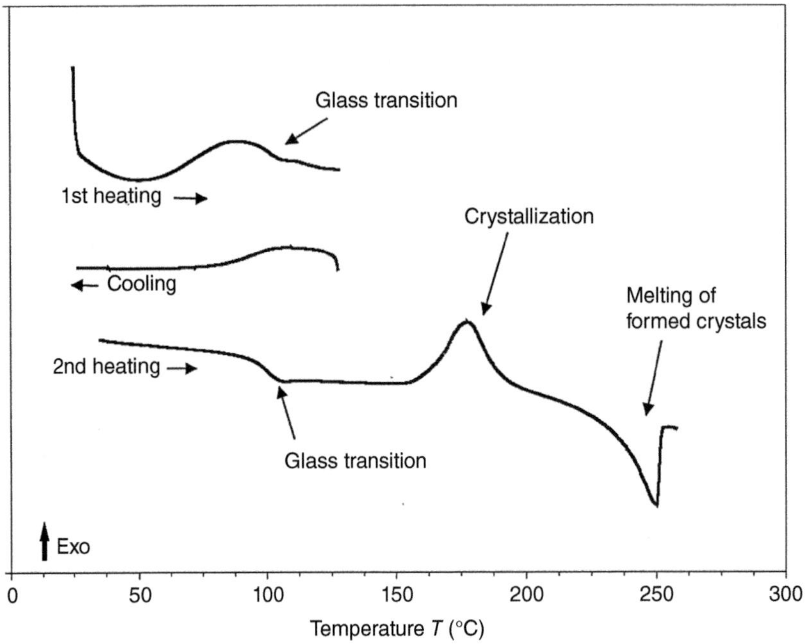

Figure 4.3.12 Cyclic DSC of the same amorphous material as the one shown in Figure 4.3.11.

Modulated DSC experiments can be applied to determine the glass transition within the first heating cycle, hence with less influence on the original sample (see Figure 4.3.13).

Using closed pans reveals the dependency of the glass transition temperature on residual solvent or water. This information is essential to define appropriate storage conditions to prevent crystallization of an amorphous product (see Figure 4.3.14). A safety margin of 50 °C was recommended by Hancock et al. [36], i.e., the amorphous product should be stored 50 °C below its glass transition temperature.

4.3.4.4 Identification of Solvate Structures

Combining evolving gas analysis with TGA is a valuable technique to identify the released gas during the weight loss step of TGA measurements. Weight loss can be caused by the loss of residual solvents or moisture, desolvation of a solvate/hydrate, or thermal decomposition of the analyzed material. When combining EGA with TGA, the weight loss and its origin can be identified at the same time.

For example, desolvation and decomposition of calcium oxalate monohydrate were analyzed using TGA coupled to a FTIR spectrometer. Figure 4.3.15 shows the decomposition steps at temperatures ranging from room temperature to 900 °C, with the first step representing dehydration of the bound water in the crystal lattice of the monohydrate, the second step representing decomposition of calcium oxalate to calcium carbonate, and the third step representing decomposition of calcium carbonate to calcium oxide (see Table 4.3.3).

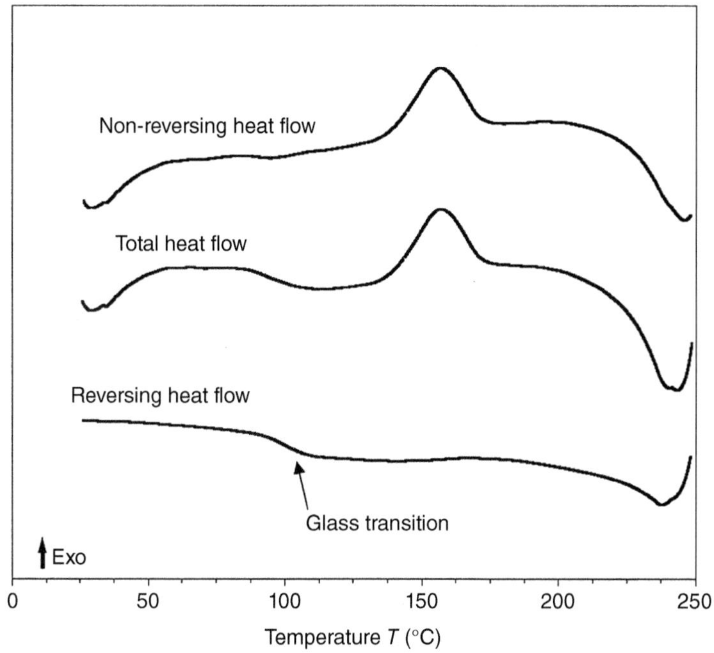

Figure 4.3.13 Modulated DSC of the same amorphous material as the one shown in Figures 4.3.11 and 4.3.12.

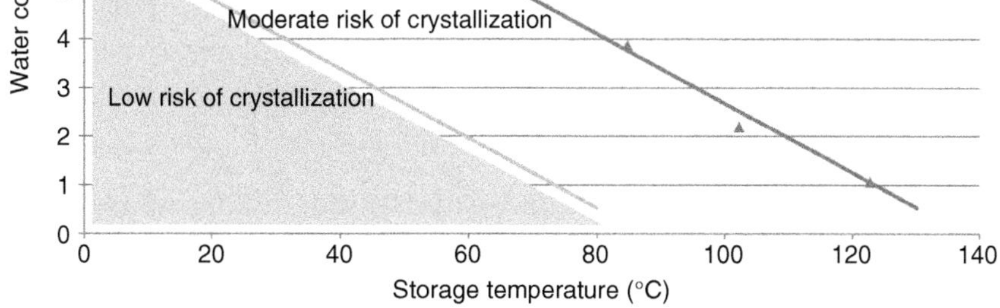

Figure 4.3.14 Example of crystallization probability depending on storage temperature and water content.

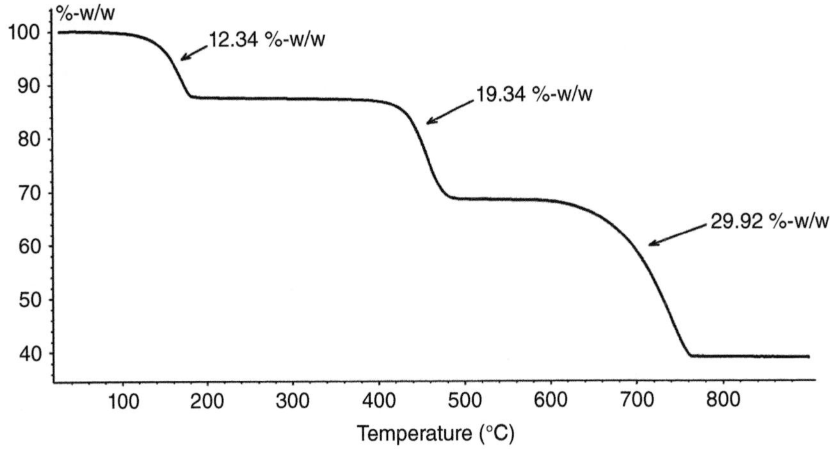

Figure 4.3.15 Thermogravimetric analysis of calcium oxalate monohydrate.

Table 4.3.3 Weight loss during decomposition of calcium oxalate monohydrate.

Step	Theoretical value (%-w/w)[a]	Measured value (%-w/w)
1. $CaC_2O_4 \cdot H_2O \rightarrow CaC_2O_4 + H_2O \uparrow$	12.3	12.3
2. $CaC_2O_4 \rightarrow CaCO_3 + CO \uparrow$	19.2	19.3
3. $CaCO_3 \rightarrow CaO + CO_2 \uparrow$	30.1	29.9

a) Weight loss of the volatile compound is calculated based on molecular mass of calcium oxalate monohydrate.

Figure 4.3.16 shows the 3D representation of the corresponding FTIR spectra at different temperatures. The observed ratio of CO (wave numbers 2181 and 2114 cm^{-1}) and CO_2 (wave numbers 2390 and 2331 cm^{-1}) in the second step was due to sample and instrumental conditions (Boudouard equation: $2\,CO \rightleftharpoons C + CO_2$). Identification of the evolved gases is based on the comparison of the measured infrared (IR) spectra with database spectra.

4.3.5 Quality and Regulatory Aspects

For quality control/release analytics, thermal analysis has to be performed in a regulated environment to meet all legal requirements issued by national health authorities. These include qualification of equipment (i.e. installation qualification [IQ], operation qualification [OQ], performance qualification [PQ], computer system validation [CSV]), meeting the requirements of CFR 21 Part 11 guidelines

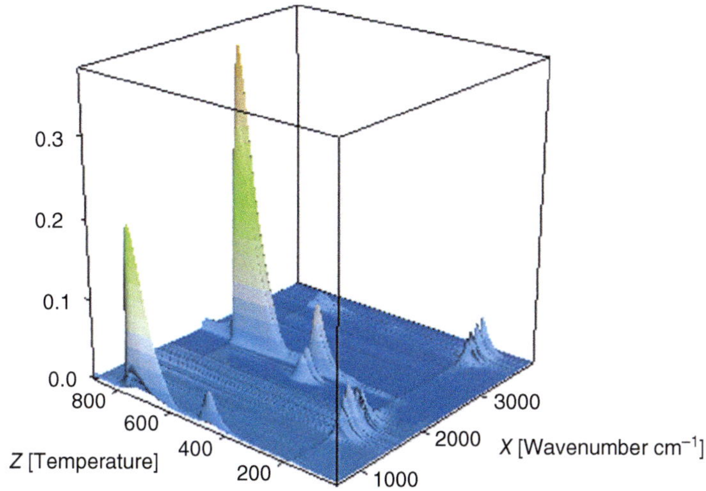

Figure 4.3.16 Three-dimensional representation of the evolved gas analysis by FTIR showing peak intensity, temperature (°C), and wave number (cm^{-1}) .

of the United States Food and Drug Administration (FDA), and adherence to good manufacturing practices (GMP) with validated methods. Guidelines to fulfill these requirements are provided by the International Council for Harmonization (ICH), FDA, European Medicines Agency (EMA), and European Commission GMP (EU-GMP).

Thermal analysis is described in both the United States Pharmacopoeia (USP <891>) and the European Pharmacopoeia (Ph.Eur. 2.2.34).

4.3.6 Outlook

Thermal analysis continues to be a cornerstone in the characterization of the solid form landscape and transition pathways of APIs. New techniques, such as fast scanning calorimetry, coupling techniques, and further refinements of thermal sensors and furnace settings have extended the boundaries of the method, thus enhancing the sensitivity and resolution.

Increased pressure on drug development timelines and the limited sample amounts available in the early stages of drug development can be counteracted by the use of fast thermal analytic techniques requiring only a few milligrams of material. In combination with coupling techniques, such measurements provide even more information.

In conclusion, drafting a holistic solid form control strategy to ensure the integrity of the solid form of drug substances during manufacture as well as shelf-life determination of drug products relies on thermal analysis.

Acknowledgments

We thank F. Hoffmann-La Roche Ltd. and the Department for Solid State Sciences for the permission to contribute to this book. Furthermore, we thank our colleagues, Mathieu Diboune, Pascal Fleischmann, Katrin Fiebich, Pirmin Hidber, Flavio Lack, Frédéric Ran, and Urs Schwitter, for the helpful discussions and their careful review of the manuscript.

List of Abbreviations

API	active pharmaceutical ingredient, drug substance
DEA	dielectric thermal analysis
DSC	differential scanning calorimetry
DTA	differential thermal analysis
DMA	dynamic mechanical analysis
EGA	evolved gas analysis
FSC	fast scanning calorimetry
FTIR	Fourier-transformed infrared
ICTAC	International Confederation for Thermal Analysis and Calorimetry
TGA	thermogravimetric analysis
TL	thermoluminescence
TMA	thermomechanical analysis
TOA	thermo-optical analyses
TSC	thermally stimulated current
XRPD	X-ray powder diffraction

Notes

1 Burger et al. [33]: "If an endothermal transition is observed at some temperature it may be assumed that there is a transition point below it, i.e. the two forms are related enantiotropically. If an exothermal transition is observed at some temperature it may be assumed that there is no transition point below it, i.e. the two forms are either related monotropically or the transition temperature is higher."

2 Burger et al. [33]: "If the higher melting form has the lower heat of fusion the two forms are usually enantiotropic, otherwise they are monotropic."

References

1 Lever, T., Haines, P., Rouquerol, J. et al. (2014). ICTAC nomenclature of thermal analysis (IUPAC Recommendations 2014). *Pure and Applied Chemistry* 86: 545–553.

2 Kofler, A. (1943). Polymorphism of organic substances: acridine, pyrocatechol, diphenylamine and suberic acid. *Berichte der Deutschen Chemischen Gesellschaft [Abteilung] B: Abhandlungen* 76B: 871–873.

3 Kofler, A. (1940). Thermal analysis with a hot-stage microscope. *Zeitschrift für Physikalische Chemie* A187: 201–210.

4 Kofler, A. (1941). Thermal analysis in the hot-stage microscope. III. Phenomena of polymorphism and isomorphism of s-trinitrobenzene, picric acid and α-trinitrotoluene. *Zeitschrift für Physikalische Chemie* A188: 201–228.

5 Kofler, A. (1940). Thermal analysis with a hot-stage microscope. II. Contact zones. *Zeitschrift für Physikalische Chemie* A187: 363–373.

6 Kofler, L., Kofler, A., and Mayrhofer, A. (1936). *Mikroskopische Methoden in der Mikrochemie*. Verlag Emil Haim & Co.

7 Maheswaram, M.P., Mantheni, D., Perera, I. et al. (2013). Characterization of crystalline and amorphous content in pharmaceutical solids by dielectric thermal analysis. *Journal of Thermal Analysis and Calorimetry* 111: 1987–1997.

8 Mantheni, D.R., Maheswaram, M.P.K., Sobhi, H.F. et al. (2012). Solid state studies of drugs and chemicals by dielectric and calorimetric analysis. *Journal of Thermal Analysis and Calorimetry* 108: 227–233.

9 Craig, D.Q.M. and Taylor, K.M.G. (1995). The use of low frequency dielectric analysis in the characterisation of metered dose inhaler formulations. *International Journal of Pharmaceutics* 114: 129–133.

10 Craig, D.Q.M. and Johnson, F.A. (1995). Pharmaceutical applications of dynamic mechanical thermal analysis. *Thermochimica Acta* 248: 97–115.

11 Sutananta, W., Craig, D.Q.M., and Newton, J.M. (1996). The use of dielectric analysis as a means of characterising the effects of moisture uptake by pharmaceutical glyceride bases. *International Journal of Pharmaceutics* 132: 1–8.

12 Aho, J., Boetker, J.P., Baldursdottir, S., and Rantanen, J. Rheology as a tool for evaluation of melt processability of innovative dosage forms. *International Journal of Pharmaceutics (Amsterdam, Netherlands)* 494: 623–642.

13 Reading, M. (1992). Thermomechanical analysis and dynamic mechanical analysis. *Special Publication – Royal Society of Chemistry* 117: 108–125.

14 Harding, L., King, W.P., Dai, X. et al. (2007). Nanoscale characterization and imaging of partially amorphous materials using local thermomechanical analysis and heated tip AFM. *Pharmaceutical Research* 24: 2048–2054.

15 Manduva, R., Kett, V.L., Banks, S.R. et al. Calorimetric and spatial characterization of polymorphic transitions in caffeine using quasi-isothermal MTDSC and localized thermomechanical analysis. *Journal of Pharmaceutical Sciences* 97: 1285–1300.

16 Collins, W. J., Dicks, M. L., Redman-Furey, N. L.et al. (2003). The role of TG/DTA in the initial evaluation of the solid state forms for pharmaceutical new chemical entities part I: evaluation of pure forms. In: Proc. NATAS Annu. Conf. Therm. Anal. Appl., 31st, 089/1-089/9.

17 Collins, W. J., Dicks, M. L., Redman-Furey, N. L. et al. (2003) The role of TG/DTA in the initial evaluation of the solid state forms for pharmaceutical

new chemical entities part II: evaluation of mixed forms. In: Proc. NATAS Annu. Conf. Therm. Anal. Appl., 31st, 090/1-090/8.

18 Grundner, C.L., Poiesz, K.B., and Redman-Furey, N.L. (2006). Development and use of a TG-DTA-microscope for evaluation of pharmaceutical materials. *Journal of Thermal Analysis and Calorimetry* 85: 91–98.

19 Burnham, L., Dollimore, D., and Alexander, K.S. (2002). Kinetic study of the drug acetazolamide using thermogravimetry. *Thermochimica Acta* 392-393: 127–133.

20 Boutonnet-Fagegaltier, N., Lamure, A., Menegotto, J. et al. (2006). The use of thermally stimulated current spectroscopy in the pharmaceutical sciences. In: *Thermal Analysis of Pharmaceuticals* (eds. D.Q.M. Craig and M. Reading). CRC Press LLC.

21 Moura Ramos, J.J. and Diogo, H.P. (2017). Dipolar reorientations in amorphous nimesulide: A TSDC and DSC study. *Current Drug Delivery* 14: 91–98.

22 Ramos, J.J.M. and Diogo, H.P. (2016). The slow relaxation dynamics in active pharmaceutical ingredients studied by DSC and TSDC: voriconazole, miconazole and itraconazole. *International Journal of Pharmaceutics (Amsterdam, Netherlands)* 501: 39–48.

23 Ibarra-Molero, B., Naganathan, A.N., Sanchez-Ruiz, J.M., and Munoz, V. (2016). Modern analysis of protein folding by differential scanning calorimetry. *Methods in Enzymology* 567: 281–318.

24 Ibarra-Molero, B. and Sanchez-Ruiz, J.M. (2006). Differential scanning calorimetry of proteins: an overview and some recent developments. In: *Advanced Techniques in Biophysics, Springer Series in Biophysics*, vol. 10 (eds. J.L.R. Arrondo and A. Alonso), 27–48. Berlin, Heidelberg: *Springer*.

25 Passot, S., Trelea, I.C., Marin, M., and Fonseca, F. (2010). The relevance of thermal properties for improving formulation and cycle development: application to freeze-drying of proteins. In: *Freeze-Drying/Lyophilization of Pharmaceutical and Biological Products*, 3e, vol. 206 (ed. L. Rey), 136–166. CRC Press.

26 Bernstein, J., Davey, R.J., and Henck, J.O. (1999). Concomitant polymorphs. *Angewandte Chemie. International Edition* 38: 3440–3461.

27 Bezerra, G.S.N., Pereira, M.A.V., Ostrosky, E.A. et al. (2016). Compatibility study between ferulic acid and excipients used in cosmetic formulations by TG/DTG, DSC and FTIR. *Journal of Thermal Analysis and Calorimetry* 127: 1683–1691.

28 Matos, A.P.S., Costa, J.S., Boniatti, J. et al. (2016). Compatibility study between diazepam and tablet excipients. *Journal of Thermal Analysis and Calorimetry* 127: 1675–1682.

29 Rosasco, M.A., Bonafede, S.L., Faudone, S.N., and Segall, A.I. (2018). Compatibility study of tobramycin and pharmaceutical excipients using differential scanning calorimetry, FTIR, DRX, and HPLC. *Journal of Thermal Analysis and Calorimetry* 134: 1929–1941.

30 Funke, S., Matilainen, J., Nalenz, H. et al. (2016). Optimization of the bake-on siliconization of cartridges. Part I: optimization of the spray-on parameters. *European Journal of Pharmaceutics and Biopharmaceutics* 104: 200–215.

31 Funke, S., Matilainen, J., Nalenz, H. et al. (2016). Optimization of the bake-on siliconization of cartridges. Part II: investigations into burn-in time and temperature. *European Journal of Pharmaceutics and Biopharmaceutics* 105: 209–222.

32 Schick, C. and Mathot, V. (2016). *Fast Scanning Calorimetry* (eds. C. Schick and V. Mathot). Springer Nature.

33 Burger, A. and Ramberger, R. (1979). Polymorphism of pharmaceuticals and other molecular-crystals 1. Theory of thermodynamic rules. *Mikrochimica Acta* 2: 259–271.

34 Burger, A. and Ramberger, R. (1979). Polymorphism of pharmaceuticals and other molecular-crystals .2. Applicability of thermodynamic rules. *Mikrochimica Acta* 2: 273–316.

35 Chua, Y.Z., Do, H.T., Schick, C. et al. (2018). New experimental melting properties as access for predicting amino-acid solubility. *RSC Advances* 8: 6365–6372.

36 Hancock, B.C., Shamblin, S.L., and Zografi, G. (1995). Molecular mobility of amorphous pharmaceutical solids below their glass transition temperatures. *Pharmaceutical Research* 12: 799–806.

4.4

Standards and Trends in Solid-State Characterization Techniques: Infrared (IR) Spectroscopy

Dagmar Lischke

GO-MSI-INT, Grünenthal GmbH, Zieglerstr. 6, 52099 Aachen, Germany

4.4.1 Infrared (IR) Spectroscopy

4.4.1.1 Introduction

As part of the powerful analysis technique group of molecular spectroscopy, infrared spectroscopy is a well-established and commonly used analytical technique in pharmaceutical research and development processes. Absorption of infrared (IR) radiation brings information about changes in molecular vibrations within molecules. Atom size, bond length, and bond strength vary in molecules, and as a consequence, the frequency at which a particular bond absorbs infrared radiation is different over a range of bonds and modes of vibration. An infrared spectrometer analyzes a material by sending radiation, over a range of different frequencies, typically in the mid-infrared region between 4000 and 400 wavenumbers (the unit is cm^{-1}), through a sample and detecting the absorption of this radiation made by each type of bond in the sample (https://en.wikipedia.org/wiki/Infrared_spectroscopy).

This produces a spectrum, which is typically illustrated as a "plot" of % transmittance against wavenumber. Because of the fact that due to a change of bond lengths and angles in a crystal structure of the same molecule, the resulting vibration bands in an infrared spectrum also change, it is possible to differentiate polymorphs by infrared radiation as long as the changes are not too small to affect the resulting infrared spectrum.

Infrared spectroscopy has been used traditionally for identification and univariate quantification methods for mostly organic substances and mixtures. Although this technique is not the gold standard to analyze the polymorphism of a substance, it is often used during the polymorph and salt screening investigation processes of a novel active pharmaceutical ingredient (API). Infrared spectroscopy is able to distinguish between polymorph and solvate forms of a substance or different salt forms and is one possibility to characterize different polymorphs. IR spectra are used for patent applications and other regulatory documents like New Drug Applications (NDA).

Solid State Development and Processing of Pharmaceutical Molecules: Salts, Cocrystals, and Polymorphism, First Edition. Edited by Michael Gruss.

Due to different measurement setups like attenuated total reflection (ATR), diffuse reflectance, and IR microscopy or imaging, the analyst is also able to identify and relatively quantify different polymorph and solvate forms in a final solid dosage form, e.g. tablets or multiparticulates.

In this chapter, the different setups and their usage during the development of novel APIs and final solid dosage forms are described. The limitations of the technique are also elucidated.

4.4.1.2 IR Spectroscopy as Identity Method for Drug Substances

Different sampling techniques exist to investigate a solid powder by means of IR spectroscopy.

4.4.1.2.1 Transmission Mode

The method has been used preferentially for a long time. The proceeding of choice was to finely grind a quantity of the sample with a specially purified salt (usually potassium bromide, KBr) that itself does not show absorption in the IR spectrum to remove scattering effects from large crystals. This powder mixture is then pressed in a mechanical press to form a translucent pellet. This pellet was then measured in transmission mode. This technique is almost completely replaced by the ATR sample technique, which is the current state-of-the-art sampling method for solid powder samples in IR spectroscopy. The main disadvantages of the transmission mode methodology in solid-state analytics are possible polymorphic transition due to the high pressure during sample preparation and potential chemical incompatibility with KBr. Salt formation or counterion exchange should be considered upon sample preparation, e.g. of hydrochloride salts [1a, b].

4.4.1.2.2 Attenuated Total Reflectance (ATR)

ATR is a sampling technique used in conjunction with infrared spectroscopy that enables sample examination directly in the present state without further preparation and with a small amount of sample material.

An ATR accessory operates by measuring the changes that occur in a reflected infrared beam when the beam comes into contact with a sample (indicated in Figure 4.4.1). An infrared beam is directed onto an optically dense crystal with a high refractive index at a certain angle. This internal reflectance creates an evanescent wave that extends beyond the surface of the crystal and protrudes a few microns (0.5–2 μm) into the sample held in contact with the crystal. Consequently, there must be good contact between the sample and the crystal surface. In regions of the infrared spectrum where the sample absorbs energy, the evanescent wave will be attenuated. The attenuated energy from each evanescent wave is passed back to the IR beam, which then exits the opposite end of the crystal and is passed to the detector in the IR spectrometer. The system then generates an infrared spectrum.

The refractive index of the crystal must be significantly larger than that of the sample. Otherwise, internal reflectance will not occur because the light will be

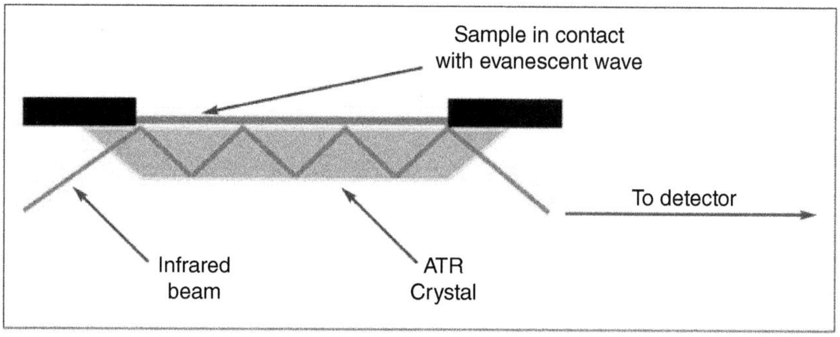

Figure 4.4.1 A multiple reflection ATR system. Source: Perkin Elmer Life and Analytical Sciences [2]. © 2019 PerkinElmer Inc.

transmitted rather than internally reflected in the crystal. Typically, ATR crystals have refractive index values between 2.38 and 4.01 at 2000 cm^{-1}. It is safe to assume that the majority of substances used in pharmaceutical development process have much lower refractive indices [2].

4.4.1.2.3 Sample preparation

ATR One of the advantages of the ATR technique is the low amount of substance (approx. 5–10 mg) that is needed to conduct an analysis. Another advantage is that there is no need for sample preparation. In contrast to measurements using KBr tablets, no milling or compression forces are applied to the sample. Therefore, it is advantageous to use this characterization method very early during the development process of an API to get information about different crystal structures potentially caused by variations of early stage synthesis conditions. Routinely applied during batch analysis, it is possible to get an early hint if a substance may form polymorphs or solvates.

IR Microscopy and Imaging One of the most important issues regarding the IR imaging techniques is an adequate sample preparation. To get a good spectral quality, the surface of the tablet or multiparticulates has to be flat and pretty smooth. To fit the needs, the tablets or the embedded multiparticulates can be prepared by different techniques. It is possible to use a milling system, e.g. Leica EM rapid or Leica TXP, to receive the optimal surface for the analysis or a kind of microtome like the Agilent sample planer [3, 4].

4.4.1.2.4 Analysis and Reporting

Regarding the phase of development, the initial IR spectra of different solid forms of a novel API can be used to build a reference database for further comparison purposes. It is advisable to confirm the identity of a polymorphic form by other techniques like X-ray powder diffraction (XRPD) or Raman spectroscopy. Later on during development, the database can be used to identify samples from the development process of the API by IR spectroscopy in a very fast and cost-efficient way. Generally, the IR instrument software includes algorithms to compare an IR spectrum with reference spectra stored in the database. Similarity between spectra is typically indicated by a correlation coefficient for the recorded spectrum and the known spectra.

It is also meaningful to export spectra into an external database and merge them with other information that is available for a sample or set of samples. This should comprise preparation details as well as results from other analytical methods.

4.4.1.2.5 Examples and Limitations

The following examples show the different IR spectra of two polymorphic forms of Oxatomide. The samples can be differentiated because one form is a hydrate form and the other is an ansolvate form. The spectra could be distinguished by the group frequencies of the hydroxide and several other bands in the fingerprint region of the particular spectrum (Figure 4.4.2). The second example shows two different polymorphs of carbamazepine (CBZ) – malonic acid cocrystals [5a] which cannot be distinguished by their IR spectra because the differences in the crystal structures do not lead to intra- and intermolecular vibrations that can be differentiated by typical IR measurements. As a consequence, the resulting IR spectra are essentially the same (see Figure 4.4.3). By using other analytical techniques (XRPD and solid-state nuclear magnetic resonance spectroscopy [NMR]), it was verified that the two cocrystal forms A and C are polymorphs [5a]. Another case where the vibrational spectra (Raman and IR) are hardly distinguishable is reported for two Alizarin polymorphs and illustrated by comparison of simulated and measured spectra [5b].

Although during the polymorph and salt screening phase of the development process, the preferred identification techniques are XRPD or Raman spectroscopy, the infrared spectra of polymorphs and solvates recorded during the early development phase can be used as additional information for the characterization of the solid phase. However, as the preceding examples showed, this information has to be considered carefully. Application of IR spectroscopy for the investigation of

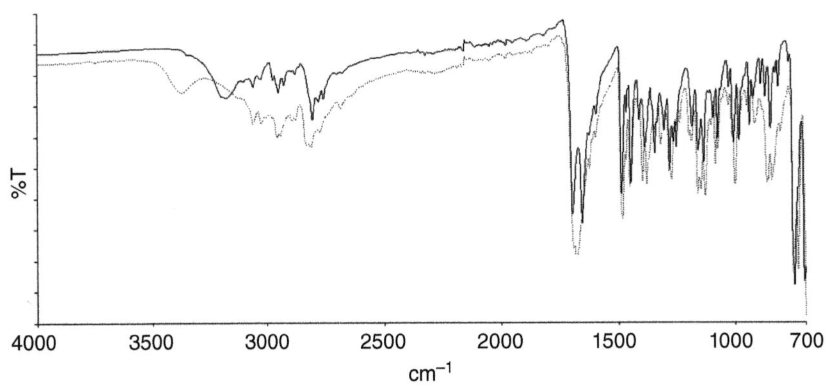

Figure 4.4.2 Overlay of ATR IR spectra of Oxatomide anhydrate (full line) and Oxatomide monohydrate (dotted line).

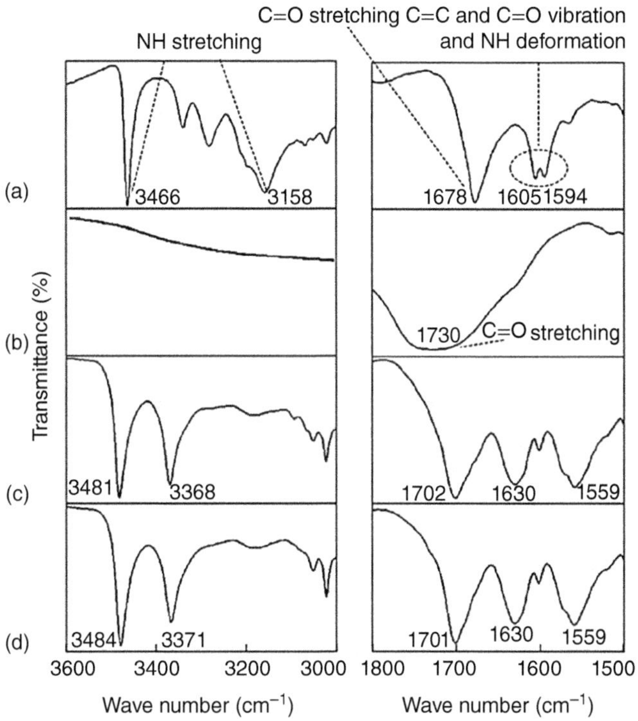

Figure 4.4.3 FTIR spectra of the carbamazepine – malonic acid system. (a) Carbamazepin; (b) malonic acid; (c) cocrystal form A; and (d) cocrystal form C. Source: Limwikrant et al. [5a]. © 2012 Elsevier.

solid phases should be decided on a case-by-case base if suitable for the intended purpose.

4.4.1.2.6 Method Validation of IR Spectroscopy Identification and Quantification Methods

There is no need to validate this identification method regarding the ICH Guidelines because IR spectroscopy is described as an identification method in all relevant pharmacopeia like USP and the European Pharm Eur. [1a]. For the current European Pharmacopoeia version 9.7, the Chapter 2.2.24 "IR spectroscopy" was updated. The method can now finally use the correlation coefficient between the sample and the reference spectrum as calculated by the comparison algorithm of the particular software. This identification criterion makes sense to be used besides the formerly used comparison method that operated on band position and the relative intensity of the bands to each other. It has to be considered that the value of this correlation factor used as limit for a positive identification of similarity has to be chosen carefully for every single substance or polymorph. Additionally, the reference substance used, which in this case could be an API, has to be approved by other techniques like ^{1}H NMR (nuclear magnetic resonance spectroscopy) or LC–MS (liquid chromatography coupled with mass spectroscopy) for chemical identity and XRPD or Raman spectroscopy to affirm identity of the physical form.

If there is a need to quantify different polymorphs in a drug substance regarding the ICH Guidelines Q6B decision tree 4 [6], techniques like XRPD or the recently upcoming Raman transmission spectroscopy [4] are methods of choice.

Due to the upcoming use of multivariate data analysis also in quantification methods using mid-infrared spectroscopy, it is possible to evaluate infrared spectroscopic methods for the quantification of different polymorphs in pharmaceutical materials. In general, the IR spectrometer supplier evaluates software tools to create such methods, e.g. the Spectrum Quant or the adulterant screen application of Perkin Elmer [7, 8]. In contrast to the use of IR spectroscopy method as identification tool, the quantification methods have to be validated regarding the ICH guidelines Q2 "Validation of analytical procedures: text and methodology" [9].

4.4.1.3 Application of IR Microscopy-Imaging Methods in Drug Development

Regarding ICH Guideline Q6A, the evaluation and validation of a quantification method for the polymorphic forms of a drug substance in a drug product are only necessary if a surrogate method like the dissolution test does not provide adequate control if polymorph ratio or polymorphic form changes [10].

State-of-the-art methods for the quantification of drug substance polymorphs in drug products are XRPD or Raman transmission spectroscopy.

Beside the possibility to monitor polymorphic changes and measure the distribution of a drug substance in a drug product like tablets or multiparticulates, the FTIR-imaging technique could also be used as a relative quantification method.

FTIR microscopy is an established method for the chemical identification and for visualizing the distribution of a certain substance in a complex sample matrix. With different detector setups like the point-by-point detection, detectors with linear array technique, and the advanced focal plane array detectors (FPA), it is a mapping or imaging technique that allows the analyst to investigate large sample areas with a high spatial resolution within a short period of time. The detectors consisting of a matrix of 128 × 128 detector elements are able to detect up to 16 384 single spectra simultaneously or map a large area line by line in a very short time. Data can be collected in reflectance, transmission, or ATR mode (Figure 4.4.4) [11].

4.4.1.3.1 Spatial Resolution

The lateral resolution is limited by the light diffraction. The smallest distance (δ) at which two points of a sample can be separated is calculated regarding the formula:

$$\delta = 0.61\lambda/\mathrm{NA} \tag{4.4.1}$$

λ is the wavelength of the light and NA is the numerical aperture of the objective.

The spatial resolution in an IR image therefore depends on the numerical aperture of the mirror objectives in the FTIR microscopes and the wavelength (λ) of the irradiated light (mid-infrared region: 2.5–25 μm). Besides that the resolution also

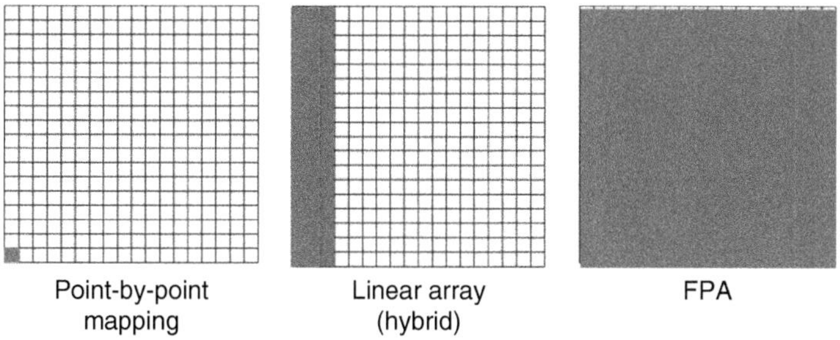

Figure 4.4.4 Detector setups. Source: Tisinger [11]. © 2016 Agilent Technologies, Inc.

depends on the specific detector and the used sampling technique. For that reasons, the resolution of an image can be better than 1 μm [11].

4.4.1.3.2 Measurement Setups

The absorptions in the spectral range of infrared spectroscopy are relatively strong. For this reason, it is important to limit the amount of material interacting with the incoming radiation. If the sample thickness is not too large, it is possible to measure the whole field of interest in transmission mode. Application is found in the plastics industry for the investigation of samples having a thickness of about 10–100 μm and the successful use depends strongly on the material's absorption coefficient for IR radiation. But in the special case of an opaque and thick sample matrix of a tablet, this is not possible.

Different instrument setups provide different approaches to handle this problem. The best way to minimize the amount of material that interacts with the infrared radiation is the ATR sample technique. Nevertheless, with respect to the size of the used ATR crystal, the image size is limited, while the spatial resolution of this relatively small image is quite good. Resolutions down to 1 μm per pixel are possible. To detect larger fields of interest, instrument suppliers provide different solutions. Some instruments provide the possibility to measure large images by near-infrared spectroscopy (NIR) in diffuse reflection sample mode, e.g. FTIR-NIR Spotlight 400 microscope by Perkin Elmer [12].

The advantage is that the absorption of near-infrared radiation is weaker than the absorption of mid-infrared radiation and no total absorption occurs. However, the interpretation of NIR spectra is only possible by multivariate data analysis, e.g. principal component analysis (PCA). That means that no direct identification of a sample spectrum by comparison with a spectral database is possible but the comparison of a sample spectrum with a set of various reference spectra. Therefore, the best way to identify a certain polymorph in a tablet matrix is to measure a large image by NIR in diffuse reflectance mode to identify fields of interest and then measure these positions subsequently by the ATR sample mode with mid-infrared spectroscopy. The resulting mid-infrared spectra can then be checked against a collection of spectra in a database in order to identify a polymorphic form or specific compound, like a salt, solvate, or a cocrystal.

Another approach is the laser direct infrared (LDIR) chemical imaging technique that enables obtaining reliable high-definition chemical images of constituents on a surface. Quantum cascade laser (QCL) technology is coupled with scanning optics that provides images and spectral data in short time. The method can be used in reflectance mode to quantify solid forms of a drug substance, like polymorphs, salts, or cocrystals in a sample [13]. Spatial resolution of images recorded in ATR sampling mode is down to a pixel size of 0.1 μm [14].

"Infrared light from the QCL is directed to the sample. The Infrared light diffuse reflected by the sample is then directed to the detector via either of the selected optical paths. In reflectance mode, infrared light from the laser is focused by a fast scanning objective system that is rapidly scanned back and forth. Concurrently, the sample is automatically moved in a perpendicular plane, and the infrared light reflected by the sample is directed back to a thermoelectrically cooled mercury cadmium telluride (MCT) detector. This process yields a two-dimensional molecular image [14].

After that certain positions of particular interest can be measured by ATR sample mode. In ATR mode infrared light from the laser is directed onto a scanning mirror that rapidly moves the light across the fixed ATR element that is in contact with the sample. Totally internally reflected light is directed to the MCT detector. The spatial resolution of the ATR images are down to a pixel size of 0.1 μm. [14]." Recorded infrared spectra can be compared with a collection of spectra in a database.

4.4.1.3.3 Case Studies

Component Detection by NIR-IR-Mapping The ICH Guideline Q6A [10] requires, besides the testing of other quality attributes of a particular solid dosage forms, the investigation of API content uniformity and the dissolution profiles during the development of the drug product. The high-pressure liquid chromatography (HPLC) test proves the homogeneous distribution of the API in the dosage form, e.g. a tablet. The dissolution test proves if the dissolution shows the expected profile. These tests are typically addressed in Phase II of the development phase.

The following example illustrates a case study of a solid dosage form, which in this case is a tablet of a nondisclosed API. Some unexpected variations in content uniformity tests and dissolution profiles were observed during the development phase. A detailed analysis should clarify if the tablet composition is homogeneous regarding the drug substance-excipient distribution and if a change in the solid form might have been occurred [15]. First, an NIR imaging of the full tablet was measured in order to get an overview (Figure 4.4.5). After that a PCA is performed and with the single score plots Figure 4.4.6, it is possible to identify regions of interest on the image, e.g. if there are any hot spots of the API in the tablet which indicate an inhomogeneity of the tablets. Scores are the coordinates of the samples in the new principal component or factor coordinate system. A score plot shows the relationship among the sample images in plots (Figure 4.4.7). In addition, it is possible to measure IR spectra by ATR at these identified positions on the tablet

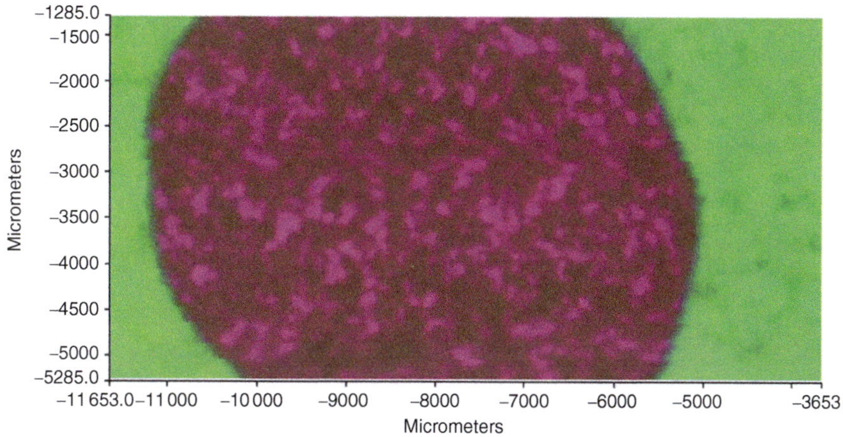

Figure 4.4.5 Example of an NIR imaging of a tablet. Source: Dagmar Lischke.

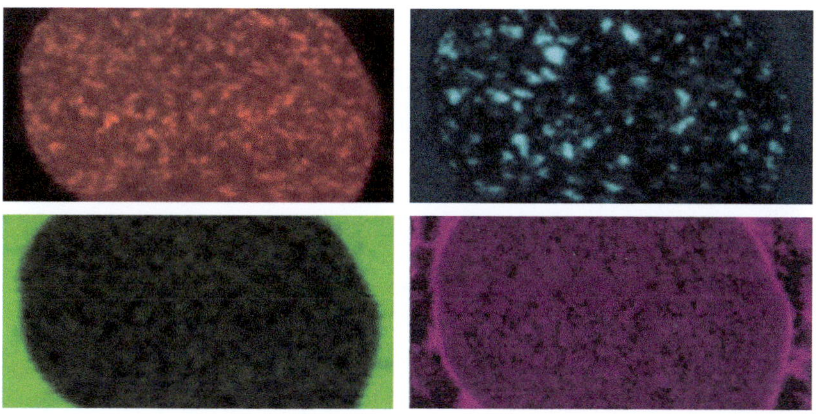

Figure 4.4.6 Single score plots of the multivariate analysis of the tablet. Source: Dagmar Lischke.

surface. The resulting mid-infrared spectra (Figure 4.4.8) can then be checked against a collection of spectra in a database in order to identify a polymorphic form or specific compound, like a salt, solvate, or a cocrystal.

Detection of Polymorphic forms of Carbamazepine in a Tablet The following example is taken from an Agilent application note [13]. It briefly explains the CBZ polymorphic system that was investigated in a tablet by using LDIR imaging. Original references can be found in [13].

> "CBZ is an anticonvulsant and mood stabilizing drug that is known to exist as different polymorphs. Of the four crystal forms (III > I > IV > II; room temperature stability order), only form III is known to have therapeutic

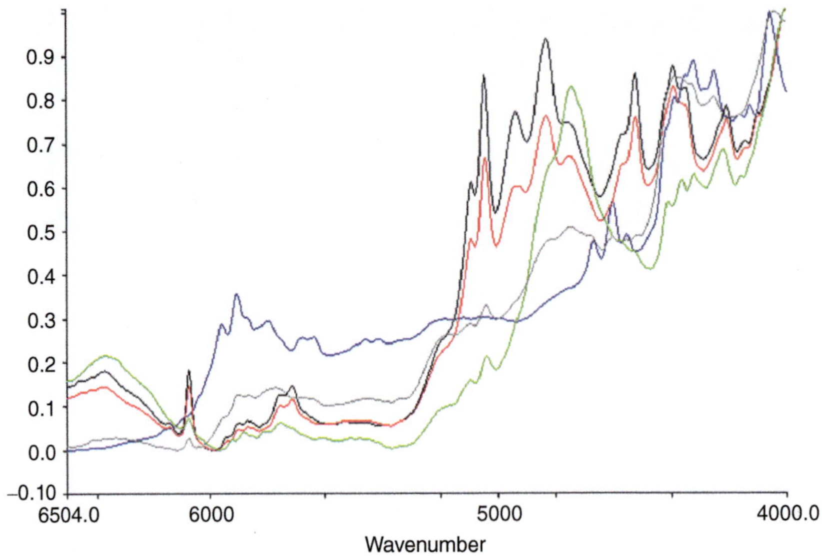

Figure 4.4.7 Corresponding NIR spectra of calculated scores.

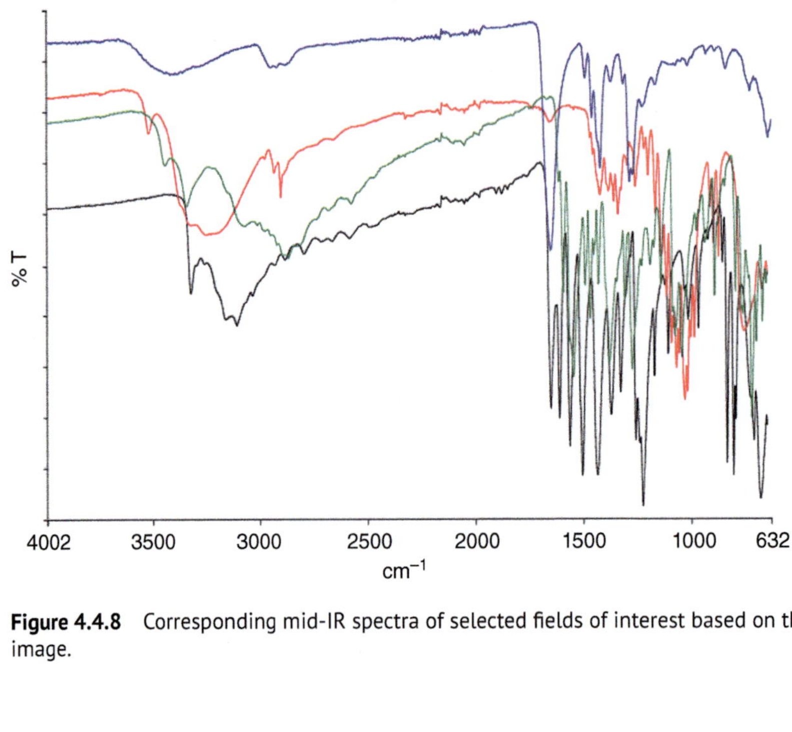

Figure 4.4.8 Corresponding mid-IR spectra of selected fields of interest based on the NIR image.

effects [13]. Detecting and understanding the formation of non-therapeutic polymorph I is essential when developing CBZ solid dosage forms. Forms I and III can be rapidly distinguished and mapped using LDIR imaging. Library spectra are first acquired for these two polymorphs and the cellulose excipient in the sample. The instrument software then generates a rapid imaging method by selecting the key diagnostic wavelengths for each of the three constituents." (Figure 4.4.9).

In a next step, the method is used to image CBZ polymorphs across an entire tablet. Images of 13 mm tablets with 10 μm pixel size were obtained in 27 minutes (Figure 4.4.10). Two formulations were studied, one containing 5.2% of form I and 15.4% of form III, while the other contained 15.3% of form I and 5.5% of form III (percentages are by weight). The remainder was cellulose in both cases. "The measured surface concentrations, in which the density of the polymorphs is not considered, showed excellent correlation with the known percentages by weight" [13]. This is shown in Figure 4.4.11, the chemical distribution of the three major constituents can be individually displayed [13].

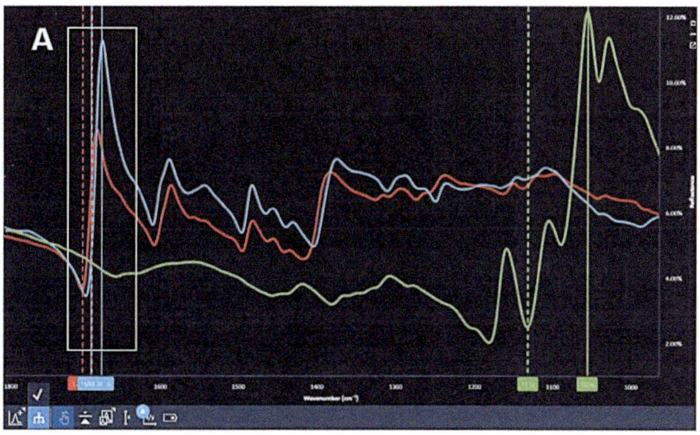

Figure 4.4.9 Diffuse reflectance infrared spectra of cellulose (green line), carbamazepine polymorph I (red) and III (blue). Source: Agilent Technologies, Inc. [13].

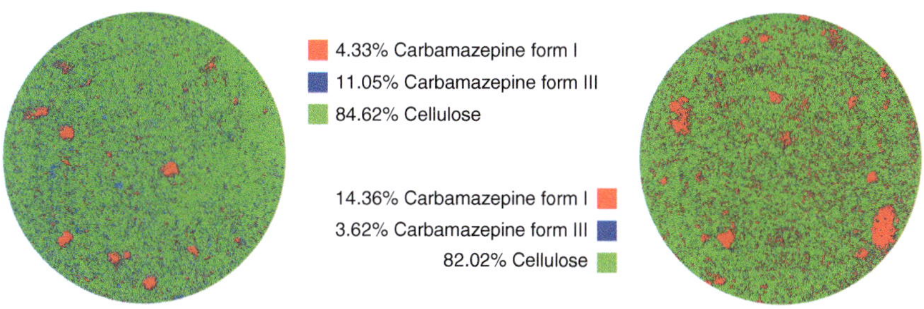

Figure 4.4.10 Classification image of a 13 mm tablet showing the distribution of cellulose, carbamazepine polymorph I and III. Source: Agilent Technologies, Inc. [13].

Figure 4.4.11 From left to right: Individual chemical maps of carbamazepine polymorph I, polymorph III and cellulose. Source: Agilent Technologies, Inc. [13].

4.4.1.4 Conclusion

IR spectroscopy is a fast and cost-efficient identification method during the development phases of new active pharmaceutical ingredients and the resulting solid drug products, although it is not the gold standard amongst the identification methods in solid state and especially polymorphism investigation due to its insensitivity to small variations in the crystal lattice. Nevertheless, the IR spectra of different polymorphs, solvate forms, salts, and cocrystals can be used to characterize new active pharmaceutical ingredients, intermediates, and excipients for patent applications or other regulatory documents. If it is possible to distinguish relevant solid forms by IR spectroscopy, the technique is a good alternative to the more time-consuming and expensive XRPD method particularly for release testing of marketed products. IR spectroscopy is a widespread identification method throughout all quality control laboratories in the pharmaceutical industry, and also non-specialist analysts are able to perform a measurement. The interpretation of the results can be simplified by providing routine procedures and databases that are implemented into daily business procedures by technical and analytical experts in the field of IR spectroscopy.

Especially in the drug development phase, IR spectroscopy is an efficient tool to test and investigate content uniformity or dissolution performance by using the FTIR-microscopy instruments. Also the monitoring of polymorphic phase transitions after or even while processing drug substance or drug product is possible.

List of Abbreviations

API	active pharmaceutical ingredient
ATR	attenuated total reflection
FTIR	Fourier transform infrared spectroscopy
HPLC	high-pressure liquid chromatography
IR	infrared spectroscopy
LC–MS	liquid chromatography coupled with mass spectroscopy
MCT	mercury cadmium telluride
NDA	New Drug Applications
NMR	nuclear magnetic resonance spectroscopy

NIR near-infrared spectroscopy
PCA principal component analysis
XRPD X-ray powder diffraction

References

1 (a) EDQM (2019). European pharmacopoeia Version 9.7, chapter 2.2.24; (b) USP 40 (2017). Chapter <854>, Physical test <854> Mid-infrared spectroscopy.

2 Perkin Elmer Life and Analytical Sciences (2005). FT-IR Spectroscopy – Attenuated Total Reflectance (ATR): Technical Note.

3 Leica, EM Rapid – Pharmaceutical Milling System for Pills (2017). Leica microsystems CMS GmbH, Ernst-Leitz-Strasse 17–37, D-35578 Wetzlar (Germany). https://downloads.leica-microsystems.com/Leica%20EM%20RAPID/Brochures/EMRAPID_Brochure_06_17_EN.pdf (accessed 07 March 2019).

4 Agilent LDIR Sample Planer https://www.agilent.com/en/product/molecular-spectroscopy/ldir-chemical-imaging-spectroscopy/ldir-accessories/ldir-sample-planer Kiefer, J. (2019). Transmission Raman Spectroscopy for Pharmaceutical Analysis, American Pharmaceutical Review (online). https://www.americanpharmaceuticalreview.com/Featured-Articles/358425-Transmission-Raman-Spectroscopy-for-Pharmaceutical-Analysis (accessed 12 August 2019).

5 (a) Limwikrant, W., Nagai, A., Hagiwara, Y. et al. (2012). Formation mechanism of a new carbamazepine/malonic acid cocrystal polymorph. *International Journal of Pharmaceutics* 431 (1): 237–240. https://doi.org/10.1016/j.ijpharm.2012.04.027. (b) Cyrański, M.K., Jamróz, M.H., Rygula, A. et al. (2012). On two alizarin polymorphs. *CrystEngComm* 14 (10): 3667–3676. https://doi.org/10.1039/C2CE06063A. Supplementary information: http://www.rsc.org/suppdata/ce/c2/c2ce06063a/c2ce06063a.pdf (accessed 12 August 2019).

6 EMA (2000). Decision tree #4 in ICH Q6A specifications: test procedures and acceptance criteria for new drug substances and new drug products: chemical substances (CPMP/ICH/367/96), Step 5.

7 Spectrum Quant – Quantitative Analysis Software (2014). Product note FT-IR and NIR spectroscopy, PerkinElmer, Inc. https://www.perkinelmer.com/lab-solutions/resources/docs/PRD_Spectrum_Quant_Quantitative_Analysis_Software.pdf (accessed 07 March 2019).

8 Adulterant Screen for Spectrum Software (2014). Product note FT-IR and NIR spectroscopy, PerkinElmer, Inc. All rights reserved. Application no-011778A_01. https://www.perkinelmer.de/lab-solutions/resources/docs/PRD_Adulterant_Screen_Spectrum_Software_011778A_01.pdf (accessed 07 March 2019).

9 EMA, ICH Guideline Q2 (R1) (1995). Validation of analytical procedures: text and methodology, step 5, note for guidance on validation of analytical procedures: text and methodology (CPMP/ICH/381/95).

10 EMA, ICH Guideline Q6A (2000). Specifications: test procedures and acceptance criteria for new drug substances and new drug products: chemical substances (CPMP/ICH/367/96), Step 5.

11 Tisinger, L. (2016). Industrial applications of FTIR imaging microscopy, Agilent Technologies, Inc. https://www.agilent.com/cs/library/eseminars/public/Chemical_Imaging_for_Industrial_Laboratories.pdf (accessed 07 March 2019).

12 Spotlight™ 400 FT-IR and 400N FT-NIR Imaging Systems (2013). 007713C_12, PerkinElmer, Inc. https://www.perkinelmer.com/lab-solutions/resources/docs/BRO%20Spotlight_400_FT_IR_400N_FT_NIR_Imaging_Systems.pdf (accessed 07 March 2019).

13 Agilent Technologies, Inc. (2018). Rapid discrimination and classification of polymorphs using the Agilent 8700 laser direct infrared (LDIR) chemical imaging system, application brief pharmaceuticals, 5991-7512EN. https://www.agilent.com/cs/library/applications/5991-7512EN_App_Brief_LDIR_Polymorph.pdf (accessed 07 March 2019).

14 Agilent 8700 LDIR Chemical Imaging System (2019). Agilent Technologies, Inc. https://www.agilent.com/cs/library/brochures/5994-0275EN_8700_LDIR_Chemical_Imaging_System_brochure.pdf (accessed 07 March 2019).

15 Lischke, D. (2017). Presentation at Perkin Elmer user meeting "Material 2017", 04.Mai 2017, Grünenthal GmbH, FT-IR/NIR Spotlight Imaging System, "Finde den Fehler", Einsatz in der pharmazeutischen Entwicklung.

4.5

Transmission Raman Spectroscopy – Implementation in Pharmaceutical Quality Control

Meike Römer

Grünenthal GmbH, Zieglerstr. 6, 52099 Aachen, Germany

4.5.1 Raman Spectroscopy – From Research to Broad Applications in Industry

4.5.1.1 Objective

The aim of this chapter is to give practical guidance on Raman applications in pharmaceutical industry not to render on theoretical topics on Raman spectroscopy. Therefore, it is focusing on explaining spectroscopic and multivariate techniques to those having experiences in univariate analysis.

4.5.1.1.1 History

The history of Raman spectroscopy dates back roughly a century to the middle of 1920, when C.V. Raman and K. S. Krishnan reported "a new radiation," which was subsequently called Raman effect [1]. Already in the 1930s, Raman effect was used for nondestructive chemical analysis but was, due to technical implications, surpassed by infrared (IR) spectroscopy. In the 1980s, after various technical improvements, such as double monochromators and notch filters, the first Fourier transform Raman instrument was developed. Raman spectroscopy finally gained more importance again. Since then new setups and areas of application have been developed, ranging from microscopes [2], fiber optic probe systems [3], the development of surface-enhanced Raman spectroscopy [4, 5], transmission [6], and spatially offset Raman systems [7], compiled in a comprehensive review focusing on pharmaceutically relevant applications [8].

4.5.1.1.2 Introduction

Raman spectroscopy, as various other techniques belonging to the field of vibrational spectroscopy, is a nondestructive and noninvasive measurement and requires minimal or no need for sample preparation. Raman spectroscopy can be used in combination with IR spectroscopy, as both techniques provide complementary information about the molecular state and its changes (Table 4.5.1). Infrared spectroscopy shows changes in dipole moment in a molecule, thus molecules are IR active if the

Solid State Development and Processing of Pharmaceutical Molecules: Salts, Cocrystals, and Polymorphism, First Edition. Edited by Michael Gruss.

Table 4.5.1 Comparison of FT-IR, NIR, and Raman spectroscopy.

	FT-IR	NIR	Raman
Technique	Absorption	Absorption	Emission
Spectral region	4000–400 cm^{-1}	12 500–4000 cm^{-1}	4000–4050 cm^{-1}
Radiation source	Polychromatic (tungsten)		Monochromatic (laser)
Spectral selectivity	High	Low	High
Detected vibration	Fundamentals	Overtones	Fundamentals
Sample volume	Small	Large	Dependent on method
Sample preparation required	Yes	No	No
Water sensitive	Yes	Yes	No

dipole changes upon vibration. The larger the change in dipole moment, the stronger the intensity of the absorption band. For being "Raman active," a molecule requires a change in polarizability thus inducing a dipole, during the vibration or rotation. This dipole emits or scatters light at the optical frequency of the incident light wave. Often infrared active molecules with a dipole are not Raman active and vice versa.

4.5.1.1.3 The Raman Effect

When a molecule absorbs energy (ν_0), in this case monochromatic light from a laser source (e.g. 785 nm or 830 nm), it may enter a higher virtual energy state. The molecule can then relax back to the ground electronic state by emitting the same amount of energy (ν_0) that was taken up resulting in elastic scattering (Rayleigh scattering). In addition, a small part returns to a different energy level than the incident one, so called inelastic scattering. Depending on the incident energy state of the molecule, this inelastic scattering can lead to either a higher energy state ($\nu_0 - \nu_v$) of the molecule, the so-called Stokes shift, or to a lower energy state ($\nu_0 + \nu_v$), the anti-Stokes shift (Figure 4.5.1). Usually, in Raman spectroscopy Stokes shifts are reported because their occurrence is of greater likelihood. 10^8 excitation photons

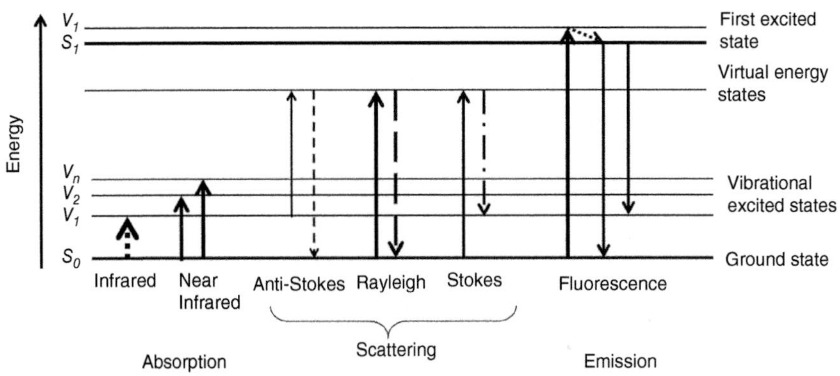

Figure 4.5.1 Jablonski energy diagram (simplified).

are required to generate a single Raman photon but depending on the structure of the molecule, fluorescence can obscure the Stokes signals. Anti-Stokes shifts do not suffer from this interference, but require molecules to be in a vibrational excited state, a situation that is negligible at room temperature.

4.5.2 Analytical use of Raman Spectroscopy for Pharmaceutical Purposes

As Raman spectroscopy is readily applicable in-line and offers the possibility for obtaining real-time data, it has been frequently used to gain insight into pharmaceutical unit operations on a molecular level [9]. Water is a poor Raman scatterer; thus, Raman spectroscopy is an ideal method for monitoring processes in which water signals may overwhelm spectra obtained with other spectroscopic methods. It can be used to detect hydrate formation [10] determining the conversion kinetics between polymorphic forms in aqueous environment [11], assessing the interaction of water with polymers [12], and detecting solid-state changes during fluid-bed drying [13]. Trace crystallinity of amorphous active pharmaceutical ingredients (APIs) has been investigated after milling [14] and after compression of a model tablet formulation to a compact [15]. The application of Raman spectroscopy has expanded to characterizing process-induced transformations of an API in final tablets [16] and quantifying API content in tablets [17]. Bulk solid dosage forms have been examined in transmission mode [6, 18], which was found to be more suitable for quantitative analysis than reflection mode due to the larger sampling volume. The coating variability was also investigated with Raman spectroscopy [19] as well as coating thickness [20]. Raman spectroscopy was found to be an effective method for detecting counterfeit pharmaceutical products [21].

Though there are various setups and different areas to apply Raman spectroscopy to, this chapter will focus on the Raman application which is of increasing importance for the pharmaceutical manufacturing, transmission Raman spectroscopy (TRS), based on fiber optic probes (Figure 4.5.2). It will also provide advice and share gained experience, how to implement such systems, and what are probable pitfalls and hurdles.

4.5.2.1 Transmission Raman Spectroscopy (TRS)

With the discovery of TRS, the typical problems arising from diffuse reflectance setups like subsampling issues (see Figure 4.5.2a) and lack of reproducibility of the measurements due to variability in distance from the laser to the sample providing difficulties to hit the focal point of the laser onto the sample.

The TRS 100 (TRS100, Agilent Technologies UK Ltd.) is a transmission Raman tabletop instrument with a high degree of precision, a defined distance between laser and detector. It is very suitable for pharmaceutical analysis due to a multi-well plate sample holder and using an excitation laser wavelength of 830 nm.

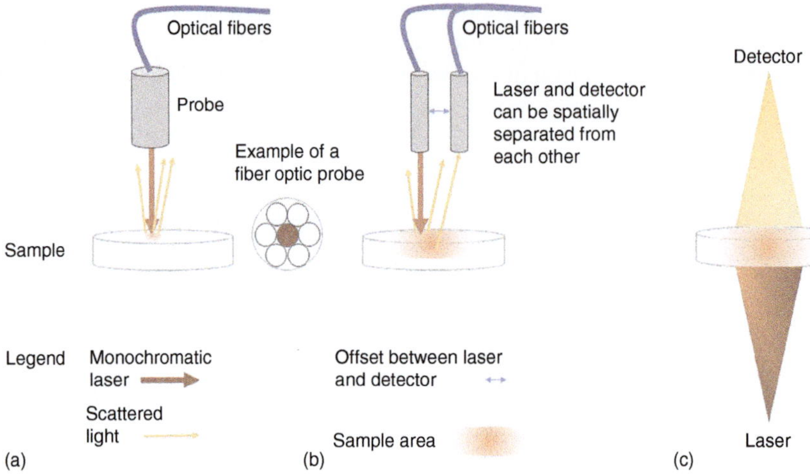

Figure 4.5.2 Raman spectroscopic configurations with (a) typical diffuse reflectance probe; (b) spatially offset Raman, and (c) transmission Raman set-up.

TRS has several advantages for the analysis of pharmaceutical samples, including

- direct measurement of bulk sample without destruction of sample, compared to high performance liquid chromatography (HPLC) analysis, which requires significant preparation steps and a suitable calibration standard
- robust process, not requiring specifically trained staff for routine analysis
- high degree of automation whereas chromatography is often manual, labor intensive and requires practical experience, making it a potential source of error and adding cost
- fast process (approx. 15 minutes per batch analysis, e.g. assay)
- does not require any solvents or consumables
- low costs per test
- high-throughput, multi-well sample holder tray
- lower maintenance costs compared to classical liquid chromatography (LC), due to the higher efficiency, one TRS100 can replace several LCs.

TRS can be used for content uniformity tests, assay quantitative analysis, and for drug product identification. Identity testing using a Raman spectrometer covers the entire drug product API and excipients in total. Whereas as in HPLC identity testing, only the respective API is detected. It is generally used for solid dose formulations, including tablets and capsules, but can also be used for powders and other form factors.

4.5.2.1.1 Principles of Transmission Raman Spectroscopy

Transmission Raman spectroscopy uses a large and adjustable spot laser beam, focusing the laser either to 2, 4, or 8 mm laser spot size, to illuminate one side of a tablet or a capsule. While the laser light travels through the sample, the light scatters through the entire sample. Occasionally, a Raman photon is generated, resulting in a signal, which is affected by the photon's pathway through the sample volume. The detector widths, on the sample side opposite to the laser, can be adjusted to fit best to the laser spot size and the nature of the sample. Due to the principle of the measurement, the single spectrum obtained is representative of the whole inner volume of that particular material. TRS is, consequently, not just a surface technique, like conventional Raman, which only illuminates one side of the tablet and collects light from the same side (see Figure 4.5.2). Thus, the benefit of transmission over conventional Raman is that it works through capsule shells and tablet coatings. This capability allows the TRS technique to work with several different types of samples and presentation methods, including powders in polyethylene bags, well plates, and various oral solid dosage forms. Typical subsampling issues of the diffuse reflectance probes do not occur.

Different types of samples, including capsules, tablets, and powders, can be loaded onto a tray into the instrument then moves the material into the sample beam, with samples analyzed sequentially. Each sample typically takes only a few seconds to be analyzed so a content uniformity test of 10 tablets only takes a few minutes.

For the development of a TRS method, a validated liquid chromatography method is required to provide measured values of a primary method to the predicted values of the secondary TRS method. Liquid chromatography relies on a standard sample detector response providing a ratio between a known reference standard and an unknown sample.

An HPLC chromatogram is separating the peaks as a function of the elution time, taking several minutes to elute the analyte. The signal intensity correlates directly to the concentration of the analyte in question.

In a Raman spectrum, the respective Raman shifts are produced simultaneously. The result is the generation of a spectrum, Figure 4.5.3. Due to the complexity of the spectrum, multivariate techniques are necessary to extract information from the data to produce quantitative information.

Though the differences of both, HPLC and TRS, methods are pronounced, the general procedure is rather similar: starting off with a solid sample, e.g. a tablet. In case of TRS, the sample is simply placed inside the instrument and within a few seconds the multivariate spectral information is generated. Whereas with HPLC, quite lengthy sample preparation is needed before the separation process and the analytical measurement can start. With Raman spectroscopy, data preprocessing is performed focusing primarily on chemical information and minimizing the effect of the physical interactions, while HPLC uses a column and separates the different sample components physically by polar interactions between a mobile and stationary phase. Subsequently, the TRS data dimensionality (i.e. complexity) is reduced by multivariate methods and illustrated as scores plots and latent variables [22]. The scores present variations in data, whereas latent variables give feedback to the wavelengths that originate the effects in the data. The HPLC data analysis approach requires peak detection and integration.

The calibration in case of TRS uses some sort of multivariate data analysis, like partial least square regression, while for HPLC, achieving a peak correlating to only

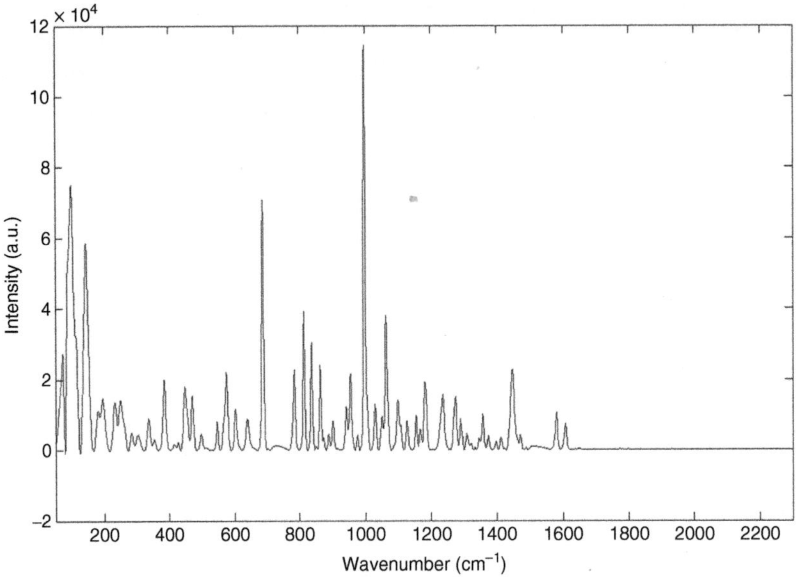

Figure 4.5.3 Transmission Raman spectrum of pure Tramadol.

the analyte in question, a linear regression is best. The measurement terminology is a little different for both techniques. For TRS, it is a called "prediction of concentration of *X*" based on a mathematical model, whereas for HPLC, it is a "measurement of concentration of *X*" where *X* is the analyte of interest such as excipient or API. However, the concentration derived from the chromatogram is also based on a mathematical model, i.e. the linear regression analysis. Even though their descriptions are slightly different, they are actually very similar because they both use regression curves, one in a multivariate and the other one in a univariate way.

Though Raman spectroscopy is not following the Beer's law, it is possible to establish a relationship between the concentration and the measured Raman intensity. The simplification of the Placzek's Equation [23, 24] for Raman scattering intensity results in the following equation:

$$I_\lambda = \sigma LCI \tag{4.5.1}$$

I_λ = Raman intensity
σ = Raman cross section
L = path length
C = concentration
I = instrument parameters

To prepare a predictive model, an experiment is designed by building a model of an output spectrum where the API and excipients change as a function of, i.e. with respect to their concentration. The benefit is that once a method is developed; there is no need to run a reference standard again, as required for LC methods.

4.5.2.1.2 A Practical Guide to a Successful Business Case

For a business case justification, some hurdles must be overcome. HPLC and UHPLC are pharmaceutically acknowledged standard methods. The requirements for method validation and method development for the chromatographic methods are well documented in the public domain, processes are established in industry, and requirements as well as acceptance criteria are known. In contrast, Raman spectroscopy, in particular TRS, is a new technique, especially for the abovementioned fields of pharmaceutical applications. Therefore, it has to be considered that new procedures need to be established, staff has to be trained in the methodologies, and experience has to be gained.

Nevertheless, the following parameters are in favor of investing into a transmission Raman system:

- The higher the costs of labor, the more savings will be generated by a switch to a rapid analytical method.
- Substituting a current HPLC method is providing more potential for savings compared to replacing a ultra-high performance liquid chromatography (UPLC) method.
- The higher the number of batches, the higher the savings.
- Applicable for assay, content uniformity, and identity testing.
- Reduction of equipment number and subsequently footprint and maintenance costs can be also considered: Agilent (formerly Cobaltlight) claims that one TRS100 can replace up to eight HPLCs [25].
- Reduction of costs for organic solvents, waste, and columns.

Some points, however, are impacting a switch to a complementary method negatively:

- Costs for regulatory fees.
- Costs for method development, calibration, and validation.
- Limit of detection does not allow to detect and quantify related substances, thus for products for which related substances, assay, and content uniformity are determined in one LC method, a switch to Raman may not be financially beneficial.

The calculation of the business case is rather complex. It is not possible to eliminate the reference method completely as it is the primary method. Additionally, the reference method is still required for model maintenance. A summary of the most important factors is shown in Table 4.5.2.

Figure 4.5.4a presents an example of a site with low volume batches (~50 per year), low labor costs, and low regulatory costs. In this situation, the payback time for TRS introduction is roughly 4–5 years. In this case, the low number of batches per product means a higher impact of method development and validation costs. This limits on the one hand side the number of products, which can be switched per year on the other hand it also means lower savings while the development costs are remaining. Realistically, if a single team has established the procedure and all functions are trained, one can calculate with 3–4 products to be switched per year from HPLC to TRS application excluding regulatory efforts and timelines.

Table 4.5.2 Comparing relevant costs of shifting products to TRS100 from HPLC analysis for assay, content uniformity, and identity testing for marketed products.

	HPLC	TRS100
Pieces of equipment	5–8 (depending on average time required for analysis)	1
Method development and validation	No costs since HPLC method should be readily available	TRS100 method
Regulatory fees	None	Average costs times number of countries
Required investment	None	Equipment and training
Routine analysis costs	Maintenance costs: costs times pieces: costs per 100 batches incl. Men hours, filters, columns, reagents/solvents, and standards	Maintenance costs + 10% of HPLC maintenance costs: costs per 100 batches incl. Men hours + 10% of the costs for HLPC routine for revalidation and lifecycle management

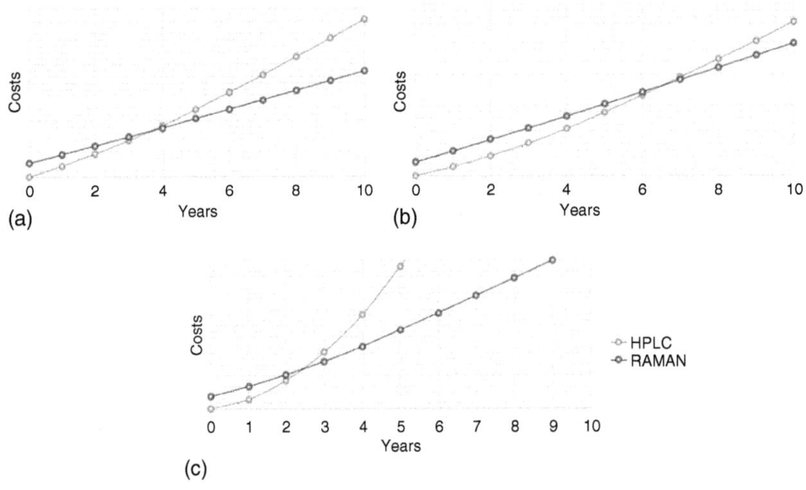

Figure 4.5.4 Return of investment for (a) low volume products, low labor costs, and low regulatory costs; (b) effect of high regulatory costs; (c) medium volume products, medium regulatory costs, and high labor costs.

In a comparable case as presented in Figure 4.5.4a but in contrast having high regulatory costs is shown in Figure 4.5.4b. The higher costs have a negative impact and may turn the business case to have a break even after six years.

If the scenario changes to medium volume products (>100 batches per year), with high labor costs, and medium regulatory costs, the payback period decreases from four years to about 2½ years (Figure 4.5.4c). The impact of labor cost is not that prominent anymore if one can reduce the men hours from 150 minutes required for an HPLC result to only 15–20 minutes for TRS.

Additional considerations contributing to a business case are as follows:

1) TRS can also analyze physical–chemical information like crystallinity, amorphous state, solvate, and polymorphic forms.
2) Physical information such as size distribution and tablet hardness can be assessed.
3) Chemical information on the API and all excipients that have been calibrated can be derived.
4) The level of expertise required for a staff member to perform a Raman routine analysis is low.
5) Highly trained laboratory staff can concentrate on more challenging and more value adding tasks.
6) Outsourced analysis may be insourced again due to freed capacity of staff and equipment.
7) Manufacturing schemes may be affected, since the number of batches per campaign may be increased.

4.5.3 Transmission Raman Spectroscopy – Another Practical Guide

Different to liquid chromatographic analytical techniques, TRS is measuring the entire tablet with all of its components. The spectral features will always be influenced by the manufacturing process. This means also a different setup of teams with respect to skills is required. The following areas of expertise are recommended within a core team:

- Spectroscopy
- Multivariate data analysis
- Analytical method validation
- Chromatographic techniques
- Manufacturing processes

The approach has to be interdisciplinary, and all involved functional experts have to trained to a defined extend, e.g.

- in the reason why specific samples have to be prepared for development and validation
- variations including their potential effect on the individual spectrum must be evaluated

These functional experts have to be involved in interdisciplinary discussions to understand potential risks of their tasks for following projects, in particular concerning the quality and robustness of the TRS method.

A first step in ensuring interdisciplinary understanding is to use clear definitions and wording, e.g. define the process of developing a spectroscopic and multivariate method as presented in Table 4.5.3. Thus, there has to be differentiation between both parts.

Table 4.5.3 Definition and terminology of the different steps in transmission Raman analytical method development and validation.

Phase	Consists of	Definition
Transmission Raman analytical method development	Spectroscopic method development	Development of suitable settings of the Raman spectrometer for the respective product
	Model development	Establishment of the model based on DoE samples covering a large range of required chemical and expected physical variability
	Model calibration	Inclusion of additional production samples to include process variability
	Model validation	Testing of unknown samples (e.g. historic production samples and artificially produced samples) to test the performance of the model
Transmission Raman analytical method validation	ICH validation	Accuracy, precision, specificity, linearity, range, and robustness

Additionally, it is important to ensure that the samples are named consistently throughout all analytical applications and software systems used. If this is not possible, sample information must be tracked in a traceability matrix, preferably also including information about the process variability of the sample.

4.5.3.1 Evaluation Phase

4.5.3.1.1 Prefeasibility Evaluation

A first and essential part of the process is the feasibility trial. Not all drug products are suitable for a switch to TRS. The difficulty is that the success of the feasibility is not only dependent on the API being Raman active but also there are various factors that have an effect.

One can establish a decision tree with all known factors affecting the technical probability of success for a drug product. These decision trees can be used for less experienced functional experts, such as interfaces from quality control or new members of the team, to evaluate the different products. An example of such a decision tree is presented in Figure 4.5.5.

First, general considerations are checked, see Figure 4.5.5, such as API concentration and colorants, e.g. in capsule shell or coating. The next step is the establishment of a decision tree concerning the physicochemical properties of the API to assess the quality of the Raman spectrum. On the one hand, if the API is known to be Raman active and therefore resulting in a good-quality Raman spectrum as shown for Tramadol in Figure 4.5.3, the chances of success are high. Still fluorescence of excipients may interfere. On the other hand, if the API is Raman active but only a

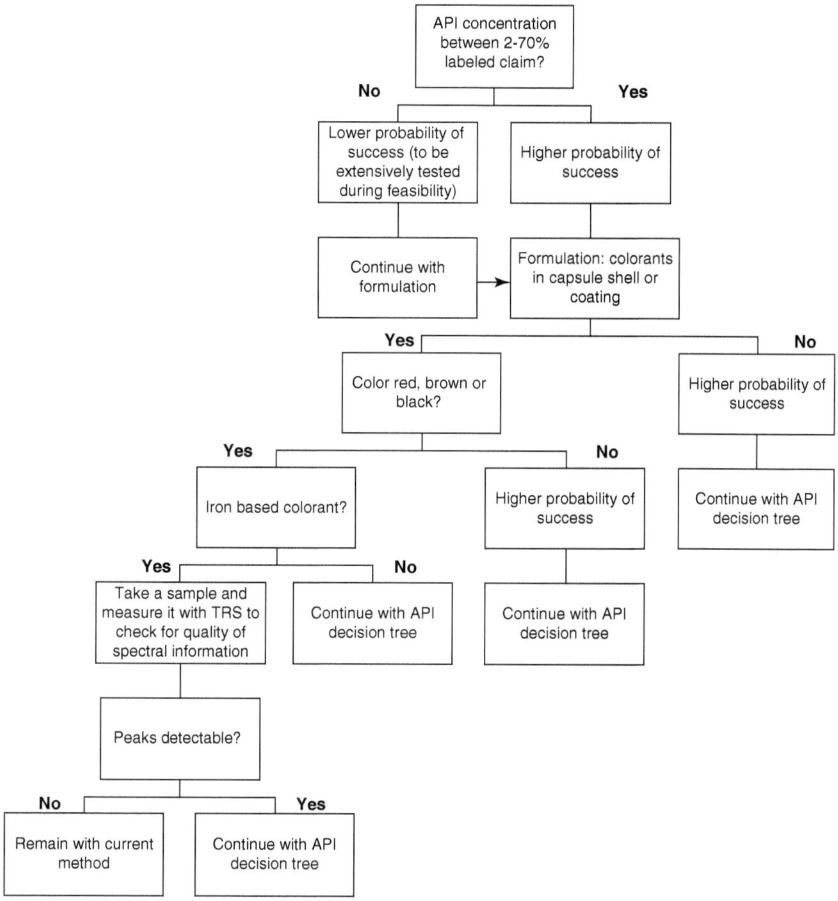

Figure 4.5.5 Decision tree for main factors: API concentration and colorants.

weak scatterer, this does not necessarily mean that the product cannot be switched. Therefore, a thorough evaluation during feasibility is essential.

The next factor to assess is the excipients.

Coming from univariate analysis, it is less known that the signal of the excipient matrix is required for robust model as is a signal of good quality of the API. Transmission Raman spectroscopy is a relative technique. This means that the robustness of the multivariate model depends on the positive correlation of the API as it depends on the negative correlation of the excipient matrix components. The lack of negatively correlating peaks results in either a less robust model or in the final conclusion that is better to remain with the current analytical method instead of switching to a TRS application. Typically, one of the many excipients is Raman active, so this is not a knockout criterion. But, if the matrix is only having a weak contribution to the spectral features, this has to be taken into the risk analysis and must be evaluated later during TRS method development and validation.

As one part of the prefeasibility study, the business case for each product to be switched should be considered.

The following questions can be assessed for this purpose:

1) Are switch of assay, content uniformity testing, and identity testing possible?
2) Are assay and impurities/dissolution testing different methods or measured in different sequences?
3) What are the current costs of the respective batch release tests in men hours, machine hours, and costs for consumables?
4) How many batches per year are produced?
5) How many market and stock keeping units are affected?
6) Is it possible to focus on European Union (EU)/US market first?
   (Pareto principle – low regulatory burden with highest amount of savings possible)
   Is it possible to split the material identification numbers to two bills of material: one for all other markets outside the EU and one for all batches produced for the EU market.

If savings are likely and theoretical considerations as well as preliminary tests on a product considered positively, the feasibility of the product can be tested.

4.5.3.1.2 Feasibility of a Product

During feasibility, the main focus is to confirm that the API signal within the formulation matrix can be identified. The next step is to carry out a design of experiment (DoE) for the calibration by varying all major components between 70% and 130%. Then, the measurements are carried out by plotting the spectra color coding with the respective concentration. This is exemplified in Figure 4.5.6, where the blue peak is equivalent to 70%, green is 100%, and the red is 130%.

The spectra of the feasibility test can already be used to for modeling using % [w/w] of API within the powder bag and also the % [w/w] for each of the excipients to find negatively correlated peaks.

It is clearly visible in Figure 4.5.7 that there is a correlation between the API concentration and the spectral information as to be seen in the measured versus predicted plot in the named figure. Additionally, there is the scores plot presented.

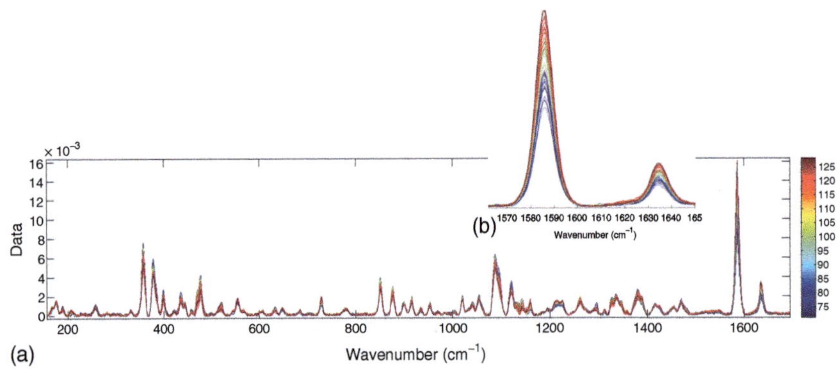

Figure 4.5.6 (a) Spectra of DoE blends of a product colored by the API concentration in % labeled claim; (b) magnification of wavenumber range 1560–1650 cm^{-1}.

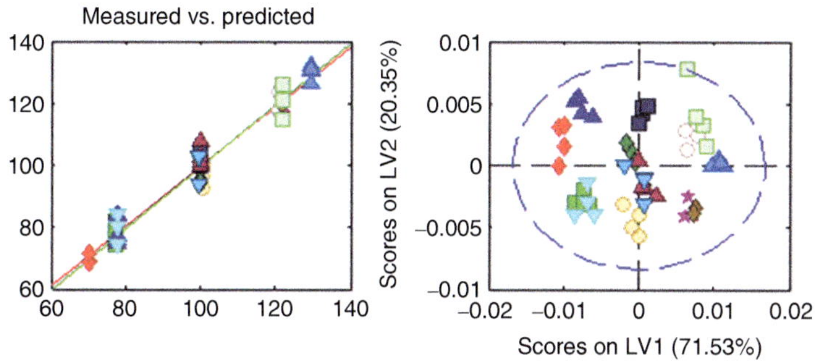

Figure 4.5.7 Example of a powder DoE model (LV: latent variable).

The scores plot shows the distribution of the sample around the mean of the data. It is based on the differences in variance between the spectra and depends on the preprocessing of the data, differences due to varying physical properties can be minimized, and chemical difference can be emphasized.

There are other method requirements to be evaluated during the TRS feasibility study:

- Can the API be detected in final bulk?
- Does fluorescence interfere?
- Is any of the components (API or excipients) heat sensitive?
- What is the correct choice of laser spot size and detector width?

If all questions are answered and the feasibility has been successfully proven, the next step is the development of TRS method.

4.5.3.2 Transmission Raman Method Development

4.5.3.2.1 Transmission Raman Spectroscopic Method Development

It is important to determine the best parameters for measuring the respective API within the bulk product. Usually Raman spectra with distinct API peaks are easily obtained, but there are several pitfalls possible and optimized Raman spectroscopic settings are crucial:

- Laser power
  - starts with the maximum and one accumulation, reduce depending on fluorescence or luminance effect
- Exposure time
  - improves signal-to-noise ratio
- Accumulations
  - make sure to stay within the linear region of detector (40 000 counts per accumulation) and optimize according to expected sample variation in advance:
    - tablets tend to vary in thickness and density: samples of higher density may result in higher peak intensities
    - coated tablets: depending on the nature of the coating peak intensities may be significantly reduced

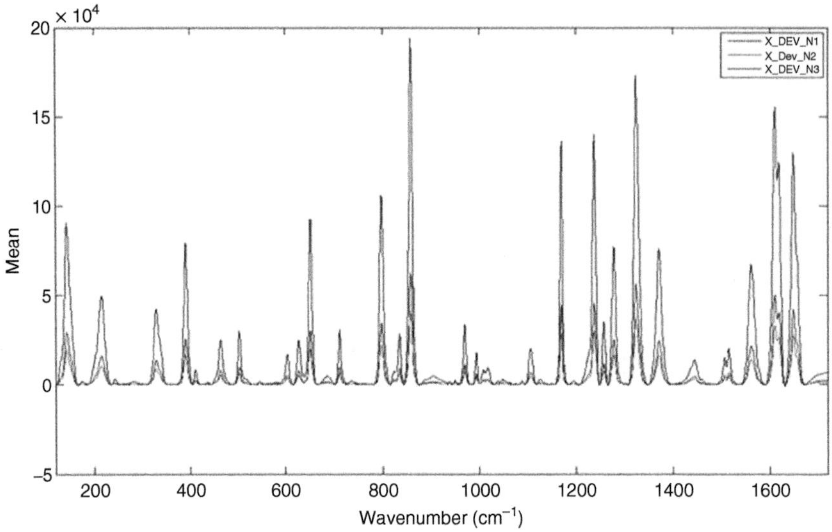

Figure 4.5.8 Different Raman settings of the same data set varying laser power and exposure time.

Table 4.5.4 Variation in Raman settings for the same matrix.

	X_DoE_N1	X_DoE_N2	X_DoE_N3
Laser power (W)	0.65	0.4	0.5
Exposure time (s)	0.1	0.1	0.5

- Measurement scheme, e.g. measurement at more than one spot may be required
  - for large, oblong tablet
  - bilayer tablets, tablets with embossing or score lines: check if both sides of the tablets produce the same spectrum.

  Effects of different settings are presented in Figure 4.5.8 with the variations explained in Table 4.5.4.

The TRS system collects raw spectra, which subsequently are preprocessed, according to the chosen settings. The following settings are possible:

- *Cast onto universal axis*: Sample peaks are shifted to the next whole wavenumber. This is a useful setting for large databases mainly build for qualitative analysis and that the matrix of all measurements (for either qualitative or quantitative analysis) has the same dimension and x-axis labels and can be used for multivariate data analysis. It also means that the error increases since small changes can result in different peak positions. Usually, this is of little impact but if the spectral information of an API is overlaid by the spectral information of the matrix, it may increase difficulties in model robustness.

- *X-axis correction*: Raman peaks are often related to as shifts. The reason is that monochromatic laser light is shifted due to interaction with a molecule to different wavenumbers. The shift is dependent on the laser performance, which is temperature dependent and generally decreases over time. Thus, confirming the correct peak position using a standard is required to enable long-term stability of each model.
- *Y-axis correction*: Peaks intensity is normalized by relating the detector answer of the measurement of the green glass standard to the detector answer of the argon-mercury lamp serving as a virtual instrument.

4.5.3.2.2 Risk Analysis

Prior to starting model development, it is recommended, according to quality by design principles [26], to perform a risk analysis. As for transmission Raman, the entire tablet is measured and thus including all naturally occurring variability of the process is crucial for a robust quantitative model. Therefore, the possibility of producing small-scale batches is important for method development. To assess the current production process, the process flow charts, the respective master batch records, and risk analysis are helpful to understand sources of variability, such as different raw material suppliers and different manufacturing equipment. An additional source of information may also be the product quality reviews, which provide not only information about the product performance but also about the robustness of the process.

To judge the risk of certain sources of variability on the robustness of a Raman model, one requires the view of the experts on the effect of process variability on the product and the knowledge of the spectroscopist on the effect on the spectral data. It is wise to include the operators into the risk analysis and the design of the trials, as first the quality of their work determines the future success and second their experience on the effects of downscaling the manufacturing process is important for the design of the trials.

Considering all this for the performance of the risk analysis, the following functional experts are required:

- Quality control representative
- Production representative
- Formulation development representative
- Raman spectroscopy expert
- Quality assurance representative
- Others: procurement for raw material suppliers

The aim of the risk analysis prior to starting method development is to check the following:

1) Are there findings from prefeasibility/feasibility study that have to be investigated during method development?
2) Are the Raman spectrometer settings sufficiently assessed?

3) Which factors have to be taken into account for setting up the DoE?
   (a) different raw material grades or suppliers
   (b) different manufacturing equipment
   (c) process variability within the specification limits that affect the physical properties of the samples.
4) What is the performance of the current release method?
   (a) method error,
   (b) preparation of samples and standards,
   (c) accuracy and precision.
5) What are the acceptance criteria?

Depending on the outcome of this risk analysis, the DoE for sample preparation will be set up. Additionally, one may not underestimate the effect of a well-thought trough labeling of the samples with the thorough documentation of the trials and the possibility to relate each sample back to its origin.

If it is not possible to mimic all variability of the manufacturing process in the DoE set, a sampling plan for production batches needs to established. This plan may also include historic batches, of which enough retention samples are available. In general, it has been proven beneficial to sample more batches than too few.

The risk analysis should be revisited at least prior to International council for Harmonization (ICH) validation and after ICH validation according to ICH guideline Q9 on quality risk management [27].

4.5.3.2.3 Transmission Raman Model Development, Calibration, and Validation

Andrews et. al. [28] and Villaumié et al. [29] have published comprehensive articles about developing and validating a transmission Raman method. They have touched all relevant topics of method development and method validation, giving also practical examples. Additional information about precision, intermediate precision, and instrument bias can be found in Reference [18].

4.5.4 Regulatory Assessment and Guidelines

Due to the increasing areas of application in pharmaceutics, the regulatory agencies have reacted on the requirements of pharmaceutical industries to provide guidance. After several years of mainly focusing on near infrared and infrared spectroscopy, which were already included in one way or the other in the European Pharmacopoeia (Ph. Eur.) and US Pharmacopoeia (Ph. Eur. 2.2.40 and United States Pharmacopoeia (USP) <1119>/<856>), Raman spectroscopy is now gaining more regulatory attention. The USP has issued a new chapter on chemometrics <1039> and is currently drafting a new chapter on Raman <858>. In Ph.Eur., Raman spectroscopy Chapter 2.2.48 has recently been updated (Ph.Eur. supplement 8.7) including handheld devices. Even though the respective regulatory agencies have increased their information about spectroscopic techniques in general and Raman spectroscopy in particular, even providing guidance about multivariate data

Table 4.5.5 Regulatory guidance documents supporting TRS method development and validation.

Topic	Reference document
Raman spectroscopy	Ph. Eur. 2.2.40, USP <858>
Near-infrared spectroscopy	Ph. Eur. 2.2.40, USP <856>, <1119>
Chemometrics	Ph. Eur. 5.21, USP <1039>
Spectroscopy and light scattering	USP 851
Guidelines	ICH Q2(R1) validation of analytical procedures: text and methodology (under revision – R2 is drafted)
	ICH Q12 technical and regulatory considerations for pharmaceutical product lifecycle management (Draft)
	ICH Q14 analytical procedure development (Draft)
	EMA – use of near infrared spectroscopy (NIRS) by the pharmaceutical industry and the data requirements for new submissions and variations (R2)
	FDA – guidance for industry: PAT — a framework for innovative pharmaceutical development, manufacturing, and quality assurance

analysis (chemometrics), there is still a lack of specific information on specification limits and life cycle management of multivariate models. The relevant documents are summarized in Table 4.5.5.

There are, however, still topics and concepts, which are not entirely described and sufficiently understood by the pharmaceutical community due to a lack of experience in this field. Being used in liquid chromatography and univariate analysis, the strengths of multivariate analysis are difficult to understand, as the manner of analysis is not comparable and the results and acceptance criteria cannot be related to each other one by one. Multivariate analysis results in predictive models, which have to be revalidated and this results in a different lifecycle management, on which guidance of regulatory authorities is scarce.

List of Abbreviations

API	active pharmaceutical ingredient
DoE	design of experiment
EU	European Union
FT-IR	Fourier-transform infrared spectroscopy
HPLC	high-performance liquid chromatography
ICH	International Council for Harmonization
LC	liquid chromatography
NIR	near infrared spectroscopy

PAT	process analytical technology
PLS	partial least squares
Ph. Eur.	European Pharmacopoeia
TRS	transmission Raman spectroscopy
UPLC	ultra-high performance liquid chromatography
US	United States
USP	United States Pharmacopoeia

References

1 Lewis, I.R. and Edwards, H.G.M. (2001). *Handbook of Raman spectroscopy: From the Research Laboratory to the Process Line*. New York: Marcel Dekker.

2 Gomes da Costa, S., Richter, A., Schmidt, U. et al. (2019). Confocal Raman microscopy in life sciences. *Morphologie* 103 (341): 11–16. https://doi.org/10.1016/j.morpho.2018.12.003.

3 Müller, J., Knop, K., Wirges, M., and Kleinebudde, P. (2010). Validation of Raman spectroscopic procedures in agreement with ICH guideline Q2 with considering the transfer to real time monitoring of an active coating process. *Journal of Pharmaceutical and Biomedical Analysis* 53 (4): 884–894. https://doi.org/10.1016/j.jpba.2010.06.016.

4 Aroca, R.F., Alvarez-Puebla, R.A., Pieczonka, N. et al. (2005). Surface-enhanced Raman scattering on colloidal nanostructures. *Advances in Colloid and Interface Science* 116 (1–3): 45–61. https://doi.org/10.1016/j.cis.2005.04.007.

5 Alula, M.T., Mengesha, Z.T., and Mwenesongole, E. (2018). Advances in surface-enhanced Raman spectroscopy for analysis of pharmaceuticals: a review. *Vibrational Spectroscopy* 98: 50–63. https://doi.org/10.1016/j.vibspec.2018.06.013.

6 Johansson, J., Sparén, A., Svensson, O. et al. (2007). Quantitative transmission Raman spectroscopy of pharmaceutical tablets and capsules. *Applied Spectroscopy* 61 (11): 1211–1218.

7 Matousek, P. (2018). Spatially offset Raman spectroscopy for non-invasive analysis of turbid samples. *TrAC Trends in Analytical Chemistry* 103: 209–214. https://doi.org/10.1016/j.trac.2018.04.002.

8 Paudel, A., Raijada, D., and Rantanen, J. (2015). Raman spectroscopy in pharmaceutical product design. *Advanced Drug Delivery Reviews* 89: 3–20. https://doi.org/10.1016/j.addr.2015.04.003.

9 Rantanen, J. (2007). Process analytical applications of Raman spectroscopy. *Journal of Pharmacy and Pharmacology* 59: 147–159.

10 Wikström, H., Marsac, P.J., and Taylor, L.S. (2005). In-Line monitoring of hydrate formation during wet granulation using raman spectroscopy. *Journal of Pharmaceutical Sciences* 94 (1): 209–219. https://doi.org/10.1002/jps.20241.

11 Tian, F., Zeitler, J.A., Strachan, C.J. et al. (2006). Characterising the conversion kinetics of carbamazepine polymorphs to the dehydrate in aqueous suspension using Raman spectroscopy. *Journal of Pharmaceutical and Biomedical Analysis* 40: 271–280.

12 Taylor, L.S., Langkilde, F.W., and Zografi, G. (2001). Fourier transform Raman spectroscopic study of interaction of water vapour with amorphous polymers. *Journal of Pharmaceutical Sciences* 90: 888–901.

13 Walker, G.M., Bell, S.E.J., Greene, K. et al. (2009). Characterisation of fluidised bed granulation processes using in-situ Raman spectroscopy. *Chemical Engineering Science* 64 (1): 91–98. https://doi.org/10.1016/j.ces.2008.09.011.

14 Heinz, A., Strachan, C.J., Atassi, F. et al. (2008). Characterising an amorphous system exhibiting trace crystallinity: a case study with saquinavir. *Crystal Growth and Design* 8 (1): 119–127.

15 Okumura, T. and Otsuka, M. (2005). Evaluation of the microcrystallinity of a drug substance, indomethacin, in a pharmaceutical model tablet by chemometric FTRaman spectroscopy. *Pharmaceutical Research* 22: 1350–1357.

16 Taylor, L.S. and Langkilde, F.W. (2000). Evaluation of solid-state forms present in tablets by Raman spectroscopy. *Journal of Pharmaceutical Sciences* 89: 1342–1353.

17 Johansson, J., Petterson, S., and Folestad, S. (2005). Characterisation of different laser irradiation methods for quantitative Raman tablet assessment. *Journal of Pharmaceutical and Biomedical Analysis* 39: 510–516.

18 Sparén, A.J., Svensson, J., Folestad, O. et al. (2009). Transmission Raman Spectroscopy for Quantitative Analysis of Pharmaceutical Solids. *American Pharmaceutical Review* 12: 62–71.

19 Romero-Torres, S., Perez-Ramos, J.D., Morris, K.R., and Grant, E.R. (2005). Raman spectroscopic measurement of tablet-to-tablet coating variability. *Journal of Pharmaceutical and Biomedical Analysis* 38 (2): 270–274.

20 Romero-Torres, S., Perez-Ramos, J.D., Morris, K.R., and Grant, E.R. (2006). Raman spectroscopy for tablet coating thickness quantification and coating characterization in the presence of strong fluorescent interference. *Journal of Pharmaceutical and Biomedical Analysis* 41 (3): 811–819.

21 Kwok, K. and Taylor, L.S. (2012). Analysis of the packaging enclosing a counterfeit pharmaceutical tablet using Raman microscopy and two-dimensional correlation spectroscopy. *Vibrational Spectroscopy* 61: 176–182.

22 Esbensen, K.H., Swarbrick, B., Westad, F. et al. (2018). *Multivariate Data Analysis. An Introduction to Multivariate Analysis, Process Analytical Technology and Quality by Design*, 6e. Oslo: CAMO.

23 Long, D.A. (1977). *Raman Spectroscopy*. New York: McGraw-Hill.

24 Lewis, I. and Griffiths, P. (1996). Raman spectrometry with fiber-optic sampling. *Applied Spectroscopy* 50: 12A–30A.

25 https://www.europeanpharmaceuticalreview.com/wp-content/uploads/Cobalt-Whitepaper-SEPT-2013.pdf (accessed 17 November 2019).

26 Yu, L.X., Amidon, G., Khan, M.A. et al. (2014). Understanding pharmaceutical quality by design. *The AAPS journal* 16 (4): 771–783. https://doi.org/10.1208/s12248-014-9598-3.

27 EMA (2015). ICH guideline Q9 on quality risk management (Step 5), September 2015, EMA/CHMP/ICH/24235/200.

28 Andrews, D., Geentjens, K., Igne, B. et al. (2018). Analytical method development using transmission Raman spectroscopy for pharmaceutical assays and compliance with regulatory guidelines—part I: transmission Raman spectroscopy and method development. *Journal of Pharmaceutical Innovation* 13: 121–132.

29 Villaumié, J., Andrews, D., Geentjens, K. et al. (2019). Analytical method development using transmission Raman spectroscopy for pharmaceutical assays and compliance with regulatory guidelines—part II: practical implementation considerations. *Journal of Pharmaceutical Innovation* 14: 245–258. https://doi.org/10.1007/s12247-018-9350-0.

4.6

Solid-state Characterization Techniques: Particle Size

Maria Paisana and Constança Cacela

Hovione Farmaciencia SA, R&D Analytical Development, Campus do Lumiar – Edifício S – Estrada do Paço do Lumiar, 1649-038, Lisboa, Portugal

4.6.1 Introduction

Particle-size distribution (PSD) of drug substances, excipients, and drug product intermediates may influence many properties of solid dosage forms, such as their performance, flowability, content uniformity, or processability. For this reason, a precise measurement of these particulate systems is of paramount importance [1, 2]. When the particle size of a certain material has a critical impact on dissolution, bioavailability, and stability of final dosage forms, testing for PSD should be carried out using appropriate procedures and acceptance criteria [3].

The acceptance criteria should be set based on the observed range of variation and should consider the dissolution profiles of the batches that showed acceptable performance *in vivo*. The potential for particle growth should also be explored during product development, and the chosen acceptance criteria should take the results of these studies into consideration [3].

There is a wide range of measurement methods for particle-size analysis, each of which can be employed over a certain range of particle size and type of products. Methods such as microscopy (optical and scanning electron), sedimentation, sieving, electrozone sensing, laser diffraction (or light diffraction), and dynamic light diffraction are used for particle-size characterization in the pharmaceutical industry [4]. Among the abovementioned methodologies, laser diffraction is one of the most commonly used techniques in this industry for measuring the PSD of pharmaceutical products due to the wide range of particle sizes that it can measure. Furthermore, precise and representative results can be obtained with this technique, which is highly valuable when dealing with materials whose particle size has a tremendous effect on their pharmaceutical performance (e.g. inhalation drug substances). Moreover, this technique is easy to work with, and it allows for the measurement of particle size of samples presented in different physical forms [5]. This laser-diffraction technique uses the assumption that particles are spherical, being the results commonly reported as equivalent-sphere size distributions. Although this approach is

Solid State Development and Processing of Pharmaceutical Molecules: Salts, Cocrystals, and Polymorphism, First Edition. Edited by Michael Gruss.

simplistic, the shapes of the particles generated by most industrial processes are such that the spherical assumption does not cause serious problems. However, for non-spherical particles, the PSD obtained using a light-diffraction approach may be different from those obtained by methods based on other physical principles, such as sedimentation or sieving. For example, in a sieve analysis, the two smallest dimension sizes (i.e. thickness and width) of a particle will allow it to pass through the sieve mesh regardless of the length. As a result, in this technique only 2 dimensions are considered for particle size description. On the other hand, a PSD profile based on the rate of fall of the particles through a viscous medium will force particles of a certain shape, such as plate-like, to orient themselves in order to maximize drag while they are sedimenting, shifting the reported particle size in the smaller direction.

In the case of image analysis, such as microscopy, particle dimensions are commonly described by their particle diameter. In the case of spherical particles, the particle-size description by just its diameter is not an ambiguous measurement; however, for irregular particles, different diameter definitions may be encountered in the literature, such as Feret's diameter, Martin's diameter, and projected area diameter. In detail, Feret's diameter measures the distance between parallel tangents on opposite sides of a randomly oriented particle, whereas Martin's diameter has in consideration the diameter of the particle "at the point that divides a randomly oriented particle into two equal projected areas" [6]. Depending on the particle orientation upon the measurement, Martin's and Feret's may differ. Thus, it is important to obtain statistically significant measurement for these diameters through the analysis of a large number of randomly oriented particles. Projected area diameter, on the other hand, measures the diameter of a sphere having the same projected area as the particle. Representative images of these diameter measurements may be encountered in United States Pharmacopeia (USP) <776> Optical Microscopy chapter [6].

Based on the above, it is not a surprise that slightly different PSDs may be obtained if a given sample is analyzed by different analytical methodologies.

PSDs are commonly represented as frequency (also referred as differential) or cumulative curves. In the first case, the distribution plot represents how frequently each particle size is observed. These distributions can be based on weight, number, surface area, or volume of the particles. These frequency plots also give interesting information about the number of populations of particle sizes in a sample; when the sample is homogeneous in terms of its particle size, a single peak in the distribution is observed. However, it is also possible that the sample presents multiple particle-size populations. For example, the frequency graph in Figure 4.6.1 shows a bimodal curve where it is possible to observe a larger particle-size population with a particle size in the micro size range and a smaller submicron population.

The plots in cumulative distribution curves are represented by the particle size *versus* the cumulative percentage at or below a given size (Figure 4.6.1). This representation is more useful for estimating statistical parameters. The x-axis in both PSD representations may be described using a logarithmic x-axis distribution

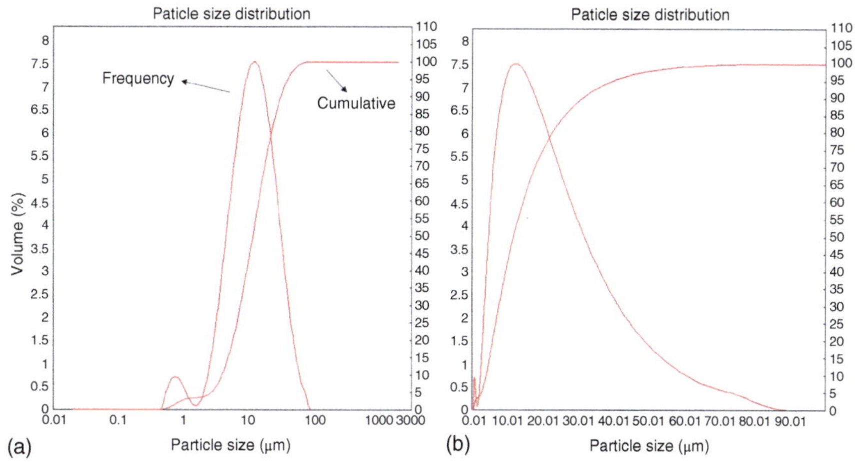

Figure 4.6.1 Representation of different particle-size distribution curves (frequency and cumulative) using different *x*-axis scales: (a) logarithmic; (b) linear.

(Figure 4.6.1a) or a linear *x*-axis distribution (Figure 4.6.1b). The first representation is usually used in the cases where the samples present a wide PSD.

To simplify the PSD interpretation, it is a common approach to report the central point of the distribution together with two other points which describe the coarsest and finest particles, for a better description of the curve width. D_{50} (or ×50) is a commonly used term to represent the median, which is the particle size where 50% of the population lies below/above this value. On the other hand, D_{90} (×90) indicates that 90% of the distribution is below this size and D_{10} (×10) presents the size where 10% of the population lies below.

4.6.2 Analytical Methodologies Used to Measure Particle Size

4.6.2.1 Sedimentation

Sedimentation methods are based on Stokes' law, which defines the velocity of particles settling in a viscous liquid when exposed to an accelerating force such as gravity. Gravitational sedimentation methods are best suited to particles in the 2–50 μm range. Limitations of this technique include the fact that Stokes' law is only valid for spheres and particles unaffected by Brownian motion (the latter limiting submicron particle measurement). Furthermore, for this technique to be applied, it requires the knowledge of the density of the powder under investigation, as well as a temperature control to avoid fluctuations in the viscosity of the liquid phase [4].

4.6.2.2 Electrozone Sensing

Electrozone sensing equipment (also known as Coulter counters) analyses powder particles as diluted suspensions in an electrolyte. In the suspension, there are two electrodes that are separated by a very small aperture and, by applying a field between the two electrodes, the particles are made to flow through the opening, producing a voltage pulse. When a particle passes through the aperture, the measurable disturbance in electric field is proportional to the volume of the particle. Since the signal is originated from the displacement of the electrolyte caused by the particle, the measurement of the particle does not depend on any optical properties. If a very large aperture is chosen, then the risk of multiple particles passing through it together and being counted as a single particle increases. This method is reliable in the micron size range and avoids the need to know the physical properties of the powder being analyzed. The limitations of the aperture result in a lower size limit of about 0.4 μm, and obviously, very wide size distributions will be more difficult to measure [4].

4.6.2.3 Sieving

Sieving is the method of choice in starting raw materials, such as excipients, and drug product intermediate characterizations involving powder blending or granulation steps. It requires large quantities of material for these measurements. Drug substances are not commonly evaluated by sieving due to their low particle size, as well as their more irregular particle shapes. Sieving is most suitable for powders whose average particle size is >30 μm. Sieve analysis does not account for particle shape effects of different particles. The two minor dimensions of a three-dimensional particle dictate whether a particle passes through a mesh opening. The major dimension does not affect the particle-size calculation or the PSD [5].

4.6.2.4 Microscopy

Optical microscopy and scanning electron microscopy (SEM) can be used to evaluate the PSD of samples. The optical microscopy method is applicable to particles in the 0.8–150 μm size range and down to 0.001 μm when using electron microscopy. These microscopy techniques use extremely small sample amounts. Thus, these analyses require an extremely sensitive step of sample collection to assure that a representative sample is analysed.

4.6.2.5 Dynamic Light Scattering

Dynamic light scattering (DLS) is a light scattering technique for particle characterization of colloidal dispersions, providing size information in the nanometer (nm) range, which is out of the primary range of static light scattering (e.g. laser diffraction). For DLS measurements, particles need to be suspended in a liquid and, by the measurement of its Brownian motion, their size is therefore determined. Brownian motion is the random movement of suspended particles in a liquid resulting from their collision with the fast-moving molecules in the fluid. Larger particles will generate slower Brownian motions. The particle diameter that is measured in DLS is a value that refers to how a particle diffuses within a fluid, and for that reason, it is referred to as a hydrodynamic diameter. In this methodology, the temperature needs to be stable during the experiment; otherwise, convection currents in the sample will cause nonrandom movements that will have a negative impact on the correct interpretation of the size [7].

4.6.2.6 Laser Diffraction

As aforementioned, laser diffraction is one of the most commonly used techniques for measuring the PSD of a wide range of particle sizes. This method consists of the analysis of the diffraction pattern produced when particles are exposed to a collimated beam of light. Due to recent advances in optics and lens and equipment design and construction, newer instruments are routinely capable of measuring particles from 0.1 μm to 8 mm. It is therefore the responsibility of the user to prove the applicability of the instrument for its intended use and to validate the method prior to its adoption for routine usage [8, 9]. In general, for a PSD evaluation, a representative sample, dispersed at an adequate concentration in a suitable liquid or gas, is passed through the beam of a monochromatic light source, usually from a laser. The light scattered by the particles at various angles is measured by a multielement detector, and numerical values relating to the scattering pattern are then recorded. These numerical scattering values are then transformed using an appropriate optical model and mathematical procedure, resulting in a volumetric PSD (e.g. $\times 50$ describes a particle diameter corresponding to 50% of the cumulative undersize distribution) [9].

4.6.3 Method Development for Precise Particle-size Measurements by Laser Diffraction

4.6.3.1 Instrumentation and Measurement

The laser diffraction technique consists of light source(s), a measurement region, optics to collect the scattered and unscattered light, multielement detector(s), and the needed hardware and software for the computations. Commonly, more than one light source may be used, which usually comes from lasers. Sources of different wavelengths are mainly used to extend the size range, especially to better characterize samples in the submicron region. The detectors used in the analysis must ensure the adequate spatial resolution needed to best define the diffraction pattern. Depending on the instrument design, different lenses may be used for specific size ranges. The light scattered by the particles at different angles is transformed, using an appropriate optical model and mathematical procedure, to yield the proportion of total volume to a discrete number of size classes, forming a volumetric PSD [9, 10] (Figure 4.6.2).

When a representative sample is dispersed, it passes through the light beam in a measuring zone by a transporting fluid (gas or liquid). The measuring zone should be within the working distance of the lens used. In liquid or gas dispersions, the interaction of the incident light beam with the dispersed particles results in a scattering

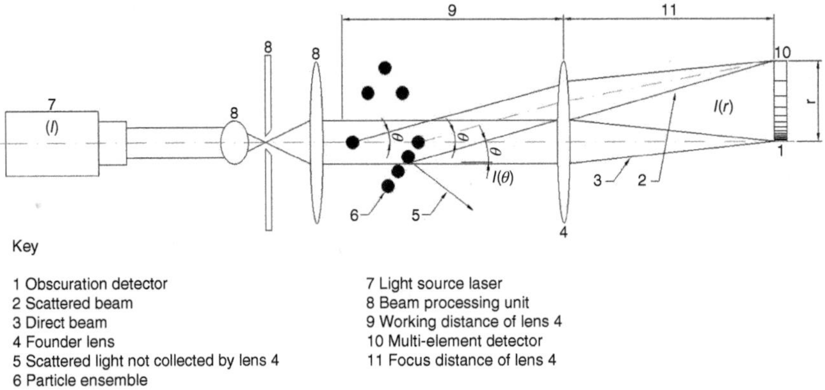

Figure 4.6.2 Illustration of common laser diffraction instrumentation configurations. Source: Adapted from Stetefeld et al. [7].

pattern with different light intensities at various angles. The total angular intensity distribution consists of both direct and scattered light [8, 9].

Before undertaking a measurement, it should be ensured that the optical part of the instrument is aligned so that the blank measurement can be performed. During sample analysis, the magnitude of the signal from each detector element depends upon the detection area, the light intensity, and the quantum efficiency. The coordinates (size and position) of the detector elements together with the focal distance of the lens determine the range of scattering angles for each element. The ratio of the intensity of a dispersed sample indicates the proportion of scattered light and, hence, the particle concentration [10]. Because of how the results are derived, the PSD from light diffraction measurement is best considered as the cumulative fraction of volume (or mass, assuming an uniform density). The particle sizes at the undersize values of 10%, 50%, and 90% (denoted as ×10, ×50, and ×90, respectively) are frequently used. ×50 is also known as the median particle size. It is recognized that the symbol *d* is also widely used to designate the particle size, thus the symbol *x* may be replaced by *d* [10].

4.6.3.2 Selection of an Appropriate Optical Model

Most laser diffraction instruments use either the Fraunhofer or the Mie theories as optical models, although other approximation theories are applied for the calculation of the scattering matrix [9, 10]. In the Mie theory, which is based on the solution to Maxwell's equation, both the light diffracted by the spherical particles and the light passing through the particles (refracted light) are taken into consideration. When applying this theory, the refractive index (RI) of the particulate system and medium where particles are going to be suspended (or the ratio between them) must be considered so that the calculation of the model matrix can be determined.

The Fraunhofer model represents a special case of the Mie theory, which assumes that all the particles are spherical, opaque, and large relative to the wavelength of the incident light. This model is therefore more applicable to particles with large particle size (>25 μm) and has the advantage of being a simple method that does not need RI values.

Mie scattering theory overcomes Fraunhofer limitations since it is sensitive to smaller sizes (wide angle scatter) and to a wide range of opacity (i.e. light absorption), allowing accurate measurement even in cases of significant particle transparency [9, 10]. It also accounts for light that refracts through the particle (i.e. secondary scatter).

The RI of the particles consists of a real and an imaginary (or absorption) part, which are both necessary for the Mie theory to be applicable [8, 9]. The real component is commonly determined by either a microscopy method (Becke line technique) or from the refractive indices of solutions of known sample concentrations. The Becke line is a band of light visible along a particle boundary in plane-polarized light. In this technique, particles are immersed in oils of known RI and observed with transmitted light. Using different oils, the RI of particles may be determined well within the needed accuracy [11]. In the second method, a refractometer is used to measure the RI of a solution in which the concentration of the material under analysis is known. Extrapolating the solution's RI from a series of known concentrations of the material to that of the pure substance yields a good approximation of the material's RI. The imaginary (or absorption) component of the optical model, on the other hand, is related to the light absorption by the particles. The degree of this absorption depends on a number of particle properties, such as surface irregularities, texture, internal reflection/refraction, color, and opacity [8]. Although this absorption value is not typically measured, it has a huge impact on particle-size measurement accuracy.

One way of roughly estimating the absorption component value of a given particulate material is to observe the sample particles in a dispersant under an optical microscope. If the particles are spherical and transparent under the microscope, then they will absorb very little light and the absorption will be reduced to a value of about 0.00 to 0.01. On the other hand, opaque, colored, and irregular surface particles are more prone to light absorption and, for that reason, the absorption value is higher [8, 12].

Another possibility for estimating both real and imaginary components of the RI of a determined sample is to use an iterative approach based upon the fit of the modeled data *versus* the actual data in the equipment. Although the Mie model results are not solely dependent on the chosen RI values, it can sometimes lead to unrealistic results. In these cases, recalculations of an existing measurement using different RIs can be made. The sensitivity of the measurement to small changes in the selected RI values should be assessed during method development. The obtained results should be compared to other techniques, such as microscopy, to assure that they realistically represent the sample.

4.6.3.3 Sample Dispersion

When developing a method to measure the PSD of a particulate system, it is important to define the desired state of dispersion for the sample, i.e. if only the primary particle size should be measured or if agglomerates/aggregates of powders should be included in the measurement once almost all powder samples exhibit some degree of agglomeration. The measurement of primary particles (fully deagglomerated particles) gives information about the particle formation process or inherent dissolution behavior of a powder. On the other hand, measuring agglomerates may be more useful for understanding issues of content uniformity and powder flow. For the measurement of primary particles, the first step is to choose an appropriate dispersant. To disperse a sample, two basic approaches may be followed, namely, dispersing the particles in a liquid (wet dispersion) or in a gas (dry dispersion). When liquids are used as dispersants, it is important to ensure that no dissolution or re-agglomeration occurs. In the case of air or gas dispersions, dissolution is not an issue but achieving the proper degree of dispersion can be. After defining the state of dispersion, method development should be focused on how to best achieve reproducible and reliable results. This can be made by defining the amount of sample that is representative of the bulk and selecting the appropriate measurement conditions.

To choose the most appropriate dispersion method using laser diffraction, it is important to consider both the physical and chemical characteristics of the samples, as well as the goal of the dispersion procedure. Both wet dispersion and dry dispersion have their own advantages. Consequently, when making the choice between wet and dry dispersion, influencing factors include the natural state of the sample, its ability to disperse, and the volume of sample to be measured. Sample types that benefit from dry dispersion include materials that are difficult to be suspended in liquids without dissolution or agglomeration or products designed to be dispersed in air (e.g. sprays and inhaled drugs). Wet measurement, on the other hand, may be preferred when handling with cohesive and or very fine particles, which can be difficult to disperse completely using dry techniques. Wet dispersion is also preferable when handling with hygroscopic materials (at risk of collecting moisture when exposed to the atmosphere), friable particles, or potent substances that are more easily controlled in the wet state with a reduced risk of inadvertent inhalation [9, 13, 14].

4.6.3.3.1 Wet Dispersion

The most suitable dispersant is the one that can wet the particles without dissolving them.

The dissolution of the samples can be minimized by using a dispersant of opposite polarity to the sample; however, this has to be balanced by the need for particle wetting [14]. The selected dispersant must have a suitable viscosity, be transparent to the laser beam, have a different RI from the sample (for the Mie calculation), be chemically compatible with the materials used in the instrument (O-ring, tubing, etc.), and not be harmful to health and meet safety requirements. Table 4.6.1 shows a list of possible dispersants in decreasing order of polarity.

As a first experiment to select an adequate dispersant for a given particulate material, it is worthwhile to test in a vial/beaker how different dispersants will wet the sample (Figure 4.6.3). This visual check is quite effective and fast. Good wetting between particles and dispersant shows a uniform suspension of particles in the

Table 4.6.1 Dispersants in decreasing order of polarity.

Dispersant	Polarity
Water	Very polar
Organic acids	⇓⇓
Alcohols (e.g. methanol, isopropyl alcohol)	
Simple alkanes (e.g. hexane, heptane, *iso*-octane)	
Long-chain alkanes and alkenes (e.g. dodecane, mineral oils, palm oil)	Very non-polar

Figure 4.6.3 Examples of different powder dispersion behaviors into different liquids.

liquid (Figure 4.6.3, vial 1). In vial 2 (Figure 4.6.3), it is possible to observe an example where the dissolution of the sample happened, which needs to be avoided. Poor wetting, on the other hand, may show droplets of liquid on top of the powder, or agglomeration and sedimentation (Figure 4.6.3, vial 3 and 4).

The wettability of the powder by the dispersant depends on the surface tension between the particles and the liquid and can be improved by using a surfactant to reduce the surface tension. However, it is important to control the concentration at which surfactants are used. In general, a very reduced concentration of a surfactant solution is enough to improve particle wetting. Too much surfactant can cause foaming and bubbles, which may be interpreted as large particles [12].

The particle size should remain stable after being dispersed. If the particle size starts to increase due to particle re-agglomeration, then an additive may be required to stabilize the dispersion. Additives such as ammonium citrate and sodium pyrophosphate can help to stabilize a suspension by adding charge to the particle surface. Typically, additives are used in a concentration of <1 w/v%. After choosing an appropriate dispersant to wet the sample, the state of dispersion of the sample in the instrument must be assessed. This assessment is performed through a dispersion titration, which generally involves a series of measurements to assess the effect of ultrasound on the sample. This includes, for instance, a ramp of increasing ultrasound potency that can be made using the ultrasound system inside or outside of the equipment. When external ultrasounds are applied, a probe

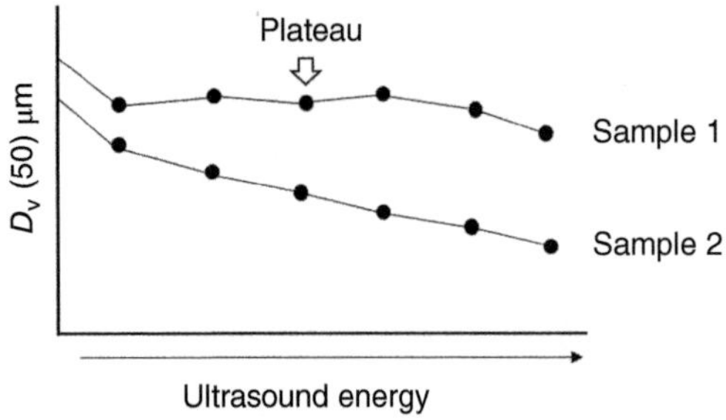

Figure 4.6.4 Example of a titration curve with increasing ultrasound energy.

that is capable of achieving higher potency values than the internal ultrasound system may be an advantage when leading with very small and cohesive materials. In both cases, a reduction of particle size and increase of obscuration are expected to be observed as the power of the ultrasound increases, as a result of a larger concentration of particles in the system as each agglomerate is broken into multiple primary particles. It is worth to mention that when the dispersant is an organic solvent, the measurement should be made only after a while so that the temperature of the dispersant could be stabilized prior to the measurement. Otherwise, false peaks at large particle sizes may be observed as a result of temperature gradients in the dispersant. The increase of ultrasound potency should continue until no further decrease in size due to ultrasound is observed. Throughout this study, the determination of the power and duration of ultrasound required to disperse the sample to its primary particle size can be achieved. If the whole dispersion titration is flat and no reduction in particle size is observed during ultrasound titration, then the sample may be fully dispersed without the use of ultrasound. In opposition, if a continual decrease in particle size is observed as the ultrasound energy increases (e.g. a stable particle size is not reached), then the primary particles may be damaged by the ultrasound. Figure 4.6.4 shows an example of two different samples when subjected to increasing ultrasound energy. It can be noted that for sample 1, a plateau is visually detected, and sample 2 shows a continuous decrease in particle size. In the case of sample 2, the verification of the state of dispersion of the sample using a microscope, where observations are made before and after ultrasound application, could be of great help during the method development. With this approach, it will be possible to detect when agglomerates are dispersed or when the particle shape/size is changed due to breakage.

4.6.3.3.2 Dry Dispersion

For a successful dispersion in this methodology, the separating force applied by, for example, a dispersing air flow, must be enough to overcome interparticle forces of attraction, such as van der Waals, liquid bridges, and electrostatics. Therefore, this method is commonly applied to materials with good flowability and low cohesiveness, so that problems of clumping and sickness can be avoided. Sample agglomeration issues can often be overcomed by drying the sample in an oven or a desiccator. Since laser diffraction is not able to differentiate between primary particles and agglomerates, the dispersion process is probably the most critical issue impacting accuracy and precision of this PSD analysis. The USP highlights the importance of having a good dispersion when developing a method: "For the development of a method, it is necessary to check that comminution of the primary particles does not occur and, conversely, that a good dispersion of the agglomerates has been achieved." [9]. Therefore, it is of great importance to study the effect of the gas pressure delivered by the dry powder feeder on the reported result. For this reason, dry particle-size method development should start with carrying out a pressure titration. For instance, particle-size measurements should be made with increasing pressures, such as 0.1, 0.5, 1, 2, 3, and 4 bar and the effect of pressure on the state of dispersion should be investigated. A plateau is expected to be obtained when total deagglomeration occurs. However, if the particles are being broken-up, a plateau may never be obtained. For this reason, a second technique should be used as an orthogonal confirmation of the PSD obtained. Running the sample in a different dispersion mode, such as wet mode, or using a static microscopic technique could be some of the approaches that can help in the identification of the most suitable pressure to be applied to the sample under study. For instance, if the results of a dry measurement are finer than the wet measurement, this indicates that the primary particles are being broken at that pressure. International Organization for Standardization (ISO) 13320:2009 states that a "pressure/particle-size" titration should, in the ideal case, identify a region where particle size is nearly constant over a range of pressures, but when this is difficult to be visualized, it becomes important to reference dry results against wet measurements [8].

4.6.3.4 Sample Representativeness and Obscuration

It is of a great importance that the sample that is put into the instrument in both wet and dry methodologies is representative of the bulk of the material. Sampling could be the largest possible source of error in the measurement of coarse particles or samples containing a broad range of sizes. To get a representative result, there is a minimum number of particles that should be measured. However, as the size of the particles increases, the mass containing enough particles increases and the minimum mass required to achieve reproducible results also increases. The adequate sample concentration must be a balance between adding enough sample to get signal and to be representative of the bulk material, but, at the same time, the possibility of adding too much sample may generate multiple scattering phenomena which should be taken into account [14]. These multiple scattering events occur when the concentration of particles in the cell is so high that the laser light is scattered by more than one particle before reaching the detector. As a result of this, the laser light is then scattered to higher angles. Higher angle scattering is associated with finer particles, and multiple scattering results in an underestimation of the particle size.

Obscuration is the parameter used to measure the sample concentration, representing the percentage loss of laser light through the sample. The particle concentration in the dispersion should be above a minimum level, which for many instruments will correspond to about 5% obscuration, to produce an acceptable signal-to-noise ratio in the detector. Additionally, it should be below a maximum level to avoid multiple scattering phenomena (e.g. 35% above 20 μm and 15% below 20 μm). The ideal concentration is influenced by the laser beam width, the path length of the measurement zone, the optical properties of the particles, and the sensitivity of the detector elements. Thus, the study of the optimum concentration range that achieves the required obscuration for any typical sample of material is important during method development [9].

4.6.3.5 Readiness for Method Validation

The particle size of drug substances may have a strong impact on their dissolution rate, bioavailability, and/or stability, and for that reason, testing the PSD should be executed using an appropriate procedure whose robustness and reliability are proven [2, 3]. The method validation will allow for a deeper understanding of the key variables associated with the variability of the results and control them as part of the measurement procedure. In these particle-size analyses by laser light diffraction, the specificity parameter as defined by the International Council for Harmonization (ICH) [15] cannot be applicable as it is not possible to discriminate between different components in a sample, nor is it possible to discriminate agglomerates from dispersed particles unless properly complemented by microscopic techniques [10]. Linear relationship between concentration and response, or a mathematical model for interpolation, is also not applicable to this procedure. Instead of evaluating the linearity, this method requires the definition of a concentration range within the result of the measurements that do not vary significantly [10]. Concentrations below that range produce an error due to a poor signal-to-noise ratio, while concentrations above that range can lead to multiple scattering. The range will depend mostly on the instrument's hardware. Accuracy should be confirmed through an appropriate instrument qualification and comparison with microscopy, while precision may be assessed throughout a repeatability determination [10]. For repeatability assessment, it is a good practice to consider a variation of not more than 10% for any central value of the distribution (e.g. for ×50). Values at the sides of the distribution (e.g. ×10 and ×90) are oriented toward less stringent acceptance criteria such as a variation of not more than 15%. Below 10 μm, these values must be doubled [9, 10].

Method robustness may be tested during the optimization of the sample dispersion so that it can be proved that the results remain unaffected by small variations in the test parameters [10]. This is important because small variations in particle size may have a strong impact on their bioavailability and manufacturability [2, 3]; thus, the proven robustness of the method would provide assurance of the method reliability during routine use. Some of the robustness parameters are common to both wet and dry methods, whereas others are specific from each technique.

In both wet and dry methods, the measurement duration should be included in the method robustness assessment. This is the variable that should be set to ensure that representative sampling has been achieved. For narrow distribution of fine particles, reduced measurement times can be used. If polydisperse or coarse PSDs are analysed, then longer measurement times may be necessary to ensure that a representative volume of large particles has been sampled. The obscuration range should also be optimized so that the difference between sample signal and background is acceptable. The results should not decrease significantly with increasing obscuration due to multi-scattering events.

In the case of methods using liquids as dispersants, suspension stability, stirring speed, ultrasound power and duration should be assessed as part of the method robustness assessment. For the suspension stability study, how long the suspension remains stable after being dispersed should be evaluated. The stirring speed, on the other hand, defines the range of speed so that the suspension of material is formed without causing air bubble formation. Power and duration of ultrasound are the most critical variables and should be assessed together with microcopy measurements. Sample suspensions should also be analyzed before and after ultrasound application to examine what effect sonication has on the particle size.

The most critical robustness variable in dry dispersion methods is the air pressure, so the range of air pressure for each particulate material should be deeply studied upon method development. Besides this variable, the feed rate should also be part of robustness studies. This variable controls the number of light scattering nuclei passing through the laser beam during measurement. An ideal feed rate is the one that will allow for a reasonable measurement integration within the defined obscuration ranges and in which, at the same time, all of the sample is consumed without segregation.

Upon definition of the appropriate measurement conditions, precision of the results should be verified. Precision can also be evaluated during robustness assessment. Preferably, method precision should be proven when sample preparation is performed by a different analyst (intermediate precision) and in a different equipment in a different laboratory (reproducibility). The required precision of the method is dependent on the characteristics of the material (milled *versus* not milled, hard *versus* fragile) and on the requirements of the application (formulation type and technique) [10].

4.6.4 Unexpected Results and Troubleshooting in Laser Diffraction Measurement

During routine analysis, it is important that the results generated by a particle-size analysis are thoroughly understood in the context of their application. Most homogenous solid particle sample types should produce a continuous PSD, so the appearance of a distinct disconnected peak should be considered suspicious and investigated (example of a disconnected peak at about 1000 μm in Figure 4.6.5). Thus, other techniques such as microscopy can be very useful in verifying the accuracy of a PSD [16].

4.6.4.1 Inconsistent Disconnected Peaks

When liquids are used as dispersants, disconnected peaks that result from bubbles or gaseous occurrences may occur. For most of the instruments, these species

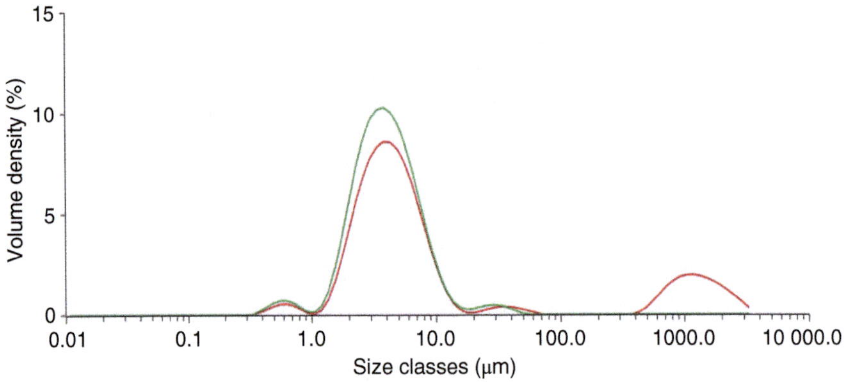

Figure 4.6.5 Example of a distribution containing a disconnected peak (red) *versus* the actual particle-size distribution (green).

cannot be distinguished from the sample and, thus, if signals are collected from both the sample and bubbles, the instrument will report both indiscriminately yielding wrong results [17]. Depending on the viscosity and the surface tension of the dispersing medium, the size range of the bubbles can vary. There are certain sample types and dispersions, such as those containing surfactants or highly viscous carriers, that might have a greater possibility to generate bubbles [14]. However, these events are not repeatable, so it is easy to discard them as part of the true PSD. Actions such as adding slowly the sample to the dispersant unit or wait a while prior to collecting the data could be possible strategies to be followed.

As bubble peaks, thermal artifact peaks are also inconsistent in their appearance. These peaks are formed when volatile dispersing media do not reach thermal equilibrium within the instrument [16]. These peaks can be avoided in most cases after ensuring that the temperature of a volatile carrier medium has reached thermal equilibrium. Equilibration is usually accomplished by allowing longer recirculation times prior to taking a background measurement. Depending on the size of the primary particle, the thermal peak is often completely detached from the "true" PSD [16].

When air is used as a dispersant, a lack of equilibration state of the air current can also generate disconnected peaks. For this artificial peak to appear, the background measurement records a lower signal in the low-angle detectors than the signal recorded during the sample measurement. This artificial peak is manifested at the coarse end of the distribution and is potentially exacerbated by lower air pressures. These occurrences may be reduced when slow steps are taken to stabilize and equilibrate the dispersion environment prior to the analysis.

Shiny and reflective particles may lead to coarse-end artifact peak due to the concentration of reflective particulate within the re-circulator of the instrument. Reducing the concentration of these particles may be helpful in avoiding this occurrence.

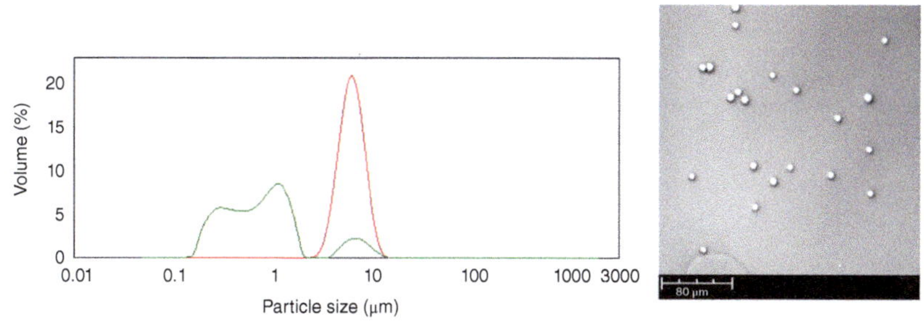

Figure 4.6.6 Example of a particle-size distribution by volume (a) using two different optical models (green and red lines) and SEM image of the particles (b).

4.6.4.2 Repeatable Artifact Peaks

Among the artifact peaks described above, optical model artifact peaks can be the most delicate ones because they can occur repeatably and consistently [16, 18]. However, an optical model-related artifact peak does not imply a complete re-analysis. A simple recalculation of the data is enough to solve this issue. Optical model artifact peaks are related to the complex RI assigned to the particles when using the Mie theory. In this theory, as opposed to the Fraunhofer theory, the optical properties of the particles are needed and so they must be entered into the instrument software. As referred above, although there are a few methods described in the literature that can be used to determine the real RI of a material, there is no commercially available method for directly measuring the absorption component. Assessing the data fit can be the fastest way to adjust the optical properties of the chosen model. Microscopy analyses of the particles, where the shape/size of the particles is assessed, can also give complementary information about the suitability of the selected imaginary RI component. Figure 4.6.6 shows an example of a SEM micrograph of a material composed by spherical particles. It is possible to observe that the particles have similar sizes and that the distribution is monomodal, so the distribution shown by the green line is not real and is the result of a wrong optical model application.

List of Abbreviations

DLS	dynamic light scattering
ICH	International Council for Harmonization
ISO	International Organization for Standardization
PSD	particle-size distribution
RI	refractive index
SEM	scanning electron microscope
USP	United States Pharmacopeia

References

1 Yu, W. and Hancock, B.C. (2008). Evaluation of dynamic image analysis for characterizing pharmaceutical excipient particles. *International Journal of Pharmaceutics* 361 (1–2): 150–157.

2 ICH-Q8(R2) (n.d.). Pharmaceutical development, international conference on harmonisation of technical requirements for registration of pharmaceuticals for human use. *ICH Harmonised Tripartite Guideline.*

3 ICH-Q6A (n.d.). Specifications: test procedures and acceptance criteria for new drug substances and new drug products: chemical substances, international conference on harmonisation of technical requirements for registration of pharmaceuticals for human use. *ICH Harmonised Tripartite Guideline.*

4 Mingard, K.P., Morrell, R., Jackson, P. et al. (2009). *Improving the Consistency of Particle Size Measurement. Measurement Good Practice Guide No. 111.* UK: National Physical Laboratory.

5 Qiu, Y., Chen, Y., Zhang, G.G.Z. et al. (2016). *Developing Solid Oral Dosage Forms Pharmaceutical Theory and Practice.* Academic Press.

6 USP42-NF37 (2019). General Chapter, <776> "Optical microscopy". https://online.uspnf.com/uspnf/document/GUID-4D510FF2-03A5-443E-BE1F-5D5BFC15F5F5_1_en-US?highlight=776 (accessed 8 September 2019).

7 Stetefeld, J., McKenna, S.A., and Patel, T.R. (2016). Dynamic light scattering: a practical guide and applications in biomedical sciences. *Biophysical Reviews* 8 (4): 409–427. https://doi.org/10.1007/s12551-016-0218-6.

8 ISO (2009). *Particle Size Analysis – Laser Diffraction Methods.* International Standards Organization, 13320:2009.

9 USP42-NF37 (2019). General chapter, <429> Light diffraction measurement of particle size. https://online.uspnf.com/uspnf/document/GUID-B3D11DFA-2021-4961-9F88-F942F7C96D5D_1_en-US?highlight=429 (accessed 8 September 2019).

10 Ph.Eur (2007). 6.6 General Chapter. 2.9.31. In: *Laser Diffraction Measurement of Particle Size*, 5103. Strasbourg, France: EDQM.

11 Nesse, W.D. (2012). *Introduction to Optical Minealogy.* New York: Oxford University Press.

12 Li, H., Li, J., Bodycomb, J., and Patience, G.S. (2019). Experimental methods in chemical engineering: particle size distribution by laser diffraction – PSD. *Canadian Journal of Chemical Engineering* 97 (7): 1–17.

13 Malvern Instruments Limited (2016). A basic guide to particle characterization. White Paper. https://www.cif.iastate.edu/sites/default/files/uploads/Other_Inst/Particle%20Size/Particle%20Characterization%20Guide.pdf (accessed 3 September 2019).

14 Malvern Instruments (2013). Wet or liquid dispersion method development for laser diffraction particle size measurements. Mastersizer Application Note: MRK1902-02. https://warwick.ac.uk/fac/cross_fac/sciencecity/programmes/internal/themes/am2/booking/particlesize/wet_dispersion_application_note.pdf (accessed 3 September 2019).

15 ICH, 2005. Q2 (R1), "Validation of analytical procedures: text and methodology". International Conference on Harmonisation of Technical Requirements for Registration of Pharmaceuticals for Human Use, ICH Harmonised Tripartite Guideline.

16 Sabin, A. (2011). Problems in particle size: laser diffraction observations. *Journal of GXP Compliance* 115 (4): 1–13.

17 Scherer, T.M., Leung, S., Owyang, L., and Shire, S.J. (2012). Issues and challenges of subvisible and submicron particulate analysis in protein solutions. *The AAPS Journal* 14 (2): 236–243.

18 Beekman, A., Shan, D., Ali, A. et al. (2005). Micrometer-scale particle sizing by laser diffraction: critical impact of the imaginary component of refractive index. *Pharmaceutical Research* 22 (4): 518–522.

4.7

Micro Computational Tomography

Susana Campos and Constança Cacela

Hovione Farmaciencia SA, R&D Analytical Development, Campus do Lumiar – Edifício S Estrada do Paço do Lumiar, 1649-038 Lisboa, Portugal

4.7.1 Tomography Imaging Techniques

Computed tomography (CT) scans are mostly known for their use in the clinical investigations of human beings as well as in living animals. However, these techniques have gained relevance within the pharmaceutical industry due to their potential to retrieve three-dimensional information of solid dosage forms, in a nondestructive way and at a molecular level [1]. Due to the increasing demands of the pharmaceutical industry, namely, the increased complexity of new formulations and the need to better understand them, these types of techniques started to be explored and used in the pharmaceutical industry, since 2003, with different purposes [2–4].[1]

When referring to CT, it is underlined the use of radiation, within the electromagnetic spectrum, as a source for imaging reconstruction. Most CT use X-rays, and so the most common technique is the micro X-ray computed tomography (XμCT or simply μCT). Although this is the most frequent technique, there are several others that are emerging like magnetic resonance imaging (MRI), imaging at terahertz frequencies, and optical coherence tomography (OCT) [1].

This chapter will focus on the use of X-ray CT for the advanced characterization of solid dosage forms in the pharmaceutical industry.

4.7.2 Micro X-ray Computed Tomography Scan

μCT is a nondestructive X-ray transmission imaging technique. In this technique, X-rays are emitted from an X-ray generator, travel through a sample, and reach a detector placed on the other end of the sample. The sample is then rotated by a fraction of a degree and the process is repeated. The process is iterated until the sample has rotated 180° or 360° producing a series of projection images. The outcome is obtained by processing all these projections and reconstructing the three-dimensional image that provides the internal structure of the object [5].[1]

Solid State Development and Processing of Pharmaceutical Molecules: Salts, Cocrystals, and Polymorphism, First Edition. Edited by Michael Gruss.

The advantage of having X-rays as the source of radiation, especially in the pharmaceutical industry, is that this high-energy (low wavelength) radiation can easily penetrate most pharmaceutical excipients and drug substances, allowing high-resolution images [1]. Moreover, using smaller X-ray focal spot sizes and finer detectors, it is possible to increase even more the image resolution. On the other hand, the use of X-rays also has some drawbacks, since the image contrast is dependent not only on the ability of X-rays to penetrate the sample but also on the different densities of solid components. Therefore, it is not uncommon to obtain poor image contrast that often leads to a loss in the power to distinguish between different components in the sample [1, 2, 5].

A CT system comprises an X-ray tube, a detector, and a sample holder apparatus. Additionally, a working station for data acquisition and data treatment is also required. To obtain 3D images, the instrument is nowadays equipped with diverging cone beams and a scintillator screen coupled with a two-dimensional CCD array detector [1]. After the scan, the image is processed applying algorithms capable of reconstructing the 3D image obtained from the cone-beam data [1]. Although these are the most common types of equipment used in the pharmaceutical industry, they still present some limitations. The most relevant being the trade-off between sample size and magnification due to the shadow of the object in the field of view. Another inherent limitation of this type of configuration is the susceptibility of having ring artifacts in the reconstituted images, generated by the shadow of the object being imaged in the array detector [1, 6]. These can be corrected using time delay integration models in the processing of the data. All these considerations should be kept in mind when applying these methodologies to quantitative analysis [6]. A detailed overview of the different types of artifacts can be found elsewhere [7] as well as some tips on how these can be overcome [8].

4.7.2.1 The Use of CT in the Pharmaceutical Industry

In the last decades, there has been an increasing number of publications exploring the use of CT for advanced characterization of solid dosage forms. That is because this technique has the potential to give some insight of the entire solid dosage form, at a molecular level. Moreover, as this is a tomography technique, information about the three dimensions of the solid dosage forms can be retrieved in a nondestructive way (Figure 4.7.1).

Hancock and Mullarney [5] in 2005 have compiled the first overview of the range of applications of this technique in the pharmaceutical industry and later,

Figure 4.7.1 3D μCT reconstructed images from a round tablet (perspective view).

in 2008, Zeitler and Gladden have made a more thorough review of its full potential [1]. The list of µCT applications includes porosity determination in tablets and granules [4, 6], defect analysis, failure analysis,[1] volumetric analysis of individual materials [6], particle size, shape, and volume [9, 10], spatial uniformity of active pharmaceutical component and also coating thickness and uniformity [2]. Some studies conducted with µCT also aimed to understand the mechanisms of drug release dynamics on the process of tablet disintegration [3, 11–13]. All the above-mentioned applications contribute to a better understanding of the final dosage form which is essential to develop new and innovative medicines.

Although this technology is not described in the most relevant pharmaceutical guidelines (Food and Drug Administration [FDA]; European Medicines Agency [EMA], United States Pharmacopeia and the National Formulary [USP-NF], or European Pharmacopoeia [Ph. Eur.]), their use is encouraged [14–19]. Better understanding of the formulation and the new drug products being developed requires the use of advanced characterization during formulation development, processing, and ultimately performance.

A more detailed overview of each application is discussed later.

4.7.2.1.1 µCT Applied to Density Distribution and Porous Characterization

The application of µCT to the characterization of tablets [3, 6, 20] and evaluation of granules porosity [4, 21] has been previously described in the literature.

X-ray tomography has been used to obtain detailed information on the structure, morphology of the pores, and its spatial distribution and connectivity [20]. However, when compared with the traditional techniques, such as gas absorption measurements, this technique has more limitations in what concerns pore size, especially if these are smaller than 4 µm [20]. This limitation is related to the inherent difficulty to unequivocally distinguish between air and final solid form constituents due to the close-like density. Nevertheless, the use of this technique to characterize compaction density (porosity, pore size, and distribution) in granules, tablets, and blends is widely described in the literature [1, 4, 6, 20–22].

Sovány et al.'s [20] work describes in detail the use of CT scan to understand how the spatial distribution of pores influences halving of tablets. This type of characterization allows a better insight of the general density and force distribution inside the tablet. The direction of the lines of force and of the pore structure enables to understand why two tablets with the same porosity show differences in halving. Moreover, the authors correlate these microstructures with tableting parameters, therefore supporting an easier scale-up or selection of equipment train. This study provided a better understanding of how the granulation and compression processes impact on the general density and force distribution within the tablet, and consequently the impact of this on the final quality attributes of halving in tablets.

Another study [4] demonstrated the benefits of using CT scans for the characterization of force distribution within a tablet when correlated with tooling and compression forces. These authors [4] also suggested that the three-dimensional

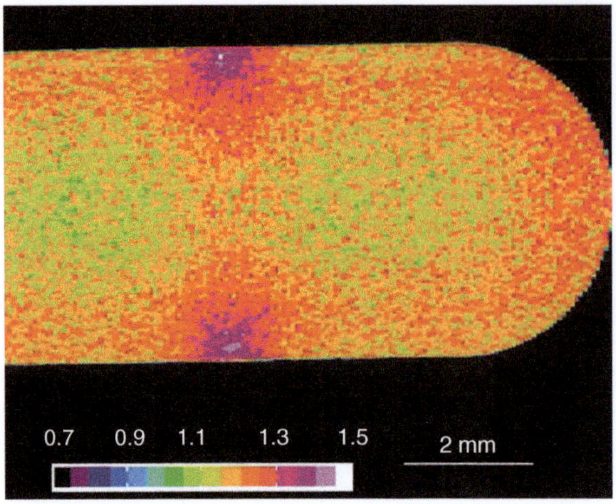

Figure 4.7.2 Density distribution in X–Y cross section by CT scan for a halving tablet showing a high-density spot in the region of the break line of the tablet. Source: Image taken form Ref. Sinka et al. [6].

distribution of density can be correlated with typical quality attributes of these types of solid dosage forms, such as friability. Additionally, Sinka et al. [6] used μCT to create quantitative density maps in tablets (Figure 4.7.2). This information was then used to optimize and understand the real effect of friction interactions between powder and the equipment, the tooling, and the process of punch motion. Although density maps were obtained in this work, the authors highlighted the difficulties inherent to this type of analysis and identified some corrections that needed to be made when collecting these maps. These included corrections to nonlinear experimental effects associated with the low X-ray attenuation experienced in the most common pharmaceutical components of a tablet.

In another study using dry coating tablets, the relationship between exfoliation between core and coating and compaction pressure was evaluated [23]. Busignies et al. [24] demonstrated that this technique can also be used to understand how different compaction forces influence local density and distribution.

All the studies aforementioned suggest that there is a correlation between density distribution and tablet compaction pressure, tooling, and friability [4, 6, 20, 23, 24].

This technique has been reported as the preferred analytical methodology to fully characterize granulation processes (e.g. wet granulation). Understanding hollowness of the cores allows a better knowledge of the impact of the process in the quality attributes of the granules [4, 21, 22]. CT has also been used to investigate the impact of compaction and blending process [20] on the formulation. A qualitative and quantitative study on the impact of different excipient particle sizes and different loading configurations in the process efficiency was therein described [20].

Crean et al. [21] have also used CT to perform a deeper study on the porosity of granules obtained via wet granulation. In this study, it was highlighted that, although CT was not able to characterize porous smaller than 3 nm, contrary to conventional mercury porosimetry, it enabled the discrimination between intra- and intergranular porosity.

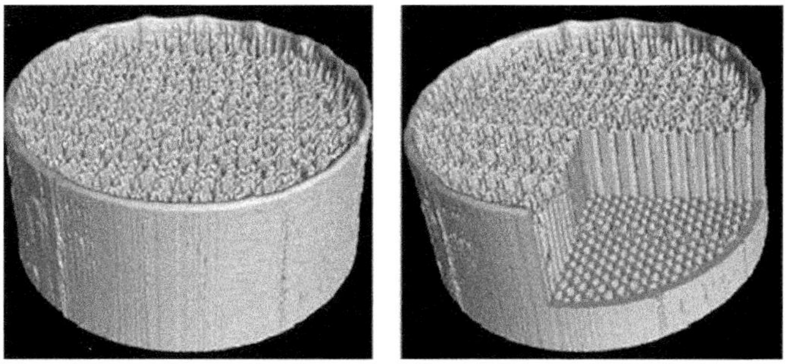

Figure 4.7.3 3D μCT reconstructed images for the honeycomb architecture tablets. Source: Taken from Ref. Kyobula et al. [10].

4.7.2.1.2 μCT Applied for Characterization of Structural Features: Size, Shape, and Dimensions and Interfaces

Like other imaging techniques, μCT can be used for the simple characterization of size and shape of final dosage forms. However, what differentiates μCT from other imaging techniques is the possibility of fully characterize regions that are not commonly accessible.

The use of this technology has been reported for the characterization of the emerging printlets [9, 10]. μCT has shown to be very useful in the characterization of a tunnel-like shape pharmaceutical dosage form produced with a 3D printer that enhanced the controlled release features [10]. Making use of μCT analysis, the detailed interior structure of the honeycombs architecture was revealed (Figure 4.7.3), which could not be done with conventional imaging techniques.

Using X-ray computed tomography, Goyanes et al. [9] were able to assess the internal structure of printlets, manufactured with a 3D printer. The printlets were produced from enteric polymers, by single filament fused deposition modeling, with different drug loads. Using this advanced characterization, the authors were able to understand the effect of drug load in the creation of cavities in the cores as well as on the density of the tablets. They were then able to correlate these characteristics with performance, i.e. dissolution behavior [9].

Finally, it has also been reported the use of this tomography technique for the characterization of defect and fractures in bilayer tablets [25].

4.7.2.1.3 μCT Applied to Coating Characterization

Another application of this technique is the characterization of coatings. Several authors previously mentioned the use of μCT for this purpose [1–3, 5]. Coatings have several functions, and solid dosage forms must comply with several quality key attributes [24, 26–29] like uniformity of color, porosity, and thickness as these parameters impact on drug performance and stability [25].

One of the advantages of CT in the characterization of coating is the fact that the coated layer can be characterized independently of the core (Figure 4.7.4).

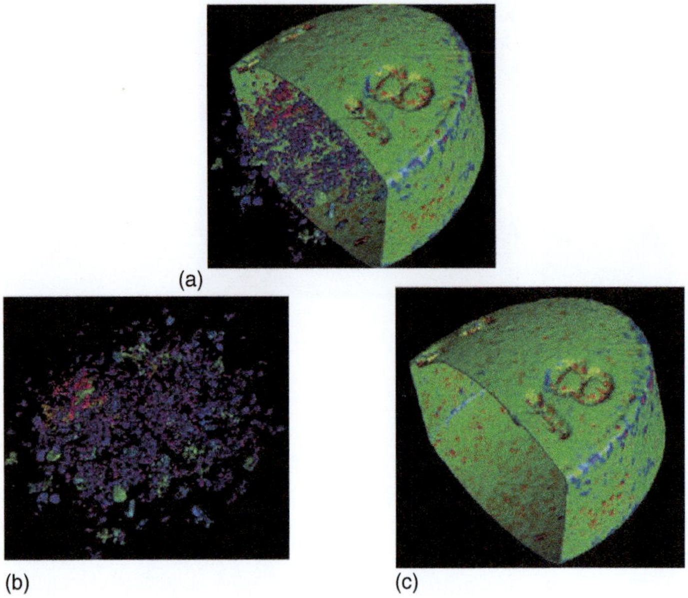

Figure 4.7.4 μCT 3D image of a tablet core and coating (a), the core only (b), and the coating only (c). Source: Image taken from Ref. Muehlhauser and DeRoo [2].

Figure 4.7.5 μCT image of a coated tablet to evaluate coating layer thickness, uniformity, and gap between multiple layers. Source: Image taken from Ref. Muehlhauser and DeRoo [2].

Another feature that has been reported is the ability to identify several layers and characterizing the gaps between those (including voids in between or in the layers) (Figure 4.7.5) [2].

The thickness of coating of tablets is a useful information when trying to understand the physicochemical properties of the dosage forms, both in a deformulation strategy and in modified release systems [3, 10].

4.7.2.1.4 μCT Applied to Performance Evaluation

For final dosage forms with controlled release applications, the understanding of the dynamic release of the drug is of high importance. There have been some studies demonstrating the suitability of μCT to the dynamic advance characterization of solid forms [3, 6, 12].

Karakosta et al. [12] were able to demonstrate that, although μCT has some limitations due to the poor contrast between water and the excipients used in a formulation, the images of tablets exposed to water, collected over time, allowed understanding how dissolution (water entrance) occurs, by looking at how voids change with time. The information provided by the entire tablet images, acquired over time, enabled the quantitative investigation of the dynamics involved in the dissolution process. The authors were able to correlate the air voids growth and their collapse with diffusion models, allowing a deeper knowledge on the performance of the solid dosage form.

The same type of study on pellets was published by Chauve et al. [13]. The authors used μCT to evaluate the three-dimensional structure of the pellets after manufacturing, and at successive time points, after being exposed to water. It was concluded that dissolution occurs in two steps: the first one consists of the pellet surface quenches, followed by water transport to the core, while the second step was the quenches of the core, achieved by the formation of a membrane that evolved into a hydrogel.

In another work, the dynamic properties of pore structure and porosity during dissolution in osmotic tablets (extended release) were evaluated [11]. As for the previous studies [12, 13], the CT scans were acquired in the whole dried tablet, and after several time points, during dissolution. The CT images revealed the swelling of the polymer at the surface of the tablet followed by the release of the drug (Figure 4.7.6). The dynamic images proved that the release of the drug was not homogeneous and was happening in the external part of the tablet rather than in the core. The work demonstrated the suitability of CT scans for mapping dissolution profiles.

Furthermore, Oliveira Junior et al. [3] highlighted the relevance of this type of studies to gather information for the deformulation process [3]. In their study, the three-dimensional morphology of an osmotic tablet was characterized, including the interior part, containing the drug, the osmotic agent, and coating layers as well as the orifice through which the drug is expelled. The authors were able to understand the release profile of the drug as well as the rate of disintegration.

It is important to highlight that in the majority of the works herein cited, the evaluation of the experimental setup for imaging the final dosage forms while dissolution is occurring has been debated. Some disadvantages can be foreseen, and in some cases, comparative studies between "as is" samples (imaging directly after removal from the dissolution medium) and samples that were freeze-dried or dried in an oven (prior to imaging) were made. The possibility of in-line CT scans is becoming a reality, and so the need for these sample preparation procedures loses relevance.[2,3]

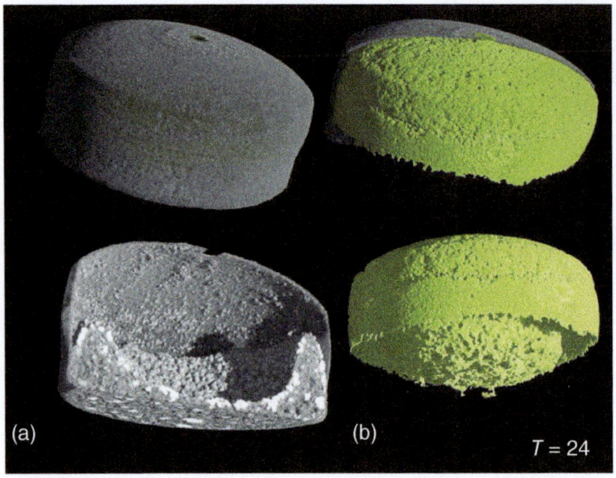

Figure 4.7.6 3D μCT reconstructed images "after 24 hours in the dissolution batch. (a) Visualization of the empty drug cavity with the collapsed swellable polymer at the bottom of the semi-permeable shell, still intact. (b) The void, previously occupied by the drug, highlighted and presented as a three-dimensional space." Source: Taken from Ref. Traini et al. [11].

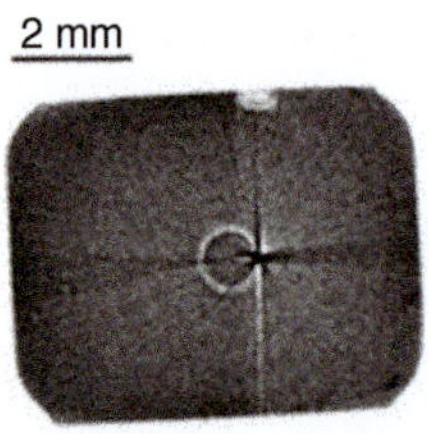

Figure 4.7.7 μCT scan showing a metallic particle in a tablet. Source: Image taken from Ref. Hancock and Mullarney [5].

4.7.2.1.5 Foreign Matter Detection by μCT

Another application of μCT in the pharmaceutical industry is related to the identification of foreign matter [2, 5]. The basic principles of this technique rely on the fact that different components are penetrated differently by X-ray due to different atomic densities of the materials. Therefore, it can be clearly understood that when foreign particulates are present, a CT image can clearly distinguish these from the remaining pharmaceutical ingredients (e.g. metallic particles have very different densities compared to the solid dosage forms constituents; Figure 4.7.7). Although this type of analysis cannot yet be done in an entire batch of final dosage forms, the suitability of this technique to clearly identify foreign particles may be pushing the application of in-line measurements and implementation of this type of technology as a PAT tool.[3,4]

List of Abbreviations

CT	computed tomography
μCT	X-ray microtomography
MRI	magnetic resonance imaging
OCT	optical coherence tomography

Notes

1 http://www.blue-scientific.com/micro-ct-in-pharmaceuticals/ (accessed 1 July 2019).
2 https://www.microphotonics.com/products/micro-ct/skyscan-1275-micro-ct/ (accessed 10 September 2019).
3 https://www.industrial.ai/inspection-technologies/radiography-ct/x-ray-computed-tomography/speedscan-ct-64 (accessed 10 September 2019).
4 https://www.tescan.com/en-us/technology/x-ray-instruments/unitom (accessed 10 September 2019).

References

1 Zeitler, J.A. and Gladden, L.F. (2008). In-vitro tomography and non-destructive imaging at depth of pharmaceutical sold dosage forms. *European Journal of Pharmaceutics and Biopharmaceutics* 71: 2–22.
2 Muehlhauser, B. and DeRoo, N. (2016). Computed tomography: applications for evaluating tablets and capsules. *Tablets and Capsules.*
3 Oliveira, Junior, J.M., Balcão, V.M., Vila, M.M.D.C. et al. (2015). Deformulation of a solid pharmaceutical form using computed tomography and X-ray fluorescence. *XXXVII Brazilian Meeting on Nuclear Physics, Journal of Physics: Conference Series* 630: 012002.
4 Farber, L., Tardos, G., and Michaels, J.N. (2003). Use of X-ray tomography to study the porosity and morphology of granules. *Powder Technology* 132: 57–63.
5 Hancock, B.C. and Mullarney, M.P. (2005). X-ray microtomography of solid dosage forms. *Pharmaceutical Technology* April.
6 Sinka, I.C., Burch, S.F., Tweed, J.H., and Cunningham, J.C. (2004). Measurement of density variations in tablets using X-ray computed tomography. *International Journal of Pharmaceutics* 271: 215–224.
7 Davis, G.R. and Elliott, J.C. (2006). Artifacts in X-ray microtomography of materials. *Materials Science and Technology* 22 (9): 1011–1018.
8 Kozul, N., Davis, G.R., Anderson, P., and Elliott, J.C. (1999). Elemental quantification using multiple-energy X-ray absorptiometry. *Measurement Science and Technology* 10 (3): 252–259.

9 Goyanes, A., Fina, F., Martorana, A. et al. (2017). Development of modified release 3D printed tablets (printlets) with pharmaceutical excipients using additive manufacturing. *International Journal of Pharmaceutics* 527 (1–2): 21–30.

10 Kyobula, M., Adedeji, A., Alexander, M.R. et al. (2017). 3D inkjet printing of tablets exploiting bespoke complex geometries for controlled and tuneable drug release. *Journal of Controlled Release* 261: 207–215.

11 Traini, D., Loreti, G., Jones, A.S., and Young, P.M. (2008). X-ray computed microtomography for the study of modified release systems. *Microtomography* 22 (1): 13–15.

12 Karakosta, E., Jenneson, P.M., Sear, R.P., and McDonald, P.J. (2006). Observations of coarsening of air voids in a polymer-highly-soluble crystalline matrix during dissolution. *Physical Research E* 74 (1): 011504.

13 Chauve, G., Raverielle, F., and Marchessault, R.H. (2007). Comparative imaging of a slow release starch excipient tablet: evidence of membrane formation. *Carbohydrate Polymers* 70 (1): 61–67.

14 EMA May 2000 CPMP/ICH/367/96 (n.d.). ICH Topic Q 6A, specifications: test procedures and acceptance criteria for new drug substances and new drug products: chemical substances.

15 FDA (2004a). *Pharmaceutical CGMPs for the 21s Century – A Risk Based Approach*. FDA.

16 ICH (2009). *Q8(R2) Pharmaceutical Development*. ICH.

17 ICH (2005). *Q9 Quality Risk Management*. ICH.

18 ICH (2008). *Q10 Pharmaceutical Quality System*. ICH.

19 ICH (2018). *Q14: Analytical Procedure Developemnt and Revision of Q2(R1) Analytical Validation*, Final Concept paper. ICH.

20 Sovány, T., Kása, P. Jr., and Vakli, K. (2009). X-ray computed microtomography for determination of relationships between structure and breaking of scored tablets. *X-Ray Spectrometry* 38: 505–509.

21 Crean, B., Parker, A., Le Roux, D. et al. (2010). Elucidation of the internal physical and chemical microstructure of pharmaceutical granules using X-ray micro-computed tomography, Raman microscopy and infrared spectroscopy. *European Journal of Pharmaceutics and Biopharmaceutics* 76: 498–506.

22 Sondej, F., Rieck, C., Schmidt, M. et al. (2019). Use of X-ray micro-computed tomography in pharmaceuticals and food industry. 6th Conference on Industrial Computed Tomography (ICT).

23 Ozeki, Y., Watanabe, Y., Inoue, S., and Danko, K. (2003). Comparison of the compression characteristics between new one-step dry-coated tablets (OSDRC) and dry-coated tablets. *European Journal of Pharmaceutics and Biopharmaceutics* 259 (1–2): 69–77.

24 Busignies, V., Leclerc, B., Porion, P. et al. (2006). Quantitative measurements of localized density variations in cylindrical tablets using X-ray microtomography. *European Journal of Pharmaceutics and Biopharmaceutics* 64 (1): 38–50.

25 Inman, S.J., Briscoe, B.J., and Pitt, K.G. (2007). Topography characterization of cellulose bilayered tablets interfaces. *Chemical Engineering Research and Design* 85 (A7): 1005–1012.

26 FDA (2015). *Size and Shape, and Other Physical Attributes of Generic Tablets and Capsules; Guidance for Industry*. FDA, June.

27 USP-NF <1151> (2020). Pharmaceutical dosage forms.

28 USP-NF <1062> (2017). Tablet compression characterization.

29 USP-NF<1217> (2019). Tablet breaking force.

4.8

In Situ Methods for Monitoring Solid-State Processes in Molecular Materials

Adam A. L. Michalchuk, Anke Kabelitz, and Franziska Emmerling

Department Materials Chemistry, Bundesanstalt für Materialforschung und -prüfung (BAM), Richard-Willstätter-Straße 11, 12489 Berlin, Germany

4.8.1 In Situ Methods for Monitoring Solid-State Processes in Molecular Materials

4.8.1.1 The Complexity of Solid Materials

The life of an active pharmaceutical ingredient (API) is complex. The molecule is first prepared in solution by some chemical synthesis, isolated as a crystalline solid, and formulated into a tablet or powder. This solid next travels to the patient, who ingests it. Once inside the body, the solid must dissolve, releasing the API back into solution to allow molecular-level interactions. The need for a solid-state intermediate product is often ascribed the enhanced stability of the molecule, which elongates shelf-life and product quality.

A solid material can be thought of as a large supramolecular entity. In this way, it is easy to draw similarities to chemical concepts. To the chemist, it is intuitive that molecular properties are defined in some way by the nature of the constituent atoms and their (intramolecular) connectivity. In this way, the substitution of atoms (or isotopes) can have a large influence on the physical properties of the molecule. It is not surprising to find that different isomers of the same atomic composition also display unique chemical properties. By analogy, the properties of a solid material are defined by the constituent molecules and their intermolecular connectivity. For discussion of solid materials one can simply replace the chemical concepts of *atoms*, *intramolecular connectivity*, and *isomers* with *molecules*, *intermolecular connectivity*, and *polymorphic forms*, respectively. While this analogy does somewhat oversimplify the true complexity of solid materials, it holds for the general case and serves to illustrate the necessary discussion of this chapter. With this simple analogy in mind, it is straightforward to identify why the correct function of a pharmaceutically active solid depends intimately on its structure.

The structure of solids is considerably more complex than that of an isolated molecule. To a first approximation, a solid comprises individual molecules, which interact via non-covalent interactions (e.g. van der Waal's interactions, $\pi \ldots \pi$

Solid State Development and Processing of Pharmaceutical Molecules: Salts, Cocrystals, and Polymorphism, First Edition. Edited by Michael Gruss.

stacking, and hydrogen bonds). These interactions are repeated through translation in three dimensions. In a real solid, some molecules may have entered the solid with incorrect geometry, or structurally similar molecules may be entered instead. This leads to *local defects*. Alternatively, the translational symmetry in one-, two-, or three dimensions may be imperfect, producing *extended defects*. Both defect types can have considerable influence on the properties of the solid, including solubility and mechanical strength. An additional complication of solids is the high proportion of surface. This extended defect-type affects properties such dissolution rate, phase stability, hygroscopicity, and friability. Hence, the production of solids with correct performance requires careful control of structure across many orders of scale.

It is an unfortunate reality that it is all but impossible to obtain a solid with all the desired physical qualities. This is because the self-assembly process is stochastic and results from an ill-understood interplay between kinetics and thermodynamics. As such, it is not yet possible to reliably target and control solid forms. Some control can be obtained through the careful selection of crystallization medium and crystallization rate, although careful monitoring is still required in order to ensure the correct solid form is being produced.

Unlike the investigation of molecular structure, analysis of solids can be performed at many different scales of size. Indeed, one may wish to identify the structure of molecular subunits that make up the solid form, for example, to prove the existence of multi-component crystals. In other cases, classification of polymorphic form may be of interest, for which the structure of intermolecular interactions is required. The presence of extended defects (e.g. dislocations and stacking faults), surface structure, and particle size or shape are also of great interest when characterizing the physical properties of solids. The correct choice of analytical method therefore depends intimately on the problem at hand, and the preparative procedure being followed.

Among the many phenomena that can occur within and between solids, perhaps the most important for pharmaceutical sciences is the formation and physical transformation of solid forms. Traditionally, these processes have occurred in batch (or more recently continuous) solution reactors. With recent pressures to design ecologically benign processes, however, many solid-state transformations have instead moved toward mechanochemical techniques. It is therefore prudent to examine which techniques are available for monitoring these new mechanochemical transformations and how they differ when considering their use in traditional solution-based techniques.

4.8.1.2 Methods to Consider

There are typically two major questions associated with producing solid products, (i) how quickly the crystallization of the solid takes place and (ii) the phase and purity of the resulting solid. To some extent, the techniques that are available to answer these questions depend on the physical state of the initial sample, solution or solid. From a technical viewpoint, the question is further divided into the type of information one hopes to obtain. In this way, techniques can be broadly divided into four categories, Scheme 4.8.1. The first two categories are pertinent only to crystallization from solution. In the first (A, see Section 4.8.1.3) the kinetics of crystallization are monitored by studying the rate at which material is lost from solution. In this case, no information is obtained about the solid product. Related are methods (B, see Section 4.8.1.4) that follow crystallization from solution by monitoring the formation of the solid itself. Typically, these techniques can offer only limited information regarding the physical state of the solid product (e.g. particle size). The final two categories are common across the solution- and solid-state, differing only in technical detail. They can (C, see Section 4.8.1.5) provide limited information regarding the solid product, or (D, see Section 4.8.1.6) intrinsic structure of the solid product.

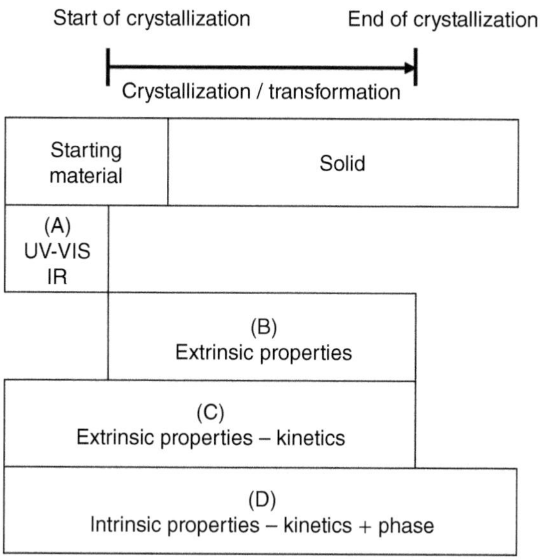

Scheme 4.8.1 General classification of analytical techniques for monitoring solid-state transformations. Starting materials refers in the solution-based crystallization to "solution" and in the mechanochemical approach to "solid."

The study of process in situ requires (i) that the probing radiation is able to enter the sample vessel, interact with the sample, and subsequently leave the vessel to be detected and (ii) that sufficient signal can be produced and detected on time scales faster than the events being monitored. These two simple requirements place enormous restrictions on the ability of most techniques to monitor transformations in situ and in real time.

In principle, any tool that captures a change in the system properties due to solid-phase processes can be used. Most tools are based on monitoring changes in vibrational, electronic, thermal, or scattering properties of the system. While we cannot hope to discuss all methods that are available, this chapter aims to introduce the most commonly employed techniques that are used to monitor solid-state processes in solution and, in particular, during mechanochemical

Table 4.8.1 Overview of major analytical techniques for in situ monitoring of solid-state processes.

Method	Mechano-chemistry	Crystallization (solution)	Solid-state information?	Classification (see Scheme 4.8.1)
UV–Vis spectroscopy	No	Yes	Indirect information	A
Infra-red spectroscopy	Yes	Yes	N	A
Light scattering	No	Yes	Extrinsic	B
Thermography	yes	yes	Extrinsic	C
Acoustic emission	Yes	Yes	N	C
X-ray diffraction	Yes	Yes	Intrinsic/extrinsic	D
Raman spectroscopy	Yes	Yes	Intrinsic	D

treatment, Table 4.8.1. A number of excellent reviews are also available in the literature, which predominantly highlight the applications of in situ monitoring of solution-phase processes[1–3]. Additional focus is aimed at identifying the technological differences that are required to monitor processes in solution vs. during mechanochemical reactions.

Prior to further discussions, it is worth making a note on common nomenclature. Within the community of solution-phase crystallization, it is common to discuss the monitoring of solution-phase transformations *off-line, at-line, on-line,* or *in-line*. These describe, respectively, manually transferring sample away from the reaction vessel for analysis at a different or the same location, automatic redirection of sample from the apparatus into an analysis compartment, and subsequent reinsertion into the reaction vessel, and analysis in real time, directly in the reaction vessel. In contrast, the mechanochemistry community discusses analytical procedures as begin ex situ, in situ, or simultaneously in situ and in real time. While the latter two are often used (erroneously) as being interchangeable, the first simply describes a set-up in which the system is monitored within the reaction vessel and is not disturbed, although the reaction is stopped. In contrast, the second describes the situation in which monitoring is performed *during* the reaction or transformation.

4.8.1.3 Methods to Monitor Crystallization Kinetics from Solution

4.8.1.3.1 UV–Vis Spectroscopy

The Basics Conceptually, the simplest way to monitor crystallization (i.e. controlled solidification) is via a simple mass balance. That is to say that the precipitation (i.e. *rapid, uncontrolled* solidification) as well as the crystallization of a target molecule is associated with a decrease in the concentration of that molecule in the solution. In this way, the crystallization itself can be monitored indirectly, by identifying the rate at which the target molecule is lost from solution. Fortunately, the Beer–Lambert relation,

$$A = \epsilon.c.l \tag{4.8.1}$$

Directly relates the absorption, A, of an incident beam to the concentration of a molecule in solution c. The molecular absorption coefficient ϵ relates the strength with which the target molecule absorbs the incident photon.

Considerations for In Situ Applications While UV–Vis measurements offer a powerful means to monitor crystallization kinetics, they are not universal. Not all molecules exhibit measurable absorption of photons within the UV–Visible region. Furthermore, some solvent systems may instead absorb strongly through the UV–Visible region. A major restriction to Eq. (4.8.1) is its dependence on the path length, l. The linearity of the Beer–Lambert relation, which is its main strength, is lost for highly concentrated samples. Hence, the typical transmission mode set-up which is common to chemistry laboratories is no longer relevant. Instead, UV–Vis probes typically work within the remit of attenuated total reflectance (ATR). One can therefore monitor solution concentration also in systems with high concentration of solid products.

Examples (ATR)-UV/Vis spectroscopy has proven indispensable for monitoring crystallization processes from solution. In particular, the simplicity of the Beer–Lambert relation has proven invaluable to assess the quantity of solute in solution, and hence monitor (super-)saturation. Although UV–Vis measurements have been previously used to monitor the saturation levels of crystallizing systems, more sophisticated applications are now developed. With knowledge of the solubility diagram, Nagy and coworkers [4] demonstrated how real-time and in situ monitoring by ATR-UV/Vis spectroscopy could be used to actively and autonomously control polymorphism in *ortho*-aminobenzoic acid (OABA), Figure 4.8.1. By monitoring supersaturation levels through ATR-UV/Vis spectroscopy, the supersaturation levels could be monitored. Through the careful control of these saturation levels via temperature, various polymorphs could be selectively prepared. Similar applications of ATR-UV/Vis spectroscopy have now been demonstrated on a large variety of systems [5–8], including the control of particle size in paracetamol through careful control of supersaturation levels, as monitored by UV/Vis spectroscopy [6].

4.8.1.3.2 Infrared Spectroscopy

The Basics *Infrared spectroscopy (IR)* is based on the absorption of radiation by quantized oscillators in the sample. As an absorption technique, IR spectroscopy is generally performed by scanning across a range of incident photon energies (c. 0–4000 cm^{-1}). When incident on the sample, the dipole moment of the oscillating radiation interacts with the dipole moment of molecular normal modes. At energies equivalent to an excitation (e.g. $\nu_1 \leftarrow \nu_0$), the incident photon is absorbed. Hence, IR spectroscopy does not monitor the vibrational modes themselves, but their excitation. This can be approximated by the harmonic expression $\Delta E = \hbar\omega(\Delta\nu)$. The intensity of absorption depends on the magnitude of the dipole moment of the normal mode and the magnitude of overlap between the ground and excited vibrational states. This imposes the restrictions that $\Delta\nu = \pm 1$ and only modes with a dipole moment are IR active.

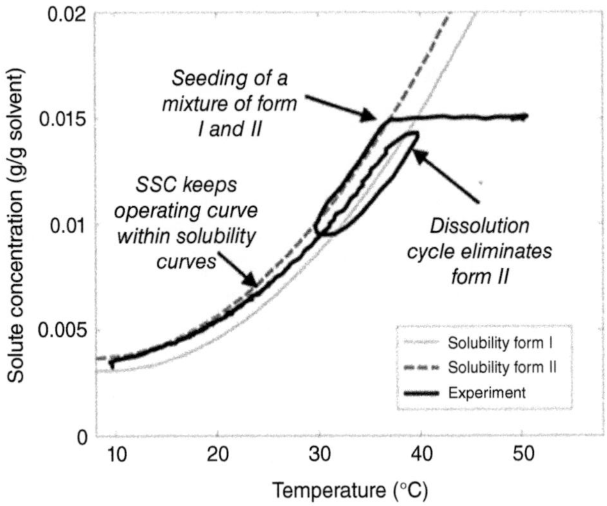

Figure 4.8.1 Controlling supersaturation levels to selectively nucleate polymorphic forms of OABA. The solubility lines are shown as gray and dotted lines, and the experimental path is highlighted as a solid black line. (SSC: supersaturation control). Source: Reprinted with permission from Ref. Simone et al. [4]. © 2014 American Chemical Society.

Considerations for In Situ Real-Time Applications As an absorption spectroscopy, applications of IR require a scanning wavelength incident beam. For this reason, focused lasers are not commonplace. Radiation cannot therefore be focused through sample vessels, and the observed IR signal becomes a function of all irradiated phases. Thick vessel walls effectively absorb all of the incident radiation, rendering non-contact methods difficult. Instead, monitoring crystallization from solution is readily done with the use of ATR-Fourier-transform infrared spectroscopy (FTIR) methods. For this purpose, an immersion optic is typically placed within the solution, and the spectra are collected from material which comes in direct contact with the probe. In this way, the observed signal is not greatly diminished by the attenuation from a large sample volume. The principles of ATR also restrict applications of this approach to capturing spectra in the liquid state; solid products are therefore not often observed. Unfortunately, given these limitations, IR spectroscopy has not been possible as a method to monitor in situ and in real-time mechanochemical transformations. Very often, process monitoring is instead performed using near-infrared spectroscopy (NIR) spectroscopy, probing between c. 780 and 2500 nm. This region corresponds to the excitation of molecular overtone and combination bands. As these transitions are forbidden according to quantum mechanics, NIR is much more poorly absorbed, and can therefore penetrate deeper into a material. The tradeoff, however, is that the spectra are poorly resolved, and often require statistical analysis in order to obtain chemical information. For this purpose, extensive use is made of multivariate calibration techniques, including principal components analysis, partial least squares methods, and artificial neural networks.

As with other spectroscopic techniques, absolute values are not realistically attainable from IR-based methods. All quantitative measurements therefore require careful construction of calibration curves, relating product quantity to vibrational band intensity. Construction of calibration curves for IR-based techniques often involves the use of multivariate modeling.

Examples While UV–Vis measurements are often regarded as "golden-standard" approach for monitoring the solution concentration during a crystallization experiment, properly calibrated IR spectroscopic techniques have also been used. In the batch crystallization of paracetamol, multivariate-based calibration models were used to correlate solution-phase paracetamol as a function of concentration and temperature [9]. This allowed the authors to establish a full solubility curves for the material and control the crystallization procedure. The pH-dependent reactive crystallization of L-glutamic acid offers an additional example where multivariate analysis has been successfully employed to monitor the rate of crystallization, here as a function of temperature and pH [10].

A much simpler calibration method was suggested for the aqueous crystallization of succinic acid [11]. Rather than constructing statistical calibrations as in the case of paracetamol, the authors instead compared the integral intensity of characteristic vibrational bands, namely that of the solvent (water) and of the solute (succinic acid). By monitoring the ratio of these two bands, the saturation profile was successfully monitored and controlled over the course of cooling crystallization. This approach to simple calibration curves has been successfully employed during the scale-up from laboratory to industrial scale [12].

4.8.1.4 Monitoring Crystallization from Solution: Following Solid Product Formation

4.8.1.4.1 Light Scattering

The Basics Light scattering involves the elastic interaction between an object (here particles or molecules) and a photon. Upon interacting with the object the photon momentum shifts, deviating from the path of the incident beam, Figure 4.8.2. The coherence of the scattered radiation varies with the motion of the scatterers, which depends on their size. It follows that the observed intensity of scattered radiation, I, also varies in time. In its simplest form, $I = I_0 \cdot e^{-\alpha\tau}$, where I_0 is the incident intensity, α is the decay rate, and τ is the time. The decay rate is proportional to the diffusion coefficient, and hence (via the Stokes–Einstein equation) to the size of the scattering object. Techniques that depend on the time dependence of the scattering intensity are known as *dynamic light scattering (DLS)* and is typically restricted to hydrodynamic radii between 1 and 1000 nm. In other applications, it can be useful to use instead the time-averaged intensities. This technique, known as *static light scattering (SLS)*, can be used to obtain information regarding the average mass or gyration radii of scatterers. SLS is often also used as a simple tool to monitor bulk transformations in crystallizing systems.

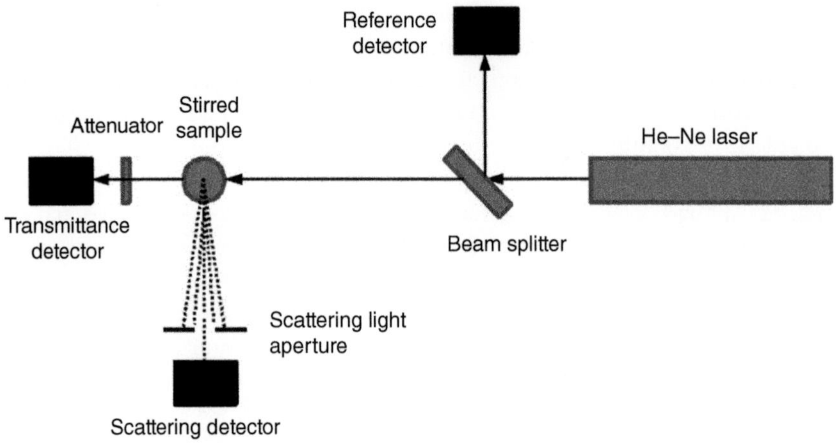

Figure 4.8.2 Schematic representation of a light scattering experiment. Source: Reprinted with permission from Ref. Sypek et al. [13]. © 2012 American Chemical Society.

A particularly popular method for in situ and time-resolved monitoring of both particle size distribution (PSD) and particle number is the focused beam reflectance measurements (FBRM) [14, 15]. In this technique, a rapidly rotating beam is focused into a slurry sample. When incident on a particle, the radiation is backscattered into the detector until the beam reaches the opposite end of the particle. Hence, FBRM measurements provide a measure of particle chord length. Each chord length can be retained, thereby producing a chord length distribution (CLD). Although it is in principle possible to relate the CLD count to the number of particles in the sample, a robust strategy to do so has not yet been developed. This owes to complexities surrounding agglomeration, statistical sampling of large and small particles, and approximations made in order to interpret particle shape from CLD.

Considerations for In situ Applications Product formation can be monitored directly in solution using light scattering methods. These techniques are typically fast (on the order of seconds to minutes), and with commercial devices, for example, the ZetaSizer™, are easy to perform. A plethora of information can be obtained in these short collection times, including solid particle size (typically on the nm scale) and cloud points (turbidity). High polydispersity can, however, lead to marked misinterpretations of scattering data. Of particular note, whereas other laser methods (e.g. Raman spectroscopy) often require large amounts of material, light scattering measurements can be performed on small volumes (c. 1 mL). The major restrictions pertain to the light–matter interactions. The experimenter must ensure that the refractive index of the solvent is different from the molecules or particles in solution. Furthermore, the sample should not absorb at the wavelength of the incident laser light.

Additional complications arise when considering CLD methods. In order to link scattering data to particle size, mathematical models are required to generalize particle shape. This is particularly problematic for aggregates, which appear to the CLD device as a single scattering object. These models are always idealized, although it is possible to increase model sophistication to varying degrees. As such, all particles for which the same chord length is measured are assumed to be identical. This problem is further amplified when particles reach the dimensions of the rotating laser beams. In such cases, the relationship between particle size and the CLD signal becomes severely compromised. As the number of large particles increases, the sensitivity toward smaller particles decreases rapidly and can lead to erroneous interpretation of the PSD [16]. Furthermore, other object, for example bubbles, can also cause issue for CLD measurements.

Most common light scattering methods require particles to be suspended in a solvent. While this is no issue for solution crystallizations, it can be problematic when applied to dry powders. The interaction of a dry powder with solvent invariably alters the surface structure, can induce agglomeration, and (where the powder is even partially soluble in the solvent) can lead to marked changes in particle size and shape. Care must therefore be taken when applying light scattering methods to dry powders, or alternative (e.g. light scattering of air-dispersed powders) must be attempted.

Examples Due to the commercial availability of light scattering instruments, these methods are commonly used for in situ detection of solid phases. For the detection of early stages of crystallization with regular shapes like spheres the DLS/SLS approach is the method of choice [17–20]. Nucleation and growth information including the onset of crystallization, size, shape and size distribution of larger irregular particles in the μm range can be reached with the FBRM approach [6–8]. In situ monitoring of drug crystallization and fermentation enable to control processes for desired polymorphs or solid-state products [21, 22]

The group of Sefcik and coworkers [13] investigated a custom-built setup for the in situ study of different polymorphs of carbamazepine (CBZ) in anhydrous ethanol. They combined both transmitted and scattered light measurements to distinguish between different pathways of crystallization. The stirring of the solution was the key factor for controlling the formed polymorphs. The crystallization of commercial carbamazepine (COM) needs longer induction times than a recrystallized carbamazepine (REC) (Figure 4.8.3a). Besides the monitoring of the induction time, it was possible to distinguish between different polymorphs due to their different morphologies. Furthermore, the recrystallized CBZ showed a metastable intermediate during crystallization which is absent in the case of commercial CBZ.

The detection of particle sizes during the size reduction of crystalline slurries in pharmaceutical processes, using high shear wet milling is crucial [23, 24]. Inline CLD data can be used to determine the size and shape distribution, as well as number of the particles from saturated suspensions. In Figure 4.8.3b the CLD and their estimated PSD during the wet milling experiment of paracetamol in 2-propanol is shown [16]. By increasing the time, the total chord counts increase due to the breaking of particles during the process. By comparison of the inline CLD data with the corresponding offline data, it was shown that inline light scattering data are suitable for monitoring changes in slurries.

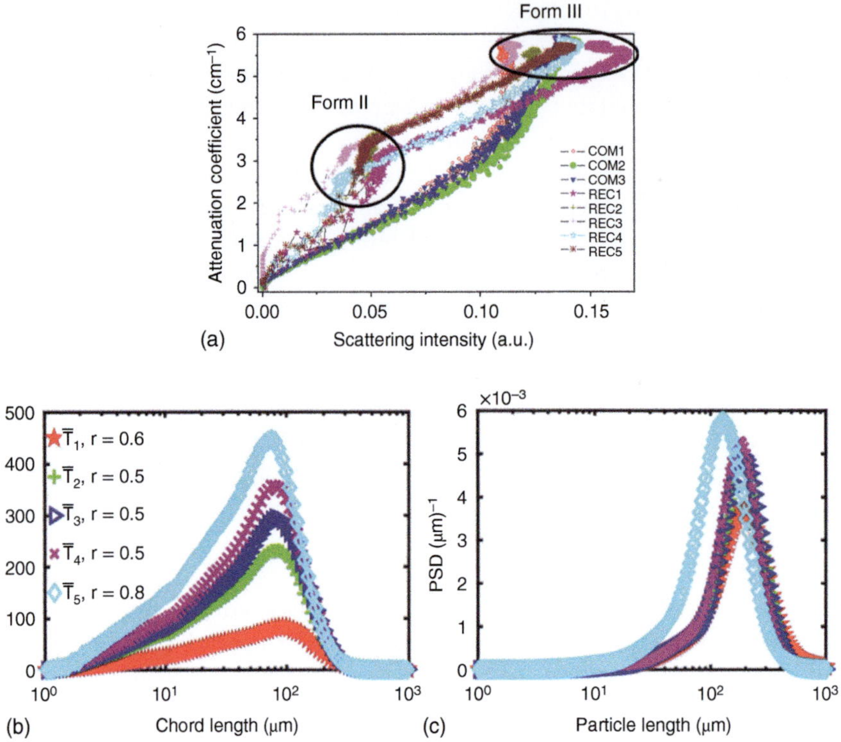

Figure 4.8.3 (a) Intensity data from attenuation coefficient against the 90° light scattering for several experimental repetitions (three times for the recrystallized CBZ (REC), five times for the commercial carbamazepine (COM). Source: Reprinted with permission from Ref. Sypek et al. [13]. © 2012 American Chemical Society. (b) Volume based data form inline chord length distribution (CLD) analysis for paracetamol during wet milling stages at different time intervals T_1–T_5. Left: The solid lines in are the estimated CLDs at the aspect ratios *r*. Right: The estimated volume-based particle size distributions (PSDs) from the CLDs shown in at the aspect ratios indicated in the left picture. Source: Reprinted with permission from Ref. Agimelen et al. [16]. © 2018 Elsevier.

4.8.1.5 Methods to Monitor Extrinsic Solid Properties

4.8.1.5.1 Acoustic Emission

The Basics The term *acoustics* refers to the generation, transmission, and detection of vibrational waves through matter. Each wave can be described by its speed, c, frequency, f, and wavelength, λ,

$$c = \lambda f = \sqrt{\frac{B}{\rho}} \tag{4.8.2}$$

where B is the bulk modulus (measure of resistance to compression) of the material and ρ is its density. The characteristics of the acoustic wave depend crucially on the mechanism of their formation. An emitted acoustic wave can be defined according to c and λ, as well as its amplitude, A. As a sinusoidal wave, the amplitude varies with time, and decreases as it travels through a medium due to attenuation. When the acoustic wave travels through fluid (liquid or gas), it is longitudinal. Hence, a traveling pressure wave is generated. This can be detected as a function of time and intensity.

Energy conservation requires that physical processes, including plastic deformation, particle-particle collisions, particle-wall collisions, and the rupture of agglomerates, emit unique acoustic waves of characteristic frequency and amplitude. These are the result of the vibrational properties of the colliding materials, which are related to their ratio B/ρ. The beating pattern is also related to the particle size. The detected signal therefore becomes a superposition of all acoustic emissions (AEs), whose frequencies are characteristic of the material and dynamics within.

Considerations for In Situ Real-Time Applications Monitoring solid-state processes by acoustic-based methods requires careful experimental considerations. Foremost, careful selection of the acoustic sensors and amplifiers must be made, such that the correct acoustic frequencies can be measured. Particularly when AE from mechanical apparatus (e.g. impellers or pumps) may introduce spurious signals, filters are required to eliminate unwanted signals. Typically, solid-state processes (including both mixing and crystallization) have been successfully monitored in the range of a 100–600 kHz [25, 26]. Depending on the applications, it may be possible to monitor transformations by variations in signal amplitude at fixed frequency [27], while other cases require broadband transducers [25]. As most acoustic transducers are based on the piezoelectric response of the sensor, careful attention must be paid to the calibration of the sensor, and its detection and temperature limits.

Although it has been demonstrated that external microphones can be used to detect AE [28], most current applications make use of surface transducers. In this case, the transducer is fixed to the surface of the sample vessel, detecting AEs that propagate through the vessel walls. This has the added benefit over microphone-based approaches as it intrinsically filters much of the background signals. Because AE transducers respond to acoustic *pressure* waves, they must be securely attached to the surface of the vessel, ensuring good contact throughout monitoring.

Perhaps the most difficult aspect of AE-based monitoring methods is the ability to resolve the acoustic contributions from a variety of processes. For example, the AE signal collected from crystallization in a stirred tank reactor will include contributions at least from the stirring itself [29], crystallization of different phases, and dissolution. Unlike many other methods, *a priori* assessment of AE from solid-state processes is not well established. Isolating individual signal contributions therefore requires thorough calibration of the AE properties of the characteristics under investigation.

Examples AEs are a promising, non-contact means to monitor solid-state processes. Nearly all physical phenomena emit acoustic waves, which are themselves dependent on the physical structure of the material system. Both the dissolution [27] and crystallization of pharmaceutical compounds have also been investigated via AE measurements [29, 30]. The crystallization of citric acid in a stirred tank reactor

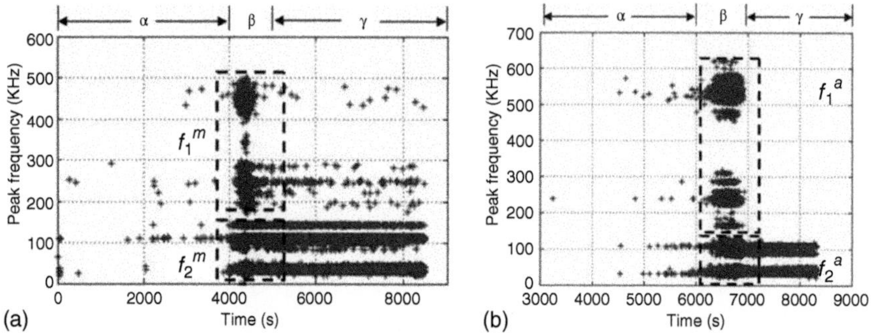

Figure 4.8.4 Acoustic emission profiles for (a) citric acid and (b) citric acid hydrate. Both profiles exhibit three stages, corresponding to (α) < 4100 s, (β) 4100–4800 s, and (γ) 4800–8000 s. Source: Figure adapted from Reprinted from Ref. Wang et al. [26]. © 2018 Elsevier.

offers a good example. In this system, the AE profile could distinguish between crystallization of anhydrous and hydrate forms [26], Figure 4.8.4, and their interconversion could be observed in real time [31]. It has been further shown that the AE signal contains information regarding particle size [25, 27], and can furthermore be used to monitor the efficiency and homogeneity of powder blending (e.g. of Aspirin and Avicel [32]). AE therefore offers a unique opportunity to probe the extrinsic properties of bulk solid phases that cannot be accessed by most other techniques.

Many additional phenomena have also been investigated in pharmaceutically relevant materials. This includes, for example, fluidized bed granulation [33], high-shear granulation [34], fluidized bed coating [35], powder compaction processes [28, 36], and even organic chemical reactions [25].

4.8.1.5.2 Thermography

The Basics Two methods are predominantly used to monitor temperature of transformations (i) thermocouples and (ii) thermographic cameras. Thermocouples comprise two wires, each made from a different metal. At one end, the wires are welded to create a junction. When the junction is exposed to a change in temperature, a voltage is created. The voltage is temperature. Many types of thermocouples exist, each having their characteristic temperature range, durability, vibration resistance, chemical resistance, and application compatibility. In contrast, thermographic cameras are a non-contact approach to temperature monitoring and are therefore favorable for many practical applications. Thermographic cameras operate by detecting the black-body infra-red radiation emitted by objects. Although calibrations for both approaches are typically performed during manufacture, in-house calibration checks are often required. These can be performed using known standards (thermocouples) and black-body emitters (thermal cameras).

Considerations for In Situ Applications Thermocouples are particularly well suited for solution-based experiments, where the probes can be inserted directly into the reaction vessel. In some cases, however, this may lead to unwanted nucleation sites.

Provided the reaction mixture is stirred, and hence heat transfer is facilitated, the positioning of the thermal probe is often of little importance. Successful attempts have also been made at using thermal couples for monitoring mechanochemical reactions. Where the risk of damage from the milling balls is minimal – i.e. during resonant acoustic mixing [36] or if sensors are embedded *within* the milling vessel walls [37], it has been possible to place the thermal couples within the sample vessel. For most cases of ball milling, however, the probes are placed externally, and therefore rely on effective heat transfer across the vessel walls [38, 39]. The use of IR cameras has also proved to be a powerful method to monitor solid-state transformations, both in solution and in mechanochemical reactors. It is important to remember that these cameras do not monitor the internal temperature of the system, but rather the temperature of the vessel walls. The temperature that is measured in this case is therefore greatly dependent on the thermal conductivity of sample jar material.

The absolute temperature of the system (i.e. that which is monitored by common techniques) is the result of many contributions. From the chemical system, this includes contributions from crystallization, mixing, dissolution, fracture, plastic deformation, and chemical reactions. The reactor can also contribute through contributions from moving components (e.g. impellers), or the impact and friction of milling balls in mechanochemical reactors. Changes in the heat capacity of the sample (including the content of atmosphere) and the rheology of solid phases can all contribute to changes in thermal properties of the system. Finally, one must also account for dynamic heat transfer between the reaction vessel and the surrounding atmosphere. All of these factors can play considerable and largely unpredictable roles in determining the thermal profile being monitored.

Examples Due to their relative availability, thermocouples are more commonly used as temperature sensors for both solution and mechanochemical solid-state processes. Earlier approaches focused on monitoring the evolution of temperature with transformation [40, 41]. This proved useful to monitor the progression of chemical and physical transformations and indeed to identify different reactivity regions during the course of the transformation [42]. More recently, thermocouple-based monitoring has been used as active feedback in order to control the overall temperature of solid-state processes, for example, during ball mill grinding [43]. In a one-pot synthesis of an amide-urea compound for 30 minutes, a thermocouple monitored the temperature increase from 20 to 80 °C directly in the wall of the polymethylmethacrylate (PMMA) jar, Figure 4.8.5b. The use of temperature data to actively direct and control crystallization processes is becoming increasing popular, particularly when coupled to additional monitoring techniques [5, 6].

While thermocouple-based techniques are now appearing in the mechanochemical literature, the intense mechanical force often ruptures the thin thermocouple wires. For this reason, thermographic cameras have proven particularly popular [44–46]. This is exemplified by the mechanochemical crystallization of pyrazinamide + oxalic acid [44, 47]. Using the mechanochemical approach, the cocrystal was formed within 20 minutes ball mill grinding at 50 Hz. The time-resolved in situ thermographic data collected during the mechanochemical crystallization,

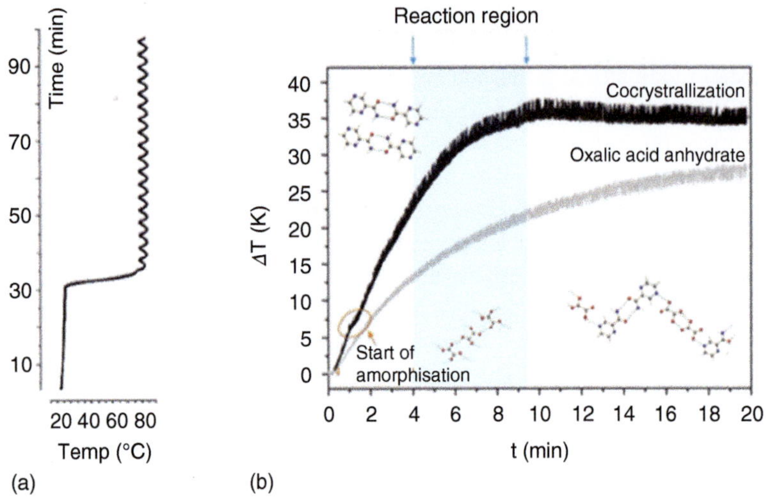

Figure 4.8.5 In situ time-resolved temperature monitoring for solid-state processes. (a) Feedback temperature control via thermocouples during mechanochemical ball mill grinding. Source: Reprinted with permission from Ref. Cindro et al. [43]. © 2019 American Chemical Society. (b) Monitoring the temperature evolution of mechanochemical cocrystallization of pyrazinamide: oxalic acid, using a thermal camera. Curves are shown for the ball milling of a stoichiometric mixture of pyrazinamide + oxalic acid at 50 Hz (black) and for oxalic acid anhydrate at 50 Hz (gray). Source: Reproduced from Ref. Kulla et al. [44]. © 2017 Royal Society of Chemistry.

Figure 4.8.5b, show a steady increase in temperature over the course of the transformation, with a maximum $\Delta T = 36$ K. Importantly, the accumulated temperature is much larger for the crystallization process than when the pure components (here oxalic acid dihydrate) were milled alone. It was suggested that this additional heat relates to the exothermic effects of crystal formation. A small shoulder in the temperature curve around 1.5 minutes was also suggested to signify an onset of amorphization. Although a powerful technique, monitoring the temperature alone is not sufficient to resolve many of the mechanistic details of interest during these solid-state transformations.

4.8.1.6 Methods to Monitor Intrinsic Solid Properties

4.8.1.6.1 X-ray Diffraction

The Basics Undoubtedly, the most popular solid-state characterization technique is X-ray diffraction. Structural analysis based on the diffraction of X-rays by single crystals offers the clearest description of the equilibrium structure of solids, down to the position and connectivity of atoms within constituent molecules. The characterization of bulk samples is instead performed by diffraction of a polycrystalline powder. A powder contains many crystallites whose orientations are distributed randomly. Scattering from such a sample is collected in all directions (Debye–Scherrer rings) using a 2D detector, Figure 4.8.6. By carefully calibrating the distance between the

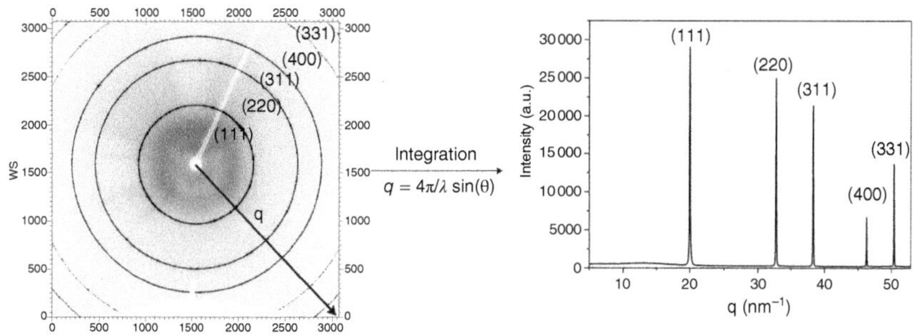

Figure 4.8.6 Representation of the 2D diffraction image of the Si standard (SRM640d) using the Debye–Scherrer rings and its conversion to a conventional 1D diffractogram. Miller indices are indicated.

scattering sample and the detector, the scattering angle θ (and hence momentum, q) can be directly obtained.

Conceptually, the scattering of X-rays with wavelength λ this is most readily defined according to the famous Bragg Law,

$$n\lambda = 2d \sin \theta \quad (4.8.3)$$

where the spacing between so-called *Bragg planes* is defined by a distance, d. The intensity of radiation scattered at each value of θ depends on the position and charge density of atoms with respect to the scattering Bragg plane. In this way, the position of Bragg peaks (Eq. 4.8.3) is defined by the *size* (and internal symmetry) of the translational unit of the crystal (i.e. its unit cell), and the intensity of the Bragg peaks is indicative of the position and type of atoms within this unit cell. With sufficient perseverance, it is therefore possible to determine the atomic structure of materials, even from powder diffraction data.

The sensitivity of X-ray diffraction to (i) unit cell shape and symmetry and (ii) atomic charge density and distribution means that even small changes in the arrangement of molecules within the solid state can be detected. Hence, X-ray diffraction is capable of resolving differences between crystalline phases of *different* molecules as well as of the *same* molecule (polymorphs). Although less common, X-ray scattering methods (e.g. the pair distribution function) are also used to monitor the formation and evolution of non-crystalline phases [48, 49].

Considerations for In Situ Applications The application of X-ray diffraction techniques for in situ monitoring of solid-state processes is accompanied by a variety of experimental restrictions. Many of these restrictions can be mitigated in various ways, although it is crucial that the experimentalist is aware of their presence and impact on resulting data. In the following, we outline the major restrictions that are associated with X-ray scattering techniques, although we note that many system-dependent factors (e.g. scattering strength) will also be important considerations.

X-ray source: There are two main sources for conducting X-ray experiments: laboratory X-ray methods and synchrotron X-ray methods. The X-rays generated within a laboratory are restricted in energy and are of low intensity. This radiation is therefore quickly absorbed by sample vessels. Furthermore, the low-intensity beam requires long collection times, on the order of tens of minutes for a fingerprint pattern. Laboratory sources are therefore often not suitable for monitoring the kinetics of solid-state transformations in real time. These methods are instead very popular for monitoring solid-state transformations under start–stop conditions, both in situ and ex situ. Notable examples of in situ and real-time monitoring by laboratory sources are transformations that occur slowly upon ageing, or when mapping of thermodynamic phase diagrams (pressure–temperature) is required. The major benefit to laboratory-based X-ray sources is their wide availability and ease of operation. In contrast, synchrotron radiation can be tuned widely in energy and is produced with very intense fluxes. There is generally no issue with the penetration of synchrotron radiation through sample vessels, with high-quality diffraction patterns obtained within seconds. Real-time monitoring of solid-state transformations thus typically makes use of these synchrotron X-ray sources. Unfortunately, obtaining time at synchrotron facilities is a difficult task, and all equipment must be transported. X-ray-based monitoring of solid-state transformations in situ and in real time is therefore not routine.

Vessel choice: Imposed by the diffraction phenomenon itself, a number of experimental considerations are required for conducting in situ measurements by X-ray diffraction. Foremost, the sample vessel must permit the penetration of X-rays. This is generally ensured by use of non-crystalline sample vessels (i.e. those which do not diffract X-rays coherently). To date, vessels composed of Perspex™, Teflon, and Polycarbonate have all been routinely employed. Assuming that the sample contained within the vessel is crystalline, the Bragg scattering is thus visible atop the amorphous background of the sample vessel, Figure 4.8.7a. The amorphous background can be minimized by using thinly walled vessels [50]. This is required to reduce the loss of broad diffraction peaks within the amorphous halo should questions surround potential amorphous phases be the subject of investigation. More difficult than permitting X-rays through the sample vessel is minimizing sample width. In accordance with Eq. (4.8.3), a given set of Bragg planes diffracts at a well-defined angle. Unfortunately, this also means that diffraction by a set of Bragg planes from individual particles within a thick (or widely dispersed) sample will reach the detector at very different positions, Figure 4.8.7b. This results in artificial broadening of the diffraction profile and loss of important structural information (e.g. particle size). Furthermore, this artificial broadening renders quantitative refinements very difficult and highly unreliable. The most common approach to reduce artificial broadening is to minimize the path length of the X-ray beam through the milling vessel. Other attempts have included preparation of specially designed vessels, with regions of very small internal diameter [50, 51].

Sampling restrictions: Unlike ex situ and in situ analysis, the beam path through the sample vessel is fixed during real-time measurements. As such, only material that passes through the beam is analyzed. It is common to average multiple diffraction patterns to account for time-dependent fluctuations in scattering intensity,

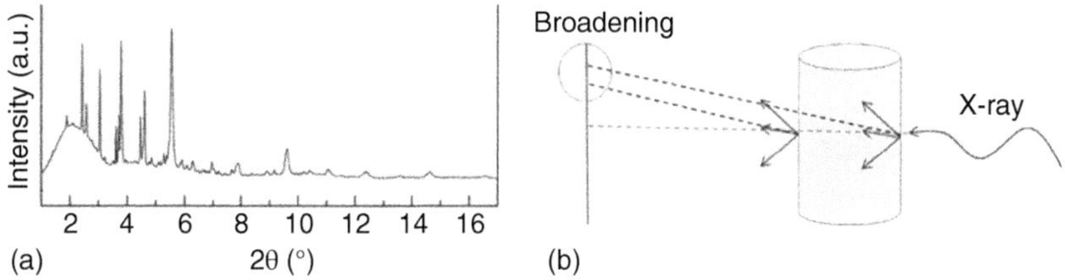

Figure 4.8.7 (a) Bragg scattering from a crystalline material within an X-ray amorphous milling vessel. Source: Adapted from Ref. Tumanov et al. [50]. © 2017 International Union of Crystallography. (b) Scattering of X-rays occurs from multiple positions within a ball milling vessel. This leads to artificial broadening of the scattering signal.

although this is not always sufficient [52]. Changes in the flowability [53] and distribution of material during the transformation can have drastic consequences on the observed transformation profile, both by changing the intensity and composition of diffracting powder. This can lead to difficulty with the interpretation of results. Attempts to mitigate these effects by inclusion of internal standards have been sought [54]. While such approaches have proved useful to varying degrees, it is important to remember that this approach requires the assumption of homogenous mixing, which is rarely the case, and the presence of external material in the mixture, alongside associated toxicity and seeding effects.

Examples The literature describing in situ monitoring of solid-state phenomena via X-ray diffraction is vast, with select example chosen to highlight the various approaches. These examples span ex situ analysis as well as in situ and real-time analysis.

Ex situ and in situ analysis: On account of its accessibility, laboratory source diffraction techniques are most common for monitoring solid-state transformations. This is generally associated with stop–start approaches to reaction monitoring, where the transformation is periodically quenched and analyzed. This approach is typically not possible for solution-phase crystallizations, where the system cannot be quenched in an intermediate state. In contrast, this approach is very common for solid-state mechanochemical reactions and has been used widely to follow transformations in organic mechanochemistry [55–59]. While disturbing the powder within a mechanochemical reactor can influence the overall transformation [60], ex situ analysis ensures that the entire powder composition is probed. This is exemplified by the organic system glycine + malonic acid, where different regions of the milling vessel produced different products [60, 61]. Furthermore, ex situ analysis allows careful and detailed investigation of powder structure. Its importance is exemplified by recent work of Belenguer et al. [57] which demonstrated the effects of particle size and surface topology [62] on the polymorphic form of mechanochemical disulfide exchange reactions and polymorphism of organic molecules, caffeine, sorbitol, and chlorpropamide.

In situ and real-time analysis: Although recent technological developments have suggested a new approach for monitoring solution-phase crystallizations using laboratory X-ray sources [63, 64], similar studies have generally been restricted to synchrotron X-ray methods. Synchrotron sources have been widely used to monitor solution and mechanochemical reactions in situ and in real time. A wide range of studies have reported the monitoring nucleation and polymorphic control within flow reactors [65], alongside batch reactors [66–68]. In this way, crystallization kinetics can be monitored as a function of experimental parameters including temperature [69, 70], solvent [71–73], and additives [74]. More recently, crystallization kinetics of solutions suspended within an acoustic field (i.e. acoustic levitation technology) have become possible. By eliminating contact surfaces, this removes potential seeding sites, and any potential background scattering from amorphous vessels. The resulting high-quality diffraction has allowed monitoring of nucleation kinetics and has, for example, allowed the identification of new amorphous intermediates phases of paracetamol via pair distribution function (PDF) analysis [75].

The strength of X-ray diffraction techniques is its ability to monitor solid-to-solid reactions: *mechanochemistry*. In principle, any mechanochemical reactor that has an accessible vessel can be modified for in situ and real-time analysis. To date, examples of real-time monitoring of mechanochemical reactions using various types of vibratory ball mill [76–78] and the Resonant Acoustic Mixer [79] have been reported. By monitoring reactions in real time, new intermediates can be identified and, as in the case of pyrazinamide-malonic acid cocrystals, isolated, Figure 4.8.8a [80]. The transformations induced by mechanochemical treatment have been monitored in real time for many phenomena [81, 82]. For example, the formation of cocrystals of many pharmaceutical compounds has been monitored, including both binary and ternary cocrystals of pyrazinamide [83]. In this way, the rate of transformation could be followed by monitoring the change in Bragg reflections during milling. In other cases, here exemplified by a Knoevenagel condensation, organic synthetic reactions can also be followed in real time by X-ray diffraction, Figure 4.8.8b. The only requirement is that crystalline phases are present. Where non-crystalline phases (liquids or amorphous phases), diffraction techniques are not suitable.

Despite its immense strength for monitoring solid-state transformations, X-ray diffraction techniques are mainly applied to crystalline phases. Many materials, both organic and inorganic [46], are known to lose their crystallinity upon extended mechanical action [48, 84–87]. As the crystallinity deteriorates, the intensity of coherently scattered X-rays does as well. Eventually, these materials no longer exhibit Bragg-reflections and are difficult to monitor by X-ray diffraction methods.

4.8.1.6.2 Raman Spectroscopy

The Basics Raman spectroscopy results from the inelastic scattering of monochromatic radiation, typically a laser, by molecules. When an electromagnetic field (i.e. the laser) is incident on a sample, its oscillating field, $\boldsymbol{E}$, induces an electric dipole, μ, on the molecule, dependent on the polarizability, α, of the molecular system, $\mu = \alpha E$. The magnitude of α varies in time as a function of the molecular vibrational frequency, ν. This interaction leads to the formation of a so-called "virtual state," which subsequently relax into either the same initial vibrational state (Rayleigh scattering;

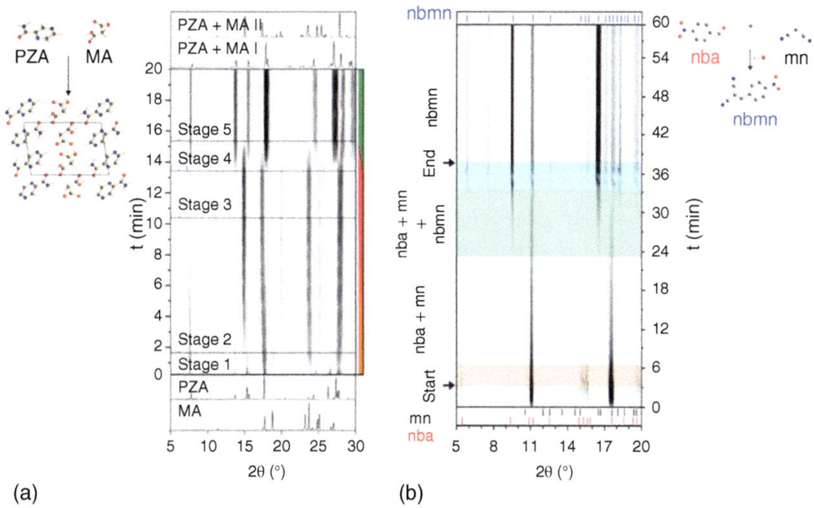

Figure 4.8.8 Monitoring mechanochemical transformations by in situ real-time synchrotron X-ray powder diffraction. (a) Cocrystallization of pyrazinamide (PZA) with malonic acid (MA). The starting materials (orange) initially form PZA: MA form II (red) and subsequently convert to form I (green). (b) Mechanochemical Knoevenagel condensation between *p*-nitrobenzaldehyde (nba) and malononitrile (mn) to form *p*-nitrobenzylidenemalononitrile (nbmn). Source: Reprinted with permission from Ref. Kulla et al. [80]. © 2017 American Chemical Society. And reproduced with permission from Kulla et al. [46]. © 2018 John Wiley & Sons.

elastic scattering) or into a neighboring vibrational state, differing by a single quantum: i.e. $\Delta\nu = \pm 1$ (+1, Stokes; −1, anti-Stokes). The intensity of the Raman band is proportional to the magnitude of the change in polarization as a function of the normal coordinate and the overlap probability of the virtual state with the final state. As a result, Rayleigh scattering dominates, and the Raman effect is small. Only vibrational modes whose polarizability is affected by the normal coordinate will exhibit Raman signal. As with other inelastic scattering processes (e.g. inelastic neutron or X-ray scattering), the incident radiation has both magnitude and direction, $k = 2\pi/\lambda$. The energy of incident photons is hence,

$$E(\mathbf{k}) = mv^2/2 = p^2/2m = (\hbar k)^2/2m; p = \hbar k \qquad (4.8.4)$$

and both energy and momentum can be exchanged with the system. For solids, one has, in principle, access to dispersion structure. However, on account of the small Raman effect, as well as the small incident momentum of photons, this is not obtained in practice. Thus, laser Raman techniques are restricted to $k \approx 0$, with important additional interactions of E with the internal Coulomb field of the crystal at $k \rightarrow 0$.

Considerations for In Situ Real-Time Applications *Excitation wavelength:* There are in principle many wavelengths possible for Raman spectroscopy, ranging from UV to near IR. The Raman effect is proportional to λ^{-4}, and spatial resolution is proportional to $\lambda/4$. Detection of shorter wavelengths is also more efficient. Shorter wavelengths therefore seem the obvious choice. In many cases, however, short wavelength (UV) excitation is accompanied by considerable fluorescence, which can drown out useful signal. One must also ensure that the organic material being studied is not reactive under excitation of the laser. Practical considerations must also be given to the cost and flexibility of the chosen laser, with UV lasers being particularly costly. In practice, most Raman spectroscopy experiments are conducted with excitation in the red or near-IR region (660–830 nm or 785 nm are very common), although the use of blue excitation is increasing (particularly 532 nm).

Sample vessel: The collection of Raman signal requires that the excitation laser first penetrates the sample vessel. With the use of visible excitation sources, this requires that the sample vessel is transparent in the visible range. Unlike diffraction, the Raman signal results from the incoherent scattering of photons and is therefore not restricted to crystalline materials. As a result, most sample vessels will exhibit a characteristic Raman spectrum. The successful use of Raman spectroscopy to monitor reactions in situ and in real time requires that these bands overlap minimally with those of the sample being monitored. To date, successful studies have coupled 785 nm excitation with the use of Perspex, Teflon, and polycarbonate sample vessels.

Sample choice: Perhaps the greatest limitation of Raman spectroscopy experiments for monitoring solid-state reactions in organic materials is its insensitivity to crystallization. This results from the weak perturbation of the lattice potential on the internal potential of the molecules. If detection is limited to > c. 200 cm^{-1}, as is commonly the case, crystallization or solid–solid transformations can only be observed where strong intermolecular interactions are changed (e.g. formation of new hydrogen bonds). With development of new technologies, the routine collection of lower frequency vibrations is becoming possible. These modes (external modes) are entirely defined by the lattice structure and are therefore very sensitive to formation and changes of solid forms. In contrast, Raman spectroscopy is an excellent tool to monitor chemical reactions, in which new covalent bonds are formed. Furthermore, as the Raman scattering intensity is low, a relatively large quantity of material is often required before useful signal intensity can be obtained. While one can in principle increase the accumulation time when dealing with small quantities of material, this greatly reduces the time resolution. This is particularly true within mechanochemical reactions, where the powder is in constant motion and distributed throughout the milling vessel and between milling bodies. The Raman signal is therefore much lower during in situ real-time mechanochemical reactions.

Sampling: The experimental set-up for in situ real-time monitoring by Raman spectroscopy requires the initial focusing of the Raman laser. Under normal circumstances, this focusing cannot be changed during the measurement. Hence, most scattering is collected from the sample which passes through the focal point of the excitation laser. If sample becomes statically distributed, coated, or otherwise hindered, it may become invisible to the beam. These effects can be mitigated to some

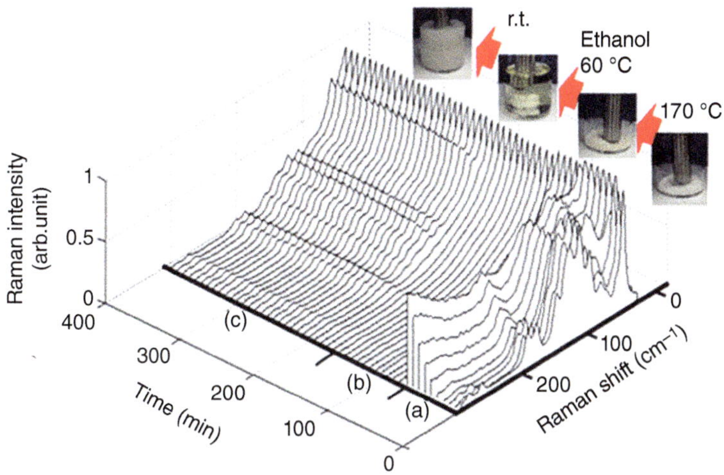

Figure 4.8.9 Low-frequency Raman monitoring of the conversion of carbamazepine polymorphic forms and dihydrate as a function of solution composition, temperature, and time. Source: Reprinted with permission from Ref. Inoue et al. [90]. © 2017 American Chemical Society.

degree by accumulating Raman signal over an extended period, or by moving the vessel within the beam (e.g. shaking during mechanochemistry).

Data processing: As with most spectroscopic techniques, it is difficult to work on the base of absolute scattering intensity. Instead, the relative intensity of the spectral intensities is obtained. This renders the direct comparison of spectra rather complex, and lengthy calibrations or statistical analyses [88] are often required. Accurate normalization requires the collection of the complete scattering. For this reason, quantitative measurements by spectroscopic means are more challenging than by X-ray scattering methods, for example.

Examples Solution phase crystallization processes cannot be monitored ex situ, as quenching the crystallization mid-way is not possible. For this reason, *in-line* (in situ real-time monitoring) is typical [89]. Although instrumental restrictions typically limit Raman spectroscopy to above c. 150–200 cm^{-1}, low-frequency Raman spectroscopy [90, 91] has also been used to monitor the crystallization and conversion of solid forms within solution. When considering polymorphism in molecular materials, for example, in CBZ, the low-frequency spectral region is of particular interest, Figure 4.8.9. Whereas the high-frequency (i.e. molecular vibrations) region remains largely unchanged, the supra-molecular (i.e. lattice) vibrations vary between polymorphs. The corresponding vibrational bands typically sit <150 cm^{-1}.

In addition to following the crystallization of materials from solution by Raman spectroscopy, the technique has also been demonstrated as a tool to monitor the evaporation of solvent during crystallization. In the case of paracetamol crystallization in an acoustically levitated droplet, for example, Raman spectroscopy demonstrated the interconversion of polymorphic forms during crystallization.

Additionally, however, the Raman spectra were simultaneously used to detect the presence of residual solvent in the levitated droplet. This proved crucial for identifying the end of the crystallization process and for differentiating between X-ray amorphous solid phases and dissolved solute [76].

Many examples of ex situ Raman analysis of mechanochemical reactions can be cited [92–95], including the use of low-frequency Raman spectroscopy to monitor the polymorphic transformation of anhydrous caffeine upon dry grinding [96]. Much less common is the use of in situ analysis. A particularly notable example describes the mechanochemical preparation of a zeolitic imidazolate framework, the ZIF-8, a biocompatible metal–organic framework, proposed for applications in pharmaceutical delivery [97]. During the mechanochemical formation of ZIF-8, the milling was intermittently stopped and a Raman spectrum obtained [98].

In situ and real-time analysis of mechanochemical transformations has become a powerful laboratory tool for studying mechanochemical transformations, and has been used for both qualitative and quantitative monitoring of mechanically induced reactions [99–101]. For example, by means of a Raman probe and optically transparent milling vessels, Raman spectroscopy was used to monitor the mechanochemical crystallization of nicotinamide and suberic acid, including the effects of milling parameters and liquid additives on the reaction kinetics [102]. Similar studies have also allowed monitoring of reaction kinetics as a function of temperature, alongside the elucidation of apparent activation barriers to crystallization and chemical transformations [100, 103].

As a final example, it is worth considering a model Knoevenagel condensation reaction, Figure 4.8.10. Within approximately 10 minutes of milling, the reaction appears complete according to visual inspection of the Raman spectra. However, the product initially forms as a monoclinic polymorphic modification and subsequently transforms into a triclinic polymorphic form. As the Raman spectra of the two forms are nearly identical, successful monitoring of this transformation was only possible by multivariate data analysis. Hence, while it is not always straightforward to use Raman spectroscopy to monitor solid–solid transformations, it can often be accomplished with the correct data processing strategies.

4.8.1.7 Benefits of Combining Methods for In Situ Monitoring

By this point it should be evident that no single tool can be regarded as the ultimate analytical probe. In fact, it is in practice very uncommon for these tools to be used in isolation, particularly when industrial processes are being investigated, Figure 4.8.11. Historically, many tools have been developed for tandem use to monitor the crystallization processes in solution, both in batch operations and under continuous control.

The tandem use of many process analytical techniques (PAT) is very well established today in the field of solution-based crystallizations. Through combinations of light scattering, UV–Vis, IR, and thermographic measurements, among other techniques, it has been possible to design entirely automated manufacturing protocols. Constructed with interconnected feedback loops, these manufacturing processes

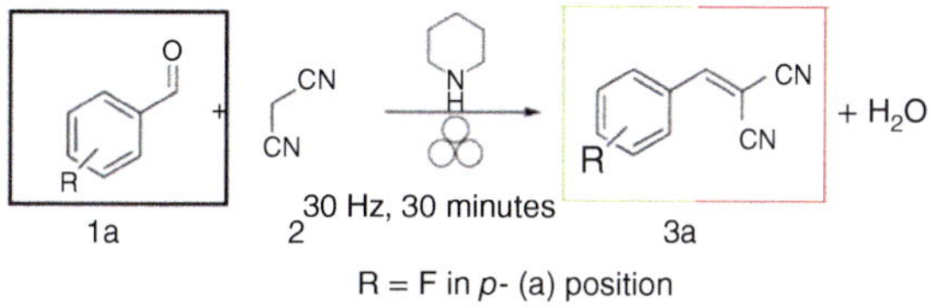

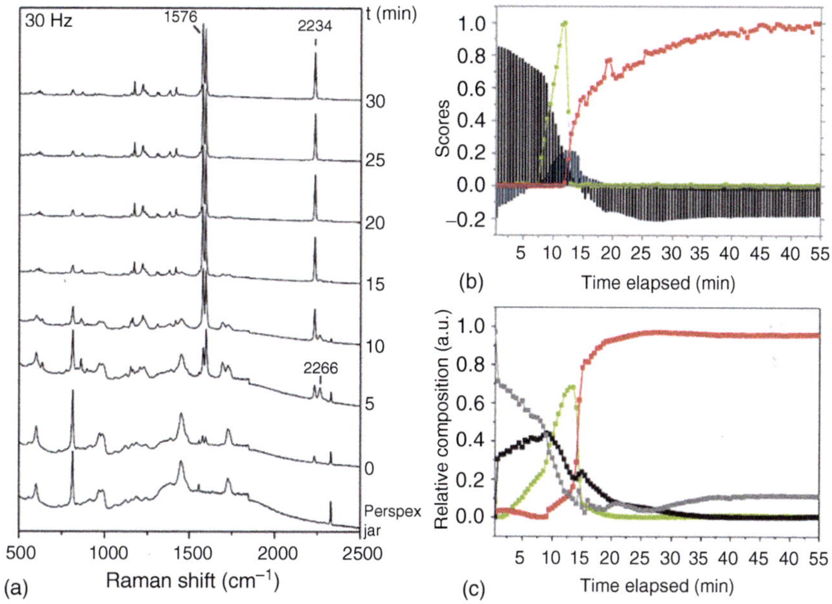

Figure 4.8.10 Monitoring the mechanochemical Knoevenagel condensation between *p*-fluorobenzaldehyde and malononitrile. (Top) Reaction scheme for the chemical transformation. (Bottom left) Time-resolved Raman spectra, collected at 5 minute intervals during ball mill grinding. (Bottom right) Multivariate principal component analysis of Raman spectra, showing contributions from two cocrystal polymorphs (red and green), and starting material (black and gray). Source: Reprinted with permission from Ref. Haferkamp et al. [104]. © 2019 Beilstein-Institut.

employ PAT methods to control and fine-tune processing parameters in real time, to ensure formation of the desired product. As an example, Nagy and coworkers [106] employed a combination of UV–Vis spectroscopy, FBRM, and thermography to control the nucleation and hence particle size distribution in the solution crystallization of paracetamol. In this adaptive crystallization control approach, FBRM data provided time-resolved monitoring of particle size, while calibrated UV–Vis spectra were obtained to monitor solution concentration. By dynamically varying the temperature of the crystallization vessel, the concentration of solution could be controlled, allowing unwanted fines to redissolve in favor of growth of larger particles. Similar adaptive control procedures have been described across an array of solution-based processes, always requiring the intimate combination of PATs [107].

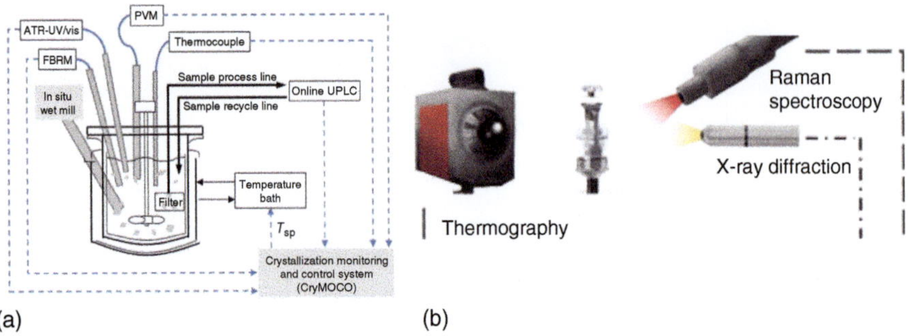

Figure 4.8.11 Realistic schematics of in situ real-time monitoring of solid-state transformations. (a) In solution, typically monitor particle size (FBRM), phase composition (ATR-FTIR), and temperature (*T*). Source: Reprinted with permission from Ref. Yang et al. [105]. © 2017 Elsevier. (b) Mechanochemical experiments can now be monitored using a tandem three-pronged approach. Source: Reprinted with permission from Ref. Kulla et al. [46]. © 2018 John Wiley & Sons.

Only recently, however, have multiple approaches been developed for tandem use during mechanochemical investigations [37, 46, 77, 83, 108, 109]. In Section 4.8.1.5.2 it was shown how thermographic measurements could be used to identify events during mechanochemical processing. As was clear above, however, thermography alone does not allow for unambiguous identification of the underlying processes. The mechanochemical synthesis of a metal phosphonate is a particularly good example, where many events appear to take place in the thermograph, Figure 4.8.12. Despite many intriguing features in the thermograph, it was not until tandem in situ X-ray powder diffraction studies were conducted that these thermal events could

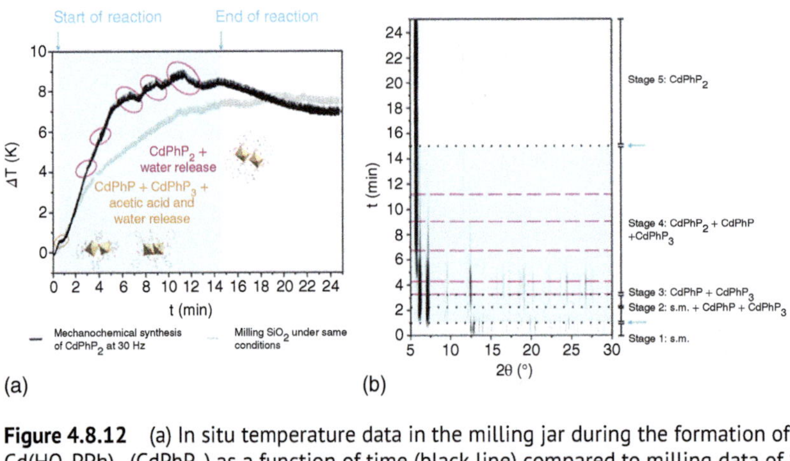

Figure 4.8.12 (a) In situ temperature data in the milling jar during the formation of $Cd(HO_3PPh)_2$ ($CdPhP_2$) as a function of time (black line) compared to milling data of SiO_2 powder under the same conditions (gray line). The formation/conversion of the intermediates and the release of acetic acid and water are shown in orange circles; red circles show the product formation and release of water. The blue area shows the reaction period. (b) In situ XRD data of the formation of $Cd(HO_3PPh)_2$ ($CdPhP_2$) where the blue area is the reaction period. Source: Reproduced from Ref. Kulla et al. [44]. © 2017 Royal Society of Chemistry.

be rationalized. Many of the thermal events correspond the chemical or physical changes in the reactants and therefore exhibits the need for the simultaneous structural analysis. Importantly, however, there are more thermal events than there are visible events in the X-ray diffraction, which suggests more subtle effects (perhaps rheological) are taking place during milling, but remain invisible in the diffraction pattern. Furthermore, the thermographic measurements allow important characterization of the bulk temperature profile, that can be crucial for rationalizing and controlling the outcome of mechanochemical reactions [38, 39, 103]. This temperature information is obviously missing from X-ray analysis. Both techniques are therefore clearly complementary to each other.

This is further exemplified by a second mechanosynthesis of manganese phosphonate, Figure 4.8.13 [46]. This example highlights the inability of X-ray diffraction to probe the reactions of non-crystalline materials. Tandem thermography, however, reveals new information in this region, suggesting important events are taking place. Tandem analysis by X-ray diffraction, Raman spectroscopy, and thermography has also been conducted for chemical reactions, for example, the model Knoevenagel condensation, Figure 4.8.13 [46]. Over the course of the transformation, both the X-ray diffraction and Raman spectroscopy confirm the general transformation mechanism. The Raman spectroscopy proved particularly sensitive to the formation of new chemical bonds and confirmed the transformation kinetics where the crystallinity proved to be poor. The thermographic measurements offered a new dimension of the reaction, exhibiting a very clear exothermic event taking place toward the end of the transformation. Each technique simultaneously reinforces the dynamics observed by the other, while providing an additional layer of process understanding.

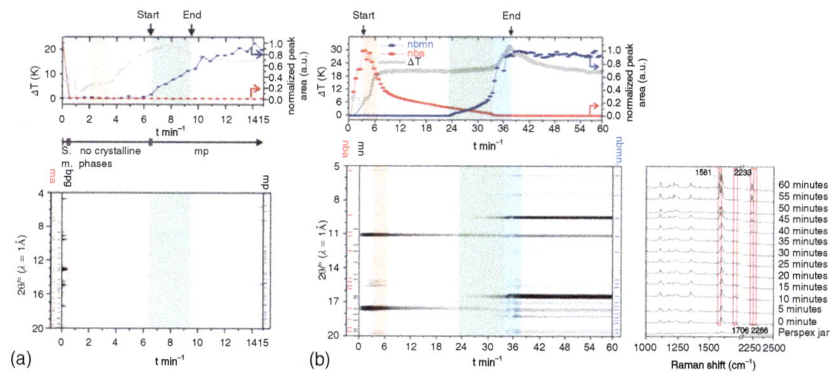

Figure 4.8.13 Combining multiple techniques to following mechanochemical reactions in situ and in real time. (a) Mechanochemical synthesis of manganese phosphonate, in which (top) the temperature is monitored over the course of the reaction (gray) and compared against integral Bragg peak intensities (blue and red). (bottom) The corresponding X-ray powder diffraction profile as a function of milling time. In both the thermographic and X-ray plots, the green indicates an X-ray amorphous region. (b) Mechanochemical Knoevenagel condensation, monitored by (top) thermography (bottom left) X-ray diffraction, and (bottom right) Raman spectroscopy. Source: Reprinted with permission from Ref. Kulla et al. [46]. © 2018 John Wiley & Sons.

The power of combining Raman and X-ray scattering data has also been highlighted as a means to quantify the relative amounts of crystalline and non-crystalline phases during mechanochemical reactions [110]. Although X-ray-based methods offer a direct route to quantifying the phase composition of a mixture, the established techniques are only valid in cases where the structures of all phases in the mixture are known. Based on knowledge of the Raman spectra for each participating phase, it was possible to quantitative extract phase compositions by Raman spectroscopy. Comparison against quantitative analysis by X-ray methods showed important differences, which were ascribed to the fact that where X-ray methods are sensitive only to bulk crystalline phases, Raman spectroscopy methods provide insight into both crystalline and non-crystalline phases. Hence, a comparison of the two approaches yields quantitation of the amorphous content of the system.

4.8.1.8 Summary

The solid form of a pharmaceutically relevant material is central to its application. One can think of a solid form as a *supra-molecular* system. By this analogy, it follows that solid material properties – for example, solubility and bioavailability – depend not only on the properties of the constituent molecules but also on the way in which these molecules interact. If these interactions change, for example, by polymorphism or defects, the solid properties also change. Furthermore, the extrinsic properties of a solid, namely, their size and shape, also greatly influences on the physical properties of the solid. Owing to its importance in controlling the physical properties of a solid material, controlling the solid attributes during crystallization (both from solution and during mechanochemical preparations) has been a central goal of industrialists.

Most often, process monitoring involves (i) identifying the rate of solid formation/transformation or (ii) identifying the physical properties of the resulting solid form. To monitor these processes, many process analytical technologies (PAT) have been developed. The choice of PAT depends on many factors, including the nature of the experimental setup, the molecule and solvent system being used, the time and sensitivity resolution required, as well as PAT compatibility and availability. Perhaps more importantly, the choice of PAT depends critically on the *type* of information that is sought. For example, whereas UV–Vis spectroscopy can be used to monitor the rate of crystallization via changes in solution concentration, it offers no information regarding the nature of the solid product phase. In contrast, X-ray diffraction techniques are ideal for monitoring the polymorphic phase of the crystalline solid form, but are not particularly well suited to monitoring particle size attributes. Owing to their niche applications, it is most common to employ tandem techniques to obtain a complete picture of the evolving system. By doing so, active feedback-driven control methods have been developed, which allow automated tuning of process conditions to drive solid phase transformations to a desired outcome.

Through this chapter, a number of common PAT techniques were introduced and are summarized in Table 4.8.1. The techniques described cover a range of common

Table 4.8.2 Type of experiment to which the technique has been applied is indicated, along with its ability to be used in real time. Furthermore, indication is given as to which other PAT methods it has been coupled in the literature.

	Type	Real time?	Combinations?
X-ray diffraction (XRD)	Used in solution and mechanochemistry	Yes	Raman spectroscopy [77, 109]
			Thermography [108]
			Light scattering [111]
			UV–Vis [112]
			ATR-FTIR [112]
Raman spectroscopy (RS)	Used in solution and mechanochemistry	Yes	Thermography [43]
			Light scattering [8]
			XRD [77, 83]
			ATR-IR [113]
			UV–Vis [8]
			Optical imaging [16]
Thermography	Used in solution and mechanochemistry	Yes	Raman spectroscopy [43, 46]
			XRD [46]
			ATR-IR [9]
			ATR-UV–Vis [106]
			Light scattering [24]
Infrared spectroscopy (ATR-IR)	Used in solution	Yes	Thermography [9]
			Light scattering [9]
			XRD [112]
			ATR-UV–Vis [8]
			Raman spectroscopy [113]
			Optical imaging [16]
UV–Vis spectroscopy (ATR-UV–Vis)	Used in solution	yes	Thermogravimetry [106]
			Light scattering [5]
			XRD [112]
			ATR-IR [8]
			Raman spectroscopy [8]
			Optical imaging [8]

(*continued*)

Table 4.8.2 (Continued)

	Type	Real time?	Combinations?
Light scattering (LS)	Used in solution	yes	Thermography [24]
			XRD [111]
			ATR-IR [9]
			ATR-UV–Vis [5]
			Raman spectroscopy [8]
			Optical imaging [21]
Acoustic emission (AE)	Used in solution and mechanochemistry	yes	XRD [114]

laboratory methods, such as Raman and UV–Vis spectroscopy, and more advanced techniques such as synchrotron X-ray diffraction. These techniques cover a range of applications, including those based on monitoring (i) changes in solution concentration (e.g. to follow kinetics of crystallization), (ii) crystallization of solids from solution (e.g. particle size), (iii) extrinsic properties of solid materials, either in solution or mechanochemically, (iv) intrinsic properties of the solid material, either in solution or mechanochemically. Although we have endeavored to cover the common techniques in each case, the list of techniques and examples discussed is by no means exhaustive. There is no single, correct choice of PAT that will be suitable for all applications. The combination of techniques chosen will depend on the system and question at hand. Knowledge of available techniques, their advantages and disadvantages, is therefore an important part of the toolbox for all involved in process monitoring of solutions (Table 4.8.2).

List of Abbreviations

AE	acoustic emission
API	active pharmaceutical ingredient
ATR	attenuated total reflectance
CBT	carbamazepine
CLD	chord length distribution
COM	commercial carbamazepine
DLS	dynamic light scattering
FTIR	Fourier-transform infrared spectroscopy
IR	infrared spectroscopy
LS	light scattering
NIR	near-infrared spectroscopy
OABA	*ortho*-aminobenzoic acid

PSD	particle size distribution
PMMA	polymethylmethacrylate
PAT	process analytical technologies
RS	Raman spectroscopy
REC	recrystallized carbamazepine
SLS	static light scattering
UV–Vis	ultraviolet and visible spectroscopy
XRD	X-ray diffraction
ZIF	Zeolitic imidazolate frameworks

References

1 Pienack, N. and Bensch, W. (2014). *Angewandte Chemie, International Edition* 50: 2014–2034.

2 Hughes, C.E., Williams, P.A., and Harris, K.D.M. (2014). *Angewandte Chemie, International Edition* 53: 8939–8943.

3 Fonteyne, M., Vercruysse, J., De Leersnyder, F. et al. (2015). *TRAC-Trends in Analytical Chemistry* 67: 159–166.

4 Simone, E., Saleemi, A.N., Tonnon, N., and Nagy, Z.K. (2014). *Crystal Growth & Design* 14: 1839–1850.

5 Bakar, M.R.A., Nagy, Z.K., and Rielly, C.D. (2009). *Organic Process Research and Development* 13: 1343–1356.

6 Saleemi, A.N., Rielly, C.D., and Nagy, Z.K. (2012). *Crystal Growth & Design* 12: 1792–1807.

7 Saleemi, A.N., Rielly, C.D., and Nagy, Z.K. (2012). *Chemical Engineering Science* 77: 122–129.

8 Simone, E., Saleemi, A.N., and Nagy, Z.K. (2015). *Organic Process Research and Development* 19: 167–177.

9 Fujiwara, M., Chow, P.S., Ma, D.L., and Braatz, R.D. (2002). *Crystal Growth & Design* 2: 363–370.

10 Alatalo, H., Hatakka, H., Kohonen, J. et al. (2010). *AICHE Journal* 56: 2063–2076.

11 Feng, L. and Berglund, K.A. (2002). *Crystal Growth & Design* 2: 449–452.

12 Helmdach, L., Feth, M.P., Minnich, C., and Ulrich, J. (2013). *Chemical Engineering and Processing Process Intensification* 70: 184–197.

13 Sypek, K., Burns, I.S., Florence, A.J., and Sefcik, J. (2012). *Crystal Growth & Design* 12: 4821–4826.

14 Yu, Z.Q., Chow, P.S., and Tan, R.B.H. (2008). *Organic Process Research and Development* 12: 646–654.

15 Dave, K., Luner, P.E., Forness, C. et al. (2018). *AAPS PharmSciTech* 19: 155–165.

16 Agimelen, O.S., Svoboda, V., Ahmed, B. et al. (2018). *Advanced Powder Technology* 29: 2987–2995.

17 Wang, X.Z., Liu, L., Li, R.F. et al. (2009). Online characterisation of nanoparticle suspensions using dynamic light scattering, ultrasound spectroscopy and process tomography. *Chemical Engineering Research and Design* 87 (6): 874–884.

18 Tobler, D.J., Shaw, S., and Benning, L.G. (2009). *Geochimica et Cosmochimica Acta* 73: 5377–5393.

19 Hoffmann, B.Y.M.A.M., Roefs, S.P.F.M., Verheul, M. et al. (1996). *Journal of Dairy Research* 63: 423–440.

20 Villari, V. and Micali, N. (2008). Light scattering as spectroscopic tool for the study of disperse systems useful in pharmaceutical sciences. *Journal of Pharmaceutical Sciences* 97 (5): 1703–1730.

21 Laffierrere, L., Hoff, C., and Veesler, S. (2004). *Crystal Growth & Design* 4: 1175–1180.

22 Helmdach, L., Schwartz, F., and Ulrich, J. (2014). *Chemical Engineering and Technology* 37: 213–220.

23 Kamahara, T., Takasuga, M., Tung, H.H. et al. (2007). *Organic Process Research and Development* 11: 699–703.

24 Yang, Y., Song, L., Gao, T., and Nagy, Z.K. (2015). *Crystal Growth & Design* 15: 5879–5885.

25 Nordon, A., Carella, Y., Gachagan, A. et al. (2006). *Analyst* 131: 323–330.

26 Wang, X., Huang, Y., and Michelitsch, T.M. (2018). *Pharmaceutical Development and Technology* 24: 1–18.

27 Serris, E., Cameirao, A., and Gruy, F. (2016). 32nd European Conference on Acoustic Emission Testing, Czech Society for NDT; European Working Group on Acoustic Emission, Prague, Czech Republic, 451–460.

28 Hakanen, A., Laine, E., Jalonen, H. et al. (1993). *Drug Development and Industrial Pharmacy* 19: 2539–2560.

29 Gherras, N., Serris, E., and Fevotte, G. (2012). *International Journal of Pharmaceutics* 439: 109–119.

30 Fevotte, G., Wang, X.J., and Ouabbas, Y. (2014). *Acoustic emission, a new sensor for monitoring industrial crystallization processes. IFAC Proceedings Volumes* 47: 2727–2733.

31 Wang, X.J., Fevotte, G., Huang, Y., and Michelitsch, T.M. (2016). *Powder Technology* 301: 70–77.

32 Allan, P., Bellamy, L.J., Nordon, A., and Littlejohn, D. (2010). *Analyst* 135: 518–524.

33 Hansuld, E.M., Briens, L., Mccann, J.A.B., and Sayani, A. (2009). *International Journal of Pharmaceutics* 378: 37–44.

34 Papp, M.K., Pujara, C.P., and Pinal, R. (2008). *Journal of Pharmaceutical Innovation* 3: 113–122.

35 Veski, P., Pedersen, J.G., Anov, D. et al. (2007). *International Journal of Pharmaceutics* 332: 90–97.

36 Hope, K. S., Lloyd, H. J., Ward, D., Michalchuk, A. A. L. and Pulham, C. R. (2015). Resonant acoustic mixing and its applications to energetic materials. New Trends in Research of Energetic Materials (NTREM) 2015, Pardubice, Czech Republic. https://doi.org/10.13140/RG.2.1.4234.8646.

37 Užarević, K., Ferdelji, N., Mrla, T. et al. (2018). *Chemical Science* 9: 2525–2532.
38 Andersen, J.M. and Mack, J. (2017). *Chemical Science* 8: 5447–5453.
39 Andersen, J., Brunemann, J., and Mack, J. (2019). *Reaction Chemistry & Engineering* 4: 1229–1236. https://doi.org/10.1039/C9RE00027E.
40 Hottot, A., Vessot, S., Andrieu, J. et al. (2005). *PDA Journal of Pharmaceutical Science and Technology* 59: 138–153.
41 Chouvenc, P., Vessot, S., Andrieu, J., and Vacus, P. (2004). *Drying Technology* 22: 1577–1601.
42 Nail, S., Tchessalov, S., Shalaev, E. et al. (2017). *AAPS PharmSciTech* 18: 2379–2393.
43 Cindro, N., Tireli, M., Karadeniz, B. et al. (2019). *ACS Sustainable Chemistry & Engineering* 7: 16301–16309.
44 Kulla, H., Wilke, M., Fischer, F. et al. (2017). *Chemical Communications* 53: 1664–1667.
45 Sarfraz, A., Schlegel, M.C., Wright, J., and Emmerling, F. (2011). *Chemical Communications* 47: 9369–9371.
46 Kulla, H., Haferkamp, S., Akhmetova, I. et al. (2018). *Angewandte Chemie, International Edition* 57: 5930–5933.
47 Kulla, H., Greiser, S., Benemann, S. et al. (2016). *Molecules* 21: 917.
48 Descamps, M. and Willart, J. (2007). *Journal of Pharmaceutical Sciences* 96: 1398–1408.
49 Nollenberger, K., Gryczke, A., Meier, C. et al. (2009). *Journal of Pharmaceutical Sciences* 98: 1476–1486.
50 Tumanov, N., Ban, V., Poulain, A., and Filinchuk, Y. (2017). *Journal of Applied Crystallography* 50: 994–999.
51 Ban, V., Sadikin, Y., Lange, M. et al. (2017). *Analytical Chemistry* 89: 13176–13181.
52 Michalchuk, A.A.L., Tumanov, I.A., Konar, S. et al. (2017). *Advancement of Science* 4: 1700132.
53 Hutchings, B.P., Crawford, D.E., Gao, L. et al. (2017). *Angewandte Chemie, International Edition* 56: 1–6.
54 Halasz, I., Friščić, T., Kimber, S.A.J. et al. (2014). *Faraday Discussions* 170: 203–221.
55 Tumanov, I.A., Achkasov, A.F., Boldyreva, E.V., and Boldyrev, V.V. (2011). *CrystEngComm* 13: 2213.
56 Fischer, F., Joester, M., Rademann, K., and Emmerling, F. (2015). *Chemistry-A European Journal* 21: 14969–14974.
57 Belenguer, A.M., Lampronti, G.I., Cruz-Cabeza, A.J. et al. (2016). *Chemical Sciences* 7: 6617–6627.
58 Hasa, D., Schneider, G., Voinovich, D., and Jones, W. (2015). *Angewandte Chemie, International Edition* 54: 7371–7375.
59 Am Ende, D.J., Anderson, S.R., and Salan, J.S. (2014). *Organic Process Research and Development* 18: 331–341.
60 Michalchuk, A.A.L., Tumanov, I.A., Drebushchak, V.A., and Boldyreva, E.V. (2014). *Faraday Discussions* 170: 311–335.

61 Michalchuk, A.A.L., Tumanov, I.A., and Boldyreva, E.V. (2013). *CrystEngComm* 15: 6403–6412.
62 Belenguer, A.M., Cruz-Cabeza, A.J., Lampronti, G.I., and Sanders, J.K.M. (2019). *CrystEngComm* 21: 2203–2211.
63 Clevers, S., Lafontaine, A., Baaklini, G., Schindler, M., Sanselme, M., Cartigny, Y., Dupray, V., and Coquerel, G. (2019). In situ X-ray for monitoring crystallization and phase transitions in suspension: In-situX®. In *BIWIC 2019 International Workshop on Industrial Crystallization*, Rayong, Thailand.
64 Takahashi, H., Iwama, S., Clevers, S. et al. (2017). *Crystal Growth & Design* 17: 671–676.
65 MacCalman, M.L., Roberts, K.J., Kerr, C., and Hendriksen, B. (2002). *Journal of Applied Crystallography* 28: 620–622.
66 Jacques, S.D.M., Pile, K., and Barnes, P. (2005). *Crystal Growth & Design* 5: 395–397.
67 Dharmayat, S., De Anda, J.C., Lai, R.B.H.Ã.X. et al. (2006). *Journal of Crystal Growth* 294: 35–40.
68 Hammond, R.B., Lai, X., Roberts, K.J. et al. (2004). *Crystal Growth & Design* 4: 943–948.
69 Haselhuhn, F., Doyle, S., and Kind, M. (2006). *Journal of Crystal Growth* 289: 727–733.
70 Bayes-Garcia, L., Calvet, T., Cuevas-Diarte, M.A. et al. (2011). *CrystEngComm* 13: 3592–3599.
71 Blagden, N., Davey, R., Song, M. et al. (2003). *Crystal Growth & Design* 3: 197–201.
72 Gnutzmann, T., Thi, Y.N., Rademann, K., and Emmerling, F. (2014). *Crystal Growth & Design* 14: 6445–3450.
73 Trobs, L., Thi, Y.N., Rump, D., and Emmerling, F. (2014). *Zeitschrift für Physikalische Chemie* 228: 493–501.
74 Nunes, C., Mahendrasingam, A., and Suryanarayanan, R. (2005). *Pharmaceutical Research* 22: 1942–1953.
75 Nguyen Thi, Y., Rademann, K., and Emmerling, F. (2015). *CrystEngComm* 17: 9029–9036.
76 Fischer, F., Greiser, S., Pfeifer, D. et al. (2016). *Angewandte Chemie, International Edition* 55: 14281–14285.
77 Batzdorf, L., Fischer, F., Wilke, M. et al. (2015). *Angewandte Chemie, International Edition* 54: 1799–1802.
78 Friščić, T., Halasz, I., Beldon, P.J. et al. (2012). *Nature Chemistry* 5: 66–73.
79 Michalchuk, A.A.L., Hope, K.S., Kennedy, S.R. et al. (2018). *Chemical Communications* 54: 4033–4036.
80 Kulla, H., Greiser, S., Benemann, S. et al. (2017). *Crystal Growth & Design* 17: 1190–1196.
81 Fischer, F., Heidrich, A., Greiser, S. et al. (2016). *Crystal Growth & Design* 16: 1701–1707.
82 Batzdorf, L., Zientek, N., Rump, D. et al. (2017). *Journal of Molecular Structure* 1133: 18–23.

83 Kulla, H., Michalchuk, A.A.L., and Emmerling, F. (2019). *Chemical Communications* 1: 2–5.
84 Oliveira, P.F.M., Willart, J., Siepmann, J. et al. (2018). *Crystal Growth & Design* 18: 1748–1757.
85 De Gusseme, A., Neves, C., Willart, J.F. et al. (2008). *Journal of Pharmaceutical Sciences* 97: 5000–5012.
86 Martinetto, P., Bordet, P., Descamps, M. et al. (2017). *Crystal Growth & Design* 17: 6111–6122.
87 Di Nardo, T. and Moores, A. (2019). *Beilstein Journal of Organic Chemistry* 15: 1217–1226.
88 Mazivila, S.J., Castro, R.A.E., Leitão, J.M.M., and Esteves da Silva, J.C.G. (2019). *Journal of Pharmaceutical and Biomedical Analysis* 169: 235–244.
89 Wang, F., Wachter, J.A., Antosz, F.J., and Berglund, K.A. (2000). *Organic Process Research and Development* 4: 391–395.
90 Inoue, M., Hisada, H., Koide, T. et al. (2017). *Organic Process Research and Development* 21: 262–265.
91 Otaki, T., Tanabe, Y., Kojima, T. et al. (2018). *International Journal of Pharmaceutics* 542: 56–65.
92 Hédoux, A., Guinet, Y., and Descamps, M. (2011). *International Journal of Pharmaceutics* 417: 17–31.
93 Wypych, A., Guinet, Y., and Hédoux, A. (2007). *Physical Review B* 76: 144202.
94 Hedoux, A., Guinet, Y., Derollez, P. et al. (2004). *Physical Chemistry Chemical Physics* 6: 3192–3199.
95 Decroix, A., Guinet, Y., Paccou, L. et al. (2011). *The Journal of Physical Chemistry B* 115: 5746–5753.
96 Hedoux, A., Guinet, Y., Paccou, L. et al. (2013). *Journal of Pharmaceutical Sciences* 102: 162–170.
97 Sun, C., Qin, C., Wang, X. et al. (2012). *Chemical Communications* 41: 6906–6909.
98 Ma, X., Yuan, W., Bell, S.E.J., and James, S.L. (2014). *Chemical Communications* 50: 1585–1587.
99 Julien, P.A., Malvestiti, I., and Friščić, T. (2017). *Beilstein Journal of Organic Chemistry* 13: 2160–2168.
100 Fischer, F., Fendel, N., Greiser, S. et al. (2017). *Organic Process Research and Development* 21: 655–659.
101 Haferkamp, S., Fischer, F., Kraus, W., and Emmerling, F. (2017). *Beilstein Journal of Organic Chemistry* 13: 2010–2014.
102 Gracin, D., Strukil, V., Friscic, T. et al. (2014). *Angewandte Chemie, International Edition* 53: 6193–6197.
103 Fischer, F., Wenzel, K.-J., Rademann, K., and Emmerling, F. (2016). *Physical Chemistry Chemical Physics* 18: 23320–23325.
104 Haferkamp, S., Paul, A., Michalchuk, A.A.L., and Emmerling, F. (2019). *Beilstein Journal of Organic Chemistry* 15: 1141–1148.
105 Yang, Y., Pal, K., Koswara, A. et al. (2017). *International Journal of Pharmaceutics* 533: 49–61.

106 Saleemi, A., Rielly, C., and Nagy, Z.K. (2012). *CrystEngComm* 14: 2196–2203.

107 Nagy, Z.K. and Braatz, R.D. (2012). *Annual Review of Chemical and Biomolecular Engineering* 3: 55–75.

108 Stroh, J., Ali, N.Z., Maierhofer, C., and Emmerling, F. (2019). *ACS Omega* 4: 7734–7737.

109 Klimakow, M., Leiterer, J., Kneipp, J. et al. (2010). *Langmuir* 26: 11233–11237.

110 Lukin, S., Stolar, T., Tireli, M. et al. (2017). *Chemistry-A European Journal* 23: 13941–13949.

111 Gicquel, Y., Schubert, R., Kapis, S. et al. *The Journal of Visualized Experiments* 134: e57133. https://doi.org/10.3791/57133.

112 Newton, M.A. and van Beek, W. (2010). *Chemical Society Reviews* 39: 4845–4863.

113 Qu, H., Alatalo, H., Hatakka, H. et al. (2009). *Journal of Crystal Growth* 311: 3466–3475.

114 Garcés, G., Máthis, K., Medina, J. et al. (2018). *International Journal of Plasticity* 106: 107–128.

4.9

Application of Process Monitoring and Modeling

Jochen Schoell[1] and Roberto Irizarry[2]

[1] *MSD Werthenstein BioPharma, Process Research and Development, Industrie Nord, Schachen, 6105, Switzerland*
[2] *Merck & Co., Inc., Data Science and Applied Mathematics, Sumneytown Pike, West Point, 19486, PA, USA*

4.9.1 In-process Solid Form Monitoring Techniques

During the development of new active pharmaceutical ingredients (APIs), a multitude of offline analytical solid-state characterization techniques is being utilized as described in Chapter 4 on Standards and Trends in Solid State Characterization Techniques. With the advent of Food and Drug Administration's (FDA) quality-by-design (QbD) initiative which stated that product quality cannot be tested into products but must be built-in or should be by-design, more and more process analytical technologies (PATs) have been applied to monitor API solid form during processing over the last two decades [1]. More recently, drug product manufacturers are aiming to implement continuous processing which requires robust and reliable PATs to support control strategies and enable real-time release. This section distinguishes between two categories of monitoring tools, i.e., direct and indirect solid-state monitoring techniques. Direct monitoring techniques continuously characterize the actual solid form at processing conditions throughout the unit operation, ideally in situ and in real time, and can be directly correlated with offline solid-state data, as long as the different forms can be clearly distinguished. Contrarily, indirect techniques monitor physical parameters which are caused by a certain solid form, e.g., solute concentration or particle shape as function of time, and without additional information the solid form present in the process cannot be definitively identified. As such, indirect techniques provide only indications on potential solid form changes and require further analytical confirmation. However, indirect techniques often provide insight to previously unknown behavior and thus can be very valuable in API process development as will be shown in the following sections.

Solid State Development and Processing of Pharmaceutical Molecules: Salts, Cocrystals, and Polymorphism, First Edition. Edited by Michael Gruss.

4.9.1.1 Direct Characterization Techniques

Over the last two decades, vibrational spectroscopy has become the main direct characterization technique for solid form monitoring in pharmaceutical processing [2]. Vibrational spectroscopy probes intermolecular interactions and thus is often suited to distinguish between different solid forms of an API. Near infrared (NIR) and Raman spectroscopy have become relatively inexpensive tools and can be easily coupled to fiber-optic probes to follow solid-state as well as solution mediated phase transformations of active pharmaceutical ingredients. Both techniques can be considered complementary: if the vibrational modes of a molecule result in a periodic change of the dipole moment, the molecule is infrared active. If the vibrational modes cause a change in polarizability, the molecule is Raman active. Thus, water is a strong IR absorber but a very weak Raman scatterer. As shown in the following sections, this sensitivity in monitoring the presence of water results in various monitoring applications in the formulation of drug products, when the API is being mixed with excipients and further processed to manufacture the final drug product.

4.9.1.1.1 Raman Spectroscopy

Raman spectroscopy is based on inelastic light scattering of monochromatic light, i.e., light with a single wavelength generally obtained from a laser source. Raman spectroscopy can be employed to probe any type of matter, from gases to liquids and solids, which generally feature distinct spectral differences depending on the solid form. Thus, Raman spectra of the active ingredient can be gathered along the drug substance/drug product manufacturing chain in the form of powders, slurries, liquids, or gels. Generally, the molecular density correlates with Raman signal strength, i.e., the Raman spectra of dry API powders are more intense than the spectra of API slurries or a drug formulation where the drug substance is diluted with excipients. One of the main limitations of Raman spectroscopy is fluorescence which often covers the much weaker Raman signal entirely, particularly in complex biological samples [3] which might require enhanced Raman spectroscopic techniques described elsewhere [4]. One possibility to reduce or eliminate fluorescence is increasing the laser excitation wavelength because the scattering intensity is inversely proportional to the fourth order of the excitation wavelength. However, Raman spectrometers with long excitation wavelength are typically more expensive and feature lower sensitivity. Thus, commercially available Raman monitoring systems often use 785 nm diode lasers which provide a good trade-off between system sensitivity, laser diode robustness, and limiting sample fluorescence. Wang et al. have been among the first to describe the monitoring of a solvent-mediated polymorph transformation during progesterone crystallization using Raman spectroscopy [5]. In this work, the authors have followed a univariate calibration approach, using the peak shift in the time-resolved Raman spectra from 1662 to 1667 cm^{-1} to monitor the turnover of the metastable Form II to the thermodynamically stable Form I. In the case of progesterone, the Raman spectroscopy signal allowed for quantitative analysis using a specific peak to obtain composition–time profiles. However, in other cases, the

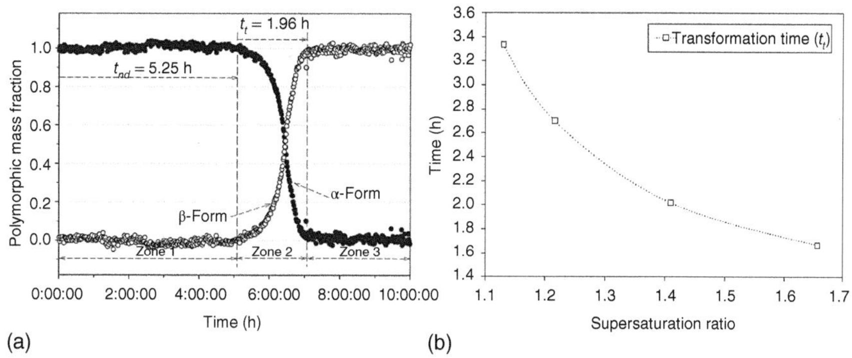

Figure 4.9.1 (a) In situ polymorphic mass fraction profiles of taltirelin through the solvent-mediated transformation process of metastable α to the stable β form. (b) Polymorph transformation time as function of the supersaturation ratio of taltirelin crystallizations at 15 °C. Source: Reproduced with permission from Nguyen et al. [7]. © 2016 John Wiley & Sons.

spectral differences between polymorphs are smaller and more sophisticated data analysis techniques need to be employed. Multivariate approaches, e.g. partial least squares (PLS) regression, utilize information from selected spectral regions or the entire spectrum and are generally considered superior to simple univariate methods. Section 4.9.2.1 provides an overview of different quantification approaches. Several publications have described solvent-mediated solid phase transformations using Raman spectroscopy in combination with other PAT tools to estimate crystallization kinetics [6–8]. Figure 4.9.1 shows the Raman profile data of the conversion from α to β taltirelin and the enhanced transformation kinetics with increasing supersaturation levels.

It is worth noting that PAT opens the door for advanced process control strategies beyond simple monitoring. Thus, the online PAT signal is fed to a process controller which compares the current state of the process with the desired process trajectory and adjusts process parameters to obtain the desired outcome. Pataki et al. employed real-time Raman spectroscopy to control polymorphic purity of the monotropically related Forms I and II of carvedilol occurring in the crystallization process [9]. Simone et al. used Raman spectroscopy in combination with UV/Vis spectroscopy to monitor solid forms and liquid phase concentration of *ortho*-aminobenzoic acid and used the signal data for feedback control of crystallization processes to isolate either of the pure solid forms [10]. Finally, PAT tools are vital for continuous process monitoring and real-time release in the pharmaceutical industry. Acevedo et al. have used Raman spectroscopy to quantify solute concentration and to monitor the ratio between Forms II and III of carbamazepine with the objective to use these data to develop control strategies for continuous crystallization in a reactor cascade [11]. Furthermore, Raman spectroscopy has been applied to monitor cocrystal formation [12, 13] as well as in drug product formulation processes to track API solid forms in mixtures with excipients. Reddy and coworkers monitored the hydration of an anhydrous drug substance during high-shear wet granulation and the subsequent dehydration during the drying process with Raman spectroscopy [14]. In this case, the monitoring data highlighted potential transformation risks which ultimately resulted in the selection of a dry formulation process to avoid any API hydration/dehydration risks at scale. Similarly, Gift et al. have used Raman spectroscopy to study the effect of certain polymeric excipients to inhibit drug substance hydration during aqueous high-shear wet granulation [15]. Two different inhibition mechanisms of the polymeric excipients have been suggested by the authors: reduction of the hydrate crystal growth rate and preferential absorption of available water. A final comparison of compaction and flow properties of formulations with and without inhibitory polymer excipients indicated that these had no adverse effects on the formulation processability. De Beer and coworkers have studied the hot-melt extrusion (HME) of an API in a polymer matrix using Raman spectroscopy. The authors not only demonstrated the importance of various process parameters on powder mixing and drug crystallinity [16] but have also scaled the HME formulation process from lab to pilot scale [17].

4.9.1.1.2 Near Infrared Spectroscopy

NIR spectroscopy is based on the absorption in the electromagnetic spectrum from about 800 to 2500 nm which corresponds to the wave number range between 12 500 and 4000 cm^{-1}. The spectral absorption bands are a combination of C—H, N—H, O—H, and S—H stretching modes and overtones which is why NIR spectroscopy is particularly useful for the quantification of polymorphs with different hydrogen bonding. Generally, NIR spectra are less resolved and feature much broader bands than Raman or mid-IR spectra and contain sample information on chemical composition and physical properties like particle size. Thus, advanced quantitation methods and spectral preprocessing are required to obtain a robust calibration model. NIR spectroscopy was employed in crystallization process monitoring of polymorphic compounds for the liquid phase concentration as well as for the dispersed solid phase. DeBraekeleer et al. were among the first to use NIR spectroscopy combined with advanced data calibration models like principal component analysis (PCA) and evolving factor analysis (EFA) [18] or soft independent modeling of class analogy (SIMCA) [19] to monitor solvent-mediated polymorph transformation during crystallization processes. Zhou and coworkers tracked the solute concentration during the crystallization of etoricoxib using a PLS calibration model to ultimately control the final solid form using a seeded process at moderate supersaturation via avoiding nucleation of the undesired form [20]. Wang et al. employed NIR spectroscopy to monitor the solvent-mediated polymorph transformation during paracetamol crystallization [21]. At manufacturing scale, Barnes and co-workers employed NIR spectroscopy with a two-level PLS calibration model for the monitoring of an API monohydrate to anhydrate conversion [22] and Schaefer et al. described the application of NIR spectroscopy with PLS modeling as online control tool for API crystallization highlighting method validation at this scale [23]. A particularly advantageous application for NIR spectroscopy is API drying, since not only the advancing drying process but also potential solid form conversions can be monitored [24, 25].

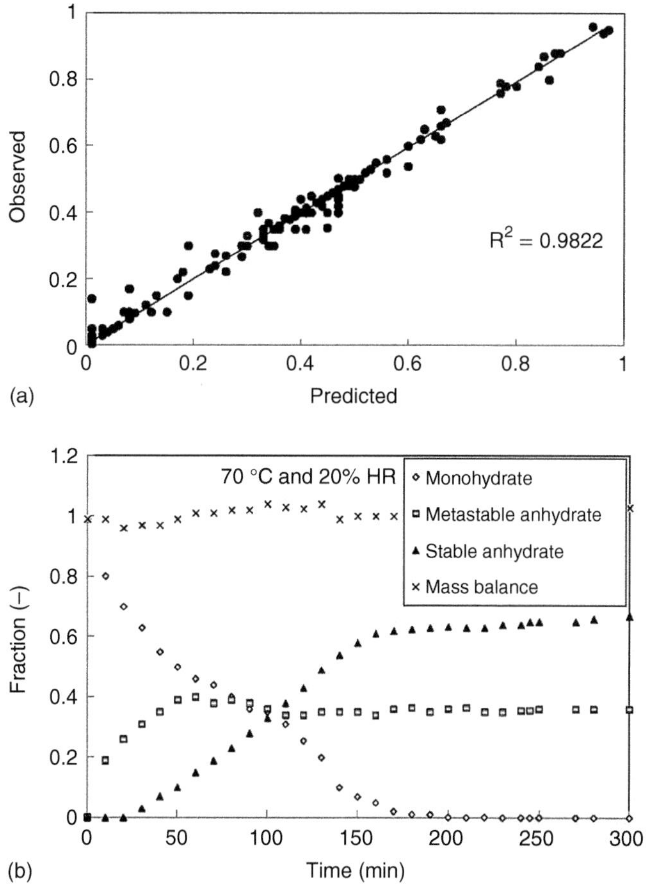

Figure 4.9.2 (a) Observed versus predicted relative content for PLS model. (b) Dehydration kinetics of theophylline monohydrate during the process at 70 °C and 20% RH. Source: Reproduced with permission from Amira et al. [24]. © 2016 Taylor & Francis.

Figure 4.9.2a shows the satisfactory agreement of observed versus predicted relative solid form content which allows to follow the drying process dynamics. The quality of the underlying PLS model is also reflected in the acceptable closure of the mass balance of the different solid forms throughout the process shown in Figure 4.9.2b. It is worth noting the relatively large number of calibration samples required for the PLS model shown in Figure 4.9.2a due to the lower specificity of the NIR bands in comparison to Raman spectroscopy.

Similar to Raman spectroscopy, NIR spectroscopy has been applied not only to drug substance solid form monitoring but also to drug product processing throughout various downstream unit operations such as high-shear wet granulation [26, 27]. Besides, NIR spectroscopy has been used for drug substance process characterization and development for powder flow [28], tablet manufacturing [29], and tablet coating processes [30] to enable QbD methodologies.

4.9.1.2 Indirect Monitoring Tools

Unlike Raman and NIR spectroscopy, other monitoring techniques do not directly characterize the solid form present in the process but can often provide relevant information on changes of the dispersed solid phase such as crystal shape or unexpected nucleation events. Thus, such indirect process information can be used to gain valuable insights in solid form transformations as will be shown in the following sections.

4.9.1.2.1 Focused Beam Reflectance Measurement (FBRM)

The focused beam reflectance measurement's (FBRM) name describes its measurement principle which is based on measuring the backscattered reflections of a focused laser beam rotating concentrically inside a probe. The probe is directly inserted in the particulate process stream and sensitively tracks changes of number, dimension, and shape of the dispersed particles as shown schematically in Figure 4.9.3. It is important to note that the FBRM signal yields CLD of any backscattering dispersed phase but does not provide any direct information on the solid form and its crystal structure. Besides, the CLD differs fundamentally from PSD data due to the nature of the FBRM measurement principle and optical reflection properties depending on the analyzed system. Several authors have tried to formulate general correlations to transform CLD to PSD data, but none of these approaches has found broad acceptance due to deviations between theoretical and experimental chord length measurements [18, 31–34]. A proposed approach to extract PSD from CLD data based on a data-driven model which incorporates the experimental particularities of a given system has shown promising results for various particle morphologies [35] and is described in detail in Section 4.9.2.2.

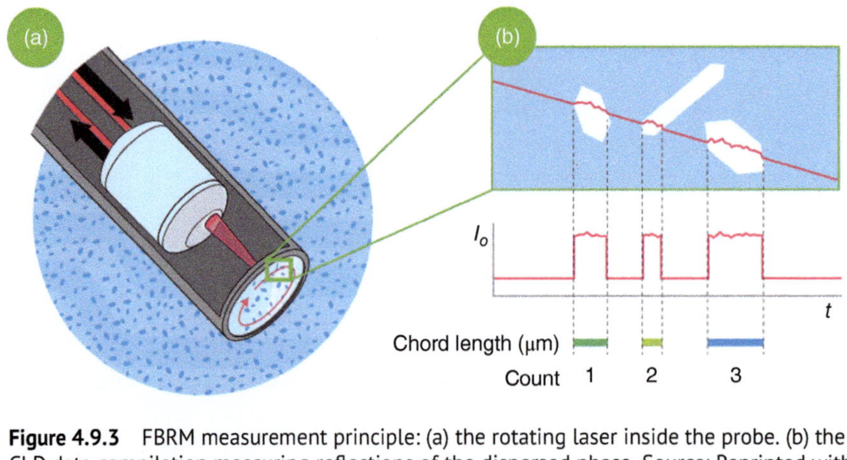

Figure 4.9.3 FBRM measurement principle: (a) the rotating laser inside the probe. (b) the CLD data compilation measuring reflections of the dispersed phase. Source: Reprinted with permission from Mettler-Toledo, AutoChem Inc., USA.

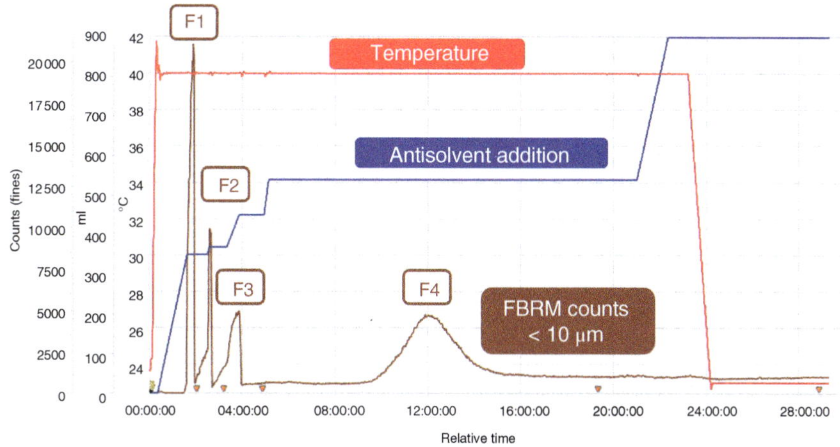

Figure 4.9.4 Temperature, antisolvent addition, and FBRM profile of an unseeded crystallization process. The spikes labeled F1–F4 highlight nucleation events of four different metastable forms.

Despite the fundamental differences between CLD and PSD data, FBRM has been routinely applied in a wide range of pharmaceutical processes, from API crystallization to high-shear wet granulation, drying and micronization processes to follow process dynamics and evolution. In process development, FBRM can provide valuable data about relative trends of fine and coarse chord counts which can be attributed to fundamental mechanisms such as nucleation, growth, or agglomeration. Distinguishing between such mechanisms with CLD data only is often rather difficult which is why the FBRM is frequently used together with in-process imaging techniques as described in the subsequent section. During scale-up and in pharmaceutical manufacturing, CLD data can be used as a real-time "fingerprint" of a particulate process which can be employed to determine product quality. Although the FBRM cannot distinguish directly between different solid forms, it is sensitive to nucleation events as shown in the cumulative number of chord counts smaller than 10 μm as a function of time in Figure 4.9.4. Unexpected nucleation events close to equilibrium conditions, i.e., without significant supersaturation with respect to the form present in the slurry, can indicate the formation of a different solid form.

In the example in Figure 4.9.4, it can be readily observed that after the initial flat line the chord counts show four distinct peaks, the first three exhibiting a rather sharp increase with subsequent decrease. All four peaks indicate distinct nucleation events of different API metastable forms which were confirmed offline using X-ray powder diffraction of wet cake samples since alternative in situ methods were not suitable due to low solids concentration in the batch. It is worth noting that the first three forms also nucleated on the FBRM probe window, resulting in the steep increase in the number of counts. Manual cleaning of the probe window quickly reduced the number of counts, before the subsequent nucleation of another metastable form induced by further antisolvent addition.

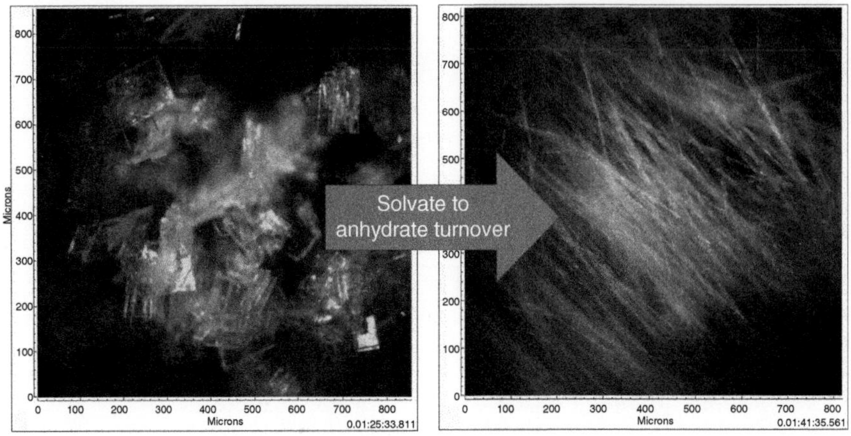

Figure 4.9.5 In-process micrographs of a solvent-mediated transformation of API solvate platelets to the anhydrous, needle-shaped form.

4.9.1.2.2 Monitoring Particle Shape Using In-process Microscopy

In-process microscopy has been used for decades to monitor process evolutions of multiphase processes. One of its main advantages is that the visual observation facilitates process understanding and data interpretation. Various sensor types have been developed ranging from reflection [8, 36] to transmission probes [37] and view-cell systems incorporated in a bypass loop [38], each of these featuring particular advantages and setbacks. Probe-based systems have the advantage that these are typically introduced directly in the reactor, thus characterizing the particles as they exist in the process. If a system requires a bypass loop, additional pumps and potentially also temperature control are needed, to avoid undesired effects in the bypass line. Reflection probes provide process images at any solids concentration, whereas transmission probes are limited by a maximum solids concentration which is a function of the PSD and the optical properties of the system. However, images from transmission probes at low solids concentration can rather easily be used for image analysis, whereas reflection probe images are hampered by overlaying particles and insufficient confocality. However, any imaging system can track solid form transformations as long as different solid forms feature significantly different particle shapes. Figure 4.9.5 shows an example of a solid form transformation in which the platelet agglomerates of a solvate (left) turn over to thin anhydrate needles (right).

It is worth noting that this technique on its own is insufficient to distinguish between different solid forms because it relies exclusively on particle shape. Particle shape and solid form are however not directly related, i.e., the same solid form can grow in different particle shapes and different solid forms can exhibit the same particle shape, which is why in-process microscopy requires additional in situ or offline solid form characterization techniques for reliable results.

4.9.1.2.3 Monitoring Solute Concentration

UV–Vis and mid-infrared (IR) spectroscopy can be used to determine the solute concentration of pure liquids via the absorption of energy. Combining these

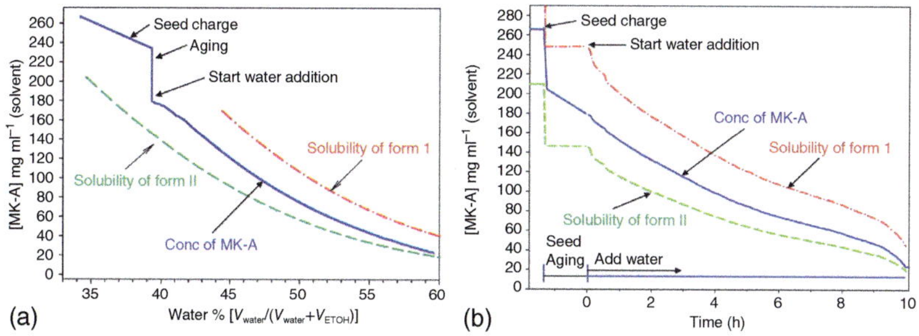

Figure 4.9.6 Solubility curves and crystallization profile of the solute concentration at 65 °C for a feedback-controlled water addition applied to maintain 33% supersaturation: (a) versus wt% water and (b) versus time. Source: From Cote et al. [45]. © 2009 American Chemical Society.

spectroscopic techniques with attenuated total reflectance (ATR) allows solute concentration quantification also in slurries because the absorption process occurs only at the surface of the ATR crystal without the influence of the dispersed phase [39, 40]. Consequently, ATR-IR spectroscopy is not able to directly characterize the solid form of suspended particles. However, the measured solute concentration data can be used together with complementary information like the solubility of various solid forms to characterize the process state. Besides, mid-IR absorbance spectra contain direct structural information due to the characteristic energy absorption of the different functional groups of a molecule. Thus, mid-IR spectroscopy has been used not only for in situ solute concentration monitoring but also for other objectives, e.g., tracking the molecular speciation as function of pH [41], automated solubility characterization [42], crystal growth kinetics characterization [43], and monitoring and control of polymorph crystallizations [44, 45]. Figure 4.9.6 shows the profile of a pharmaceutical crystallization process using a feedback-controlled water addition to ensure the generation of the stable Form II of a nondisclosed compound called MK-A [45].

The proposed methodology was combining ATR-IR spectroscopy and feedback control of the supersaturation level throughout the process to generate exclusively the most stable form. This has been achieved by keeping the supersaturation below the solubility of the metastable Form I which prevented any Form I formation. Keeping the solute concentration between the solubility lines of the two forms effectively ensures the crystallization of the thermodynamically stable Form II. Moreover, Raman spectroscopy and FBRM were employed in this example to monitor polymorphic purity and CLDs throughout the process.

4.9.1.3 Advantages and Challenges of In Situ Solid Form Monitoring Techniques

This paragraph summarizes the advantages and challenges of the various monitoring techniques in the subsequent Table 4.9.1.

Table 4.9.1 Advantages and challenges of solid form monitoring techniques.

Technique	Advantages	Challenges
Probe-based monitoring techniques in general	• Less analytical labor intensive, particularly at manufacturing scale • Improved operator safety via reduced exposure to process • Enables process control, continuous processing, and real-time release	• Although first solutions have become available for drug product processing, probe fouling remains an issue • No unified interfacing protocols with process control systems at manufacturing scale • Often, offline techniques feature an improved limit of detection which is not attained with in situ probes
Raman spectroscopy	• Directly distinguishes between forms • Monitors liquid and solid phase	• Fluorescence can limit applicability
NIR spectroscopy	• Directly distinguishes between forms • Monitors liquid and solid phase	• Overtones in NIR spectra require extensive calibration models
Focused beam reflectance measurement (FBRM)	• Statistically robust signal • Characterizes number and dimension of dispersed phase	• No direct form information • Advanced data-driven model required for PSD estimation
In-process microscopy	• Representative images without sampling and sample preparation	• Robust image analysis requires highly confocal optics • No direct form information
ATR-IR spectroscopy	• Able to monitor liquid phase only	• Equipment sensitive to vibrations which sometimes is an issue in manufacturing settings • Only indirect form information via the knowledge of the solubility levels of different forms

4.9.2 Quantification Methods and Application to Solid Form Transformation Modeling

The monitoring techniques described in the previous sections require calibration models to translate the measurement signal into quantifiable values. Univariate calibration is based on Beer–Lambert's law which is commonly used to relate absorbance A with the attenuation coefficient ε, the solute concentration c of an absorbing solute, and the path length d of the measurement beam in the sample:

$$A = \epsilon c d \tag{4.9.1}$$

However, Beer–Lambert's law applies at low concentrations and other limitations of this relation have been discussed by Mayerhoefer et al. [46]. Thus, multivariate data analysis is recommended to build reliable calibration models which are not limited to a single band in the spectrum but capture more information of broader spectral parts.

4.9.2.1 Multivariate Data Analysis

Multivariate calibration models allow for the correlation of vibrational spectroscopic data for direct solid form monitoring with the concentration in multicomponent mixtures. Different techniques can be combined for this purpose like PCA and PLS regression. PCA is a powerful tool to compress variables that are highly correlated, e.g., an NIR spectrum containing from hundreds to thousands of absorbance values. Typically, these absorbance values are correlated with a certain extent such that the spectra can be explained by a certain number of principal components, ideally only 2–3. In this way, the vectors containing a large number of correlated variables are compressed into a smaller number of uncorrelated variables (scores) by projecting the vectors on rotated axes, i.e., the principal components. Thus, most of the information contained in the original vector of variables can be expressed in terms of only a few principal components. A detailed description of this method can be found elsewhere as it has become a standard method in the fields of chemometrics, statistical analysis, and machine learning [47]. Once PCA has been performed on the spectral data, a regression model can be established between the principal components and the corresponding concentration values. In this way, PCA is performed as a first step to generate a reduced set of features and then these features are used to develop a regression model. One advanced approach is the PLS method which generates a regression model via the projection of input and output onto optimal rotated axes [48]. One example of such a combination is the quantification of the polymorph ratio in slurries using Raman spectroscopy. Hu and coworkers studied the effect of particle size on Raman spectra employing a multivariate calibration to monitor polymorph composition [49]. The authors found deviations of up to 20% when polymorph mixtures of large particles were analyzed with calibration models of small particles and vice versa as shown in Figure 4.9.7. This phenomenon is

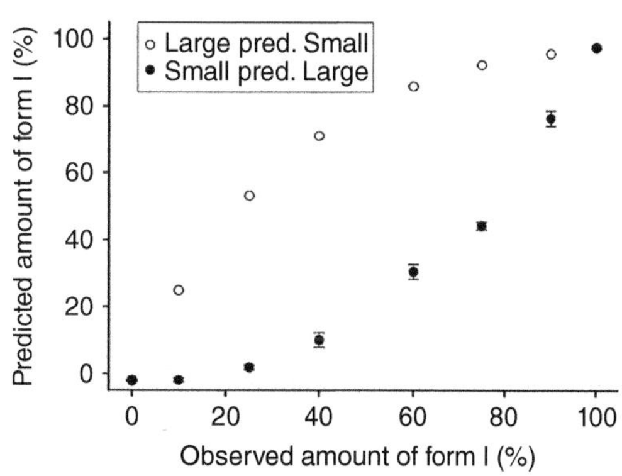

Figure 4.9.7 Predicted versus observed amount of Form I particles depending on the particle size used for measurement and calibration. Source: From Hu et al. [49]. © 2006 SAGE Publications.

due to the dependence of particle surface available for Raman scattering on particle size, i.e., the smaller a certain particle population of a given solid form, the higher is the corresponding surface for Raman scattering. Thus, the measured Raman spectra depend not only on the mass fraction of the dispersed polymorph but also on the PSD of this form. This becomes particularly complex in dynamic processes like crystallization, in which the mass fraction and the PSD of different forms dynamically vary over time. Consequently, Hu et al. highlighted that data preprocessing techniques such as standard normal variate (SNV) transformation and orthogonal signal correction (OSC) were required to take spectral features related to particle size into consideration and reduce the prediction error of the quantification model to acceptable ranges.

This has been confirmed by Simone et al., who suggested a "good calibration practice" approach, which systematically studied the impact of any dynamic parameter during a crystallization process using a design of experiments (DoE) approach, including the PSD [50].

4.9.2.2 Data-driven Model for CLD–PSD Prediction

Unlike direct solid form monitoring techniques, indirect techniques are typically not employed for quantitative solid form monitoring. However, the in-process signal of indirect techniques can still be used for process control (as discussed in Section 4.9.1.2.3) and process modeling. A recently developed data-driven approach allows the prediction of PSDs from in situ measured CLD data. Previous approaches for generating such estimates have been mainly based on the use of geometric models in combination with deconvolution algorithms to extract the underlying PSD [51, 52]. However, this methodology produces large deviations compared to the actual size distribution, particularly during dynamic processes such as crystallizations. An example is shown in Figure 4.9.8, which compares experimental PSD data with the predicted PSDs from CLD measurements at different solids concentrations using deconvolution of a geometrical model for spherical particles. Two main conclusions can be drawn from these results. Firstly, there is a strong impact of concentration on the measured CLD which results in a significant change to the PSD estimate using the geometrical model alone. Secondly, the geometric deconvolution results in a rather inaccurate PSD estimate with deviations in median size up to 100 μm. The inaccurate PSD estimate of the geometrical model is due to the simplification of the actual measurement optics and the subsequent data processing algorithm which ignores the impact of varying particle concentration on the measured chord lengths [53].

Thus, a data-driven model has been developed that can produce quantitative estimate of PSDs with one (1D) or two characteristic dimensions (2D) from measured CLD and solids concentration data, suggesting an effective solution to the abovementioned limitations of the FBRM. The particle size measurement of two characteristic dimensions of a population is particularly advantageous for non-compact particles such as needles or platelets which can cause production issues in downstream unit operations such as filtration, washing, and drying.

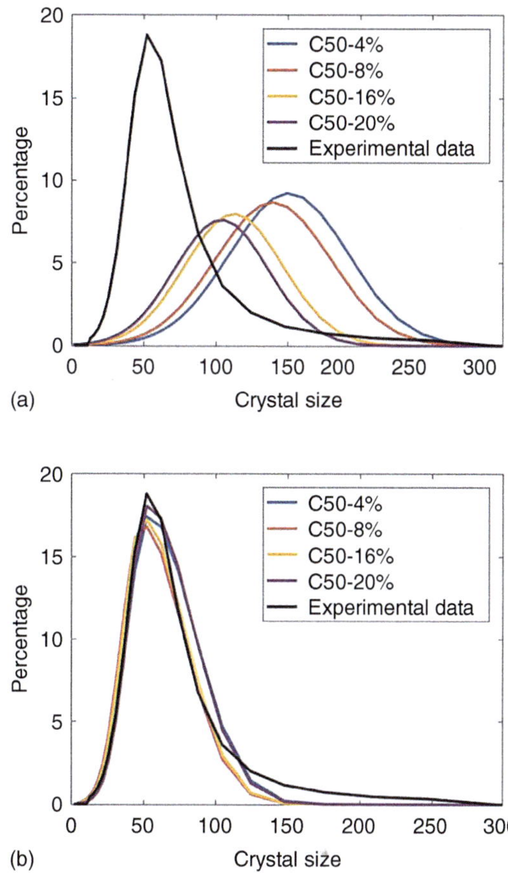

Figure 4.9.8 Comparison of PSD data estimated from CLD measurements: (a) using a geometrical model and (b) using a data-driven model. The nomenclature used in the legend refers to the experimental sample average size measured by laser diffraction (C50 for 50μm) and the different slurry concentrations (4, 8, 16, and 20 wt%).

Furthermore, it has been shown that the model can be used to predict PSD regardless of particle morphology, as long as the morphology does not change fundamentally during the process. The model has been successfully implemented for PSDs generated by laser diffraction and by image analysis for crystals of different morphologies at concentrations typical of industrial crystallization processes [35]. The model architecture is shown in Figure 4.9.9, based on three sequential steps:

- First, the entire CLD data are compressed into a small set of CLD low order moments with the moment describing the integral of the distribution, e.g., the moment of the number distribution is the total number of chord counts and the moment of the length distribution is the sum of all chord lengths of this sample.
- Second, the resulting CLD moments are mapped into a small number of PSD moments using a regression model.
- Finally, the PSD moments are expanded into a full PSD using a two-layer network model. An important aspect of this step is the definition of the generating function, i.e., a parametrized distribution, to reduce the number of model parameters.

Since this is a data-driven model, experimental CLD and PSD data are needed for every system to train the model and gather the system's optical properties in

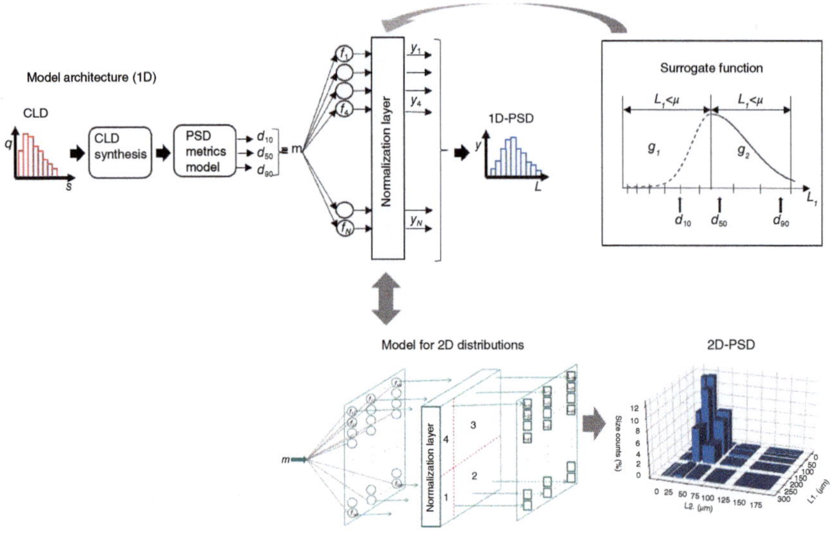

Figure 4.9.9 Model architecture to predict 1D and 2D PSDs from CLD measurement data. The first two axes refer to the minor and major particle size and the third axis refers to the number-based density. Source: From Irizarry et al. [53]. © 2017 Elsevier.

order to determine the model parameters. The experimental set featuring different PSD for model training can be generated via a combination of milling and thermal annealing and is ultimately slurried at different concentrations in saturated solutions to measure CLD as function of solids concentration. A positive feature of this model architecture is that accurate and robust results can be obtained from training sets with only a limited number of experimental distributions. This can be seen in Figure 4.9.8b which compares the experimental PSD with accurately estimated PSD data at different solids concentration using the data-driven model. It can be readily observed that the model eliminates the concentration effect which enables this approach to estimate PSD data from CLD monitoring data of crystallization processes. Moreover, this model has also the potential to be used for multi-dimensional size distributions which were generated via image analysis of scanning electron micrographs as shown by the comparison to image analysis data in Figure 4.9.10.

4.9.2.3 Process Modeling of Polymorph Transformation Processes

Solid form transformation processes can be modeled using population balance equation (PBE) accounting for the different solid forms involved [44, 54]. In the case of a perfectly mixed batch process with constant volume, size independent growth, and the absence breakage, the PBE can be formulated for the solid form i as follows [55, 56]:

$$\frac{\partial n_i}{\partial t} + G_i \frac{\partial n_i}{\partial L_i} = B_i - D_i \tag{4.9.2}$$

where n_i denotes the number density of particles, t is the time, G_i is the growth rate, L_i is the characteristic particle size, B_i and D_i are the birth and death terms due to

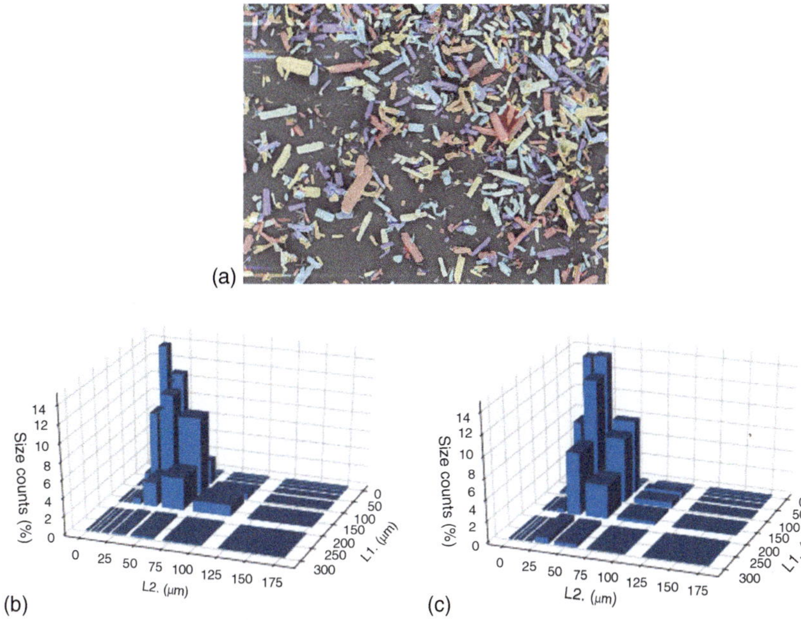

Figure 4.9.10 PSD determination by image analysis characterizing minor and major sizes measured in microns. The PSD data are shown number distributions: (a) particle segmentation of an SEM, (b) 2D-PSD established from data-driven model, and (c) 2D-PSD determined via image analysis as described in [53]. Source: From Irizarry et al. [53]. © 2017 Elsevier.

agglomeration. The material balance of the process is described in Eq. (4.9.9.3) using the solute concentration c:

$$\frac{\mathrm{d}\,c(t)}{\mathrm{d}\,t} = \sum_{i=1}^{n} \left(-3k_{v_i}\rho_i G_i \int_0^{\infty} n_i L_i^2 \mathrm{d}L_i \right) \tag{4.9.3}$$

where ρ_i and $k_{V,i}$ are the solid density and the volumetric shape factor of the ith solid form, respectively. The initial and boundary conditions for the PBE and the material balance are denoted as follows in the case of seeded transformation experiments:

$$n_i(0, L_i) = n_{i,0}(L_i) \tag{4.9.4}$$

$$n_i(t, 0) = \frac{J_i}{G_i} \tag{4.9.5}$$

$$c(0) = c_0 \tag{4.9.6}$$

where c_0 represents the initial concentration, whereas J_i is the nucleation rate and $n_{i,0}(L_i)$ denotes the initial PSD of the ith form, respectively. The supersaturation S_i is typically defined as a dimensionless ratio of the current concentration c divided by the solubility of the corresponding solid form c_i^*.

$$S_i = \frac{c}{c_i^*} \tag{4.9.7}$$

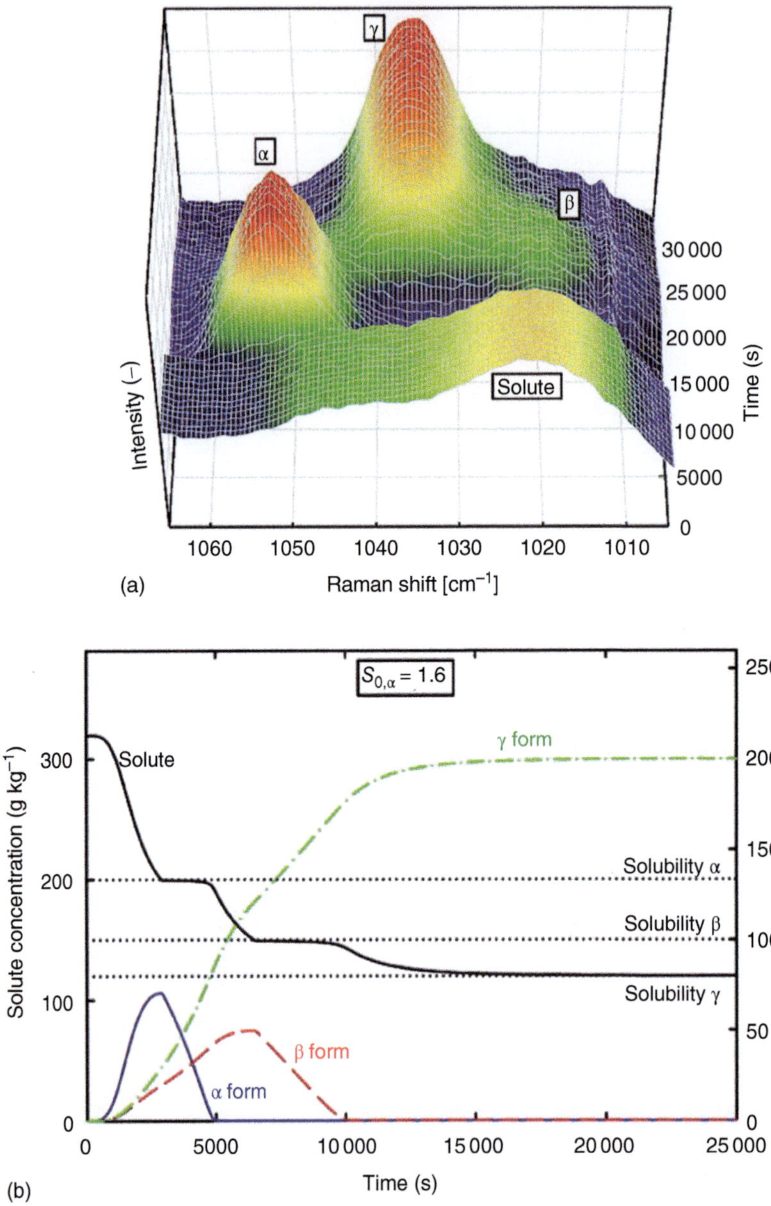

Figure 4.9.11 (a) Waterfall plot of time-resolved Raman spectra data of the unseeded polymorph transformation of D-mannitol at 10 °C with an initial supersaturation of 1.27 indicating the formation of α, β, and γ form. (b) Simulated solute and solid concentrations for the precipitation of D-mannitol at an initial supersaturation of 1.6. Source: From Cornel et al. [61]. © 2010 American Chemical Society.

Such a multiform PBE model can be solved numerically using various alternative methods, such as high-resolution finite differences [57] and Monte Carlo methods [58–60]. However, applying such a modeling strategy to actual crystallization processes requires the characterization of the unknown kinetic parameters describing the fundamental processes of nucleation, growth, and agglomeration. Several authors have combined online and offline monitoring data with PBE modeling to characterize polymorph transformation kinetics of different systems [44, 61] and to optimize continuous crystallizers while ensuring form purity [62]. Figure 4.9.11 shows the unseeded crystallization of D-mannitol as an example: in the waterfall plot of the time-resolved Raman spectra shown in (a), it can be readily observed how the characteristic peak of the solute concentration disappears after an initial lag time and the Raman spectra are subsequently dominated by the corresponding peaks of the metastable α and β forms, before the peak of the stable γ form appears. Figure 4.9.11b shows the simulation results using a population balance model after estimating the kinetics parameters for the D-mannitol system for the crystallization at a higher supersaturation of 1.6 as described in [61]. It is worth noting that the kinetics of solvent-mediated phase transformations of inorganic systems, which had been studied in the 1970s and 1980s, have been accurately analyzed and modeled before the advent of in situ process monitoring techniques by Davey and coworkers [63].

A practical review on polymorphism in crystallization processes featuring a useful table on possible strategies to isolate the stable or metastable form depending on the system's ability for spontaneous nucleation of either form can be found in the literature [64].

List of Abbreviations

API	active pharmaceutical ingredient
ATR	attenuated total reflectance
CLD	chord length distribution
DoE	design of experiments
EFA	evolving factor analysis
FBRM	focused beam reflectance measurement
HME	hot-melt extrusion
HR	relative humidity
IR	infrared
NIR	near infrared
OSC	orthogonal signal correction
PAT	process analytical technologies
PBE	population balance equation
PCA	principal component analysis
PLS	partial least square
PSD	particle size distribution
RH	relative humidity
SIMCA	soft independent modeling of class analogy

SNV	standard normal variate
UV/vis	ultraviolet/visible wavelength range (spectroscopy)
QbD	quality by design

References

1 Wu, H., Dong, Z., Li, H., and Khan, M. (2015). An integrated process analytical technology (PAT) approach for pharmaceutical crystallization process understanding to ensure product quality and safety: FDA scientist's perspective. *Organic Process Research and Development* 19 (1): 89–101.

2 Erxleben, A. (2016). Application of vibrational spectroscopy to study solid-state transformations of pharmaceuticals. *Current Pharmaceutical Design* 22 (32): 4883–4911.

3 Lieber, C.A. and Mahadevan-Jansen, A. (2003). Automated method for subtraction of fluorescence from biological Raman spectra. *Applied Spectroscopy* 57 (11): 1363–1367.

4 Buckley, K. and Ryder, A.G. (2017). Applications of Raman spectroscopy in biopharmaceutical manufacturing: a short review. *Applied Spectroscopy* 71 (6): 1085–1116.

5 Wang, F., Wachter, J.A., Antosz, F.J., and Berglund, K.A. (2000). An investigation of solvent-mediated polymorphic transformation of progesterone using in situ Raman spectroscopy. *Organic Process Research and Development* 4 (5): 391–395.

6 Starbuck, C., Spartalis, A., Wai, L. et al. (2002). Process optimization of a complex pharmaceutical polymorphic system via in situ Raman spectroscopy. *Crystal Growth & Design* 2 (6): 515–522.

7 Nguyen, D.L.T. and Kim, K.-J. (2016). Solvent-mediated polymorphic transformation of α-taltirelin by seeded crystallization. *Chemical Engineering and Technology* 39 (7): 1281–1288.

8 Nicoud, L., Licordari, F., and Myerson, A.S. (2019). Polymorph control in batch seeded crystallizers. A case study with paracetamol. *Crystal Engineering Communications* 21 (13): 2105–2118.

9 Pataki, H., Csontos, I., Nagy, Z.K. et al. (2013). Implementation of Raman signal feedback to perform controlled crystallization of carvedilol. *Organic Process Research & Development* 17 (3): 493–499.

10 Simone, E., Saleemi, A.N., Tonnon, N., and Nagy, Z.K. (2014). Active polymorphic feedback control of crystallization processes using a combined Raman and ATR-UV/Vis spectroscopy approach. *Crystal Growth & Design* 14 (4): 1839–1850.

11 Acevedo, D., Yang, X., Mohammad, A. et al. (2018). Raman spectroscopy for monitoring the continuous crystallization of carbamazepine. *Organic Process Research and Development* 22 (2): 156–165.

12 Lee, M.J., Chun, N.H., Kim, M.J. et al. (2015). In situ monitoring of antisolvent cocrystallization by combining near-infrared and Raman spectroscopies. *Crystal Growth & Design* 15 (9): 4385–4393.

13 Soares, F.L.F. and Carneiro, R.L. (2017). In-line monitoring of cocrystallization process and quantification of carbamazepine–nicotinamide cocrystal using Raman spectroscopy and chemometric tools. *Spectrochimica Acta A* 180 (5): 1–8.

14 Reddy, J.P., Jones, J.W., Wray, P.S. et al. (2018). Monitoring of multiple solvent induced form changes during high shear wet granulation and drying processes using online Raman spectroscopy. *International Journal of Pharmaceutics* 541 (1–2): 253–260.

15 Gift, A.D., Luner, P.E., Luedeman, L., and Taylor, L.S. (2009). Manipulating hydrate formation during high Shear wet granulation using polymeric excipients. *Journal of Pharmaceutical Sciences* 98 (12): 4670–4683.

16 Van Renterghem, J., Kumar, A., Vervaet, C. et al. (2017). Elucidation and visualization of solid-state transformation and mixing in a pharmaceutical mini hot melt extrusion process using in-line Raman spectroscopy. *International Journal of Pharmaceutics* 517 (1–2): 119–127.

17 Almeida, A., Saerens, L., De Beer, T. et al. (2012). Upscaling and in-line process monitoring via spectroscopic techniques of ethylene vinyl acetate hot-melt extruded formulations. *International Journal of Pharmaceutics* 439 (1-2): 223–229.

18 DeBraekeleer, K., Cuesta Sanchez, F., Hailey, P.A. et al. (1998). Influence and correction of temperature perturbations on NIR spectra during the monitoring of a polymorph conversion process prior to self-modelling mixture analysis. *Journal of Pharmaceutical and Biomedical Analysis* 17 (1): 151–152.

19 DeBraekeleer, K., De Maesschalck, R., Hailey, P.A. et al. (1998). On-line application of the orthogonal projection approach (OPA) and the soft independent modelling of class analogy approach (SIMCA) for the detection of the end point of a polymorph conversion reaction by near infrared spectroscopy (NIR). *Chemometrics and Intelligent Laboratory Systems* 46 (2): 103–116.

20 Zhou, G.X., Crocker, L., Xu, J. et al. (2006). In-line measurement of a drug substance via near infrared spectroscopy to ensure a robust crystallization process. *Journal of Pharmaceutical Sciences* 95 (11): 2337–2347.

21 Wang, I.C., Lee, M.J., Seo, D.Y. et al. (2011). Polymorph transformation in paracetamol monitored by in-line NIR spectroscopy during a cooling crystallization process. *AAPS PharmSciTech* 12 (2): 764–770.

22 Barnes, S.E., Thurston, T., Coleman, J.A. et al. (2010). NIR diffuse reflectance for on-scale monitoring of the polymorphic form transformation of pazopanib hydrochloride (GW786034); model development and method transfer. *Analytical Methods* 2 (12): 1890–1899.

23 Schaefer, D., Clicq, D., Lecomte, C. et al. (2014). A process analytical technology (PAT) approach to control a new API manufacturing process: development, validation, and implementation. *Talanta* 120: 114–125.

24 Amira, T., Roman, P., and Fethi, Z. (2016). On-line monitoring of vacuum drying of theophylline using NIR spectroscopy: solid-state transitions, water content and semi-empirical modeling. *Drug Development and Industrial Pharmacy* 42 (11): 1825–1832.

25 Lamberto, D.J., Diaz-Santana, A., and Zhou, G. (2017). Form conversion and solvent entrapment during API drying. *Organic Process Research & Development* 21 (11): 1828–1834.

26 Li, W., Worosila, G.D., Wang, W., and Mascaro, T. (2005). Determination of polymorph conversion of an active pharmaceutical ingredient in wet granulation using NIR calibration models generated from the premix blends. *Journal of Pharmaceutical Sciences* 94 (12): 2800–2806.

27 Otsuka, M., Kanai, Y., and Hattori, Y. (2014). Real-time monitoring of changes of adsorbed and crystalline water contents in tablet formulation powder containing theophylline anhydrate at various temperatures during agitated granulation by near-infrared spectroscopy. *Journal of Pharmaceutical Sciences* 103 (9): 2924–2936.

28 Alama, M.A., Shic, Z., Drennen, J.K. 3rd, and Anderson, C.A. (2017). In-line monitoring and optimization of powder flow in a simulated continuous process using transmission near infrared spectroscopy. *International Journal of Pharmaceutics* 526 (1–2): 199–208.

29 Tabasi, S.H., Fahmy, R., Bensley, D. et al. (2008). Quality by design, part I: application of NIR spectroscopy to monitor tablet manufacturing process. *Journal of Pharmaceutical Sciences* 97 (9): 4040–4051.

30 Tabasi, S.H., Fahmy, R., Bensley, D. et al. (2008). Quality by design, part II: application of NIR spectroscopy to monitor the coating process for a pharmaceutical sustained release product. *Journal of Pharmaceutical Sciences* 97 (9): 4052–4066.

31 Ruf, A., Worlitschek, J., and Mazzotti, M. (2000). Modeling and experimental analysis of PSD measurements through FBRM. *Particle and Particle Systems Characterization* 17 (4): 167–179.

32 Li, M. and Wilkinson, D. (2005). Determination of non-spherical particle size distribution from chord length measurements. Part 1: theoretical analysis. *Chemical Engineering Science* 60 (12): 3251–3265.

33 Pandit, A.V. and Ranade, V.V. (2016). Chord length distribution to particle size distribution. *AIChE Journal* 62 (12): 4215–4228.

34 Unno, J., Umeda, R., and Hirasawa, I. (2018). Computing crystal size distribution by focused-beam reflectance measurement when aspect ratio varies. *Chemical Engineering and Technology* 41 (6): 1147–1151.

35 Schoell, J., Irizarry, R., Sirota, E. et al. (2019). Determining particle-size distributions from chord length measurements for different particle morphologies. *AIChE Journal* 65 (5): e16560.

36 Zhou, G., Moment, A., Cuff, J. et al. (2015). Process development and control with recent new FBRM, PVM, and IR. *Organic Process Research & Development* 19 (1): 227–235.

37 Presles, B., Debayle, J., Fevotte, G., and Pinoli, J.C. (2010). Novel image analysis method for in situ monitoring the particle size distribution of batch crystallization processes. *Journal of Electronic Imaging* 19 (3): 1–7.

38 Schorsch, S., Ochsenbein, D.R., Vetter, T. et al. (2014). High accuracy online measurement of multidimensional particle size distributions during crystallization. *Chemical Engineering Science* 105 (1): 155–168.

39 Groen, H. and Roberts, K.J. (2001). Nucleation, growth, and pseudo-polymorphic behavior of citric acid as monitored in situ by attenuated total reflection Fourier transform infrared spectroscopy. *The Journal of Physical Chemistry B* 105 (43): 10723–10730.

40 Schoell, J., Lindenberg, C., Vicum, L. et al. (2007). Precipitation of α L-glutamic acid: determination of growth kinetics. *Faraday Discussions* 136: 247–264.

41 Scholl, J., Vicum, L., Muller, M., and Mazzotti, M. (2006). Precipitation of L-glutamic acid: determination of nucleation kinetics. *Chemical Engineering and Technology* 29 (2): 257–264.

42 Cornel, J., Lindenberg, C., and Mazzotti, M. (2008). Quantitative application of in situ ATR-FTIR and Raman spectroscopy in crystallization processes. *Industrial and Engineering Chemistry Research* 47 (14): 4870–4882.

43 Schoell, J., Lindenberg, C., Vicum, L., and Mazzotti, M. (2007). Antisolvent precipitation of PDI 747: Kinetics of particle formation and growth. *Crystal Growth and Design* 7 (9): 1653–1661.

44 Schoell, J., Bonalumi, D., Vicum, L. et al. (2006). In situ monitoring and modeling of the solvent-mediated polymorphic transformation of L-glutamic acid. *Crystal Growth & Design* 6 (4): 881–891.

45 Cote, A., Zhou, G., and Stanik, M. (2009). A novel crystallization methodology to ensure isolation of the most stable crystal form. *Organic Process Research and Development* 13 (6): 1276–1283.

46 Mayerhoefer, T.G., Mutschke, H., and Popp, J. (2016). Employing theories far beyond their limits-the case of the (Boguer-) Beer–Lambert law. *ChemPhysChem* 17 (3): 1948–1955.

47 Jolliffe, I.T. (2002). *Principal Component Analysis*, 2e. New York, NY: Springer.

48 Wehrens, R. (2011). *Chemometrics with R – Multivariate Data Analysis in the Natural Sciences and Life Sciences*. Berlin: Springer.

49 Hu, Y., Wikström, H., Byrn, S.R., and Taylor, L.S. (2006). Analysis of the effect of particle size on polymorphic quantitation by Raman spectroscopy. *Applied Spectroscopy* 60 (9): 977–984.

50 Simone, E., Saleemia, A.N., and Nagy, Z.K. (2014). Application of quantitative Raman spectroscopy for the monitoring of polymorphic transformation in crystallization processes using a good calibration practice procedure. *Chemical Engineering Research and Design* 92 (4): 594–611.

51 Kempkes, M., Eggers, J., and Mazzotti, M. (2008). Measurement of particle size and shape by FBRM and in situ microscopy. *Chemical Engineering Science* 63 (19): 4656–4675.

52 Agimelen, O.S., Jawor-Baczynska, A., McGinty, J. et al. (2016). Integration of in situ imaging and chord length distribution measurements for estimation of particle size and shape. *Chemical Engineering Science* 144: 87–100.

53 Irizarry, R., Chen, A., Crawford, R. et al. (2017). Data-driven model and model paradigm to predict 1D and 2D particle size distribution from measured chord-length distribution. *Chemical Engineering Science* 164: 202–218.

54 Nicoud, L., Licordari, F., and Myerson, A.S. (2019). Polymorph control in MSMPR crystallizers. A case study with paracetamol. *Organic Process Research and Development* 23 (5): 764–806.

55 Mersmann, A. (2001). *Crystallization Technology Handbook*. New York, NY: CRC Press.

56 Ramkrishna, D. (2000). *Population Balances – Theory and Applications to Particulate Systems in Engineering*. San Diego, CA: Academic Press.

57 Qamar, S., Elsner, M.P., Angelov, I.A. et al. (2006). A comparative study of high resolution schemes for solving population balances in crystallization. *Computers and Chemical Engineering* 30 (6–7): 1119–1131.

58 Liffman, K. (1992). A direct simulation Monte Carlo method for cluster coagulation. *Journal of Computational Physics* 100 (1): 116–127.

59 Maisels, A., Kruis, F.E., and Fissan, H. (2004). Direct simulation Monte Carlo for simultaneous nucleation, coagulation, and surface growth in dispersed systems. *Chemical Engineering Science* 59 (11): 2231–2239.

60 Irizarry, R. (2008). Fast Monte Carlo methodology for multivariate particulate systems – I: point ensemble Monte Carlo. *Chemical Engineering Science* 63 (1): 95–110.

61 Cornel, J., Kidambi, P., and Mazzotti, M. (2010). Precipitation and transformation of the three polymorphs of D-mannitol. *Industrial and Engineering Chemistry Research* 49 (12): 5854–5862.

62 Lai, T.T.C., Cornevin, J., Ferguson, S. et al. (2015). Control of polymorphism in continuous crystallization via mixed suspension mixed product removal systems cascade design. *Crystal Growth & Design* 15 (7): 3374–3382.

63 Davey, R.J., Cardew, P.T., McEwan, D., and Sadler, D.E. (1986). Rate controlling processes in solvent-mediated phase transformations. *Journal of Crystal Growth* 79 (1-3): 648–653.

64 Mangin, D., Puel, F., and Veesler, S. (2009). Polymorphism in processes of crystallization in solution: a practical review. *Organic Process Research and Development* 13 (6): 1241–1253.

4.10

Photon Density Wave (PDW) Spectroscopy for Nano- and Microparticle Sizing

Lena Bressel and Roland Hass

PDW Analytics GmbH, Geiselbergstr. 4, 14476 Potsdam, Germany

4.10.1 Classification of Particle Sizing Technologies

Particle sizing technologies can be classified into single particle or particle ensemble measurement approaches. Characterization of single particles with respect to size, shape, or other features provides best access to histograms, e.g., high resolved particle size distributions. However, counting statistics are relevant, i.e., how many particles need to be characterized to provide reliable distribution data. In contrast, particle ensemble measurements typically assume a certain distribution function and fit such a function to the actual experimental result of the various sizing techniques. It applies to all sizing techniques that "particle size" itself is derived only indirectly from other experimental quantities, e.g., angle-dependent scattering intensities, diffusion coefficients, attenuation data, or image analysis algorithms. Particle size as experimental result per method of measurement may be obtained based on theoretical principles or by correlative methods needing calibration with particles of defined size. However, methods of measurement that differ in their physical measurement principle typically yield differing particle sizes.

For all techniques, the operational suitability with respect to lower and upper particle size limit as well as lower and upper solid fraction limit is a relevant concern. In addition, characterization of size-distributed particles might be challenging for ensemble measurements if smaller or larger particles in a given distribution are screened due to the other particles of the ensemble and thus are not detected by the sizing technique.

Without claim to be complete, Table 4.10.1 categorizes common methods for particle sizing into single or ensemble measurements, correlative or "first principle" basis, and their suitability for process analytics, i.e., online or inline measurement capability. Here, "inline" is defined as dilution-free measurement within the actual processing environment, i.e., a reactor or a transfer pipeline, and "online" is defined as dilution-free measurement in bypass installations or dilution-free material feed into the analyzer. All dilution-based approaches would be referred to "offline" or "at-line." Still, in literature, the term "online" is often used in conjunction with particle

Solid State Development and Processing of Pharmaceutical Molecules: Salts, Cocrystals, and Polymorphism, First Edition. Edited by Michael Gruss.

Table 4.10.1 Classification of particle size characterization methods into single or ensemble measurement, correlative or first principle analysis, and inline/online capability.

	Single	Ensemble	Correlative	First principles	Inline/online capability
Dynamic light scattering		x		x	
Static light scattering		x		x	
Single particle light scattering	x			x	
Laser diffraction		x		x	
Fiber-optical quasi elastic light scattering		x	x		x
Ultrasound attenuation		x		x	x
Imaging/light microscopy	x		x		x
Electron microscopy	x		x		
Flow cytometry		x	x		
Impedance measurement		x	x		
Centrifugation/ sedimentation		x		x	
Chord length measurement	x		x		x
Turbidimetry		x		x	
Turbidity		x	x		x
Photon Density Wave spectroscopy		x		x	x

size analyzers providing continuous measurement but being based on automated dilution.

Table 4.10.1 depicts only a limited number of particle size characterization technologies. Various literature describes different analytical approaches in a much more extensive and holistic way [1, 2].

4.10.2 Particle Size and Solid Fraction Ranges

The actual particle size range and the solid fraction range, being experimentally accessible per sizing method, depend strongly on the material under investigation. Still, Figure 4.10.1 presents generalized ranges per method, again without claim to be complete.

The boundaries with respect to lower and upper size as well as solid fraction, often provided by the various instrument suppliers, should be evaluated in detail

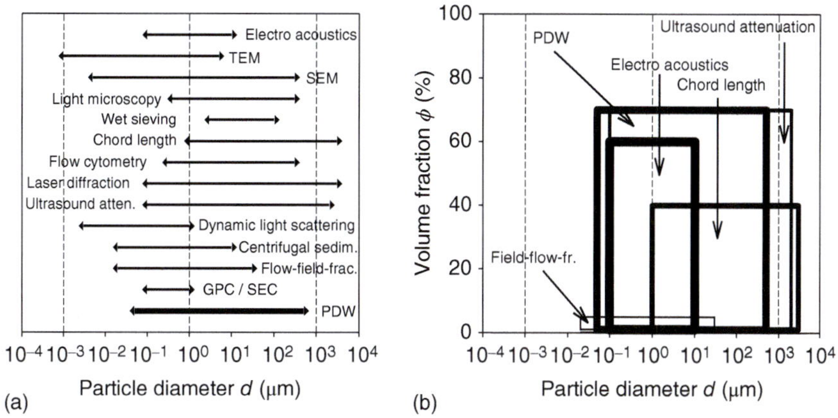

Figure 4.10.1 Sizing ranges (a) and volume fraction ranges (b) for common particle sizing methods. Data are generalized and need to be verified for a defined material under investigation. Source: Based on Merkus [1], PDW spectroscopy added, figures taken from [3].

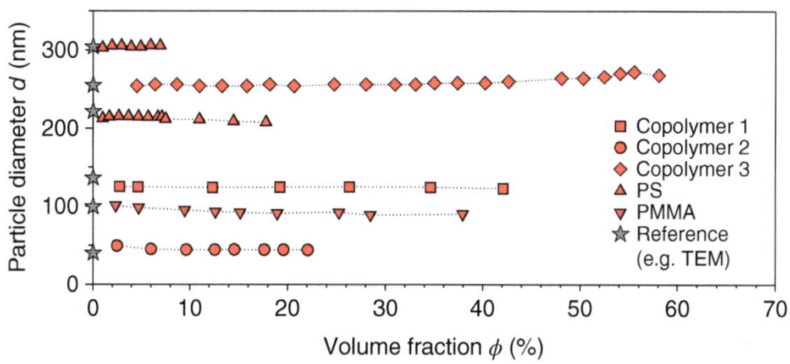

Figure 4.10.2 Particle diameter as function of volume fraction for different aqueous homo- and copolymer particle suspensions obtained by PDW spectroscopy in comparison to transmission electron microscopy (TEM) as reference.

per material of interest. A simple test for method suitability to higher particle fractions is a dilution series and comparison to dilution-based reference analysis. In case of particles not changing their size or shape by dilution, coherent sizes should be obtained across the whole solid fraction range, including the material in its original state. As an example, in Figure 4.10.2, dilution series of aqueous suspensions of homo- and copolymers investigated by Photon Density Wave (PDW) spectroscopy are represented along with dilution-based transmission electron microscopy (TEM) data as reference. The upper solid fraction limit was defined by the provided samples. The highest volume fraction was approximately 60%.

For particle sizing technologies not suitable for high solid fractions, artifacts may be observed. For example, applying dynamic light scattering (DLS) to such concentrated aqueous polymer suspensions or to aqueous nano-sized oil emulsions, deviations in size are observed toward larger volume fractions (Figure 4.10.3). Here, significant changes in diameter are found by DLS, in contrast to the consistent results found by PDW spectroscopy. In fact, also for dilution-based measurement

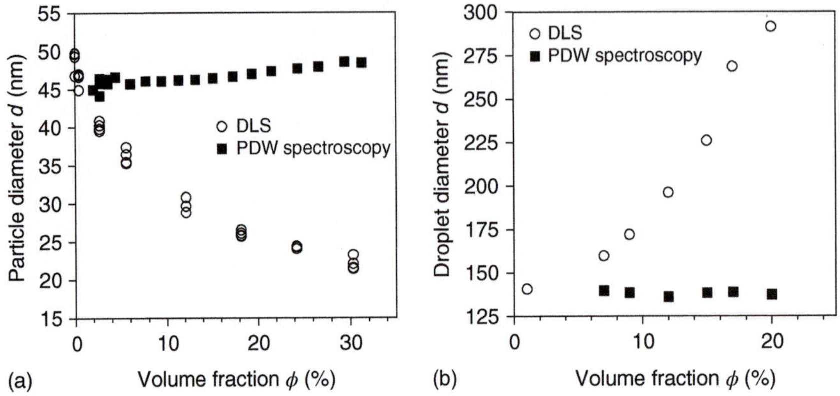

Figure 4.10.3 Particle diameter as function of volume fraction for an aqueous polystyrene suspension (a) and an aqueous nano-sized oil emulsion (b) as comparison between dynamic light scattering (DLS) and PDW spectroscopy.

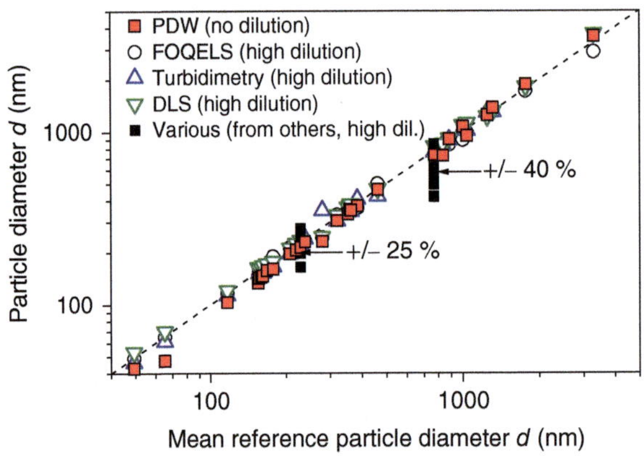

Figure 4.10.4 Experimental particle diameter of various aqueous monomodal, monodisperse, and spherical polystyrene particle suspensions determined by different methods as function of mean reference particle diameter and results of a round robin test for two of the suspensions (experimental result range of ±25% and ±40% of mean particle diameter), (FOQELS, fiber optic quasi elastic light scattering).

technologies, the actual working range for reliable sizing results should be determined per material.

In addition, method comparison based on well-defined particles (here polystyrene) may indicate in what ranges particle size measurements deviate from expectations (Figure 4.10.4). In this example for particles of around 50 nm diameter, PDW spectroscopy tends to underestimate their size. Figure 4.10.4 also represents the outcome of a round robin test for two of such polystyrene suspensions. Even though the two suspensions provided straightforward size characteristics (monomodal, monodisperse, and spherical), various laboratories using different methods including varying operators generated strongly differing experimental results with respect to particle size.

4.10.3 Photon Density Wave (PDW) Spectroscopy – Theory, Instrumentation, and Application Examples

PDW spectroscopy is based on photon transport theory in multiple light scattering environments. Here, it is assumed that a point source emits light (photons) in all directions into an infinitely large material. When hitting a particle (crystal, polymer particle, etc.), the photons are scattered and the light hence is distributed in all directions. Due to these consecutive scattering events, individual photons follow a kind of random path through the material until they are absorbed by absorbing components of the material or are coupled out of a confined sample of the material, e.g., due to a beaker wall or a detector as light sink. The distribution of photons in the material strongly depends on absorption and scattering properties of the material, expressed as absorption coefficient μ_a and as reduced scattering coefficient $\mu_s{}'$, respectively. This can be understood as counting the number of photons in a certain volume, i.e., defining a photon density ρ. To derive a mathematical expression for the photon density, the Boltzmann transport equation has to be solved. In PDW spectroscopy, this is done by applying the so-called P1-approximation. Details can be found in [4]; an excerpt is provided in the following. The resulting mathematical expression for the photon density in a multiple scattering material with a time-independent light source is given as follows:

$$\rho(r) = \frac{\rho_{\mathrm{DC}}^0}{r} \mathrm{e}^{-k_{\mathrm{DC}} r} \tag{4.10.1}$$

Here, ρ_{DC}^0 is the intensity at the source and r is the distance from the light source. The coefficient k_{DC} comprises information about the absorption coefficient μ_a and the reduced scattering coefficient $\mu_s{}'$ of the material

$$k_{\mathrm{DC}} = \sqrt{(\mu_a^2 + 3\mu_a\mu_s')} \tag{4.10.2}$$

and can be determined by measurement of the photon density, e.g., as function of distance from the light source. However, the optical properties are still coupled here. If one of the optical properties, μ_a or $\mu_s{}'$, would be known, the other one could hence be determined. If both coefficients are unknown, only their combination is obtained. To separate the absorption and scattering properties in PDW spectroscopy, the light source is additionally modulated in its intensity over time (typically with a sinusoidal modulation) with a certain modulation frequency. Modulation of the light source leads to a time-dependent modulation of the photon density in the sample, i.e., a PDW is created. Solving the Boltzmann transport equation for this time-dependent problem in the P1-approximation leads to:

$$\rho(r,t) = \underbrace{\frac{\rho_{\mathrm{DC}}^0}{r} \exp[-k_{\mathrm{DC}} r]}_{\text{time-independent part}} + \underbrace{\underbrace{\frac{\rho_{\mathrm{AC}}^0}{r} \exp[-k_{\mathrm{I}} r]}_{\text{intensity I}} \underbrace{\exp[\mathrm{i}k_{\Phi} r]}_{\text{phase } \Phi} \exp[-\mathrm{i}\omega t]}_{\text{time-dependent part}} \tag{4.10.3}$$

Here, the first summand is still the time-independent photon density from the constant light source and the second summand is the time-dependent PDW which results from the intensity modulation. The first part of the second summand can be interpreted as the amplitude or intensity I and the second part as the phase Φ of the PDW. By measurement of I and Φ of the PDW as a function of distance r, the so-called intensity and phase coefficients k_I and k_Φ can be determined. They are related to μ_a and $\mu_s{}'$ as follows:

$$k_{I/\Phi} = \sqrt{\frac{3}{2}\left\{\sqrt{\left[\left[\frac{\mu_a}{3}+\mu_s'\right]^2+\frac{\omega^2}{c^2}\right]\left[\mu_a^2+\frac{\omega^2}{c^2}\right]} \pm \mu_a\left[\frac{\mu_a}{3}+\mu_s'\right] \mp \frac{\omega^2}{c^2}\right\}} \tag{4.10.4}$$

Here, ω is the angular modulation frequency and c is the speed of light in the material. As two coefficients, k_I and k_Φ, are determined, the two unknown variables, μ_a and $\mu_s{}'$, can be obtained independently. In PDW spectroscopy, this is done by a nonlinear global analysis over all measurement parameters. As Eq. (4.10.4) requires the speed of light in the material, the only property necessary to analyze the optical coefficients is the total refractive index n_{Disp} of the material.

In practice, PDW spectroscopy uses optical emission and detection fibers to guide light from a laser diode (LD) of a certain wavelength (typically between λ = 400–1000 nm) into the sample and to guide light of the PDW to the detector (Figure 4.10.5). The fiber ends each act as point-like light sources and sinks (typical fiber core diameter of 200–1000 μm) which create or probe the PDW. To achieve the light modulation, the LD receives a constant current (DC) from a laser driver and an additional sinusoidally modulated current (HF) with a frequency between typically f = 10–1300 MHz from a vector network analyzer (VNA) resulting in a time-modulated light intensity emitted by the LD. The resulting PDW, created in the sample and received via the optical detection fiber by an avalanche photodiode (APD), is converted into a voltage, amplified by a high frequency amplifier (HFA) and fed into the VNA for analysis of the phase and intensity of the signal. To determine the intensity and phase coefficients from the intensity and phase of the PDW, the distance between the fiber tips in the sample is varied, e.g., by a motorized translation stage or a specialized probe design (cf. Section 4.10.5). To obtain high precision and accuracy of the optical coefficients, the modulation frequency is varied additionally.

PDW spectroscopy is especially suitable for highly turbid liquids ($\mu_s{}' \gtrsim 0.05\,\text{mm}^{-1}$) [5]. Due to the model used for solving the Boltzmann transport equation (P1-approximation, etc.) it requires significantly stronger light scattering than light absorption of the sample ($\mu_s{}' \gg \mu_a$). For example, white crystallization slurries like lactose or many pharmaceutical small molecules can be easily characterized, while black inorganic oxide slurries are not suitable. In the latter case, the photons emitted into the suspension are simply absorbed. Because of the assumption of an infinitely large material while solving the Boltzmann transport equation, absorption of the material should also not be too low ($\mu_a \gtrsim 10^{-5}\,\text{mm}^{-1}$) and sample sizes should typically have a volume in the range of at least 0.5–1 L to prevent

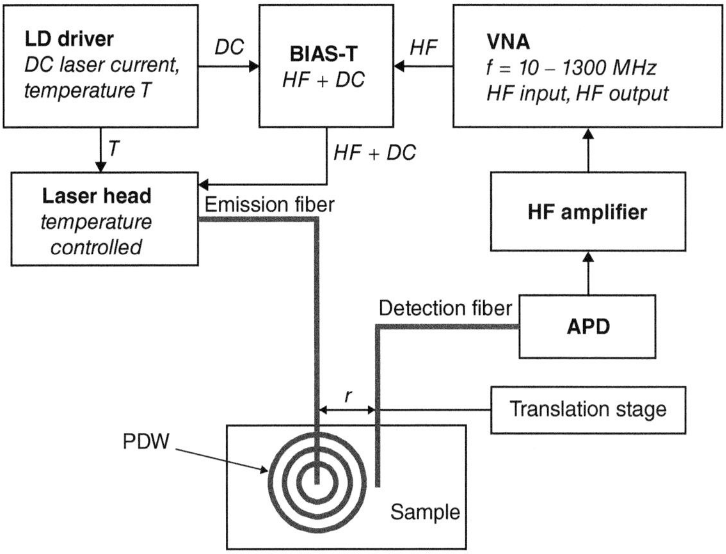

Figure 4.10.5 Schematic experimental setup of a PDW spectrometer.

substantial loss of light at the sample boundaries. As main experimental result, the reduced scattering coefficient μ_s' and the absorption coefficient μ_a are obtained at the experimental wavelength. They represent the absolute optical properties of the characterized material. For many physical, chemical, or biotechnological processes, these two coefficients are sufficient to monitor changes of the material as function of time. Examples include the oxygenation/deoxygenation of blood [6], the melting/crystallization of milk fat [7], the thermo-responsible swelling or aggregation of poly(*N*-isopropylacrylamide) (PNIPAM) polymer particles [8], zeolite synthesis [9], and the biotechnological production of polyhydroxyalkanoate (PHA) [10]. Here, process understanding is obtained without the interpretation of μ_s' with respect to particle size.

4.10.4 Particle Sizing by PDW Spectroscopy

The reduced scattering coefficient μ_s' can be interpreted with respect to particle size. Based on Mie theory [11] and theory for so-called dependent light scattering [4], particle sizes can be obtained from highly concentrated dispersions (cf. Figure 4.10.2). Size analysis in PDW spectroscopy can be carried out in different ways. In its simplest form, it is assumed that all particles possess the same size (monodisperse case). Here, the reduced scattering coefficient is calculated as follows:

$$\mu_s' = \frac{3\phi Q_s[1 - g]}{2d} \tag{4.10.5}$$

where ϕ is the volume fraction of particles, d is the particle diameter, and Q_s and g are the scattering efficiency and anisotropy factor, respectively, which can be calculated

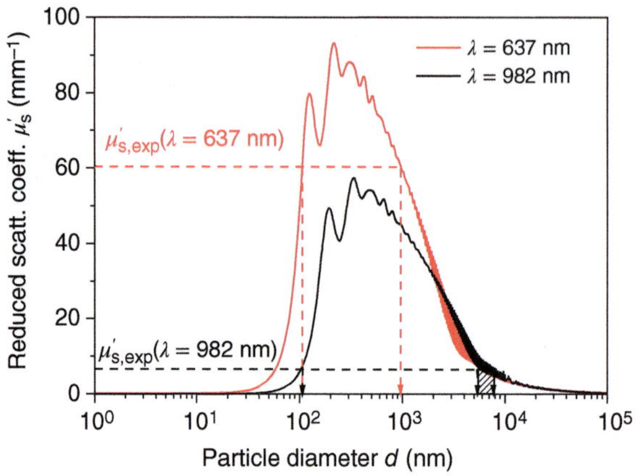

Figure 4.10.6 Reduced scattering coefficient calculated by Mie theory for a polystyrene dispersion with a volume fraction of $\phi = 0.3$ and for two different wavelengths ($\lambda = 637$ nm, 982 nm) as function of the particle diameter.

via Mie theory. Figure 4.10.6 displays calculated reduced scattering coefficients as function of particle diameter for two different experimental wavelengths. Accordingly, for each reduced scattering coefficient at least two particle diameters are possible. A comparison between experimental and calculated reduced scattering coefficients thus leads to either a small or large particle diameter. This ambiguity can be overcome by measuring $\mu_s{}'$ with at least two different wavelengths [12]. The results from both measurement wavelengths match each other either for the smaller or larger diameter.

At high particle volume fractions (typ. $\phi > 0.05$) spatial ordering of the particles takes place due to their interaction potential. This leads to so-called dependent light scattering, i.e., interference of the scattered light from different particles due to spatial ordering which finally influences the absolute value of the reduced scattering coefficient. For proper calculation of the particle size, this particle interaction has to be taken into account during the calculation of the reduced scattering coefficient. The interference effect can be expressed via the static structure factor S:

$$\mu_s' = \frac{3\phi}{2d} \int_{-1}^{1} 2\pi\, q_s(\theta) S(\theta)[1 - \cos(\theta)] \mathrm{d}\cos(\theta) \tag{4.10.6}$$

Here, q_s is the angle-dependent scattering efficiency and θ the scattering angle. In PDW spectroscopy, two different models of particle interaction are considered. The particles interact either via a hard sphere potential and the static structure factor is calculated applying the Percus–Yevick (PY) approximation [4] or the particles interact due to additional electrostatic forces via a Yukawa potential and the structure factor is calculated applying the mean spherical approximation (MSA) [4]. In the latter case apart from the particle size, electric charges on the particles can be additionally probed via PDW spectroscopy in their natural, dilution-free environment. For such an application, the analysis of the wavelength dependency is crucial and hence measurement of $\mu_s{}'$ at different wavelengths becomes necessary.

In case of polydisperse samples, Eq. (4.10.6) changes to:

$$\mu_s' = \frac{\pi}{k_m^2} \int_{-1}^{1} \sum_{\nu}^{N_c} \sum_{\mu}^{N_c} \sqrt{\rho_\nu \rho_\mu} [A_{1,\nu}(\theta) A_{1,\mu}^*(\theta) + A_{2,\nu}(\theta) A_{2,\mu}^*(\theta)] S_{\nu\mu}(\theta)[1 - \cos(\theta)] \mathrm{d}\cos(\theta) \tag{4.10.7}$$

where k_m is the absolute of the wave vector in the medium, ρ_ν and ρ_μ are the number densities of particles with sizes d_ν and d_μ, respectively, $A_{1,\nu}$, $A_{1,\mu}$, $A_{2,\nu}$, $A_{2,\mu}$ are the scattering amplitudes for incident light of parallel or perpendicular polarization of particles with sizes d_ν or d_μ (a * denotes complex conjugates), $S_{\nu\mu}$ is the partial static structure factor accounting for interference of the scattered light from particles with sizes d_ν and d_μ, and N_c is the number of different particle sizes. The partial static structure factures are calculated in PDW spectroscopy assuming hard sphere potentials between all particles applying again the PY approximation [13].

To obtain the size distribution, an assumption of the type of distribution is necessary. For example, often a volume-based log-normal distribution is assumed. The only unknown parameters in this case are the mean and standard deviation of the size distribution. These parameters are obtained during a nonlinear fitting algorithm (Levenberg–Marquardt) from the wavelength dependency of the reduced scattering coefficient.

Typically, four to eight experimental wavelengths are used in a PDW spectroscopy experiment to obtain particle size distributions. Fewer wavelengths still allow for a size distribution estimate. To facilitate the analysis under these circumstances, a monodisperse approach is taken as described above. In this case for each measured wavelength a different particle size is obtained. The average and standard deviation of these values can be used as an estimate for the size distribution in the sample. With this type of analysis, a fast estimation of size distributions can be achieved, e.g., during reaction processes. It is currently ongoing research to evaluate under which circumstances particle size distribution measurements find limits.

Mie theory assumes that the particles are spherical, possess a homogeneous refractive index and optionally a nonzero absorption index, and are embedded in a non-absorbing medium. The models for dependent light scattering used in PDW spectroscopy assume hard spherical particles. Therefore, due to the use of Mie theory and the hard sphere model with optionally additional electrostatic interactions particle sizing with PDW spectroscopy works best for compact, hard, spherical particles. Nonspherical, porous, or soft particles will lead to slightly or strongly different behavior of the reduced scattering coefficient and thus also might lead to deviations in the determined particle size. It has to be pointed out, however, that the experimental reduced scattering coefficient remains the absolute optical parameter of the material under investigation. Its interpretation with respect to size might need specialized theories for a certain particle type.

Particle sizing with PDW spectroscopy is also strongly dependent on the input parameters like volume fraction ϕ or the refractive indices of the particles n_p and the surrounding medium n_m, which are necessary for Mie theory and dependent scattering. Hence, these properties should be known or measured with good precision

($\Delta\phi/\phi \approx 0.1$, $\Delta n_p/n_p \approx 0.001$, $\Delta n_m/n_m \approx 0.001$) [3]. For systems undergoing time-dependent changes, typically a model is needed to provide these reference data as function of time. Currently, the software of PDW spectrometers includes a database for many material properties and additionally an interface to implement reference data models.

4.10.5 Sample Versus Process Measurements

PDW spectroscopy can be used for dilution-free sample measurements and for probe-based process characterization. For the first approach, a two-fiber concept (one emission and one detection fiber) is best (cf. Figure 4.10.5). PDW spectroscopy requires no sample purification or similar preparation steps. In an ideal case, the sample is stirred to avoid separation effects like sedimentation and is covered to prevent evaporation losses and hence concentration changes during the measurement. The sample should have a volume of at least 0.5–1 L depending on the optical properties. The lower the turbidity, the larger the sample volume should be. The optical fibers are placed in the middle of the sample, to minimize light interaction with surrounding materials (beaker wall, air, etc.). With the help of a motorized translation stage, a distance variation between emission and detection fiber is realized. All aspects of such a PDW spectroscopy experiment are computer controlled and automated. Such a beaker experiment in conjunction with the two-fiber concept allows also for the characterization of simple processes, e.g., Ostwald ripening of an emulsion, phase separation, homogenization, or titration experiments, since PDW spectroscopy can be executed continuously, providing a time resolution in the sub-minute regime.

For processing conditions requiring closed systems, e.g., within chemical reactors, the two-fiber concept has the drawback of the positional change of one of the two fibers. For the characterization of processes in such closed reactors, instead fiber-optical process probes are implemented which consist of more than two fibers. The distance variation between the fibers is then defined by the probe geometry. Currently, standard PDW spectroscopy process probes are made of stainless steel with a length of >250 mm and have a diameter of 25 mm. They can be implemented in processes by weld-on adapters (e.g., for TriClamp, VariVent, etc.) or with flexibility in their insertion depth by compression fittings.

In contrast to process probes of many other optical process analytical technologies (PAT), the process probes of PDW spectrometers are comparably of low cost. This is due to the absence of complex components like lenses, motors, filters, and such, which are often found in other process probes. This allows for the use of PDW process probes in a single use case. In addition, due to the optical fibers acting only as waveguides, a PDW spectroscopy process probe provides significant flexibility in its conduit length. Typical length is 10 m; up to 50 m should be technically feasible.

A fundamental benefit particularly for crystallization or precipitation processes is the robustness of the PDW spectroscopy method towards so-called probe fouling.

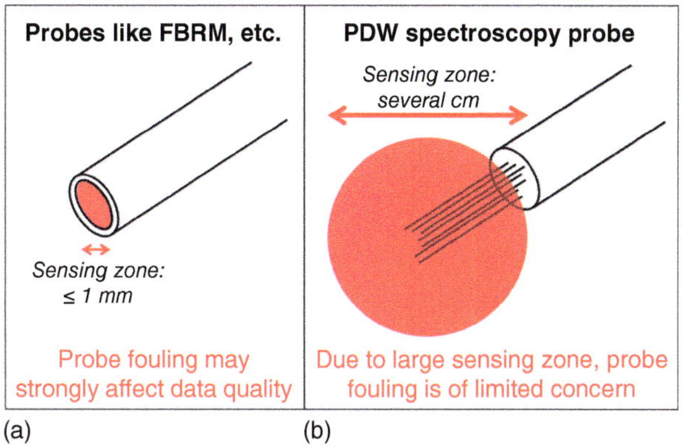

Figure 4.10.7 Schematic representation of the sensing zone close to the probe window for many optical process analytical techniques like Raman, process microscopy, FBRM, ATR-FTIR (a) and for PDW spectroscopy scaling over larger distances (b). Source: From Hartwig and Hass [14]. © 2018 John Wiley & Sons.

Often, process probes consist of a probe window and measurements are executed in close proximity (e.g., focused beam in Raman spectroscopy or focused beam reflectance measurement (FBRM), imaging-based process probes, and attenuated total reflectance (ATR) measurement principles). Material deposition on the probe window might defocus the actual measurement light and/or add to light attenuation. As a result, data from process probes exhibiting probe fouling may be of limited quantitative nature.

In PDW spectroscopy however, light is expected to be scattered multiple times, interacting with particles in zones of several centimeters diameter. In addition, the individual optical fibers have a diameter of less than a millimeter. Even if particles adhere to the fiber tips, the light will be simply scattered by these particles. Multiple light scattering within the sensing zone is, however, the basis for this method (Figure 4.10.7).

4.10.6 Technical Implementation and Data Access

A PDW spectrometer requires 220 or 110 V as power source and a computer for its operation. Connection of the spectrometer to the computer is realized by universal serial bus (USB) or local area network (LAN). Data export is realized by open platform communications unified architecture (OPC UA). The spectrometer should be placed in a dry and dust-free environment. Due to the possibility of using long optical fibers between the spectrometer and its process probe (typically 10 m, feasible up to 50 m), the instrument itself can be decoupled from the implementation place of the process probe. The process probe only requires the laser light emitted from the spectrometer, being guided by the fiber-optical conduit. During manufacturing, the probe itself is tested for its operational capability. At the present, the spectrometer includes routines for performance self-check to some degree. In addition, all relevant operating parameters are saved for later evaluation or as basis for reproducibility experiments. However, currently, the technique is yet at a technology readiness level for active process research and monitoring. Controlling of production processes, which is in principle possible on basis of the data derived from PDW spectroscopy, should only be executed based on a dual redundancy approach. The implementation of a PDW spectrometer in a laboratory or processing environment should be planned well ahead, anticipating periods for half a year as minimum.

The technical operation of a PDW spectrometer is straightforward. The operator needs to define an experiment name and needs to start a measurement. Older experimental data are not overwritten since new experiments receive their own time stamp. Care, however, needs to be taken for the correct selection of materials from the database and the correct definition of the mass or volume fraction of the disperse material, considering time-dependent changes by model development. These material parameters can be changed post-process for improved raw data analysis. It is important to stress that the raw data itself are not affected by changing data analysis parameters.

As a result, the absorption and the reduced scattering coefficient per experimental wavelength as well as the mean particle size and the distribution width are exported as function of time in a non-proprietary file format. Raw data analysis is executed in real time by the control computer, e.g., allowing for the data export to a distributed control system (DCS). In addition, data analysis can be performed post-process.

4.10.7 Examples for Process Analysis with PDW Spectroscopy

Various chemical, physical, or biotechnological processes have been characterized by PDW spectroscopy in the recent past. Examples include starved feed polymerization reactions [15], suspension polymerization reactions [12], solvent–borne copolymers [16], processes in milk [6, 7], beer mash [6], fruits and optical phantoms for fruits [17, 18], erythrocyte suspensions [6], biotechnical PHA synthesis [10], micro droplet emulsification [19] and nanodroplet emulsification [20], PNIPAM particle suspensions as "smart," temperature-dependent structures [8], zeolite synthesis in high pressure reactors [9], or synthesis of formazine as turbidity standard [5]. For crystallization or precipitation reactions, mainly glutamic acid, lactose, zeolites, and barium sulfate have been investigated so far. In general, the optical coefficients during such reactions can be obtained with good quality particularly in concentrated systems. Challenge, however, is the correct assessment of the volume fraction of the formed particles during periods of supersaturation. Here, particularly in periods of early nucleation, PDW spectroscopy may yet not correctly determine the particle size. In the following, some results for the crystallization of lactose and precipitation of barium sulfate are shown.

4.10.7.1 Crystallization of Lactose

In an industrial context, lactose is crystallized from highly concentrated solutions, in processes often facing process probe fouling. Thus, the applicability of PDW spectroscopy was evaluated during cooling crystallization for solid fractions of up to 60 wt% (see [14] for details, here only PDW spectroscopy data are discussed). First, it has to be pointed out that even under conditions of nucleation on process probe surfaces, PDW spectroscopy provided reproducible data at different cooling rates and at different mass fractions (Figure 4.10.8). In addition, particles size of lactose crystals before their dissolution is determined at identical values across the mass fraction range of 25–60 wt% (by suspending different amounts of lactose from the same lactose powder source) (Figure 4.10.9). After crystallization, the particle size must have changed, since the reduced scattering coefficient has changed when comparing the initial suspension before dissolution and the final suspension after crystallization.

However, during crystallization, at the period of nucleation (i.e., during initial increase of μ_s') particle sizes are found to be decreasing, which is counterintuitive (data not shown). As stated above, while the reduced scattering coefficient absolutely and reliably depicts the increasing number and size of the lactose crystals (two repeating experiments are shown), the theories used for the interpretation towards particle size seem yet not to interpret particle growth correctly during these early stages. Most likely, this is caused by supersaturation and results in the incorrect estimation of the volume fraction of the particles and the incorrect estimation of the refractive indices of the continuous and the solid phase.

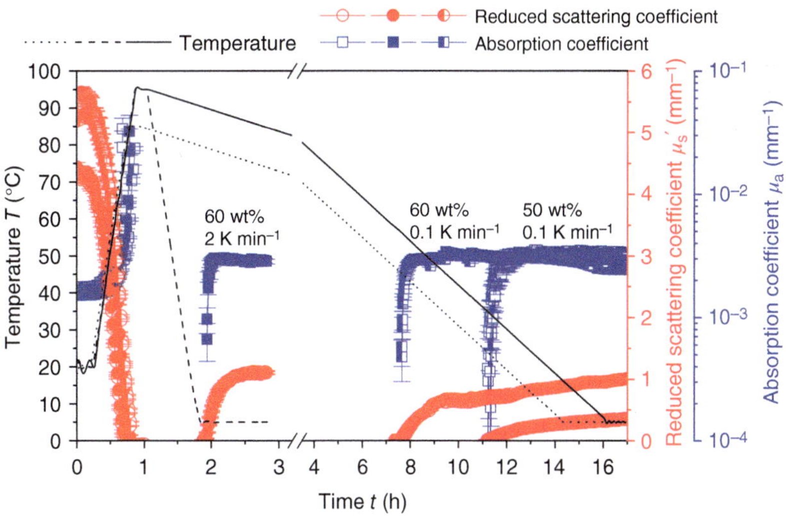

Figure 4.10.8 Absorption coefficient, reduced scattering coefficient, and temperature during dissolution under heating (2 K min^{-1}) and cooling crystallization of lactose monohydrate suspension with w = 50 wt% (0.1 K min^{-1}) and w = 60 wt% (0.1 and 2 K min^{-1}), as function of time for two repetition experiments each. Source: From Hartwig and Hass [14]. © 2018 John Wiley & Sons.

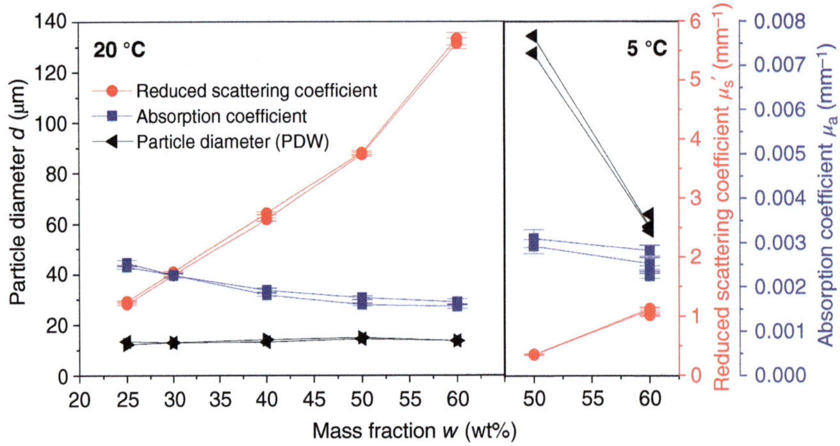

Figure 4.10.9 Optical coefficients μ_a and $\mu_s{}'$ as well as particle diameter based on PDW spectroscopy for the lactose suspensions at 20 °C (before dissolution under heating at 2 K min^{-1}) and 5 °C (after crystallization under cooling at 0.1 and 2 K min^{-1}) as function of the mass fraction for two repetition experiments each. Source: From Hartwig and Hass [14]. © 2018 John Wiley & Sons.

4.10.7.2 Precipitation of Barium Sulfate

The aim of investigating the precipitation of barium sulfate was the assessment to what extent systematic changes to the concentrations of barium and sulfate ions influence the particle formation. Thus, a design of experiments (DoE) approach was executed (see [3] for details, here only an excerpt is discussed). Figure 4.10.10 displays the reduced scattering coefficient for different precipitation conditions as function of time. All processes were investigated under stirring, to avoid sedimentation of the precipitated particles. Since PDW spectroscopy provides a time resolution in the sub-minute regime, it is a good process analytical technique to investigate processes lasting longer than 10 min, which may be a contradiction to precipitation processes. However, the results obtained indicate clearly that the barium sulfate suspensions undergo significant changes of the reduced scattering coefficient as function of time while being stirred. For a reliable DoE, randomized repetition experiments need to be included. Figure 4.10.10a shows the process trends for four different experimental conditions executed twice. Despite being a rapid precipitation by quickly mixing two solutions of each approx. 0.5 L, the temporal development of $\mu_s{}'$ is very reproducible in each of the four cases. Figure 4.10.10b shows the reduced scattering coefficient and the absorption coefficient for two of these experiments executed for more than 15 h. While the absorption remains nearly constant (no change to the chemical composition), the reduced scattering coefficient continuously changes over the course of more than 10 h, indicating a permanent alteration of the barium sulfate particles.

The so-called specific reduced scattering coefficient τ', which is a quantity not depending on the volume fraction ϕ of the barium sulfate particles, indicates very strong process deviations as function of time for all 50 precipitation experiments

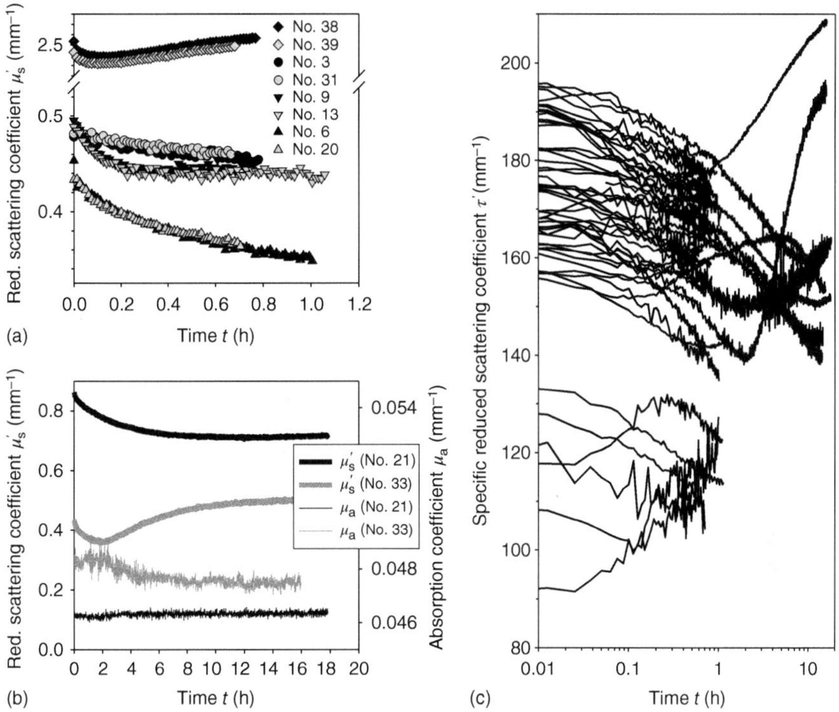

Figure 4.10.10 Reduced scattering coefficient at 982 nm as function of time of repeated precipitation reactions according to the DoE plan (no. indicates experimental run) (a), longtime development of the optical coefficients of two precipitations from the DoE plan (b) and specific reduced scattering coefficients of all 50 precipitations of the DoE plan as function of time (c). Source: Adapted from Hass [3].

executed for the DoE (Figure 4.10.10c). The precision of the data obtained by PDW spectroscopy plus the temporal resolution helps to build up a time-dependent response surface (see [3] for details). Even though particle size is not yet investigated here, it helps to build up understanding about the influencing factors on the precipitation of barium sulfate in concentrated systems.

4.10.8 Summary

PDW spectroscopy is a novel approach towards sensing of changes of particles, droplets, or cells in concentrated liquid systems. It is applicable in liquids exhibiting strong light scattering. With its fiber-optical probe design, it can be implemented in batch or continuous processes. Due to its time resolution in the sub-minute regime, it is of high relevance as process analytical technology, mainly in application areas where the suspended material is of analytical interest.

In contrast to many other particle sensing technologies working on single particle characterization, the theoretical background of PDW spectroscopy is based on light matter interaction in concentrated dispersions. Here, in contrast to other technologies, too low concentrations of particles pose limits, whereas high concentrations are beneficial.

As primary data, the absorption coefficient μ_a and the reduced scattering coefficient μ_s' are obtained. They are the absolute optical properties of the investigated material and are obtained without a need for calibration or model development. Since the quantitative separation of light absorption and light scattering is challenging in optical spectroscopy, the validation of the experimental results of PDW spectroscopy with respect to the optical coefficients is best done by theoretical calculations of these optical properties applying light scattering theories. In systems where the suspended particles can be well described, e.g., aqueous suspensions of spherical polymer particles, an excellent agreement between theoretical calculation and experimental findings is observed.

For particle size determination based on the reduced scattering coefficient, model development is needed. Volume fraction and the refractive indices of the continuous phase and the disperse phase are required. Validation of the results with respect to particle size is best done in comparison to other sizing technologies. However, it has to be stressed that different measurement principles for particle sizing yield differing results, quite often even when the same measurement concept is applied, but from different technology suppliers. In practice, PDW spectroscopy provides a new type of particle size data, which not necessarily will be in exact agreement to other particle sizing approaches. Closest comparable technique with respect to particle sizing might be static light scattering.

The strength of PDW spectroscopy is the ability to provide information about changes of the particles, droplets, or cells in real time and dilution-free directly inline from a process. As a process analyzer, PDW spectroscopy should currently still be regarded as an advanced monitoring and research tool. If its implementation and use of process data aims at process control, technological redundancy is required. Its current cost scales in the range of typical high-class optical process spectrometers like Raman, ATR–FTIR, or process microscopy. However, due to the simplicity of the process probes of PDW spectrometers, probe multiplexing, e.g., for spatial resolution is of limited cost, allowing for multiple measurement points with one spectrometer.

List of Abbreviations

APD	avalanche photodiode
ATR	attenuated total reflectance
DC	direct current
DCS	distributed control system
DoE	design of experiments
FBRM	focused beam reflectance measurement
FOQELS	fiber optic quasi elastic light scattering
FTIR	Fourier-transform infrared spectroscopy
GPC	gel permeation chromatography

HF	high frequency (additional sinusoidally modulated current)
HFA	high frequency amplifier
LAN	local area network
LD	laser diode
MSA	mean spherical approximation
OPC UA	open platform communications unified architecture
PAT	process analytical technology
PDW	Photon Density Wave (spectroscopy)
PHA	polyhydroxyalkanoate
PNIPAM	poly(*N*-isopropylacrylamide)
PY	Percus–Yevick approximation
SEC	size exclusion chromatography
TEM	transmission electron microscopy
USB	universal serial bus
VNA	vector network analyzer

References

1 Merkus, H.G. (2009). *Particle Size Measurements: Fundamentals, Practice, Quality*. Springer Science & Business Media.

2 Allen, T. (2003). *Powder Sampling and Particle Size Determination*. Elsevier.

3 Hass, R. (2011) Angewandte Photonendichtewellen Spektroskopie. Dissertation, Potsdam University.

4 Bressel, L., Hass, R., and Reich, O. (2013). Particle sizing in highly turbid dispersions by photon density wave spectroscopy. *J. Quant. Spectrosc. Radiat. Transf.* 126: 122–129. https://doi.org/10.1016/j.jqsrt.2012.11.031.

5 Münzberg, M., Hass, R., Dinh Duc Khanh, N., and Reich, O. (2017). Limitations of turbidity process probes and formazine as their calibration standard. *Anal. Bioanal. Chem.* 409: 719–728. https://doi.org/10.1007/s00216-016-9893-1.

6 Hass, R., Munzke, D., Vargas Ruiz, S. et al. (2015). Optical monitoring of chemical processes in turbid biogenic liquid dispersions by photon density wave spectroscopy. *Anal. Bioanal. Chem.* 407: 2791–2802. https://doi.org/10.1007/s00216-015-8513-9.

7 Vargas Ruiz, S., Hass, R., and Reich, O. (2012). Optical monitoring of milk fat phase transition within homogenized fresh milk by photon density wave spectroscopy. *Int. Dairy J.* 26: 120–126. https://doi.org/10.1016/j.idairyj.2012.03.012.

8 Werner, P., Münzberg, M., Hass, R., and Reich, O. (2017). Process analytical approaches for the coil-to-globule transition of poly(*N*-isopropylacrylamide) in a concentrated aqueous suspension. *Anal. Bioanal. Chem.* 409: 807–819. https://doi.org/10.1007/s00216-016-0050-7.

9 Häne, J., Brühwiler, D., Ecker, A., and Hass, R. (2019). Real-time inline monitoring of zeolite synthesis by photon density wave spectroscopy. *Micropor. Mesopor. Mater.* 288: 109580. https://doi.org/10.1016/j.micromeso.2019.109580.

10 Gutschmann, B., Schiewe, T., Weiske, M.T.H. et al. (2019). In-line monitoring of polyhydroxyalkanoate (PHA) production during high-cell-density plant oil cultivations using photon density wave spectroscopy. *Bioengineering* 6: 85. https://doi.org/10.3390/bioengineering6030085.
11 Mie, G. (1908). Beiträge zur Optik trüber Medien, speziell kolloidaler Metallösungen. *Ann. Phys.* 330: 377–445.
12 Hass, R., Münzberg, M., Bressel, L., and Reich, O. (2013). Industrial applications of photon density wave spectroscopy for in-line particle sizing. *Appl. Opt.* 52: 1423–1431.
13 Bressel, L. (2016) Bedeutung der abhängigen Streuung für die optischen Eigenschaften hochkonzentrierter Dispersionen. PhD Thesis, Universität Potsdam.
14 Hartwig, A. and Hass, R. (2018). Monitoring lactose crystallization at industrially relevant concentrations by photon density wave spectroscopy. *Chem. Eng. Technol.* 41: 1139–1146. https://doi.org/10.1002/ceat.201700685.
15 Hass, R. and Reich, O. (2011). Photon density wave spectroscopy for dilution-free sizing of highly concentrated nanoparticles during starved-feed polymerization. *ChemPhysChem* 12: 2572–2575. https://doi.org/10.1002/cphc.201100323.
16 Kutlug, Ö., Hass, R., Reck, S., and Hartwig, A. (2018). Inline characterization of dispersion formation of a solvent-borne acrylic copolymer by photon density wave spectroscopy. *Coll. Surf. Physicochem. Eng. Asp.* 556: 113–119. https://doi.org/10.1016/j.colsurfa.2018.08.011.
17 Adebayo, S.E., Hashim, N., Hass, R. et al. (2017). Using absorption and reduced scattering coefficients for non-destructive analyses of fruit flesh firmness and soluble solids content in pear (Pyrus communis 'Conference') – An update when using diffusion theory. *Postharvest. Biol. Technol.* 130: 56–63. https://doi.org/10.1016/j.postharvbio.2017.04.004.
18 Zude-Sasse, M., Hashim, N., Hass, R. et al. (2019). Validation study for measuring absorption and reduced scattering coefficients by means of laser-induced backscattering imaging. *Postharvest. Biol. Technol.* 153: 161–168. https://doi.org/10.1016/j.postharvbio.2019.04.002.
19 Reich, O., Bressel, L., and Hass, R. (2011). Sensing emulsification processes by photon density wave spectroscopy. *Proc. SPIE* 7753: 77532J1-4.
20 Münzberg, M., Hass, R., and Reich, O. (2013). In-line characterization of phase inversion temperature emulsification by photon density wave spectroscopy. *SOFW J.* 139: 38–46.

5

Impact of Solid Forms on API Scale-Up

Sophie Janbon[1], Clare Mayes[2], and Amy L. Robertson[2]

[1] *Early Chemical Development, Pharmaceutical Sciences, R&D, AstraZeneca, Macclesfield, UK*
[2] *Chemical Development, Pharmaceutical Technology & Development, Operations, AstraZeneca, Macclesfield SK10 2NA, UK*

5.1 Introduction

The solid form of an Active Pharmaceutical Ingredient (API) must be considered during all aspects of development and scale-up of crystallization processes. This is because the API material attributes, specifically solid forms (hereafter referred to as form), must be controlled to achieve the required biopharmaceutical profile for patient safety and drug efficacy. In addition, it is important that the process can be manufactured efficiently, robustly, and safely; these aspects are particularly important during scale-up. The need to scale up a process must be considered throughout development as failing to do so is likely to result in batch failures and delays to clinical supply, in bringing the API to market, and in supplying to patients postlaunch.

In this chapter, crystallization process development is introduced from early-phase development through to commercial manufacture highlighting the understanding and actions required to ensure the process is successfully delivered. This chapter is written from an idealized crystallization perspective, in practice throughout development compromises need to be made to accommodate the needs of upstream and downstream processes.

Figure 5.1 gives an overview of the crystallization development process, how it fits within clinical phases and some key collaborating skill sets. This chapter explains this overview in more details and is divided into four main sections:

- Small-scale crystallization development,
- Scale-up,
- Technical skill sets, and
- Regulatory requirements.

The approach outlined here will help ensure the risks are identified and mitigated and opportunities are exploited. This should result in a robust and scalable crystallization process, delivering the required form with high purity, yield, and

Solid State Development and Processing of Pharmaceutical Molecules: Salts, Cocrystals, and Polymorphism, First Edition. Edited by Michael Gruss.

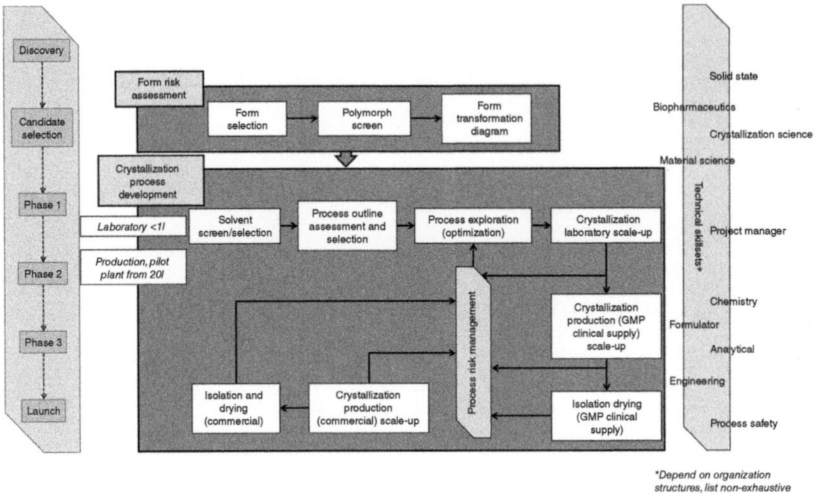

Figure 5.1 Overview of the crystallization development process alongside approximate development phase and key interacting skill sets.

the required material attributes. These aspects are referred to in this chapter as manufacturability and are fundamental for successfully delivering medicines to patients over the lifetime of the products.

The focus of this chapter is API batch crystallization. Similar approaches can be applied to the understanding of isolated intermediates and for continuous crystallization, neither of which are discussed in detail. The requirements on the crystallization for APIs are higher than for isolated intermediates and a risk-based approach can be applied. Continuous crystallization is a growing area of interest for API, and many of the activities discussed below will transfer directly to a continuous process [1–3].

5.2 Background

The pharmaceutical industry is highly regulated to ensure patient safety. As a result of worldwide variations in quality standards, the International Council for Harmonization (ICH) was created in 1990. The ICH's mission is *to achieve greater harmonisation worldwide to ensure that safe, effective, and high quality medicines are developed and registered in the most resource-efficient manner* [4]. The council consists of a panel of industry experts and regulators, which facilitates the production of a comprehensive set of harmonized guidelines on quality, safety, and efficacy. This has shaped the ways of working across the pharmaceutical industry.

To work effectively in this highly regulated environment, process development follows a stepwise and risk-based approach to ensure that each decision is rationally taken; and as a result, the outcome should be more robust and consistent. This approach means the process is well understood and small variation does not result in large changes in the output in terms of chemical and material attributes of the crystalline compound.

Developing a process suitable for the launch of an API and associated drug products requires significant collaboration between cross-functional teams and skill sets. A closely-knit team within a chemical development function is typically responsible for the delivery of the API and the technical elements of the drug substance components of the regulatory dossier [5]. This team typically consists of:

- Synthetic and process chemists whose responsibilities include design and development of the synthetic route.
- Analytical chemists whose responsibilities include the development of appropriate analytical methods to test for impurities and to support impurity tracking exercises.
- Process engineers whose responsibilities include design of work-ups, isolation, and drying stages as well as understanding how the process will perform at different scales.
- Crystallization scientists whose responsibilities include input into form selection, development of the crystallization process, and often providing the link between the technical activities in the drug substance and drug product areas. They interact with solid-state scientists to understand solid-state aspects such as form purity.

All of these groups alongside more specific specialisms in, for example, statistics, control strategy, and catalysis and as a team take smart risks and make compromises to ensure the best possible process is delivered to the manufacturing plant which delivers the required API to the drug product organization and ultimately to the patient.

5.3 Small-Scale Crystallization Development

The development of a crystallization process in the laboratory takes place through several different phases. This starts with the selection of a form, solvent, and process outline; the latter is then evaluated, and the process is further explored and optimized. Development is an iterative process, at times it is necessary to revisit earlier decision points as developing understanding may result in changed demands on the API and process. The different phases of the process development are described in more detail below.

5.3.1 Form Selection

Form selection is achieved through risk assessment comparing the different potential forms against predefined criteria. The aim of form selection is the identification of a form which has the required biopharmaceutical profile, with chemical and solid-state stability and the potential to be manufacturable.

More important at this stage of development is gaining an understanding of form stability; this is gathered through polymorph and solvent screening. A form

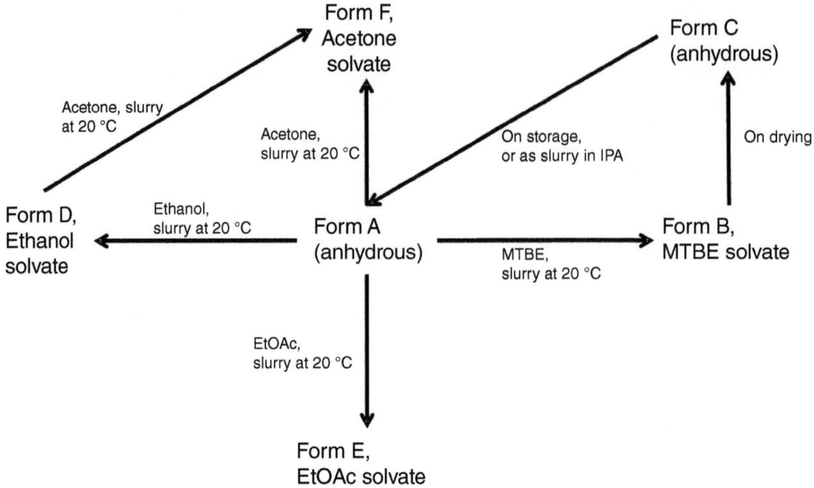

Figure 5.2 Example of a form transformation diagram showing relationship between forms, each form is identified through usual solid-state characterization tools and their crystalline state and crystallinity should be described.

transformation diagram should be initiated as soon as more than one form is identified. The form landscape is presented via a transformation diagram, e.g. solvent, temperatures an example is shown in Figure 5.2. This schematic should be a living document through the life cycle of the medicine and updated when further knowledge is gained.

To enable the development of a robust process and aid form stability of the API, the most thermodynamically stable form under manufacturing conditions should be selected; ideally, this should be a non-solvated crystalline form. The selection of a metastable form may result in problems later in the development cycle, as chemical processes will be improved and the impurity profile of the input material into the crystallization will change; this may result in a metastable form no longer having the kinetic stability to facilitate efficient manufacture [6]. Solvated forms are often challenging to maintain during drying, and there is the potential for solvent losses during storage limiting the potential shelf life. During early stages of development, manufacturability will be considered as acceptable if the initial small-scale crystallization experiment (ideally around 500 mg–1 g) shows that the selected crystalline compound can be produced as a pure crystalline form with minimal challenges.

Crystal structure data should be collected as soon as possible on the proposed form, since these data provide an insight into how the molecules are packed. Structural assessments can then help establish if further, more stable forms are likely to exist [7, 8]. For this reason, efforts to grow a quality single crystal should be initiated as early as possible. However, if it is not possible to grow a suitable single crystal, then structures can also be established via X-ray powder pattern, although calculating the structure this way can be more complex or not possible depending on the size and flexibility of the molecule. The crystal structure data can be used to predict properties and guide experimental work which may result in reduced development time.

The understanding gained during form selection and contained in the transformation diagram will also assist the identification of the most appropriate process conditions for form control during the development of a process.

5.3.2 Solvent Selection

Once the form has been selected, work can begin on crystallization process development. The first stage of this is the identification of solvents that are suitable for the crystallization. When selecting solvents for crystallization process development, both the solvent safety guidelines and an understanding of the solubility need to be considered.

The most relevant guideline for solvent safety is ICH Q3C, Guideline for Residual Solvents (R6) [9]. This guideline classifies how well common solvents are tolerated by humans (in ppm per day). They are categorized into levels of risk, from high risk (class 1) to low risk (class 3). Generally, solvents used in the API stage are selected from ICH classes 3 and/or 2. For intermediates (depending on the position within the chemical synthesis route), solvents can be selected from ICH class 1; however, this requires a strong rationale.

Some solvents are avoided for reasons other than safety. The following examples of the reason's solvents might be avoided:

- environmental impact, e.g. dichloro-methane (DCM) [10, 11]
- shortages of supply, e.g. acetonitrile [12]
- physical properties, e.g. dimethyl sulfoxide (DMSO) due to a low melt temperature of 19 °C and low volatility. Low volatility solvents are avoided as they need high temperatures for drying which could promote chemical degradation
- solvent stability, e.g. esterification of acids in alcohols
- cost, e.g. acetonitrile in 2009.

5.3.2.1 Solvent Screening

Preferred solvents for API crystallization are alcohols, ketones, esters, ethers, and alkanes and a range of these are used during the initial solubility screen. Solubility data must be collected with a temperature as solubility is temperature dependent [13]. During the solvent screen, additional data should be generated and include chemical and form stability (summarized in Figure 5.3).

Chemical stability is important to ensure that the compound of interest does not significantly degrade in situ. If degradation is observed, these impurities must be purged by the crystallization process (i.e. remain in the liquors). The presence of degradants can alter crystallization kinetics and particle habit and increases the demand on the crystallization to reduce impurities and negatively impacts yield.

The desired form must be maintained in the selected solvent system and any new learning incorporated in the form transformation diagram. It is essential to have thermodynamic understanding and stability of the selected crystalline form of the API in the solvent.

Predictive modeling is often used to rank solvents and infer their temperature dependence. They often need some experimental data, such as solubility in a limited

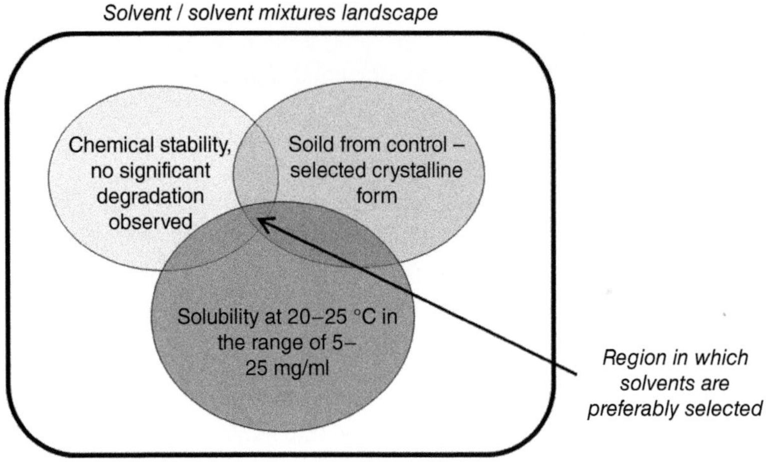

Figure 5.3 Criteria for solvent selection.

range of solvents and physico-chemical properties, such as melting point and heat of fusion. This approach enables a wider range of solvents to be assessed without additional laboratory work. However, it is currently less reliable for multicomponent solids such as salts and cocrystals due to challenges in modeling solvation in these systems. Subsequently, a more tailored solvent screen can be performed using solvents predicted to have the most suitable solubility. Both experimental and predictive methods should be used in conjunction for a faster and more reliable crystallization process development [14].

5.3.2.2 Solubility Diagram

Solubility data are graphically represented as the thermodynamic solubility curve on a solubility diagram. This diagram can also include kinetic phenomena such as the metastable zone. Figure 5.4 shows an example of a solubility diagram. Understanding the different areas in the solubility diagram is key to designing a robust and scalable crystallization process.

In the metastable zone, spontaneous nucleation occurs in a time-dependent way. Once this metastable zone is exceeded, nucleation is immediate. The time taken for spontaneous nucleation to occur is known as the induction time. This time is driven by kinetics and depends on external factors. These external factors include agitation rate, extraneous particles, impurities, scale, equipment geometry, and materials of construction (MoC). In organic pharmaceuticals, the metastable zone is typically very wide requiring very high supersaturation to generate immediate nucleation [16, 17].

The metastable zone can be further subdivided into the seed proliferation zone where secondary nucleation dominates, the seed growth-only zone where growth is the dominating mechanism, and the dead zone where no activity is observed. These subdivisions are recognized from within the sugar and dairy industries [18].

If crystallization occurs beyond the metastable zone limit, the crystallization is likely to be uncontrolled and can give rise to different kinetic forms. Crystallizing in

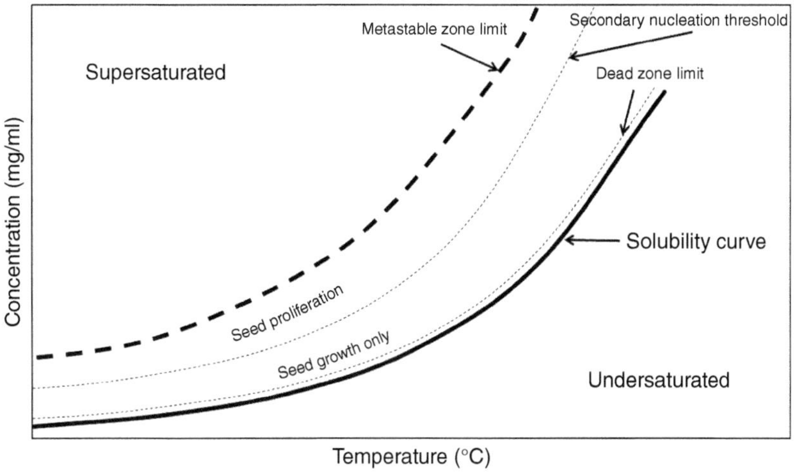

Figure 5.4 Solubility curve. Source: Threlfall and Coles [15].

this region promotes uncontrolled nucleation over growth favoring small particles and may result in a wide particle size distribution (PSD). Working at high supersaturations can also increase the risk of "oiling out." This relates to phase de-mixing where a product-rich phase separates as a liquid prior to crystallization [19]. Crystallization in these conditions generates undesirable outcomes, from batch-to-batch variability, poor impurity purging, and fines formation. Fines are small crystalline particles, which may lead to filtration issues as they can blind filter media and can give rise to a broad or a bimodal PSD as shown in Figure 5.5. The size fraction considered as fines will be dependent on the full distribution. These will all result in challenges when scaling up (see also Section 5.4.3). If oiling out occurs, this can also be recorded on the solubility diagram.

5.3.2.3 Solubility Measurement

Solubility curves can be generated experimentally or estimated by using a range of theoretical methods. Experimental solubility data can be measured in numerous ways; these methods fall into two broad categories, dynamic and equilibrium methods, and are discussed in detail below. In these small-scale experiments, magnetic stirrer bars are often used which tend to grind the solid. Therefore, at this stage, there is little to be gained from obtaining PSD and shape data.

Theoretical methods include rules of thumb, e.g. solubility doubles every 20 °C [20], use of an ideal solubility equation, e.g. Van't Hoff [13], and predictive modeling using solvent properties databases and bespoke algorithms, e.g. Aspen and Cosmotherm.

For multicomponent crystals such as salts and cocrystals, the solubility data of the free molecule, additive (e.g. counterions and co-formers), and the resulting entity need to be measured separately and then considered in combination when selecting a solvent.

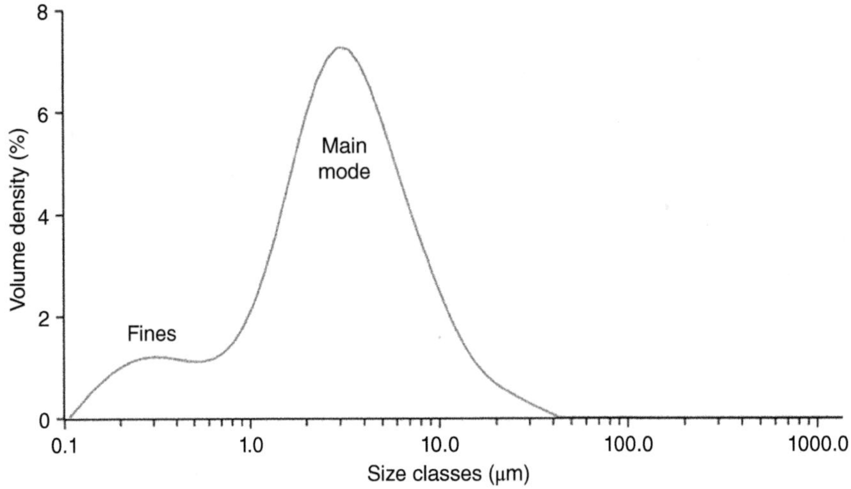

Figure 5.5 Particle size distribution (PSD) showing bimodality as a result of fines.

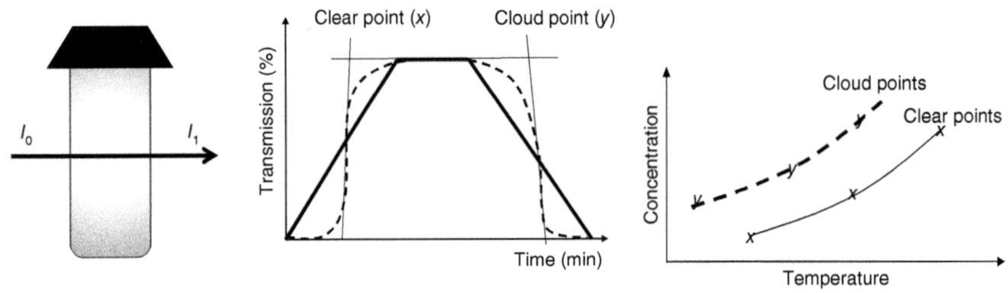

Figure 5.6 Dynamic solubility measurement.

Dynamic Solubility Measurement Dynamic solubility measurement is performed by taking a known mass of compound in a known volume of solvent. This is heated at a slow rate (typically 0.03–0.05 °C min^{-1}) until a clear solution is achieved. The temperature at which the slurry becomes clear is called the clear point and is the solubility. Using this approach, the metastable zone width (MSZW) can be determined by cooling the solution after the clear point has been observed. Until a suspension is formed, this point is called the cloud point. The cloud point temperature defines the edge of the metastable zone.

There are several technologies available to aid and automate solubility measurement. These typically have controlled agitation, accurate and precise temperature control, as well as in-built analysis. Turbidity is currently one of the most common approaches used to measure clear and cloud points at small scale (typically 1–5 ml). Figure 5.6 gives a schematic of dynamic solubility measurement.

Figure 5.6 shows how the solubility can be measured at small scale using a transparent vial in which a known mass of solute is in contact with a known volume of solvent. The system is sealed, agitated, and temperature controlled. The transmission of light through the vial is monitored upon heating. When all the solute has dissolved, the light is transmitted through and this is the clear point (x). When the system is cooled and crystallization/precipitation occurs, the light is no longer fully transmitted and this is assigned as the cloud point (y). The results obtained for a set of concentrations are then plotted and represent the solubility curve and the metastable zone.

The main advantages of dynamic solubility measurement are the ease of automation, the potential to carry out repeat measurement on the same sample, and the ability to measure the metastable zone; however, this method has the following limitations:

- As the system is continually heated, albeit slowly, the measurement will tend to underestimate the solubility of the system. There are two main sources of this discrepancy:
  1) At small scale, the temperature of the sample is not typically directly measured. So, there is a temperature difference between the point of temperature measurement and the slurry temperature.
  2) Dissolution is a kinetic process; therefore, the heating ramp may not allow time for the system to fully equilibrate before the temperature increases.

  As there is no solid remaining at the end of the solubility measurement, it is not possible to check if form change has occurred by offline methods. To confirm that form change has not occurred during the measurement, a suitable in situ monitoring technology can be used.
- If the solute, solvent, and/or solute–solvent system have limited thermal stability, degradation may occur. This could result in changes in the apparent solubility of the system.
- If the compound has an impurity, which is less soluble than the API in the solvent, the dissolution of the API could be masked.
- As nucleation is a stochastic (random) event, small-scale measurements tend to overestimate the MSZW [21–23].

Equilibrium Solubility Measurement Equilibrium solubility measurement is performed by agitating a slurry over an extended time period (for example, 24 hours) at a controlled temperature. This time enables an equilibrium to be achieved between the crystalline solid and the solute. Samples of the solution are analyzed by chromatographic or gravimetric methods to establish the solution concentration, i.e. the solubility.

Chromatography provides increased understanding of chemical degradation occurring during the experiment and more accurately quantifies low solubilities. Gravimetric methods are simpler to perform and reliable when the solubility is higher, i.e. greater than 2–5 mg ml^{-1}, and both purity and chemical stability are understood.

Equilibrium measurement has the advantage that retained solids can be further characterized by solid-state techniques such as X-ray powder diffraction (XRPD) to confirm no form change has occurred. If a new diffraction pattern is found, further analysis is required to characterize the new form. This should be captured in a form risk assessment and transformation diagram, even though this is not part of the formal polymorph screen activity. The main disadvantage of this method is measurements at elevated temperatures as the hot solution can rapidly cool down and spontaneous crystallization may occur or solvent can evaporate resulting in inaccuracies in the solubility measurement.

A combination of equilibrium and dynamic methods is often most suitable. This can be achieved by an initial solvent screen at 20 °C using an equilibrium method followed by dynamic measurements of solvents selected for further development.

5.3.3 Crystallization Process Selection

Once the solubility is understood, crystallization processes can be proposed and investigated. In principle, any method by which supersaturation can be generated could be used for crystallization. Common methods for crystallization are as follows:

- Cooling crystallization, which uses the temperature of the solution to create a supersaturated environment.
- Anti-solvent crystallization, which uses the addition of a low-solubility solvent to create the supersaturation.
- Reactive crystallizations, in which a reagent is added to another chemical entity in solution. The reaction generates a product with lower solubility which then crystallizes. The reagents can be:
  - chemically reacting to form a new molecule
  - acidic or basic salt forming agents
  - co-formers
- Distillative or evaporative crystallizations, in which solvent is removed, increase the concentration, generating supersaturation.

Where possible, a cooling or anti-solvent crystallization should be used as supersaturation control is more straightforward, with cooling crystallization being the preferred option due to being most scalable. The solubility curve (in Figure 5.4) is key data for a scalable cooling crystallization process. The same type of solubility data can be generated for anti-solvent processes. In this instance in Figure 5.4, the temperature is replaced with % volume per volume of a solvent in an anti-solvent. Where multicomponent and multi-solvent systems are used, it may be necessary to use ternary phase diagrams to further describe the system.

The other types of crystallization are investigated when the above preferred options are not feasible or favorable as part of the chemical synthesis route. For example, reactive crystallizations might be more appropriate with multicomponent compounds. If possible, distillative or evaporative processes are avoided as they are less controllable at scale due to the inherent variation in rates of distillation while changing scale and equipment. This is discussed in more detail in Section 5.4.6 of this chapter.

5.3.3.1 Process Outline Selection

To select potential processes, the solubility data are used to propose a number of process outlines to be scoped for further development. Table 5.1 provides a

Table 5.1 Selection guide for crystallization process type.

Scenario (solvent or solvent mixture)	Solubility gradient – greater than twofold every 20 °C	Solubility at 20 °C (mg ml^{-1})	Process outlines (recommendations)
A	Yes	5–20	Cooling crystallization
B	Yes	>20	Not suitable for cooling crystallization Can be used as a solvent in an anti-solvent process (using A or C as an anti-solvent at 20 °C), or a cooling crystallization in combination with solvent C
C	No	5–20	Anti-solvent in an anti-solvent crystallization (with either B or D as solvent), or in a cooling crystallization in combination with solvent C
D	No	>20	As a solvent in an anti-solvent crystallization (mixed with either A or C at 20 °C)

guideline to the selection of a cooling or anti-solvent crystallization process from the solubility data.

5.3.3.2 Process Outline Evaluation

Once the potential processes have been identified, the options need to be evaluated to identify the process which is most suitable for development and scale-up. This evaluation involves a combination of different equipment scales coupled with in-line, at-line, and offline analysis. In recent years, the development of smaller scale equipment with overhead stirrers and enhanced Process Analytical Technology (PAT) capability has enabled fast and effective evaluation of multiple processes simultaneously [24]. PAT is a suite of analytical technologies that allow the real-time collection of analytical data; this includes well-established technologies such as temperature and conductivity probes as well as a wide range of analytical tools including ultraviolet (UV), infrared (IR) and Raman spectroscopy, optical microscopy, and particle size monitoring. All of these techniques have their limitations and the results should be collectively considered alongside offline techniques. Deciding on how and what to monitor in the crystallization is dependent on the system. Data interpretation can often be complex, requiring statistical approaches and specialist input.

Seeding is important to initiate crystallization and should wherever possible be included in a process as it increases process control. The enhancement of control especially in terms of form and particle comes from the reduction of nucleation events. Seeding with the desired form should be included at a point within the metastable zone established during solubility screening and added to the bulk of the batch.

Once the number of potential processes has been reduced, the most promising processes should be trialed at a larger scale using ideally an overhead stirrer in a <250 ml vessel. This larger scale work should include remeasurement of the metastable zone so the appropriate temperature window for seeding to be identified more accurately [25, 26]. A PAT or visual observation approach can also be used to measure the clear and cloud points.

In these experiments, an increased understanding of how the crystallization is progressing and the crystallization kinetics can be achieved. For example, using in situ particle sizing technology, it is possible to monitor nucleation and growth during crystallization. Further process development needs can be established from this initial evaluation and should be used to inform future work packages; these are discussed in Sections 5.3.3.3 and 5.4.6.

In addition to in situ monitoring, offline analyses are performed; these include XRPD for checking the form, chromatography, and nuclear magnetic resonance (NMR) spectroscopy methods for measuring solution concentration and assay, and mass of the resulting solid for measuring experimental yield. A preliminary assessment of particle shape and size can be performed at this development stage using microscopy and laser diffraction methods.

This evaluation will highlight potential risks and opportunities in the processes. It is often found that solvent type and supersaturation cause a change in particle size and shape. The maximum theoretical yield based on the solubility is often not obtained. The main reasons are often related to slow crystallization kinetics and mechanical losses to the vessel.

Once the processes have been evaluated, systematic analysis is used to compare the process outlines. Kepner-Tregoe [27] methodology is a well-established tool which is often used for this purpose. To carry out this analysis, all requirements for specification and manufacturability should be listed and ranked. The outcome should be the selection of the most promising process outline.

5.3.3.3 Process Exploration

Once the solubility is understood and process outline selected, the crystallization process can be progressed to the next stage, process exploration. The aim of process exploration is to maximize impurity purging and yield, robustly deliver the selected form, produce particles with suitable material attributes for downstream processing, and minimize scale dependence.

To achieve the control required, crystallization should be initiated via seeding with the desired solid-state form in the metastable zone region established in the process outline evaluation. This should control nucleation by preventing random or stochastic events from occurring. This should be followed by the slow and controlled generation of supersaturation to promote crystal growth. Fast generation of supersaturation can lead to the process running in a zone with uncontrolled nucleation. As solubility is not linear, nonlinear cooling rates and solvent additions such as cubic profiles can be used to aid better control while not increasing the overall processing times. By separating nucleation and growth stages, a process will be less scale dependent. This is important as the different crystallization mechanisms will vary in different ways at different scales. This deconvolution will also facilitate risk identification and mitigation during scale-up and equipment change.

Various analytical techniques are used to support the development process. PAT is routinely used to monitor growth, solution concentration, and form; this information is used to feed back into further process design work packages. The selection of PAT tools is dependent on the setup and the system under investigation [24]. Currently, the range of PAT probes is wide; spectroscopic probes such as Raman, IR, and UV are used to monitor solid and solution changes including form and concentration, laser diffraction probe for in situ size monitoring, and optical system for video imaging. The PAT tools are used in conjunction with offline techniques such as high-performance liquid chromatography (HPLC) to track purity and concentration, microscopy, X-ray diffraction, particle habit monitoring, and PSD measurements. At larger scales, the use of probes is less common due to cost and difficulty of implementation in the plant equipment. If appropriate offline samples as well as in-line data have been collected during the smaller scale development, these data are still useful to infer process changes without the use of PAT.

During the development of the crystallization process, downstream processes such as isolation, drying, particle size reduction, storage, and formulation need to be considered. The development of the isolation and drying processes are started at this stage of development and involve collaboration between crystallization scientists and engineers. Problems during manufacture are often observed in these stages as a result of the solid-state properties and material attributes. Problems include filter cloths blocking upon batch isolation, formation of amorphous solid during micronization, and poor flow during formulation. In recent years, this has led to an increased focus on API material attributes. A feedback loop between the drug substance and the drug product teams is critical to develop the landscape of material attributes which can effectively and efficiently be manufactured in both the API and formulation stages [28–30].

To generate the material attributes landscape and evaluate their impact on downstream processes, the process parameters which impact them should be identified. Factors impacting particle properties often include supersaturation generation profile, type of seed, e.g. micronized or output from a previous crystallization (i.e. daughter seed), amount of seed, and seeding temperature. The impacting parameters should then be used as part of a design of experiment (DoE) [31]. DoE is a statistical approach to experimental design which enables, in an efficient manner, the reduction of the number of experiments needed and increased output in terms of process understanding. Specialist software, such as MODDE™, Design Expert™, and SIMCA™, are used to produce designs and facilitate data interpretation. When a DoE is appropriately designed, it will result in the production of a model showing how different factors interact and impact properties. If the model is accurate enough, the material attributes can be predicted and therefore this model can be used to enhance control. Poor design of a DoE will result in poor models and data which cannot be relied upon; when setting up a DoE, input should be gained from experts in DoE alongside crystallization scientists.

If the simple factors described above do not yield the required material attributes, other approaches can be investigated to engineer the particle shape and size. The following techniques are commonly used during crystallization:

- Temperature cycling, the theory here is the smaller particles are preferentially dissolved during heating followed by growth of the larger crystals during cooling. Over time this should lead to reduced small particles and overall larger particles.
- Wet milling, this can be used in a number of different ways and can be coupled with temperature cycling. The use of a wet mill involves a high shear mill being incorporated in a recirculation loop on the main vessel to primarily break particles. Use of the mill at specific points during the crystallization can, for example, change morphology from needles to rods. Wet milling can also be used to generate seed material in situ with a larger specific surface area. This provides an alternative to micronized seed and simplifies the supply chain (i.e. removing the need for a specific seed preparation step).

PAT can also help with particle engineering. Increasingly automated feedback control is used to achieve specific particle properties [32].

During process exploration, the regulatory requirements also need to be considered. At this stage, critical quality attributes (CQA) and critical process parameters (CPP) start to be determined. The requirements of the API and minimum specification are outlined in the ICH guidelines [33], of importance to a crystallization scientist are the material attributes (i.e. crystalline form and the particle properties) and chemical purity. Additionally, throughput and yield should be maximized and are of equal importance to each other; this will ensure the API is produced in an economical and environmentally sound way.

5.3.4 Process Development Conclusions

This section has presented a workflow for developing crystallization processes at laboratory scale and highlighted some of the challenges. Following the systematic approach outlined here will give the best chance to scale up and commercialize a process with limited problems, though challenges will often be encountered when scaling up. Knowledge of the form landscape, solubility and process conditions, and particle requirements will enable better trouble shooting activities.

The development of the crystallization itself cannot be considered on its own but must be considered as part of the wider picture, taking into account upstream and downstream processes and the requirements of regulations and environmentally responsible manufacture. It should be noted that successful laboratory development sets the foundation for a good manufacturing process.

5.4 Crystallization Scale-Up

A process is normally scaled up in increments, first to large-scale laboratory vessels up to a maximum of about 100 l, through pilot plant to manufacturing scale where vessels vary from 3 or 4 cubic meters to 10's of cubic meters depending on the amount of product required. It is common for processes to be accommodated to plants designed for multiproduct use. Therefore, it may not be possible to achieve an ideal setup for a process, with compromises need to be made, for example, vessel size, shape, and agitator type. Safety is not an area in which compromise is appropriate.

To successfully scale up a crystallization process, there is a minimum amount of process understanding required which is gained in the laboratory as discussed in the previous section. This includes the transformation diagram, solubility data, and process knowledge gained at small scale (see Section 5.3.1).

In this section, parameters impacting how a process scales up are discussed along with risks and common problems associated with form and particle control which may be encountered during scaled up.

5.4.1 Crystallization Process Accommodation

When a crystallization is accommodated at a new scale or in a different vessel, there are a range of parameters which inherently vary. Figure 5.7 summarizes the key parameters which should be considered when changing scale. Typically, when increasing the scale of a process, the time taken for operations will increase and fill volume and head space will vary depending on vessel geometry and batch size. The different aspects of mixing, such as tip speed, micro and macro mixing, will change independently of each other and depend on the vessel size, baffling, agitator type, and stirrer speed used. Heat transfer will depend on MoC, heat transfer fluid type (this is the liquid contained in the vessel jacket, examples include glycol or brine), and heat transfer capacity. If any additions take place, the addition point, the temperature differential between the vessel contents and the addition stream, and the addition rate will impact the process. All these factors need to be considered when accommodating a process and are discussed in more detail below.

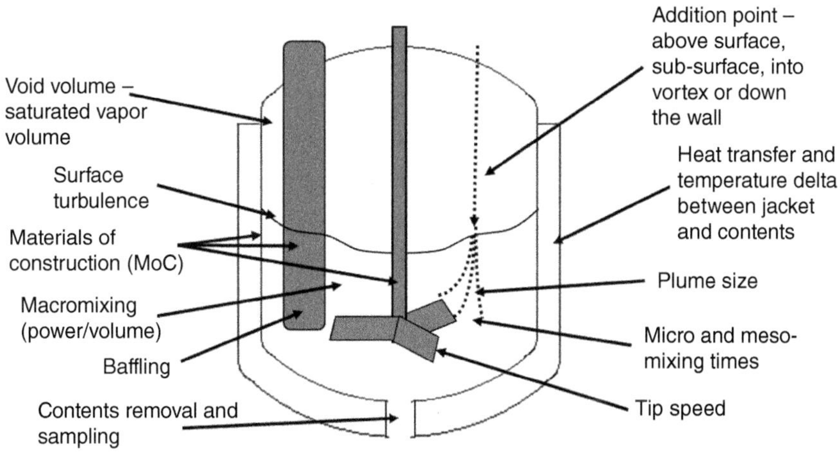

Figure 5.7 Potential factors to consider when scaling up a crystallization process.

5.4.1.1 Vessel Size and MoC

The first stage of scaling-up is the identification of the vessel requirements in terms of volume and MoC.

Vessels are typically designed to work most effectively at or close to their maximum working volume; therefore, when all additions are complete, a process should be carried out at this volume. For crystallizations involving an addition, the initial volume must be taken into account, the volume must be sufficient to reach the agitator in the vessel and ensure efficient mixing [34]. Running a process where the agitator is just covered can lead to splashing of solution onto the walls of the vessel. If this occurs and the vessel walls are heated, this may lead to rapid evaporation of the solution and encrustation occurring on the walls of the vessel. This is of particular importance where the system is polymorphic as this increases the risk of the batch being seeded with a metastable or undesired form.

The MoC of the vessel is also an important part of the selection. The internal surfaces of the vessels are commonly glass lined or made of Hastelloy or stainless steel; the selection of these is dependent on:

- *Chemical compatibility:* For example, some strong acids are not compatible with stainless steel.
- *Heat transfer:* For example, heat transfer is more efficient across a metal vessel compared to a glass lined vessel.
- *Interaction between the API and the walls*: For example, an API which has affinity to glass could have an increased propensity to crystallize on glass-lined walls than on stainless steel walls.

5.4.1.2 Agitation

All types of crystallization rely on effective mixing to guarantee a better control. This is of greater importance for crystallizations involving additions where a homogenous mixture throughout a large vessel is critical to achieving control.

Agitator Type There are a wide variety of different agitator types. All have different advantages and limitations, and these are widely discussed in other publications [35]. The MoC of the agitator is of equal importance as the MoC of the vessel, and care should be taken when selecting this for the reasons discussed in Section 5.4.1.1.

Agitation Regime When setting an agitation regime, several parameters need to be balanced:

- *Rapid integration of solutions added to the vessel:* Slow mixing can result in areas of localized high supersaturation which may lead to crystallization of undesired metastable forms.
- *Effective suspension of the crystals:* If the crystals are not effectively suspended, settling may occur in the vessel resulting in caking on the bottom which may cause agglomeration, blockages of base valves, and yield loss.
- *Minimize particle attrition:* Attrition is more often encountered when dealing with needle morphology but should be evaluated for all crystallization processes.

These parameters can often lead to competing agitation requirements during the process. These can be overcome by having an initial higher agitation rate to achieve homogenous solution at the point of seeding and during the addition, followed by a reduction in rate when additions are completed, and growth of the crystals is more important. The impact of agitation rate on the crystallization process should be investigated throughout the development from laboratory to production scale to understand the impact on the crystallization mechanism and the resulting particles. Predictive modeling can be used to investigate mixing for a given system at different scales and facilitate more effective scale-up of mixing parameters. Modeling ranges from the use of simple equations and algorithms to computational fluid dynamic (CFD) which uses computerized simulation of flows to predict mixing [36].

5.4.1.3 Heat Transfer

The cooling capacity of the vessel at production scale may vary between designs so having an appreciation of the heating and cooling capacity of the manufacturing vessel is important when designing a cooling crystallization process at laboratory scale. As said earlier, the time taken for operations tends to significantly increase upon scale-up. Fast cooling rates can be achieved in a laboratory vessel; these may not be possible on scale and will be dependent on the cooling capability of an individual vessel. Typical maximum cooling rates achievable on a production scale are approximately 0.5 °C min^{-1}. As scale increases to generate a fast cooling rate, the differential between the jacket and content is likely to increase. This large differential is likely to result in encrustation and uncontrolled crystallization on the vessel walls. If this is of a risk in the process being developed, a slower cooling profile should be implemented.

5.4.1.4 Solution Addition

Solutions such as anti-solvents and counterions can be added to different locations in a vessel, for example, above the surface into the vortex or subsurface using a dip pipe. The location of addition and the mixing regime drive the time taken for a homogenous solution to be achieved [37]. The requirements of the crystallization process need to be considered when deciding the point of addition. The solution is ideally dispersed throughout the batch quickly to reduce localized areas of high supersaturation. Areas of high supersaturation can lead to uncontrolled nucleation in parts of the vessel, generating potentially metastable forms. In this context, additions should not be carried out down the walls of the vessel as mixing is poor at the edges of the vessel. CFD modeling can be used to gain greater understanding of the mixing of the liquids. CFD images in Figure 5.8 show the difference in dispersion predicted between two different vessel configurations. In both cases, the addition point is subsurface.

5.4.1.5 Solid Addition

Solid additions such as seed need to be added to a vessel safely and be dispersed effectively and rapidly. To achieve this, solids are ideally added as a slurry. This approach

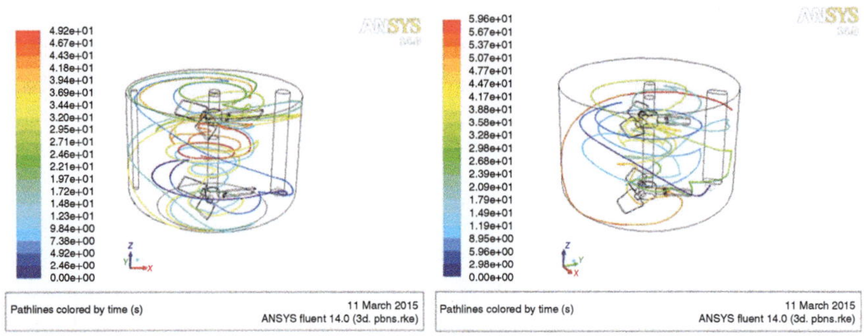

Figure 5.8 CFD models of addition to two different vessel configurations. The point of addition is at the base of the baffle (i.e. subsurface) in both vessels. The line color shows the time after addition in seconds.

enables more efficient dispersion within the vessel and limits the amount of solid material adhering to the walls, charging ports, or agitator shaft during addition.

The propensity for nucleation and crystal growth and, therefore, MSZW will vary with change of vessel and scale [25, 38]. If, on a small scale, seeding has not demonstrated any measurable benefits, it should still be included as if seeding is not used, the following may be observed:

- Differences between the first batch and subsequent ones. This is often referred to as the "first batch effect." This results from a clean vessel being used in the first batch, but in following batches residual materials being present and acting as seed.
- Impurity purging may change as well as variation in inclusion of residual solvents.
- Metastable forms may become more accessible.
- Other physical attributes of the API may vary.

5.4.1.6 Alternative Technologies

There are other technologies that can be used within vessels to deliver particles otherwise unachievable through "traditional" crystallization processes. Technologies include sonication and high shear wet milling.

Sonication can be used in the form of ultrasonic probes or horns within the vessel or externally on a recirculation loop. Sonication can promote nucleation and particle breakage and reduce encrustation on vessel walls. However, the use of ultrasonic probes in vessels results in significant heating of the solution which increases the requirements for effective heat transfer. Sonication is difficult to scale up but can be useful in early development if nucleation is challenging [39–41].

High shear wet milling can be used in situ to reduce the particle size and remove the need for dry milling processes post-isolation, or as part of a more complex temperature program, e.g. temperature cycling to aid morphology modification [42, 43]. The drawback of reducing particle size by this approach is isolation, and drying times are likely to be increased.

Both technologies have the potential risk of shedding metals leading to contamination of the batch and therefore require careful monitoring.

5.4.2 Risks and Common Problems

Different solid-state forms carry different levels of risk. As mentioned previously, a non-solvated stable form should ideally be selected for development. However, this is not always possible and special considerations are required for the development of a metastable form.

5.4.2.1 Metastable Forms

The crystallization of metastable forms can be challenging as there are inherent risks of transformation to a more stable form. If a metastable form is required to improve biopharmaceutics (e.g. better exposure), greater effort and attention needs to be applied [44] and the following information should be acquired:

- Accurate solubility and MSZW data, with an understanding of how the MSZW changes with scale (see solubility curve, Figure 5.4).
- A form risk assessment with a detailed understanding of the accessible forms via the use of a transformation diagram (Figure 5.2).
- A detailed understanding of the potential for form transformation to occur. This includes the impact of impurities, temperature, and solvents on transformation kinetics. In addition, the propensity of the material to transform in different environments, e.g. solvent- or vapor-mediated transformation.

When designing the process, the following need to be considered:

- Ideally, the process should be seeded with the desired form. However, sometimes seeding needs to be avoided, for example, when the metastable form can template the stable form [45, 46].
- Hold times should be minimized to limit the time available for nucleation of the stable form.
- A fast isolation and drying process may be required to prevent the transformation of forms [6, 47–49].
- Residual solvent at the end of drying. The presence of solvent in the dried material may be enough to facilitate the crystallization of a more stable form.

The kinetic stability of a metastable form is often strongly dependent on the impurity profile and solvent system in use. Changes in either of these aspects need to be carefully considered prior to being implemented.

5.4.2.2 Amorphous

The precipitation of amorphous forms has similar challenges to metastable forms with the risk that a more stable form may crystallize [50]. In addition, the glass transition temperature (Tg) of the dry material and an understanding of how this transition changes with solvent content are essential. Precipitation of the API below the Tg in slurry may result in the formation of fine brittle particles and at temperatures above the Tg the material will be more sticky and rubbery and tend to agglomerate and stick to vessel surfaces. There is currently no established technique for measuring the Tg in a slurry state; however, research is ongoing in this area [51].

As with metastable forms, there may be a requirement to have a fast precipitation, isolation, and drying process to prevent crystallization occurring. In the case of true amorphous solids defined as those where no crystalline phase exists, the risk of crystallization does not exist, and therefore, controlling the precipitation conditions for improving particle generation for downstream processes success is the only activity to investigate.

5.4.2.3 Salt Stoichiometry

If a salt form is being used, the different possible stoichiometries need to be understood and a speciation diagram can facilitate the visualization of the most appropriate conditions in terms of pH. It is important to note that the pK_as change with different solvent systems and are largest for an acid moving from water to an organic solvent [52]. Specific information required to develop a salt process includes knowledge of the pH range observed in the vessel during the addition of the counterion solution and understanding of the changes in solubility due to pH. As with all crystallizations, seeding of the process is recommended and should be combined with efficient mixing to ensure a homogenous solution.

5.4.2.4 Oiling and Phase Separations

Phase demixing, commonly known as oiling, is a problem which is often encountered during a crystallization of organic molecules like those seen in the pharmaceutical industry [53–56]. Oiling is where an API-rich liquid phase separates from the bulk of the solution, crystallization then often occurs in this phase. This is more commonly a problem in high supersaturation environments, for example, reactive crystallizations. The risks arising from crystallization occurring via oil include the crystallization of undesired forms and inclusion of impurities, and problems on scale-up such as encrustation and unpredictable variation of the crystalline solid phase attributes.

Crystallization from oils often leads to spherically agglomerated crystals similar to those shown in Figure 5.9. If spherical agglomerates are observed, oiling out as a potential cause should be investigated. The oil may only be present for a few seconds and therefore could be missed when offline measurements are taken. PAT such as in situ-particle size monitoring and imaging can allow the detection of transient oil phases as shown in Figure 5.10.

Where possible it is recommended that oiling should be identified and avoided. This can be achieved by reducing the supersaturation at the start of crystallization by seeding and during a crystallization process by reducing the rate of supersaturation generation.

5.4.3 Isolation and Drying

Once crystallized, the API needs to be separated from the crystallization liquors. Typically, in early development isolation and drying will be carried out using simple filtration and drying platforms. In the laboratory and on small-scale pilot plant facilities, systems such as Buchner funnels coupled with static drying in a vacuum

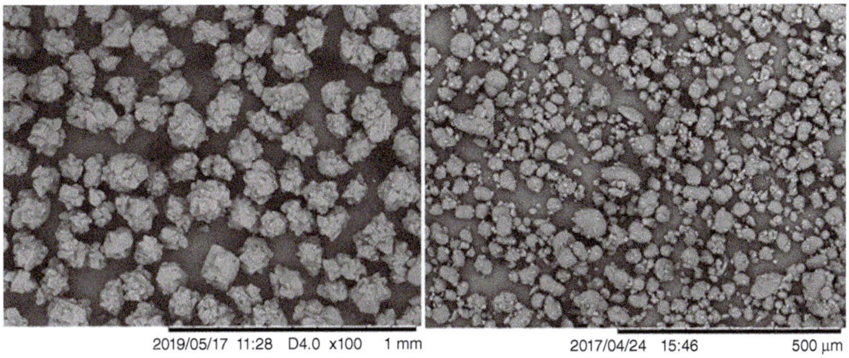

Figure 5.9 Examples of spherical agglomerates resulting from oiling out during the crystallization process.

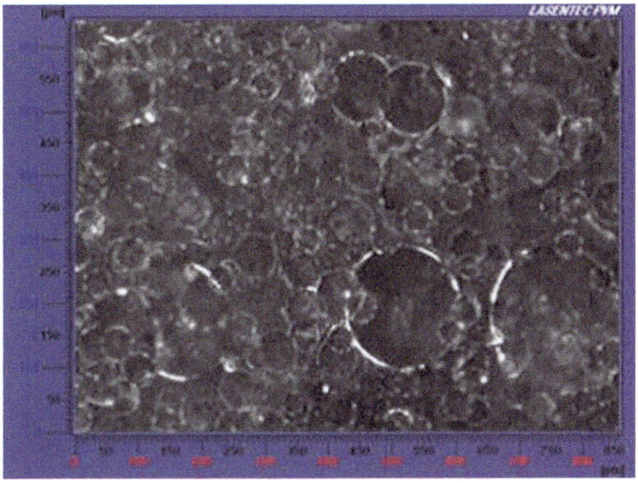

Figure 5.10 Example of video imaging showing a transient oil phase occurring prior to nucleation.

oven are commonly used. As scale increases, these types of equipment are no longer suitable and larger scale filters or centrifuges are used coupled with separate agitated drying equipment. Alternatively, filtration and drying can be completed in a single piece of equipment using agitated filter dryers. These all result in different levels of damage to the particles; an understanding of the impact of the equipment on the solid state of the API is required to ensure the right particle is delivered for formulation and ultimately the patient [57].

5.4.3.1 Isolation

The aim of isolation is not only to remove the solids from the liquid but also to remove all dissolved impurities and provide a suitable solvent environment for drying. An understanding of cake properties is essential for an effective and scalable

Figure 5.11 Image on left shows a crystals which have been slurry washed, sample, or right shows crystals which have been displacement washed.

isolation process, preventing phenomena such as cracking and over compression of the cake.

During development filtration time, wash efficiency and compressibility of the cake should be monitored. These data not only provide information on how these processes might scale but also highlight potential changes in the crystallization process, for example, changes in polymorphic form or particle size can lead to differences in filtration properties. At all stages of filtration development samples should be analyzed through the process to monitor changes in form, particle size, and morphology. Solvents with the potential to form solvates should be avoided to ensure the desired form is maintained.

A typical filtration and washing process will consist of:

1) An initial filtration to remove mother liquors. Depending on the cake properties, it is often desirable to not fully deliquor the cake at this point as deliquoring may result in cake cracking and inefficient washing.
2) The mother liquors are then displacement washed with solvent of the same composition. This present should effectively displace the mother liquors without crashing out the impurities which they contain.
3) Finally, a wash of a single volatile non-solvent is used. The use of a volatile solvent will aid rapid drying. The low solubility ensures API losses are minimized and reduces the residual solubility from the cake. This reduction in solubility lowers the propensity for formation of persistent agglomerates with solid bridges.

While displacement washing is the most efficient at removing mother liquors, it may be desirable to reslurry the cake. This has the advantage of increasing the contact time of the API with the wash and can aid the dissolution of impurities from the surface. In addition, it can help to remove fines from the surfaces of crystals which may be important to ensuring a stable API. Smaller particles have an inherently larger surface area and therefore energy; this increases the opportunity for agglomeration. Figure 5.11 shows an example of the same material with and without slurry washing; it is possible to see a visible difference in the amount of fines in the sample.

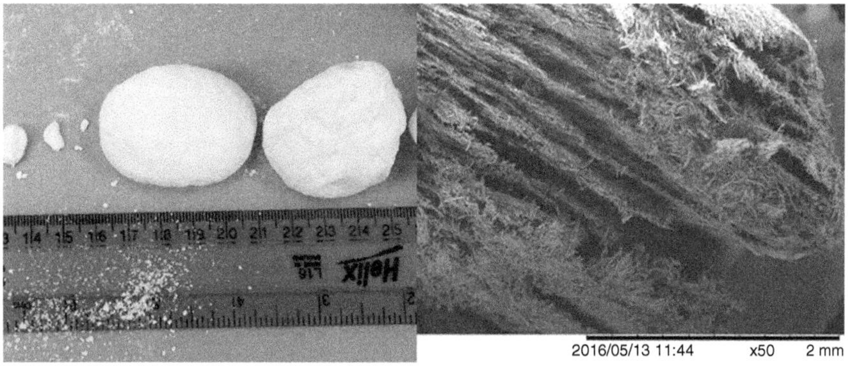

Figure 5.12 Lumps formed as a result of compression on the centrifuge.

Common problems observed during filtration and washing include [58]:

- Breakage of particles due to agitation of the cake or excessive compression.
- Formation of agglomerates due to compression. Figure 5.12 shows an example of the type of agglomerates that can result from compression.
- Formation of agglomerates due to ineffective removal of solvent resulting in a high residual solubility in the cake, allowing recrystallization and the formation of solid bridges between particles.
- Blinding of the filter with fines. The same filter is often used for multiple batches of API and not necessarily cleaned between batches. This can lead to a buildup of fines in the filter mesh which block the pores, preventing solvent removal.
- Cracking of the cake as a result of deliquoring leading to inefficient washing as a result of preferential passing of washes through cracks rather than the cake.

These problems all have the potential to impact the solid-state properties of the crystal.

5.4.3.2 Drying

The aim of this stage is to remove the solvent without causing change to the crystals and to produce a free-flowing powder ready for formulation. At small scales, drying is typically carried out in a static environment (e.g. vacuum tray dryer), at larger scales agitated dryers are used. There are a wide range of different dryer technologies available at larger scale including conical, spherical, and paddle dryers [58]. Scale-up and transfer between different drying technologies are poorly understood and are often significant challenges in API manufacture [59].

There is limited information that can be obtained from small-scale laboratory development experiments. Technology including mix torque rheometry, powder rheometry, attrition cells, and small-scale dryers can provide insight into material changes during the drying process [60, 61]. During the drying process samples should be taken and analyzed for form, particle size, and morphology to monitor changes and enable better understanding and control.

Though the process of agitation can impart significant shear leading to particle breakage [62, 63], shear force depends on the type of agitation, rate of agitation,

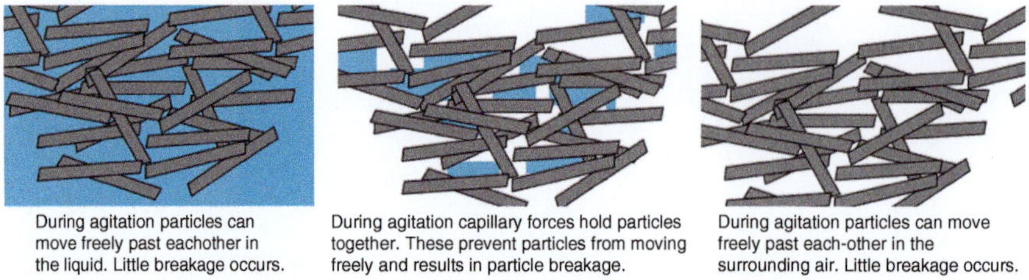

Figure 5.13 Impact of water levels on breakage forces seen during drying.

and loss on drying (LoD) (see Figure 5.13). Typically, at high LoD, particles are able to move past each other and the blade with limited breakage occurring. As LoD reduces the capillary forces between the particles become significant increasing particle breakage. As the cake dries, capillary forces are reduced and particles move freely reducing breakage.

Shear and breakage can result in the formation of impurities. Risks can be identified by an understanding of the crystal structure [64]. This can highlight proximity of functional groups which may react or reactive surfaces which can be exposed, leading to impurity formation if crystal defects are introduced. The formation of impurities as a result of shear is also observed when formulating the drug product, collaboration across the drug substance/drug product interface will allow any potential problems with agitated drying to be highlighted.

Agglomeration is often encountered during drying [63, 65, 66]. Most cakes will form lumps in the presence of solvent above a critical solvent content; this will happen irrespective of the solid-state form. The form and solubility in the drying solvent will decide the persistence of any agglomerates. Solubility in the drying solvent increases the risk of solid bridges forming between the particles and thus the formation of persistent agglomerates. If no solid bridging is present, the strength of agglomerates formed will be decided by the level of interlocking and other weaker forces such as van der Waals. Where the API has a tendency to auto-agglomerate, agitation of the dry cake bed can result in dry agglomeration of particles. Figure 5.14 shows an example of agglomerates formed during agitated drying. At solvent levels below the critical solvent content, agglomerates and lumps will tend to be broken,

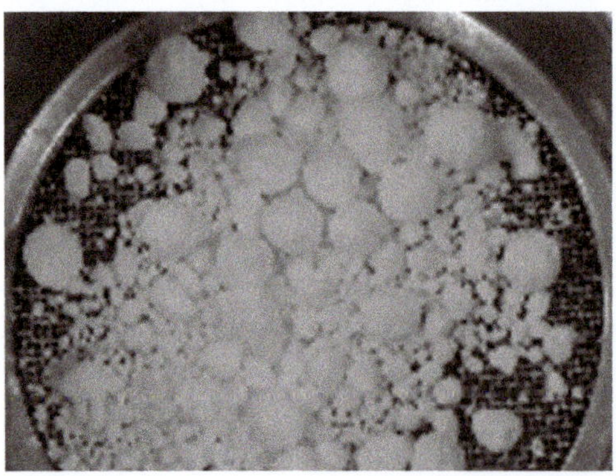

Figure 5.14 Image of lumps which have formed during agitated drying.

Figure 5.15 Image of a crystal fractured as a result of desolvation during drying.

and agitation at this point can help to break up agglomerates which were present from the isolation stage.

Form transformations can occur in the dryer. In particular, the drying of solvates and hydrates need to be performed in a way to avoid desolvation. This requires knowledge of the stability of the solvate as well as careful control of the drying process. It is possible to dry in humid environments to preserve the hydrated form [48, 67]. Figure 5.15 shows the fracturing of a crystal as a result of desolvation during drying. In some cases, the dryer is used to induce form transformation [68]. This is undesirable but sometimes necessary. In these cases, a clear understanding of the form transition at different pressures and temperatures needs to be gained. On scaling up, it is critical to have knowledge of the absolute pressure in the dryer rather than gauge pressure which is commonly measured to ensure the desired form transformation is achieved.

The ideal drying program will vary between APIs as they all exhibit different chemical and mechanical properties affecting how they dry. Ideally, drying preserves the crystal morphology and delivers a free-flowing nonagglomerated powder. Often agglomerates and lumps are formed upon drying due to solvent/solute forming bridges between particles. Those larger particles need to be broken up to ensure the cake is homogeneously dried and no pockets of solvent remain. To maintain crystal morphology, agitation should be reduced. These two things tend to require opposite actions. It is therefore common to have drying programs which consist of multiple phases, for example, an initial static phase, an intermittent mixing phase, and continuous mixing at the end of drying. The length of the phases will ideally be defined on the basis of solvent content of the cake (LoD).

5.4.4 Agglomeration

Agglomerates and lumps have been mentioned in the above sections. The nature and strength of the agglomerates is dependent on the solid-state properties of the API and the environment where they have been created [69–71]. The agglomerate strength can vary depending on the property of the bridges binding the particles. These can be weak, where only van der Waals forces are involved. Stronger agglomerates can be formed due to liquid and solid bridges [70].

Agglomerates and lumps can cause a variety of problems in drug product manufacturing. For example, if large lumps are present, it may not be possible to charge the API into the equipment as pipework may be blocked. Most formulation equipment contains various delumping steps so if these larger lumps can be charged they are often not a risk to quality. Hard lumps which have not been broken may cause problems with, for example, homogeneity of blends and content uniformity of tablets. It is therefore important to understand both the origin of lumps to try to avoid their formation and their fate in downstream processes [30].

5.4.5 Particle Size Reduction

Particle size reduction can be split into two broad categories: delumping, where agglomerated particles are broken down into either smaller agglomerates or primary particles, and milling where the primary particle is broken. During the development of a particle size, reduction process samples should be taken and analyzed for form, particle size, and morphology to ensure desired properties persist.

5.4.5.1 Delumping

Delumpers are often used to address the problem of large agglomerates in API resulting from the manufacturing process. Large-scale dryers often have milling equipment such as cone mill and the discharge pipework to carry out the delumping in situ. These are essentially automated sieves in which material can be gently milled or "delumped" to break these agglomerates. This process is gentle, and any damage to the primary particles should be minimal; care needs to be taken during the development of the process to ensure only delumping is happening, and where high-temperature and pressure solid forms are accessible, the process does not result in the formation of these.

5.4.5.2 Milling and Micronization

Milling or micronization may be required to obtain the desired particle size for the formulation, for example, for bioavailability or delivery mechanism, e.g. inhalation. There are a range of different mills which can be selected and tuned to deliver a specific PSD, e.g. jet mill and impact mills [72]. As strong mechanical forces are applied to crystalline particles, the introduction of disorders is likely. A common practice to improve crystallinity of these size-reduced particles is the use of a solvent/vapor-mediated process; for inhaled formulation, this is called conditioning.

5.4.5.3 Storage and Packing

API and intermediates need to be stored after manufacture before using them in the next process. Multiple locations of manufacture and variable product demand may lead to storage over an extended period. The packaging and storage conditions need to consider the stability of the solid-state form as well as the chemical and physical stability. Specific considerations need to be taken to ensure form is maintained for hydrates, metastable forms, amorphous, and hygroscopic APIs; common practices are refrigerated storage conditions and addition of desiccant.

5.4.6 Scale-up Conclusions

In this section, we have discussed various aspects of crystallization scale-up. During development, the crystallization will evolve as chemical synthesis and formulation routes are optimized. Development is iterative; improved understanding and findings will result in the need to revisit activities completed earlier in the development to gain greater understanding.

Changing upstream and downstream processes mean competing challenges for the form and particle forming stages to deliver specific and consistent outputs with often variable inputs. This challenge is also an opportunity to test a range of different API particle attributes to the formulation process. The assessment of particle variability enables the development of a more robust formulation with the ability to accommodate a wide range of particles properties.

5.5 People and Skill Requirements

In addition to the technical aspects, successful crystallization process development requires input from a wide range of skill sets to ensure a robust scalable process is designed. The selection of the form requires information from biopharmaceutical, solid state, crystallization, and formulation scientists. Once the form has been identified, process development then occurs within multiskilled teams consisting of crystallization scientists, process chemists, chemical engineers, and analysts. As the process is developed and scaled up, the role of the chemical engineer is critical to the understanding of the equipment dependency and the risk of transfer on API properties between equipment. Effective team working is required to ensure delivery relying on individuals possessing a wide range of soft skills including communication, problem solving, leadership capabilities, influencing, negotiation, and networking.

As scientific understanding and technology evolve, continued professional development is required. In industry, this is achieved through collaborations with other pharmaceutical companies, suppliers, and academia.

5.6 Regulatory Requirements

The regulations applied to the pharmaceutical industry ensure the safety of patients and workers. To support these regulations, there are several paper-based exercises that are required or should be considered during the scale-up of a crystallization process. The ICH guidelines form the basis of the activities which need to be undertaken. These activities relate to how the process is run to ensure safety, quality, and manufacturability and make sure learning is captured and implemented.

5.6.1 Process Documentation

There are a range of documents which are useful or necessary to have in place before a process can be scaled up or moved.

- A process description is the starting point for scale-up and provides the basis for subsequent documentation.
- A batch record is best practice for all scale-up and is required for good manufacturing practice (GMP). This details how each of the process steps will be carried out, including masses and vessels to be used. This level of planning allows the appropriate safety, quality, and manufacturability assessments to be completed.

These are the basic technical documents for scale-up and should be considered for non-GMP manufacture. As with the other elements of scale-up, this work requires cross-functional involvement from groups such as regulatory, quality assurance, and supply chain.

5.6.2 Safety

The level of safety assessment required will depend on several different factors such as scale of operation and local legislation requirements. Currently United Kingdom/European Union practices include:

- A hazard and operability study (HazOp) – this is used to understand the environmental, health, and safety (EHS/SHE) associated with a process.
- A control of substances hazardous to health (COSHH) assessment.
- A process risk assessment (PRA) can also be carried out.

The assessments should be performed by appropriately trained chemists and engineers who work on the plant where the process will be accommodated. These risk assessments do not consider the impact of the process change on the product quality and manufacturability. It is therefore necessary to carry out additional risk assessments to provide this understanding.

5.6.3 Quality and Manufacturability

The quality and manufacturability framework is defined by the ICH guidelines [73]. ICH Q9 Quality Risk Management utilizes failure modes and effects analysis (FMEA) which is a method that can be used to identify, understand, quantify, and mitigate risks [74]. FMEA can be used to understand all aspects which are important to the quality and manufacturability of crystallization processes. A quality risk assessment (QRA) is used to identify and evaluate the CQA and CPP.

This will also facilitate the identification of process stages which are sensitive to change and steps where greater understanding is required.

As with the QRA, a scale-up risk evaluation can be carried out by following an FMEA type approach and enables understanding of scale-up risks [75]. This should consider plant specific variables such as cooling ramps, temperature overshoots, or agitation regimes. This evaluation is performed to look at risks of a process in general or specific changes in a process as a result of equipment changes or process modification.

If the manufacture is to GMP regulations, a GMP risk assessment is required. This studies the risks to GMP of running the process in the selected equipment and accommodation.

Guidelines are constantly evolving together with new tools being developed to improve best practice and respond to new learning (e.g. Britest [76]). As a result, learning from historical issues is assimilated and patients, workers, and the environment are safeguarded.

5.7 Closing Remarks

In a crystallization process, the objective is to deliver the API to meet the requirements for both quality and manufacturability. This is achieved by having a well-controlled and scalable process. A well-controlled crystallization is one in which the different crystallization mechanisms are deconvoluted as far possible and the mechanisms understood and managed.

The crystallization process development workflow starts with experiments performed on a small scale to identify a suitable solvent and process, e.g. cooling or anti-solvent crystallization. This is followed by small-scale optimization to identify process parameters including seed addition and cooling profile. In an ideal process, nucleation is initiated by seeding, these seeds then grow without further nucleation, and the material is then isolated and dried without change to the properties imparted during the crystallization. This is the ideal situation and simplifies process understanding and subsequent scale-up. However, this is rarely observed, and development is required to identify the relationship between the process and the equipment. This should include investigation of vessel choice, mixing, additions, isolation, and drying. The challenge of scale-up increases where nonideal solid-state forms such as metastable or amorphous have been selected. The structured approach described in this chapter facilitates the understanding of crystallization mechanisms enabling a robust process which can be effectively scaled up.

The delivery of a robust scalable process requires cross skill set working between crystallization scientists, solid-state scientists, process chemists, engineers, and analysts. This work takes place within a highly regulated environment with clear guidelines on documentation and risk assessments required to ensure quality, efficacy, and patient safety. The level of understanding required and the degree of regulation increase as the drug moves through the process of development and commercialization.

By collaborative working, ensuring common understanding of goals, challenges, and potential problems, the right decisions and compromises can be made to ensure the best possible process is developed for all aspects of the API, enabling successful delivery of new medicines to patients.

List of Abbreviations

API	Active Pharmaceutical Ingredient
COSHH	control of substances hazardous to health
CPP	critical process parameters
CQA	critical quality attribute
DoE	design of experiment
EHS (or SHE)	environmental, health, and safety
EtOAc	ethyl acetate
FMEA	failure modes and effects analysis
GMP	good manufacturing practice
HazOp	hazard and operability study
ICH	International Council for Harmonization
IPA	isopropanol
IR	infrared (spectroscopy)
LoD	loss on drying
MoC	materials of construction
MSZW	metastable zone width
MTBE	methyl-tert-butyl ether (2-methoxy-2-methylpropane)
NMR	nuclear magnetic resonance (spectroscopy)
PAT	Process Analytical Technology
PRA	process risk assessment
PSD	particle size distribution
QRA	quality risk assessment
Tg	glass transition temperature
UV	ultraviolet (spectroscopy)
XRPD	X-ray powder diffraction

References

1 Brown, C.J., McGlone, T., Yerdelen, S. et al. (2018). Enabling precision manufacturing of active pharmaceutical ingredients: workflow for seeded cooling continuous crystallisations. *Molecular Systems Design & Engineering* 3 (3): 518–549.

2 Brown C, McGlone T, Florence A Continuous crystallisation Peter Kleinebudde, Johannes Khinast, Jukka Rantanen. *Continuous Manufacturing of Pharmaceuticals* 2017: Wiley Hoboken, United States 9781119001324.

3 Zhang, H., Lakerveld, R., Heider, P.L. et al. (2014). Application of continuous crystallization in an integrated continuous pharmaceutical pilot plant. *Crystal Growth & Design* 14 (5): 2148–2157.

4 ICH (2019). Mission. https://www.ich.org/page/mission (accessed 9 October 2019).

5 ICH (2002). The common technical document for the registration of pharmaceuticals for human use: quality – M4Q(R1). https://database.ich.org/sites/default/files/M4Q_R1_Guideline.pdf (accessed 23 October 2019).

6 Lee, A.Y., Erdemir, D., and Myerson, A.S. (2011). Crystal polymorphism in chemical process development. *Annual Review of Chemical and Biomolecular Engineering* 2: 259–280.

7 Feeder, N., Pidcock, E., Reilly, A.M. et al. (2015). The integration of solid-form informatics into solid-form selection. *Journal of Pharmacy and Pharmacology* 67 (6): 857–868.

8 PTA, G., Pidcock, E., Wood, P.A. et al. (2016). Navigating the solid form landscape with structural informatics. In: *Computational Pharmaceutical Solid State Chemistry* (ed. Y.A. Abramov). Hoboken, United States: Wiley. ISBN: 9781118700747.

9 ICH (2016). Impurities: guideline for residual solvents Q3C(R6). https://database.ich.org/sites/default/files/Q3C-R6_Guideline_ErrorCorrection_2019_0410_0.pdf

10 Scottish pollutant release inventory – methylene chloride, 2019. http://apps.sepa.org.uk/spripa/Pages/SubstanceInformation.aspx?pid=72 (accessed 21 October 2019).

11 Directive 2000/60/EC of the European Parliament and of the Council, 2000. https://eur-lex.europa.eu/legal-content/EN/TXT/PDF/?uri=CELEX:02000L0060-20141120&from=EN (accessed 21 October 2019).

12 Bonilla, R. (2010). Cost-effective solutions to the worldwide acetonitrile shortage. R&D Magazine. https://www.rdmag.com/article/2010/04/cost-effective-solutions-world-wide-acetonitrile-shortage (accessed 9 October 2019).

13 Mullin, J.W. (2002). Solutions and solubility. In: *Crystallization*, 4e (ed. J.W. Mullin), 86–134. Oxford: Butterworth-Heinemann.

14 CosmoLogic (2019). CosmoLogic – predicting solutions. http://www.cosmologic.de (accessed 9 October 2019).

15 Threlfall, T.L. and Coles, S.J. (2016). A perspective on the growth-only zone, the secondary nucleation threshold and crystal size distribution in solution crystallization. *CrystEngComm* 18 (3): 369–378.

16 Davey, R. and Garside, J. (2001). *From Molecules to Crystallizers. An Introduction to Crystallization*. New York: Oxford University Press.

17 Mullin, J.W. (2002). Metastable zone widths. In: *Crystallization*, 4e (ed. J.W. Mullin), 201–205. Oxford: Butterworth-Heinemann.

18 Srisanga, S., Flood, A.E., and White, E.T. (2006). The secondary nucleation threshold and crystal growth of α-glucose monohydrate in aqueous solution. *Crystal Growth & Design* 6 (3): 795–801.

19 Brandel, C., Gbabode, G., Cartigny, Y. et al. (2014). Crystal growth, structure, and polymorphic behaviour of an ionic liquid: phthalate derivative of N-butyl, N-methylimidazolium hexafluorophosphate. *Chemistry of Materials* 26 (14): 4151–4162.

20 Black, S. and Muller, F. (2010). On the effect of temperature on aqueous solubility of organic solids. *Organic Process Research and Development* 14 (3): 661–665.

21 Xiao, Y., Tang, S.K., Hao, H. et al. (2017). Quantifying the inherent uncertainty associated with nucleation rates estimated from induction time data measured in small volumes. *Crystal Growth & Design* 17 (5): 2852–2863.

22 Brandel, C. and ter Horst, J.H. (2015). Measuring induction times and crystal nucleation rates. *Faraday Discussions* 179 (Nucleation – A Transition State to the Directed Assembly of Materials): 199–214.

23 Davey, R.J., Back, K.R., and Sullivan, R.A. (2015). Crystal nucleation from solutions - transition states, rate determining steps and complexity. *Faraday Discussions* 179 (Nucleation – A Transition State to the Directed Assembly of Materials): 9–26.

24 Bakeev, K.A. (ed.) (2010). *Process Analytical Technology - Spectroscopic Tools and Implementation Strategies for the Chemical and Pharmaceutical Industries*, 2e, 557. Wiley.

25 Kadam, S.S., Kramer, H.J.M., and ter Horst, J.H. (2011). Combination of a single primary nucleation event and secondary nucleation in crystallization processes. *Crystal Growth & Design* 11 (4): 1271–1277.

26 Liang, K., White, G., Wilkinson, D. et al. (2004). Examination of the process scale dependence of L-glutamic acid batch crystallized from supersaturated aqueous solutions in relation to reactor hydrodynamics. *Industrial and Engineering Chemistry Research* 43 (5): 1227–1234.

27 Kepner-Tregoe (2019). Kepner-Tregoe. https://www.kepner-tregoe.com (accessed 9 October 2019).

28 Airaksinen, S., Karjalainen, M., Raesaenen, E. et al. (2004). Comparison of the effects of two drying methods on polymorphism of theophylline. *International Journal of Pharmaceutics* 276 (1–2): 129–141.

29 Zhang, Y. and Johnson, K.C. (1997). Effect of drug particle size on content uniformity of low-dose solid dosage forms. *International Journal of Pharmaceutics* 154 (2): 179–183.

30 Stauffer, F., Vanhoorne, V., Pilcer, G. et al. (2018). Raw material variability of an active pharmaceutical ingredient and its relevance for processability in secondary continuous pharmaceutical manufacturing. *European Journal of Pharmaceutics and Biopharmaceutics* 127: 92–103.

31 Eriksson, L., Johansson, E., Kettaneh-Wold, N. et al. (2006). *Multi- and Megavariate Data Analysis Part I Basic Principle and Applications*, 2e, 425. Umea, Sweden: Umetrics AB

32 Rajagopalan, A.K., Botschi, S., Morari, M., and Mazzotti, M. (2018). Feedback control for the size and shape evolution of needle-like crystals in suspension. II. Cooling crystallization experiments. *Crystal Growth & Design* 18 (10): 6185–6196.

33 ICH (1999). Specifications: test procedures and acceptance criteria for new drug substances and new drug products: chemical substances Q6A. https://database.ich.org/sites/default/files/Q6A_Guideline.pdf (accessed 21 October 2019).

34 Janbon, S.L.M., Parsons, A.R., Gavi, E., and Reynolds, G.K. (2019). Effects of scale, equipment, and operation on agglomeration during a reactive crystallization. *Organic Process Research and Development* 23 (3): 302–308.
35 McConville, F.X. and Kesslet, S.B. (2011). Scale-up of mixing processes: a primer. In: *Chemical Engineering in the Pharmaceutical Industry R&D to Manufacturing* (ed. D.J.A. Ende), 249–268. Hoboken, NJ: Wiley.
36 Schmidt, B., Patel, J., Ricard, F.X. et al. (2004). Application of process modelling tools in the scale-up of pharmaceutical crystallization processes. *Organic Process Research and Development* 8 (6): 998–1008.
37 Barrett, M., O'Grady, D., Casey, E., and Glennon, B. (2011). The role of meso-mixing in anti-solvent crystallization processes. *Chemical Engineering Science* 66 (12): 2523–2534.
38 Gillon, A.L., Steele, G., Nagy, Z.K. et al. (2006). PAT investigations into the crystallization of caffeine. In: *BIWIC 13th Industrial Workshop on Industrial Crystallisation* (eds. P. Jansens JtH and S. Jiang), 35–42. Delft, The Netherlands: IOS Press.
39 Nguyen, T.T.H., Khan, A., Bruce, L.M. et al. (2017). The effect of ultrasound on the crystallisation of paracetamol in the presence of structurally similar impurities. *Crystals* 7 (10): 294.
40 Guo, Z., Zhang, M., Li, H. et al. (2005). Effect of ultrasound on anti-solvent crystallization process. *Journal of Crystal Growth* 273 (3–4): 555–563.
41 Raman, V. and Abbas, A. (2008). Experimental investigations on ultrasound mediated particle breakage. *Ultrasonics Sonochemistry* 15 (1): 55–64.
42 Wilson, D., Bunker, M., Milne, D. et al. (2018). Particle engineering of needle shaped crystals by wet milling and temperature cycling: optimisation for roller compaction. *Powder Technology* 339: 641–650.
43 Ahmed, B., Brown, C.J., McGlone, T. et al. (2019). Engineering of acetaminophen particle attributes using a wet milling crystallisation platform. *International Journal of Pharmaceutics (Amsterdam, Netherlands)* 554: 201–211.
44 Su, W., Hao, H., Barrett, M., and Glennon, B. (2010). The impact of operating parameters on the polymorphic transformation of D-mannitol characterized in situ with Raman spectroscopy, FBRM, and PVM. *Organic Process Research and Development* 14 (6): 1432–1437.
45 Ferrari, E.S. and Davey, R.J. (2004). Solution-mediated transformation of α to β L-glutamic acid: rate enhancement due to secondary nucleation. *Crystal Growth & Design* 4 (5): 1061–1068.
46 Nicoud, L., Licordari, F., and Myerson, A.S. (2019). Polymorph control in batch seeded crystallizers. A case study with paracetamol. *CrystEngComm* 21 (13): 2105–2118.
47 Tian, J., Dalgarno, S.J., and Atwood, J.L. (2011). A new strategy of transforming pharmaceutical crystal forms. *Journal of the American Chemical Society* 133 (5): 1399–1404.
48 Cypes, S.H., Wenslow, R.M., Thomas, S.M. et al. (2004). Drying an organic monohydrate: crystal form instabilities and a factory-scale drying scheme to

ensure monohydrate preservation. *Organic Process Research and Development* 8 (4): 576–582.

49 Stephenson, G.A., Groleau, E.G., Kleemann, R.L. et al. (1998). Formation of isomorphic desolvates: creating a molecular vacuum. *Journal of Pharmaceutical Sciences* 87 (5): 536–542.

50 Kavanagh, A., McConvey, I., McCabe, J. et al. (2012). Developing amorphous pharmaceuticals: opportunity and necessity. *American Pharmaceutical Review* 15 (5): 119521/1–119521/11.

51 Zarkadas, D., Pridgen, C.S., and Liotta, V. (2018). Process development and robust control of physical attributes of an amorphous drug substance. *Journal of Pharmaceutical Sciences* 107 (1): 217–230.

52 Jones, H.P., Davey, R.J., and Cox, B.G. (2005). Crystallization of a salt of a weak organic acid and base: solubility relations, supersaturation control and polymorphic behaviour. *Journal of Physical Chemistry B* 109 (11): 5273–5278.

53 Deneau, E. and Steele, G. (2005). An in-line study of oiling out and crystallization. *Organic Process Research and Development* 9 (6): 943–950.

54 Takasuga, M. and Ooshima, H. (2014). Control of crystal size during oiling out crystallization of an API. *Crystal Growth & Design* 14 (11): 6006–6011.

55 Daver, S., Rodeville, N., Pineau, F. et al. (2017). Process development and crystallization in oiling-out system of a novel topical antiandrogen. *Organic Process Research and Development* 21 (2): 231–240.

56 de Albuquerque, I. and Mazzotti, M. (2017). Influence of liquid-liquid phase separation on the crystallization of L-menthol from water. *Chemical Engineering and Technology* 40 (7): 1339–1346.

57 Ottoboni, S., Price, C.J., Steven, C. et al. (2019). Development of a novel continuous filtration unit for pharmaceutical process development and manufacturing. *Journal of Pharmaceutical Sciences* 108 (1): 372–381.

58 Murugesan, S., Sharma, P.K., and Tabora, J.E. (2011). Design of filtration and drying operations. In: *Chemical Engineering in the Pharmaceutical Industry R&D to Manufacturing*, 1e (ed. E. DJA), 315–346. Hoboken, NJ: Wiley.

59 Conder, E.W., Cosbie, A.S., Gaertner, J. et al. (2017). The pharmaceutical drying unit operation: an industry perspective on advancing the science and development approach for scale-up and technology transfer. *Organic Process Research and Development* 21 (3): 420–429.

60 Lamberto, D.J., Cohen, B., Marencic, J. et al. (2011). Laboratory methods for assessing API sensitivity to mechanical stress during agitated drying. *Chemical Engineering Science* 66 (17): 3868–3875.

61 am Ende, D., Birch, M., Brenek, S.J., and Maloney, M.T. (2013). Development and application of laboratory tools to predict particle properties upon scale-up in agitated filter-dryers. *Organic Process Research and Development* 17 (10): 1345–1358.

62 Remy, B., Kightlinger, W., Saurer, E.M. et al. (2015). Scale-up of agitated drying: effect of shear stress and hydrostatic pressure on active pharmaceutical ingredient powder properties. *AIChE Journal* 61 (2): 407–418.

63 Lekhal, A., Girard, K.P., Brown, M.A. et al. (2004). The effect of agitated drying on the morphology of L-threonine (needle-like) crystals. *International Journal of Pharmaceutics* 270 (1–2): 263–277.
64 Krishna, G.R., Devarapalli, R., Prusty, R. et al. (2015). Structure-mechanical property correlations in mechanochromic luminescent crystals of boron difluoride dibenzoylmethane derivatives. *IUCrJ* 2 (Pt 6): 611–619.
65 Papageorgiou, C.D., Langston, M., Hicks, F. et al. (2016). Development of screening methodology for the assessment of the agglomeration potential of APIs. *Organic Process Research and Development* 20 (8): 1500–1508.
66 Lim, H.L., Hapgood, K.P., and Haig, B. (2016). Understanding and preventing agglomeration in a filter drying process. *Powder Technology* 300: 146–156.
67 Feth, M.P., Nagel, N., Baumgartner, B. et al. (2011). Challenges in the development of hydrate phases as active pharmaceutical ingredients - an example. *European Journal of Pharmaceutical Sciences* 42 (1–2): 116–129.
68 Lamberto, D.J., Diaz-Santana, A., and Zhou, G. (2017). Form conversion and solvent entrapment during API drying. *Organic Process Research and Development* 21 (11): 1828–1834.
69 Brunsteiner, M., Jones, A.G., Pratola, F. et al. (2005). Toward a molecular understanding of crystal agglomeration. *Crystal Growth & Design* 5 (1): 3–16.
70 Rossetti, D. and Simons, S.J.R. (2003). A microscale investigation of liquid bridges in the spherical agglomeration process. *Powder Technology* 130 (1–3): 49–55.
71 Hsieh, D.S., Lindrud, M., Huang, M. et al. (2018). Mechanistic elucidation of hard agglomerate formation from drying kinetics in the integrated sorption chamber. *Organic Process Research and Development* 22 (5): 608–617.
72 FDA (2014). SUPAC: manufacturing equipment addendum. https://www.fda.gov/media/85681/download (accessed 20 October 2019).
73 ICH (2019). ICH guidelines. https://www.ich.org/products/guidelines.html
74 ICH (2005). Quality risk management Q9. https://database.ich.org/sites/default/files/Q9_Guideline.pdf (accessed 21 October 2019).
75 Muller, F.L. and Latimer, J.M. (2009). Anticipation of scale up issues in pharmaceutical development. *Computers & Chemical Engineering* 33 (5): 1051–1055.
76 Britest (2019). Britest. www.britest.co.uk (accessed 20 October 2019).

6

Impact on Drug Development and Drug Product Processing

Susanne Page and Anikó Szepes

F. Hoffmann-La Roche Ltd., Grenzacherstr. 124, Basel CH-4070, Switzerland

6.1 Introduction

Structured product development is a current trend in the pharmaceutical industry, which includes the concept of designing quality into the formulations (quality by design [QbD]) and targeting robust manufacturing processes [1]. As outlined in International Conference on Harmonization (ICH) guideline Q8, Health Authorities expect pharmaceutical companies to design a quality product during pharmaceutical development and to develop a manufacturing process that is capable to consistently deliver the intended performance of the product [2].

The following elements are key for a structured development approach:

- Early focus on solid state properties of the drug candidate, which includes activities, such as pharmaceutical profiling, salt- and polymorph screening, and test of process-induced transformations (PIT) of the solid form that is selected for drug product development [1].
- Early compatibility testing of the drug substance with different excipients, which can minimize the risk of failure in long-term stability testing [1].
- Structured formulation development based on the physico-chemical characteristics and material attributes of the active pharmaceutical ingredient (API) and the formulation needs that are described in the quality target product profile (QTPP), such as indication and target population, route of administration, targeted dosage form, dose strength, and release characteristics [3, 4].
- Process development and modeling with the target to select a process technology based on the required quality attributes of the drug product [1].

Sun published the concept of materials science tetrahedron (MST) in 2009. The MST outlines the relationship among the structure, properties, performance, and processing of a drug [5]. The galenical tetrahedron (Figure 6.1) depicts the general considerations, which ensure that a drug product fulfills the three main requirements, which are quality, safety, and efficacy. These general requirements also represent the fundamental considerations which are closely linked to the QTPP of the final drug product.

Solid State Development and Processing of Pharmaceutical Molecules: Salts, Cocrystals, and Polymorphism, First Edition. Edited by Michael Gruss.

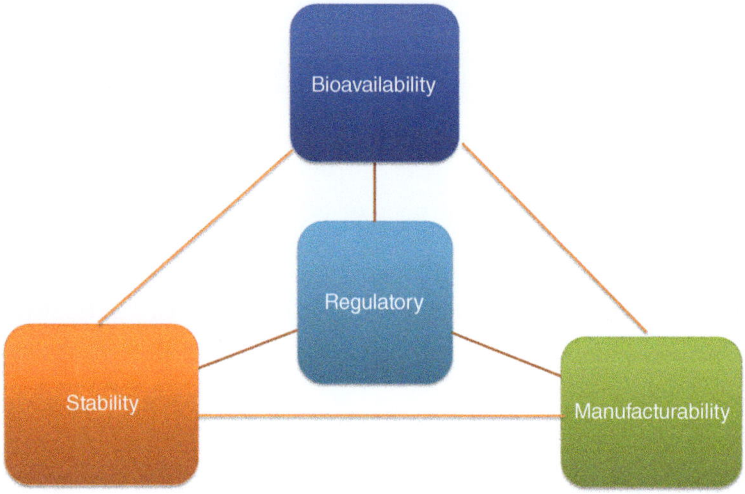

Figure 6.1 The galenical tetrahedron.

Pharmaceutical formulations have to exhibit adequate in vivo pharmacokinetic properties including bioavailability (BA), which is an important prerequisite for efficacy and an acceptable safety profile [6]. Adequate physical and chemical stability of the drug product ensure that product quality, safety, and efficacy can be guaranteed during the entire product shelf life. Technical formulation aspects are referred to as manufacturability (processability), which is considered to achieve robust manufacturing processes and quality can be built into the drug product, according to QbD principles. Finally, drug products have to be designed and developed which not only possess adequate quality and in vivo performance but also meet global regulatory standards.

Solid form screening of drug candidates is a standard procedure in pharmaceutical development and a regulatory requirement for new pharmaceuticals. Accordingly, solid form selection is an important milestone in drug development and plays a crucial role in product life cycle management [7].

A wide variety of solid forms of an API can be identified during solid form screening, e.g. crystalline and amorphous forms, solvate/hydrate forms, salts, or co-crystals. Different solid forms of a chemical entity can significantly differ in physico-chemical characteristics that can have an impact on product quality, performance, and processability [7, 8].

The goal of solid form screening is to find the suitable form of the drug candidate that possesses the best characteristics for development. Selection of the most suitable solid form of the API is usually a compromise between physical, chemical, and biopharmaceutical properties (physical and chemical stability, crystallinity, thermal behavior, hygroscopicity, mechanical properties, solubility/dissolution, permeability, etc.) [7]. On the other hand, formulation requirements also need to be considered, since the selection criteria of a solid form might completely differ depending on the indication, estimated daily dose or route of administration (i.e. oral, parenteral, inhalation). Generally, the thermodynamically most stable modification (at ambient conditions) of the parent compound is preferred, due to its low tendency to solid phase transformations. However, certain physical and chemical properties of the identified solid forms of an API may require a compromise and justify the selection of a solid form with a lower generic preference, e.g. a metastable form can have better solubility [7].

Regarding drug product design, the appropriate selection of the formulation composition is essential, since API-excipient interactions can induce solid state changes such as hydrate formation, salt disproportionation, or co-crystal formation/complexation with excipients [9–11]. It has to be noted that excipients can also stabilize the preferred solid form of the API and prevent undesired phase transitions, which may significantly affect dissolution and BA and facilitate manufacturability [12–14].

The stability of the selected solid form of the API has to be monitored during scale-up and manufacturing, since PITs can occur, which are well known but difficult to predict and to control (e.g. polymorphic transition or solvent-mediated transformations [hydrate formation/dehydration, salt/parent conversion]) [10, 15].

The pharmaceutical development of a new drug product can be divided into different stages:

- Pharmaceutical profiling
- Formulation development
- Process development and transfer to commercial manufacturing
- Development of control strategy and regulatory submission

Figure 6.2 provides an overview of the different phases during pharmaceutical development and key aspects of each phase. In each of these stages, certain decisions are taken and knowledge is generated, which finally supports the new drug application (NDA) submission. Important key aspects that have to be considered during pharmaceutical development are outlined in ICH Q8(R2) [2].

6.2 Pharmaceutical Profiling

Early on in the development of a new drug product, a large number of molecular entities are screened and selected based on the interaction with the biological target(s). In many cases, this is done by performing high throughput screenings using the API in solution, for instance dissolved in dimethyl sulfoxide (DMSO). This means that the intrinsic activity of the molecular entity is determined at this point in time by its chemical structure. Based on the lead structure found in the screening, further optimization of the chemical structure is performed and, in parallel, preformulation activities are initiated. They are in particular important as the ability to deliver the drug to the patient in a safe, efficacious, and cost-effective manner depends largely on the physico-chemical properties of the API in the solid state, which is either the solid form present in the final drug product or the form present as an intermediate in the manufacturing process.

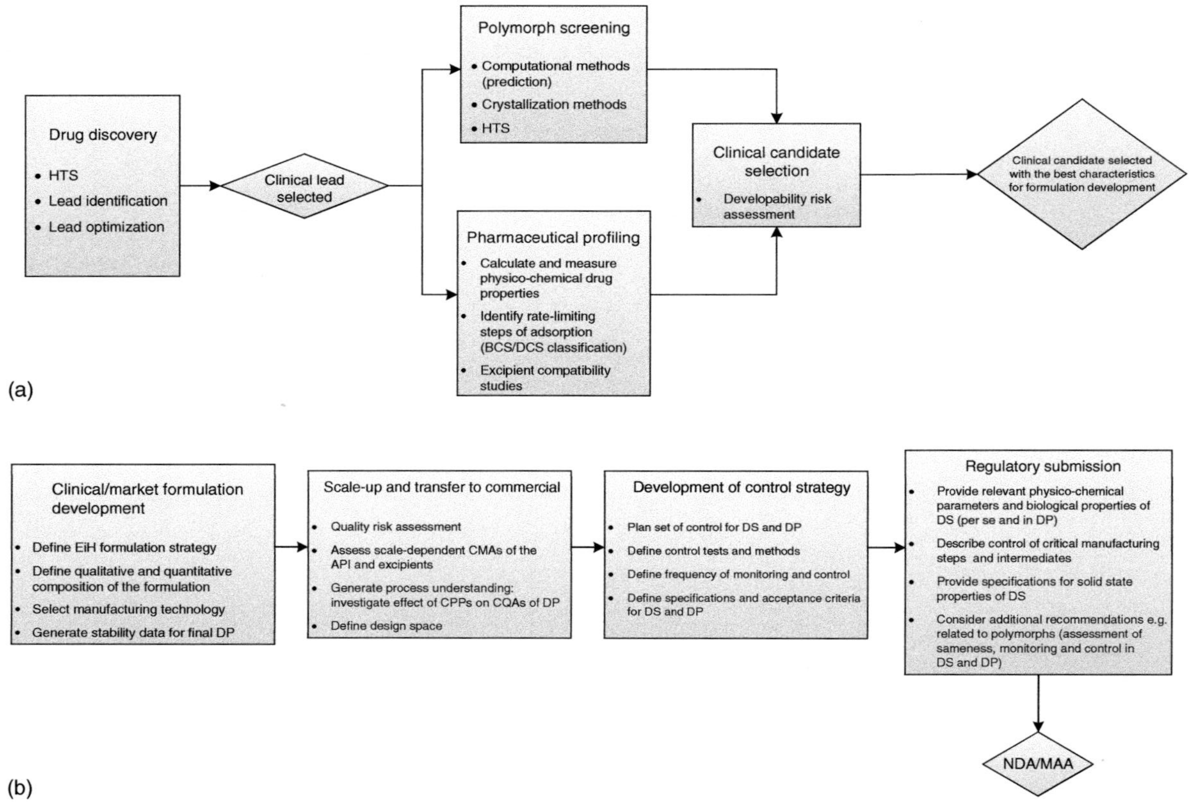

Figure 6.2 Pharmaceutical development – key aspects which are covered during the development (a), drug substance; (b), drug product. HTS, high throughput screening; BCS, biopharmaceutics classification system; DCS, developability classification system; DS, drug substance; DP, drug product; EiH, entry into human; NDA, new drug application; MAA, marketing authorization application.

One of the key deliverables in this phase of development is the selection of the clinical candidate, which is supported by the pharmaceutical profiling, that includes the determination of the physico-chemical properties of a number of APIs and their solid forms. The final selection of the chemical structure of the API, which is referred to as clinical candidate selection, is mainly triggered by the activity at the target(s), whereas the selection of the final solid form used for further development is mainly based on its solubility, which has an impact on BA, and sometimes based on other developability aspects, such as stability of the drug substance. A number of miniaturized testing methods were developed over the years to enable determination of the physico-chemical properties using mg amounts of the drug substance. Two examples of solubility assays are the partially automated solubility screening test (PASS), which bears a minor risk to generate new physical forms during dispensing of the drug substance in heptane [16], and the miniaturized assay for solubility and residual solid screening (SORESOS assay), an assay that allows simultaneous screening of drug solubility in different vehicles and identification of changes in solid form [17]. In addition, the paper, published by Alsenz and Kansy in 2007, provides an overview and comparison of high throughput assays for the determination of the kinetic and thermodynamic (equilibrium) solubility [18]. Additional assays were developed to determine the permeability of the API [19] and the intrinsic dissolution [20]. Even though the results of this early investigations can differ due to the different impurity level of API batches prepared in the medicinal chemistry lab compared to API batches manufactured for clinical studies, they provide useful information and allow ranking of different drugs and their solid forms.

In case the drug substance shows superior interaction with the target, but the investigated solid form has inferior physico-chemical properties, such as low solubility, alternative solid forms of the drug substance can be evaluated. The solid form selection process can be divided into three different parts:

- decision on the solid form composition (free form, salt, co-crystal, or combination thereof),
- decision on the physical form (polymorph, hydrate, solvate, amorphous, or liquid/oil), and
- selection of additional parameters related to particle characteristics, such as crystal habit, shape, or size of the particles).

As mentioned earlier, it can be stated that there is a clear preference for the thermodynamically stable modification (at ambient conditions) of the parent compound. However, sometimes certain physical and chemical properties of the identified solid forms of an API may require a compromise and justify the selection of a different solid form [7]. During pharmaceutical profiling, the clear focus is the selection of the solid form composition and the physical form.

Computational prediction and a number of different assays have been published in recent years to support the salt and co-crystal screening [21–33]. Once the decision on the solid form composition is taken, the final physical form, e.g. polymorph, hydrate, has to be selected. A first understanding of the polymorphic landscape can be obtained nowadays by using crystal structure predictions (CSPs)

[34–36]. The sixth blind test of organic CSP methods [37] has shown that despite many remaining challenges, these methods are becoming more applicable to a wider range of systems, including salts, hydrates, and larger flexible molecules. Furthermore, the results highlighted the potential for CSP calculations to complement and augment experimental studies of organic solid forms. A similar statement was done by Price et al. [35], who concluded that these methods can be useful in order to find and characterize practically important solid forms and to understand the solid form landscape. Based on the outcome of the CSP, additional polymorph screening experiments can be initiated in order to find the most stable polymorph and to access its physico-chemical properties [7, 21, 38, 39]. In addition, several approaches were tested to predict the formation of hydrates, and the calculation of the multi-differential hydrogen propensity score was superior to solution thermodynamics or single-differential hydrogen bond propensity in differentiating between hydrate-forming and non-hydrate-forming compounds [40].

Finally, the clinical candidate, its solid form composition and physical form, is selected and the preclinical development, which includes preparation of formulations for the animal and first in human clinical studies, is initiated. Early on during this phase, a biopharmaceutical assessment is performed in order to determine the rate limiting step for absorption and to decide if changes in the particle size distribution of the API can improve the rate of absorption, which is the case if dissolution limited absorption is expected. It needs to be taken into consideration that the estimated particle size distribution is linked to the applied doses, which can lead to different particle size distributions required for toxicology studies and first in human clinical studies compared to later phases [41–43].

In order to support the selection of a suitable formulation approach, Alsenz [44] proposed a four-step plan, which comprises the following steps:

- Calculation and measurement of the most important physico-chemical drug properties,
- Identification of the rate limiting step of absorption,
- Definition of the formulation strategy based on the findings and by further considering the development phase, therapeutic area, target dose, species, and resources, and
- Selection of the formulation approach considering the overall strategy for entering into humans and paying attention to excipient tolerability at given concentrations (and regulatory status).

6.3 Formulation Development

Nowadays, QbD principles are used throughout drug product development, and as a starting point of each development a QTPP is defined as it relates to quality, safety, and efficacy [2]. Safety and efficacy of the drug product are directly linked to the BA of the product, whereas the quality aspect is strongly related to the stability as well as manufacturability (e.g. control and robustness of the manufacturing process).

The importance of BA, stability, and manufacturability are reflected in the galenical tetrahedron (Figure 6.1).

BA is one of the key aspects regarding in vivo performance of a drug product. The biopharmaceutics classification system (BCS) was established in the late 1990s [45], which can be considered as an important milestone in formulation development of immediate-release oral dosage forms. Butler and Dressman published the concept of the developability classification system (DCS) in 2010, which provided a more appropriate classification system according to the factors limiting oral absorption. The BCS and the DCS provided a scientific framework to consider dose, solubility, permeability, and dissolution as critical factors that determine in vivo performance [46]. It must be noted that characterization of solubility in biorelevant media is important to determine DCS class and predict in vivo absorption, while solubility of the solid form in processing liquids is an aspect of manufacturability.

Drug product stability and manufacturability are multifactorial aspects of formulation development, since a wide variety of solid form characteristics and attributes of the API and the interaction thereof have to be considered to develop a final dosage form. Chemical stability of the API in solid state and solution, physical stability of the solid form during manufacturing and storage, crystallinity, particle shape/crystal habit, particle size distribution, and further parameters that are related to them, such as specific surface area, bulk density, flowability, particle adhesion/cohesion, electrostatic behavior; melting point (MP), mechanical properties, wettability, and hygroscopicity are all solid state or physico-chemical characteristics of the drug, which can be considered as potential critical material attributes (pCMA) during formulation development and drug product manufacturing.

It is widely-known that different polymorphic modification might significantly differ in the above-mentioned material attributes [7]. Therefore, selection of the appropriate solid form is essential. Furthermore, the selected solid form of the drug substance has to be controlled during manufacturing and storage of the final drug product to ensure that the required material attributes remain unchanged and, thus, sufficient product performance can be achieved. Therefore, it is of vital importance to understand phase transformations of the selected drug candidate during processing and storage, such as polymorphic changes, salt formation and disproportionation, co-crystal formation or (de)solvation during manufacturing, storage, and dissolution [47, 48].

It must be emphasized that the final drug product contains not only the API but also further excipients that support processability and drug product performance. Therefore, besides the solid state characteristics of the API, special attention has to be paid to the qualitative and quantitative composition (dose strength and drug load of the formulation) of the formulation and to potential interactions between the API and the excipients regarding possible change of the solid state form. In addition, the selected manufacturing process and the process parameters and factors, which might have an impact on the drug's solid state characteristics, need to be considered.

Based on physico-chemical and other properties of the drug substance, the formulator selects the type of formulation which has to be developed, in order to meet the requirements in the QTPP. In the following sections, the main characteristics of the drug substance and key formulation attributes, that are important when developing the different formulations, are captured, and insights are provided into the development of liquid dosage forms, solid dosage forms, and solubility enhanced dosage forms.

6.3.1 Liquid Formulations: Solutions and Suspensions

Liquid dosage forms can be applied for different routes of administration, such as oral, parenteral, topical, or inhalation application [49] and comprise a wide variety of pharmaceutical preparations, such as oral solutions and drops, syrups, elixirs, mixtures, dermal solutions and suspensions, nasal and eye drops, ophthalmic preparations, injections, infusions, or aerosol suspensions [50]. Oral liquid formulations are widely used for the medical treatment of the pediatric population including infants and toddlers (one month to two years) and young children (two to five years) since they are easy to swallow, reduce the risk of choking and ensure high dosing flexibility [51, 52].

Liquid dosage forms can be prepared by dissolving the active pharmaceutical substance in an aqueous or nonaqueous solvent (solutions), dispersing the API in appropriate liquid vehicle (suspensions) or by incorporating the drug into an oil or water phase (emulsions). This section focuses on the influence of the API solid form on the formulation aspects of solutions and suspensions.

As defined by Edman, a true molecular solution is a mixture of two or more components forming a homogeneous molecular dispersion in a one-phase system. In an aqueous solution, the solid form of the API is completely dissolved in water, which can be regarded as the primary solvent for pharmaceutical liquid formulations [49]. In order to achieve a physically stable formulation, the concentration of the solid API form should be below the equilibrium solubility in the applied solvent or vehicle to prevent crystallization/precipitation during shipment or storage.

In case of pH-dependent solubility in water (weak acids or bases), solubilization of the active substance can be promoted by pH control using buffer systems, while solubilization of nonpolar solutes can be achieved with co-solvents, such as alcohols, glycerols, or glycols [49, 53]. However, it must be noted that control of pH is a critical issue when dealing with aqueous liquid formulations, since pH might also impact chemical stability of the active substance [49]. Special attention should be paid to drugs that are applied as salts in liquid formulations, since salts can disproportionate in aqueous media and revert to the free form resulting in precipitation or pH changes and, as a consequence, chemical degradation [54, 55].

Surfactants are widely used as solubilizing agents which support wetting of the drug substance by the solvent [49, 53]. Furthermore, cyclodextrins are functional excipients that can interact with poorly water-soluble drugs resulting in an increase in their apparent water solubility via formation of dynamic inclusion complexes [56]. The increase in solubility, that can be achieved by complexation, was reported to also increase dissolution rate and hence improve oral BA of BCS II or IV compounds.

As described earlier, different drug polymorphs can significantly differ in their apparent solubility. A general rule of thumb of solid form selection is that the thermodynamically most stable form of a compound is preferred for drug product formulation purposes, which also applies for liquid dosage forms. However, metastable forms frequently exhibit better aqueous solubility and improved oral BA. A metastable form of celecoxib could be produced and stabilized in suspension by selecting the right manufacturing technology (precipitation) and adding a polymer and a surfactant in an appropriate ratio to the formulation [57]. A faster dissolution rate and a significantly higher BA were detected for the suspension containing the metastable solid form.

Besides the solid form, particle size, and crystal quality are also reported to significantly influence dissolution rate and solubility of solids. According to the Noyes–Whitney equation (Eq. (6.1)), small particles have a large surface area, which has a positive effect on dissolution rate, while the Freundlich–Ostwald equation (Eq. (6.2)) describes that small particle radii and large number of high energy sites of fine particles provide increased solubility. Grinding and micronization break the particles and increase their crystal defects that act as weak points in the crystal lattice from which the molecules can be easily removed [53].

$$\frac{\mathrm{d}m}{\mathrm{d}t} = A\,\frac{D}{d}\,(C_\mathrm{s} - C_\mathrm{b}) \tag{6.1}$$

where $\frac{\mathrm{d}m}{\mathrm{d}t}$ is the dissolution rate in kg s^{-1}, A is the surface area of the solute particle in m^2, D is the diffusion coefficient in m s^{-1}, d is the thickness of the diffusion layer in m, C_s the saturation concentration, and C_b the concentration in the bulk solution in kg l^{-1} [58].

$$S = S_\infty \exp\left(\frac{2\gamma M}{r\rho RT}\right) \tag{6.2}$$

where S is the saturation solubility of the drug, S_∞ is the saturation solubility of an infinitely large API crystal, γ is the crystal-medium interfacial tension, M is the compound molecular weight, r is the particle radius, ρ is the density, R is the universal gas constant, and T is the temperature [59].

As mentioned above, solution formulations usually contain excipients, such as co-solvents, buffers and pH stabilizers, surfactants/wetting agents, and preservatives. Pediatric formulations need taste masking by using sugars (e.g. sucrose, fructose, maltose, dextrose, fructose) and sugar alcohols (e.g. sorbitol, mannitol, xylitol), artificial sweeteners, and flavors [60].

Excipients can also be used to promote or inhibit solution-mediated solid state transformations. Surfactants were reported to affect the anhydrate (AH) to hydrate transformation of carbamazepine in aqueous solution by promoting or inhibiting crystal growth depending on their solubilizing capability, while polymers adsorbing to crystal surfaces showed an inhibition effect by reducing the growth phase of the transformation profile and changing the crystal habit of the hydrate crystal [61].

A pharmaceutical suspension is a dispersed system, in which insoluble solid particles are dispersed in a liquid vehicle, which is often water [49]. However, it must be noted that even a compound, which is poorly soluble in the suspension vehicle, can partially dissolve and recrystallize over time. Since small particles usually have higher rate of solubilization, the small particles have the tendency to dissolve and recrystallize subsequently on the surface of the larger particles resulting in the growth of the larger particles. This phenomenon is referred to as Ostwald ripening. Increasing size of the dispersed solid particles during storage due to Ostwald ripening might result in problems regarding physical stability (e.g. sedimentation, flocculation, caking), redispersability, and content uniformity and a drop in BA [50, 60]. It is impossible to obtain an absolutely uniform particle size distribution of the API and, thus, real suspensions tend to ripening. However, ripening can be minimized if the suspended solid particles are similar in size. Therefore, suspensions are most stable if the initial particle size of the suspended solid is uniform [53].

Controlling the size of particles is on one hand very important from different formulation aspects, as described related to Ostwald ripening. On the other hand, specific particle size limitations can apply for different application routes. Suspensions containing large dispersed solid particles will be gritty and unsuitable for parenteral or ophthalmic preparations due to, for example, clogging of needles and possible irritations [50].

Besides the suspending agent, suspensions also contain other excipients, such as wetting agents to promote dispersing the API particles and viscosity enhancers to prevent sedimentation and agglomeration of the dispersed solid particles and to ensure homogeneous distribution of the drug (content uniformity), which is a prerequisite of sufficient dosing accuracy [60, 62]. Extemporaneously prepared oral suspensions are frequently selected as clinical formulations in early phases. Oral compounding can be easily performed by using commercially-available ready-to-use suspension vehicles containing different types of excipients, such as suspending and buffering agents, antifoam agents, preservatives, sweeteners, and flavors [63–65].

The formulation composition has to be selected carefully, since the API might interact with the excipients present in the formulation and a new solid form can be formed in a suspension. Co-crystal formation of a free base was reported with a preservative incorporated in a suspension vehicle resulting in improved aqueous solubility of the parent compound and higher exposure in animal studies [54]. Theophylline co-crystals were formed with artificial sweeteners in suspension [66]. The formation of co-crystals led to the change of drug release rate and improved sweetness of the suspension formulations. Metronidazole benzoate undergoes AH to monohydrate conversion in aqueous suspension, which results in crystal growth due to lower solubility of the monohydrate. Monohydrate formation could be prevented using a combination of microcrystalline cellulose (MCC) and carboxymethylcellulose sodium (Avicel® RC-591) as suspending agent, indicating that selection of the right excipient can exert a positive effect on metronidazole benzoate suspension stability [67].

Pressurized metered dose inhalers (pMDIs) can contain aerosol suspensions consisting of two main components: the concentrate of the active compound(s) and the propellant gas mixture. The following formulation considerations are essential to achieve a stable and efficient suspension: to decrease the rate of sedimentation and flocculation adding surfactant or dispersing agents, reducing the particle size to less than 5 μm, matching the densities of the API and the propellant mixtures and minimizing the moisture content to optimize the respirable fraction [50]. According to Sheth et al., particle size and formulation composition often have significant impact on quality and performance variables, such as delivered dose uniformity (DDU), aerodynamic particle size distribution (APSD), plume geometry, spray pattern, and dissolution [68].

Liquid dosage forms are reported to be less stable compared with solids. Chemical stability of liquid dosage forms is often compromised by hydrolysis, oxidation, and photodegradation of the API. The challenges regarding physical stability are described above. The stability of aqueous formulations is easily affected by microorganisms, especially if they contain ingredients that may be nutritive to bacteria and fungi (e.g. carbohydrates, organic acids, and vitamins) or possess a pH value that is preferred by microorganisms [50]. Therefore, liquid formulations often require special packaging (e.g. brown or amber glass) and special storage conditions (e.g. refrigeration).

Since equilibrium solubility is defined at a given temperature, it must be noted that exposure of a pharmaceutical solution or suspension to elevated temperatures during manufacturing, transport and storage (e.g. temperature excursions during transport and storage) or to refrigerated conditions during storage can have a significant impact on physical stability of liquid dosage forms via inducing recrystallization of the active agent from the solution or suspension [51, 64]. Therefore, storage temperatures have to be selected, which are a good compromise between physical, chemical, and microbiological stability of a liquid drug product. Furthermore, heat sterilization cannot be used for parenteral formulations to prevent crystal growth and the modification of the protecting colloids. Therefore, all operations have to be carried out in sterile environments using aseptic techniques [50].

6.3.2 Solid Dosage Forms

Solid dosage forms, such as powders and granules filled into capsules, sachets and stick packs, tablets, and also dry powder inhalers (DPIs), represent the most commonly used formulations for small molecules nowadays. These dosage forms, besides the DPIs, are mainly designed to deliver the drug in the gastrointestinal tract, where it is dissolved and absorbed. The dissolution rate, which is described by the Noyes–Whitney equation (Eq. (6.1)) [58], is therefore one of the key performance indicators for BA. The equation clearly shows, that the saturation concentration as well as the surface area of the API particles have an impact on the dissolution rate. As outlined earlier, consistent delivery of the same solid form (e.g. keeping the saturation concentration the same) and suitable particle size distribution are key API properties, which impact the BA.

While the in vivo performance of oral dosage forms depends on the amount of API that is dissolved and absorbed in the gastrointestinal tract, in vivo performance of inhaled drug products is strongly influenced by the amount of API that is delivered into the lung, which is referred to as aerosolization performance [69, 70]. On the other hand, dissolution rate should be also considered for inhaled products, since they mainly act locally [69, 71]. The fine particle dose that penetrates the lung in vivo is generally considered as the mass of API particles with aerodynamic diameter <5 μm. Particles larger than 5 μm are generally considered nonrespirable [69]. Therefore, micronization is commonly used to achieve the required particle size. Besides the particle size (aerodynamic diameter), particle shape and surface properties (surface roughness and rugosity) [72] were found to influence product performance of DPIs [69]. Needle-like crystals showed better deposition profile in the lung and, as a consequence, improved product performance [69, 73]. The surface interaction of API and the excipient carrier is described as the cohesion–adhesion balance (CAB) that compares the cohesive forces within the API and excipient and the adhesive force between drug and excipient carrier particles [69, 74].

In addition to the biopharmaceutical aspects, also manufacturability and stability aspects need to be considered when selecting the solid form and particle size distribution for development. The manufacturing classification system (MCS), which is proposed by an industrial working group for the selection of the manufacturing technology for oral solid dosage forms [75, 76], provides some context and guidance with respect to interrelation of the API properties, formulation needs and the manufacturing technologies. Manufacturing technologies included in the classification system are, in order of increasing complexity: Direct Compression (MCS Class 1), Dry Granulation (MCS Class 2), Wet Granulation (MCS Class 3), and Other Technologies (MCS Class 4), such as active coating, liquid-filled capsules, or melt granulation. Furthermore, the MCS introduced the concept of developability and defines six API characteristics that are considered as developability parameters and four additional properties that can affect developability (Table 6.1).

Based on the projected dose and the envisaged size of the dosage form which is defined in the QTPP, a certain drug loading is required in the formulation. A comparison of the drug loading with the percolation threshold (the critical concentration of the API at which the drug characteristics significantly impact the properties of the finished drug product) can be used to assess in how far the API properties have an

Table 6.1 Developability and other parameters according to the MCS (ideal properties are reflected in brackets).

Developability parameters	Other parameters which can affect developability
Projected dose (dose volume < 250 ml)	Flowability
Particle size (50–500 μm)	Segregation
Particle morphology (aspect ratio 1 : 1 : 1)	Compression assessment
Surface area	Surface adhesion
Solid state form (stable under standard manufacturing environmental conditions)	
Bulk density (>0.3 g ml^{-1})	

Source: Based on Leane et al. [75].

impact on the critical quality attributes (CQAs) of the drug product [75]. Accordingly, the higher the drug load, the more pronounced effect the API characteristics have on manufacturability and further CQAs of the drug product.

The critical concentration of the API in a solid dosage form is not constant but it is reported to strongly depend on bulk API characteristics, such as particle size, morphology, and mechanical properties [75]. When considering the critical API concentration, it has to be emphasized that particle size distribution is essential not only regarding manufacturability but also from a biopharmaceutical point of view, since these two parameters are critical, especially when developing formulations with high drug load or dose strength to ensure sufficient dissolution and BA of poorly soluble drugs [77].

This illustrates the impact of the selected solid form in formulations with higher drug loads, but also the development of low dose formulations, which contain less than 2 mg or 2% drug loading (w/w) of the API according to the British Pharmacopeia [1, 78, 79], is challenging. In case a low dose formulation is targeted, an appropriate particle size is essential to achieve content uniformity [80–82].

Due to the importance of the solid form of the API for BA and process robustness during manufacturing, the solid form, e.g. the polymorph, needs to be maintained in the formulation during processing as well as throughout the shelf life of the product. AH to hydrate transformations are described in literature to occur in two different ways, either as solid–solid transformation, where the water molecules are incorporated into the crystal lattice while remaining in solid state, or via the solvent-mediated process, where the anhydrous form dissolves and creates a supersaturated state, followed by nucleation and crystal growth of the hydrate form [83]. Excipients and water, either introduced by the excipients or as processing liquid into the formulation, are known to have an impact on the phase transformation behavior [47, 84]. Therefore, in addition to the chemical compatibility of the drug substance with common pharmaceutical excipients, which can be tested early on in the development using miniaturized assay [85], also the likelihood for phase transformations in presence of excipients and water should be investigated. Based on this knowledge, Raijada et al. [47] proposed a high throughput platform to select suitable excipients for wet granulation and tested the platform using binary mixtures of sodium naproxen, theophylline, amlodipine besylate, and nitrofurantoin in combination with 10 different excipients and adding water to the samples. Even so phase transformations as well as one API-excipient incompatibility could be detected, one drawback of the proposed screening method is the absence of high shear in the test set-up, which represents an important stress factor that is linked to the manufacturing process and is expected to have an impact on the solid form [47]. The sample preparation method was further optimized to accommodate high shear stress [48] and to investigate phase transformations during dissolution. Finally, it was tested using naproxen sodium, which formed a stable co-crystal hydrate with lactose, disproportionated in acidic environment during wet massing and resulted in a dissolution rate that was even lower than that of the acid form [48]. Another miniaturized assay was developed based on X-ray powder diffraction in order to quantitatively determine the kinetics of solvent-mediated phase transformations [86].

These miniaturized assays can on one hand be used to investigate the likelihood of phase transformations in presence of excipients under simulated process conditions, and on the other hand to screen for excipients which alter the kinetics of the phase transformation or avoid it.

A number of papers have been published, which investigated the effect of excipients on solvent-mediated phase transformation, such as AH to hydrate transformation of theophylline [87–89], berberine chloride [90], olanzapine [91], thiamine hydrochloride [92], caffeine, carbamazepine and sulfaguanidine [93], AH to hydrate, and hydrate to AH transformation of naproxen sodium during wet granulation and drying [94, 95].

Airaksinen et al. [88] and Jørgensen et al. [89] investigated the impact of lactose monohydrate and silicified MCC on the AH to hydrate transformation of theophylline during high shear wet granulation. While lactose monohydrate with minimal water absorbing slightly enhanced hydrate formation compared to theophylline alone, silicified microcrystalline cellulose (SMCC) was able to initially inhibit formation of the monohydrate by taking up the added water, but could finally not prevent this transformation at water levels required for the granulation [88]. Similarly, Chakravarty et al. [92] investigated solvent-mediated phase transformation of thiamine hydrochloride during wet massing in presence of MCC or povidone. None of the added excipients could prevent the transformation to hemihydrate (HH), but the kinetics of the process was altered [92].

An extensive excipient screen including diluents, binders, disintegrants, solubilizers, and buffers was performed using anhydrous theophylline in slurry experiments at low shear as well as in a high shear wet granulation experiment [87]. The results showed that the AH to hydrate transformation kinetics was unchanged for the majority of excipients tested, and that the formation of the monohydrate could be retarded by adding methyl cellulose and prevented by adding small amounts of hypromellose. Based on this outcome, further investigations were done by comparing the addition of the polymer in solid state vs. dissolved as binder including the effect of hydration/age of the binder solution, the impact of polymer concentration, the polymer molecular weight, and substitution grade [87].

In particular, polymeric excipients were found to be suitable to inhibit the transformation to the hydrate. They prevent nucleation and slow down crystal growth by absorbing water and thus removing an excess of water from the system. Furthermore, other effects, such as formation of hydrogen bonding and hydrophobicity, were used to explain the different behavior of hydroxypropyl cellulose and povidone in comparison to polyethylene glycol (PEG) on the hydration of olanzapine [91]. The insufficient hydrophobicity of PEG could explain why PEG could not prevent hydrate formation, even though all tested polymers showed hydrogen bonding [91]. Cross-linked poly(acrylic) acid inhibited hydrate formation of caffeine, while cross-linked poly(acrylic) acid and hypromellose inhibited AH to hydrate formation of carbamazepine during high shear wet granulation [93]. In addition, povidone delayed, whereas MCC facilitated the transformation to the AH during drying of naproxen sodium dihydrate [94]. Tong et al. [90] investigated the effect of lactose monohydrate, MCC and povidone on the conversion of berberine chloride hydrates during wet massing and drying, and concluded that phase transformations between two hydrates can be related to the water activities of the granulated material during high shear wet granulation and drying.

In addition to the excipient impact on solvent-mediated transformations, the effect of the drug load in the formulation and the use of milled or unmilled API has been investigated. While the drug load did not affect the AH to hydrate formation of theophylline during high shear wet granulation, the use of the ball-milled AH showed faster conversion to the monohydrate compared to the unmilled API, which was mainly attributed to changes in the surface area and surface properties due to milling [96]. Figure 6.3 summarizes the factors that were found in literature to impact solvent-mediated transformations.

Besides solvent-mediated transformations, also pressure-induced transformations and the impact of various formulation parameters on these transformations have been investigated in previous studies, such as polymorphic transition of caffeine [97], crystallization of griseofulvin [98], and amorphization of posaconazole [99].

Juban et al. investigated the caffeine Form I to Form II transformation during direct compression, and observed that the transformation was more pronounced in tablets compared to the uncompressed powder and impacted by the type of tablet filler as well as the drug load [97]. The amorphous to crystalline transformation of griseofulvin was studied in tablets containing the neat amorphous griseofulvin alone or in combination with various excipients [98]. Although none of the tested excipients (silica, hypromellose acetate succinate, MCC, and PEG) could prevent crystallization of griseofulvin, they still had an impact on the crystallization rate. The research of Huang et al. [99] focused on the effect of MCC as plastic material, dibasic calcium phosphate anhydrous as brittle material and magnesium stearate as lubricant, and posaconazole drug load on the degree of amorphization during direct compression. The authors found that the drug load as well as both fillers significantly contributed to amorphization of posaconazole. Furthermore, neither external (tooling surface prelubricated with magnesium stearate) nor internal (magnesium stearate added to the tablet blend) lubrication with magnesium stearate could fully prevent the transformation [99]. The potential of carrageenans and polyethylene oxides to protect drugs from polymorphic transformation during direct compression was shown by Schmidt et al. [100, 101]. Figure 6.4 summarizes the factors that were found in literature to impact pressure-induced transformations. Overall, the results indicate that the transformation of the physical form of the drug can, at least in some cases, be modulated by selecting the right excipients. The impact of process parameters on phase transformation is discussed in Section 6.4.

6.3.3 Solubility Enhanced Formulations

6.3.3.1 Lipid-Based Formulations and Drug Delivery Systems

Lipid-based formulations comprise different types of drug delivery systems ranging from simple oily solutions, intravenous emulsions, and liposomal formulations to complex SEDDS (self-emulsifying drug delivery systems) and SMEDDS (self-microemulsifying drug delivery systems) [102]. They are often liquids but

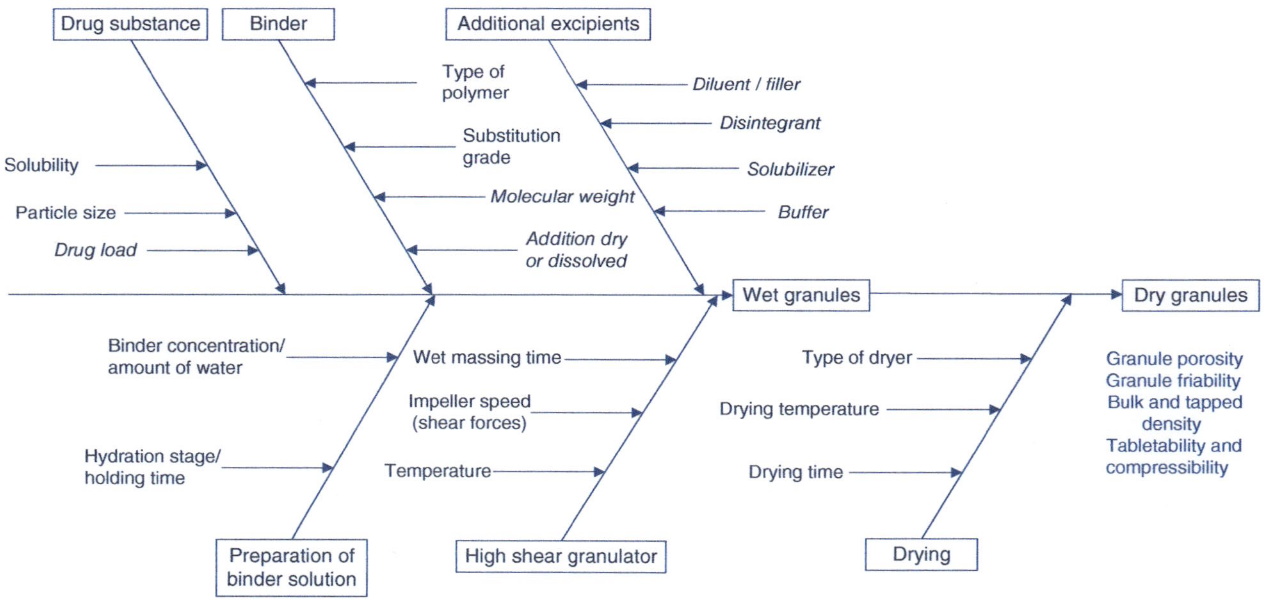

Figure 6.3 Summary of factors that proved to have an impact (bold) and that were investigated with respect to their impact (italic) on solvent-mediated phase transformations during wet granulation. Source: Based on Mirza et al. [83].

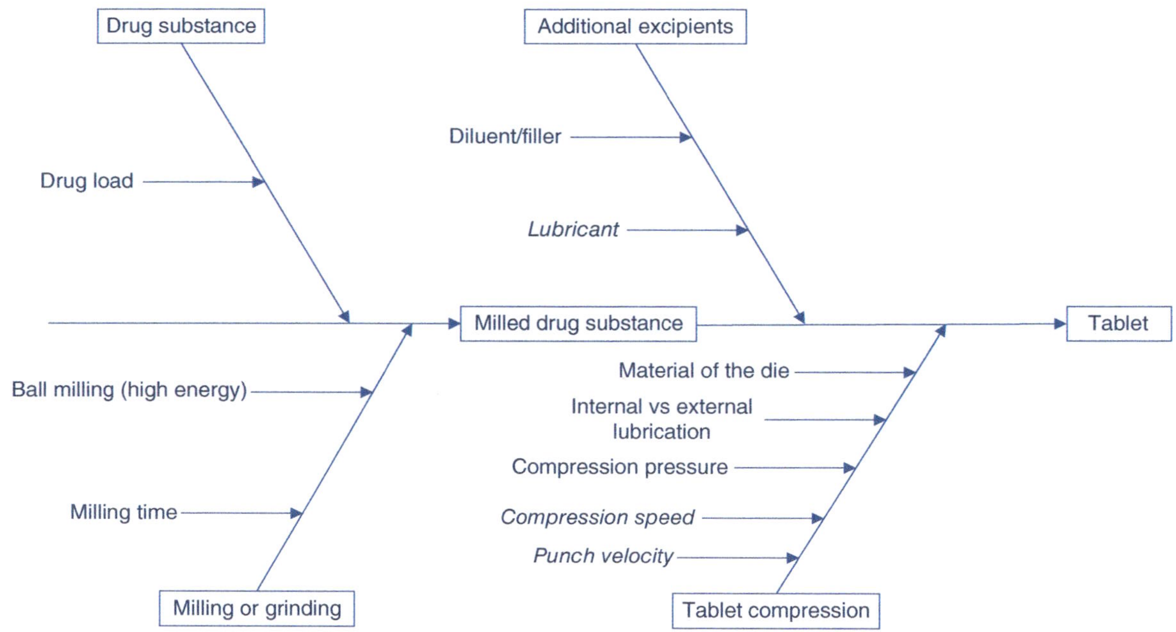

Figure 6.4 Summary of factors that proved to have an impact (bold) and that were investigated with respect to their impact (italic) on pressure-mediated phase transformations during milling and direct compression. Source: Based on Juban et al. [98].

may also be solid or semi-solid at room temperature. According to the analysis of Savla et al., soft gelatin capsule is the most frequently applied dosage form for the administration of lipid-based formulations [103, 104].

The major types of excipients, which are applied in lipid-based drug delivery systems (LBDDS), are lipids, water insoluble surfactants (hydrophilic–lypophilic balance [HLB] < 12), water soluble surfactants (HLB > 12), and hydrophilic co-solvents [103]. Pouton introduced the lipid formulation classification system (LFCS) that classifies lipid-based formulations into four main categories based on the proportion of included lipids, surfactants, and co-solvents [102, 105]. The type and relative percentage of excipients influence drug solubility, formulation characteristics, and dispersibility of the drug delivery system. Type I LBDDS contain triglyceride oil alone or mixture with its partial glycerides, while Type IV LBDDS does not contain any oil and consists of co-solvents and surfactants [106].

The primary benefit of lipid-based formulations is of biopharmaceutical nature: they increase dissolution rate and apparent intestinal solubility for lipophilic, poorly water-soluble drugs that are classified as Class II and Class IV compounds according to the BCS [46, 103, 107]. Furthermore, the excipients present in lipid-based formulations may reduce presystemic clearance and enhance permeation through the intestinal cell membrane [108, 109]. The improved absorption of poorly water-soluble compounds is based on the generation and maintenance of supersaturation in vivo for an appropriate time period. Incomplete solubilization of the active substance can result in precipitation upon dilution in the gastrointestinal tract, which has a negative impact on BA [110, 111].

Lipid-based formulations can be also used for the delivery of water soluble Class I or III drugs to improve content uniformity of high potency/low dose drugs, increase drug permeability, and to achieve modified release or taste masking [112].

The general solubility equation of Yalkowsky (Eq. (6.3)) describes the relationship between solubility (S), MP, and the partition coefficient ($\log P$) of a substance [104, 113]:

$$\log S = 0.8 - \log P - 0.01\,(\mathrm{MP} - 25) \tag{6.3}$$

Based on the above equation, solid state limited solubility is a consequence of a densely-packed, tight crystal lattice resulting in high MP or of high affinity for the lipid phase, which is reflected by a high $\log P$ value. The two extremes of poorly soluble drugs are the "brick dust" compounds possessing a MP above 200 °C and the so-called "grease balls" with a $\log P$ value above 3 [114].

Regarding the key properties of APIs that are incorporated in lipid-based formulations approved by the US Food and Drug Administration (FDA), they exhibit large variability in terms of MP (330–623 K), $\log P$ value (0.8–7.5), and molecular weight < 200 to >500 g mol^{-1} [104, 115].

Alskär et al. suggested a MP of 150 °C as a base line for the selection of drug candidates intended for lipid-based formulations, since a MP below this value indicates reasonable solubility in glyceride-based lipid vehicles [104, 116].

The solubility of brick dust compounds may be not high enough in lipid excipients. Therefore, co-solvents and surfactants might be added, in case a lipid-based formulation is targeted for drug product development. No correlation was found between MP and solubility in co-solvents. Therefore, a solubility screening in different lipid vehicles containing co-solvents and surfactants is regarded as a reasonable first approach for formulation development of brick dust compounds [104].

Lipid systems are excellent vehicles for highly lipophilic drugs which exhibit a log *P* value above 5 [103]. According to Chen et al., drugs possessing the following physico-chemical properties are ideal candidates for lipid formulations: poor aqueous solubility (e.g. $<10\,\text{mcg ml}^{-1}$); high lipophylicity (e.g. $\log P > 5$); good solubility in oils and lipids (e.g. $>25\,\text{mg ml}^{-1}$); relatively low MP and acceptable chemical stability [117].

If no appropriate solubility can be achieved in lipids, the physico-chemical properties of the drug candidate are to be modified [104]. The formation of prodrugs [118] or the preparation of lipophilic salts of the drug using lipophilic counter ions [119] could be potential options or the drug candidate could be transferred into an ionic liquid [112, 120].

6.3.3.2 Solid Solutions and Amorphous Solid Dispersions

In the last decades, the growing number of poorly water-soluble substances has led to an increased interest of researchers in the development of solid solutions with the drug being molecularly distributed in a matrix, and amorphous solid dispersions. It has to be clearly stated that in these formulations the transformation of the drug substance from its crystalline to the amorphous form is done with the purpose to enhance solubility and BA of the compound. As the amorphous form is inherently thermodynamically instable, it needs to be stabilized in the pharmaceutical preparation. This is often done by using polymers as the few marketed products show [121]. In recent years, a number of publications discussed the use of mesoporous silica for stabilization of the amorphous form [122], or the ability of other small molecules including amino acids in forming co-amorphous systems [123] or therapeutic deep eutectic systems [124]. Traditionally, spray drying is used as solvent-based and hot melt extrusion as fusion-based manufacturing technology to prepare amorphous solid dispersions and solutions. Besides these two processes, other technologies, such as microprecipitation [125], supercritical fluid technology [126, 127], electrospinning [128], and the Kinetisol® technology [129], were developed in order to fulfill specific needs.

In the initial stage of the development of an amorphous formulation, the stabilizing excipient as well as the manufacturing technology need to be selected based on the API characteristics and the formulation needs. Afterwards, the amorphous solid dispersion/solid solution can be manufactured and characterized by a number of analytical methods. It is extremely important to ensure complete conversion of the crystalline drug substance to the amorphous form during the manufacturing process as the BA is directly linked to the physical state of the drug substance. Furthermore, the amorphous state of the drug substance needs to be maintained during downstream processing and shelf life of the drug product. A number of publications have focused in the recent years on downstream processing of amorphous systems, and investigated in particular compression-induced phase separation [130–135]. For more detailed information on the development of amorphous solid dispersions and solid solutions in general, the interested reader is referred to the two following books [136, 137].

6.4 Process Development and Transfer to Commercial Manufacturing

Process development activities are usually initiated once the quantitative composition of the market formulation is defined and the manufacturing process is selected. An integral part of the process development efforts is focused on the identification of material attributes and process parameters that can have an effect on the CQAs of the drug product as outlined in ICH Q8(R2) [2]. As an initial starting point, a quality risk assessment (QRA) using prior knowledge can be performed as a first step. Based on the pCMA and potential critical process parameters (pCPP) identified in the QRA, an experimental plan is set-up in order to determine the functional relationships that link material attributes and process parameters to product CQAs. Finally, the enhanced product and process understanding generated during process development is used to define the CMA and CPP. Prior to transfer of the drug product into commercial production another QRA focusing on scale-dependent effects should be performed. Based on this, a small number of experiments might be done at commercial scale in order to confirm or disprove the relationship found during process development between CMAs/CPPs and CQAs. Besides classical approaches such as design of experiments (DoE), nowadays a number of scale-up and process models are used in order to facilitate the process development [138–143].

With regard to the solid state of the drug substance, the following aspects are important during process development: (i) influence of the particle size on the processability, and (ii) investigation of possible phase transformations depending on the selected process parameters. Even though the particle size distribution and the particle shape were defined quite early in the development, during drug product process development the impact of changes, which occurred due to scale up of the drug substance process, should be assessed. Furthermore, a good knowledge on the existing solid forms and potential transitions among them is key. Table 6.2 lists potential phase transformations and the underlying mechanisms.

The kinetics of these phase transformations depends on a number of different factors. For transformations via the solid state, kinetics is influenced by environmental factors, such as temperature, pressure, humidity as well as the presence of crystalline defects, the particle size distribution, and impurities. Melt-based transformations can occur when heating or cooling takes place, and impurities and excipients present can have an impact on transformation kinetics. In case of solution as mechanism, the drug is at least partially dissolved and the new form appears upon solvent removal, whereas solution-mediated transformations happen when the metastable phase is in contact with the saturated solution. While the first one can occur from stable to metastable and metastable to stable phases, the second one can only happen from the metastable phases to the stable phases [15].

Table 6.2 Phase transitions and their underlying mechanisms.

Phase transition	Mechanism			
	Solid-state	Melt	Solution	Solution-mediated[a)]
Polymorphic transition	X	X	X	X
Hydration/dehydration	X		X	X
Amorphous crystallization	X		X	X
Vitrification	X	X	X	

a) Only from the metastable phases to the stable phases.
Source: Based on Zhang et al. [15].

In order to provide sufficient clarity on the solid form changes that can occur during processing and to allow easy navigation, the following section is structured by unit operations.

6.4.1 Particle Size Reduction

Particle size reduction is often done as the first step during manufacturing in order to facilitate downstream processing (e.g. via improved morphology/flow properties, minimized segregation) and/or to enhance product performance (e.g. enhanced uniformity, increased surface area). Typically, particle size reduction is done by milling, which involves shearing/cutting, compressing, impacting, or attrition of drug particles. The mechanical stress and heat generated during milling can induce phase transitions, such as polymorphic transitions, dehydration, or vitrification via solid-state or melt mechanisms. It is therefore quite obvious that the rate and extent of these phase transitions depend on the type of mill and the milling conditions used [15]. Fast transformation of caffeine form I to form II was for instance observed during grinding (partially within 1 min and full transformation after 3 min) [144].

6.4.2 Blending

As a next step in the manufacturing process, the drug substance is typically blended with a number of excipients to facilitate further processing and product performance. In addition, homogeneity of the drug substance in the drug product is often achieved by mixing for solid oral dosage forms, such as capsules or tablets. As differences for instance in the surface properties, crystallinity and particle size of the drug substance as well as the excipients may result in segregation, homogeneity achieved by the mixing process should be investigated during process development and confirmed during validation [145].

6.4.3 Granulation

Due to the fact that the powder blend itself seldom possesses suitable properties for tablet compression or capsule filling, granulation is carried out to improve properties, such as flowability and compressibility. Commonly, granulation is performed by wet granulation and subsequent drying or by dry granulation.

6.4.3.1 Wet Granulation and Drying

In wet granulation processes, either solution or solution-mediated phase transformations can occur and should be investigated in detail. Formulations containing extremely low drug loads are often manufactured by dissolving the drug substance and spraying it onto a carrier using fluid bed granulation or Wurster process (in case of pellet layering) in order to achieve the required content uniformity. Whether transformation via the solution mechanism occurs during processing, it largely depends on the method of drying and the type of excipients used in the formulation. Independent of the drug load of the formulation, wet granulation processes are critical if the drug substance can exist in different AH and hydrate forms. As conversion of the anhydrous to the hydrate phase can occur via a solution-mediated transformation during wet granulation. Overall, it is important to show that the process is capable to consistently deliver the drug product with the same polymorph.

A large number of publications have reported AH to hydrate formations, in particular during high shear wet granulation, for example for thiamine hydrochloride [92, 146], theophylline [87, 96], amlodipine besylate [147], naproxen sodium [94, 95], caffeine, carbamazepine, and sulfaguanidine [93]. The behavior of thiamine hydrochloride, which can exist as an AH, a HH, and as nonstoichiometric hydrate (NSH), was extensively studied by Chakravarty et al. [92, 146]. While wet massing of a blend consisting of NSH and MCC resulted in complete transition to HH after ~150 min, fluid bed granulation followed by drying resulted in a mixture of NSH and HH in the granules when using the same NSH–MCC blend [92]. Wikström et al. investigated the theophylline AH to monohydrate transformation during high shear wet granulation, and could show that the mixing speed had a huge impact on water distribution in the granulator and thus on transformation time, which decreased with increasing impeller speed [96]. Similar observations were made for amlodipine besylate, where shear forces facilitated nucleation and overall transformation of the AH to dihydrate during high shear wet granulation. Subsequent drying of the obtained granules resulted in an increasing amount of AH and a decreasing amount of dihydrate with drying time. Furthermore, this process was more pronounced at a drying temperature of 70 °C compared to 40 °C [147]. Reddy et al. investigated the phase transitions of compound A during high shear wet granulation and subsequent drying in dependence of the amount of water, the temperature during granulation, the wet massing time, and the drying technique using DoE. While the rate of transformation from the AH to the HH was temperature-dependent (higher temperature led to faster transformation), the extent did depend on the amount of water used and the wet massing time (more water and longer wet massing time resulted in higher proportion of the HH). Finally, the drying process impacted the proportion of AH, HH and "apparent" amorphous form in the granules [148]. The impact of the drying process was also investigated for naproxen sodium, which forms a tetrahydrate during wet granulation. Subsequent drying under vacuum led to a mixture of dihydrate and tetrahydrate at room temperature, and to a mixture of the monohydrate and tetrahydrate at 40 °C. This finally resulted in different properties of granules, such as higher porosity and friability for the granules dried at 40 °C, different bulk/tapped density and compression behavior [95]. Dehydration of naproxen sodium dihydrate was additionally tested at elevated temperature under ambient air, in flow of an inert gas, under low pressure environment, and under high pressure in closed environments. The fastest transformation to the AH was observed under flow of inert gas, while the slowest change took place in a closed environment [94].

In most of these cases, Raman spectroscopy [87, 93, 96, 147–149] and/or near infrared (NIR) spectroscopy [147, 149, 150] were used to monitor the phase transition during granulation. NIR proved to be a good method to reveal the state of water, whereas Raman spectroscopy provided information on the molecule itself [147, 149, 151]. In addition, the use of torque measurements for investigating phase transitions was compared with NIR spectroscopy leading to the conclusion that NIR spectroscopy was superior to torque measurements as it allowed analysis of the state of water and reflected the property of the wet mass independent of the process parameters of the granulator equipment [150].

6.4.3.2 Dry Granulation/Roller Compaction

In comparison to the wet granulation process, in which solution or solution-mediated phase transformations can occur, the probability of phase transitions during dry granulation is reduced. Nevertheless, the applied mechanical stresses during dry granulation may lead to phase transformation via the solid state or melt mechanisms. Examples for solid state transformations, which can potentially also occur during roller compaction, are provided in the section about tablet compression.

6.4.4 Tablet Compression

A number of different publications mention compression-induced transformations, which can occur during tablet manufacturing. Different types of transformations are described: crystalline to amorphous [99, 152], amorphous to crystalline transformations [98, 130], polymorphic transformations as well as disorder of the crystal lattice and polymorphic transformations [153]. Huang et al. [99] investigated the compression-induced amorphization of crystalline posaconazole, while Thakral et al. [152] studied the compression-induced crystalline to amorphous transformation for chlorpropamide. Both research groups investigated the impact of shear stress in comparison to the axial stress or hydrostatic stress, and concluded that the amorphization is at least partially attributed to shear stress due to friction between the tablet surface and die wall or punches. In particular, at low compression pressures, the degree of amorphization is pronounced on the radial surface of the core. In addition, Huang et al. [99] investigated the effect of punch velocity, which was shown to have a negligible effect on the degree of amorphization using ^{19}F solid-state nuclear magnetic resonance (ssNMR).

Mah et al. [98] investigated the amorphous to crystalline transformation for griseofulvin in tablets containing the neat amorphous griseofulvin alone or in combination with various excipients using sum frequency generation microscopy. The authors found that compression-induced crystallization occurred predominantly on the tablet surface, and only minimal crystallinity was detected in the core of the tablet [98]. Amorphous to crystalline transformation can also occur unforeseen during the development of solid dispersion formulations. Singh et al. investigated the effect of compression (pressure and dwell time) for an Itraconazole–Soluplus solid dispersion, and proved that compression-promoted phase separation occurred [130].

The effect of the applied compression pressure on crystallographic properties was studied for carbamazepine, tolbutamide, and chlorpropamide as a function of distance from the tablet surface using grazing incidence X-ray diffraction. Compression-induced disorder, e.g. amorphization, crystal defects, and occurrence of small crystallites, was observed on the tablet surface for tolbutamide and chlorpropamide (amorphization), and both on the surface and in the core for carbamazepine (crystal defects) [153].

Polymorphic transformation of anhydrous caffeine was detected during tablet compression [97, 144, 154]. While fast transformation of form I to form II was observed during tablet compression already at low compression pressures (50 MPa) [144], increasing the compression pressure or changing the compression speed had no additional impact on the transition [97, 154]. Chakravarty et al. [146] investigated the impact of the NSH to HH transformation on tablet properties, such as microstructure, physical properties, and performance, for thiamine hydrochloride in stored tablets manufactured by direct compression or wet granulation. In comparison to the powder blends and granules, NSH to HH transformation was faster in the tablets. Complete conversion to HH was detected after 30 h storage at 40 °C/75% RH in tablets, whereas transformation was still incomplete in granules and powder blends after 144 h. Furthermore, changes in the porosity, hardness, and disintegration time of the tablets were measured for tablets prepared by wet granulation [146].

6.4.5 Film Coating

During film coating, an aqueous or solvent-based polymer system is applied to the pellets or tablets in a coating pan or in a fluid bed. Due to a highly efficient air exchange in these equipment, the time between the impingement of coating liquid onto the surface and the subsequent solvent evaporation is quite short, which minimizes the interaction between the core material and the coating dispersion. Therefore, a phase transition via the solution mechanism is highly unlikely during film coating [15].

6.5 Control Strategy

Based on the process and product knowledge and process and product monitoring conducted throughout development, a control strategy can be established for manufacturing toward the end of the process development phase. It consists of a planned set of controls, which should assure sufficient and consistent process performance and product quality, and can include drug substance, drug product, material attributes, operating conditions for the facility and the equipment, in-process controls and finished product specifications, respectively. Furthermore, the associated methods and frequency of monitoring and control are mentioned. Additional monitoring performed during scale-up activities can provide some indication of the performance of the commercial manufacturing process, and the knowledge gained during transfer and scale-up can be useful to further enhance the control strategy [155].

As it was mentioned earlier, the solid state of the drug substance can have an impact on the BA, manufacturability, stability, and quality of the final drug product. Nevertheless, this section highlights the interrelation between material properties and galenical/formulation aspects (BA, manufacturability, etc.) again using polymorphs as an example. Polymorphic forms of a drug substance can have different chemical and physical properties, including apparent solubility and dissolution rate, which have a direct impact on BA. Furthermore, they often differ with respect to mechanical properties and density, which can have a direct effect on the suitability for processing and/or manufacture of the drug substance and drug product [156]. Therefore, the control strategy has the aim to ensure that the solid form remains unchanged throughout the manufacturing process, or is under control, e.g. low batch-to-batch variability of the drug substance particle size distribution; amorphous form of the drug substance during manufacturing, downstream processing, and storage of amorphous solid dispersions.

Depending on the nature of the drug substance and the impact on BA and product quality, specifications for the polymorphic form and particle size distribution need to be defined. Guidance, in which cases specifications need to be set as part of the overall control strategy, is provided in ICH Q6A [157]. Figure 6.5 presents a modified decision tree from ICH Q6A for setting particle size distribution specifications for the drug substance.

A similar decision tree in ICH Q6A for setting acceptance criteria for polymorphs starts with the question, whether other polymorphic forms are known, and if yes, whether they differ with respect to their properties so that safety, efficacy, or quality are impacted. If an impact is expected, an acceptance criterion should be defined for the drug substance. In addition, the impact on drug product performance needs to be assessed and the polymorph should be monitored during release and the entire shelf life. This can either be done using a drug product performance test, e.g. dissolution (if this provides adequate control) or other analytical test methods, such as X-ray powder diffraction, thermoanalysis, and spectroscopic methods. Independent of the selected analytical method, it is important to establish acceptance criteria which are consistent with safety and/or efficacy.

6.6 Regulatory Submissions

Due to the importance of the solid form of the drug substance per se and in the drug product, regulatory advice is provided for both the NDAs and the abbreviated new

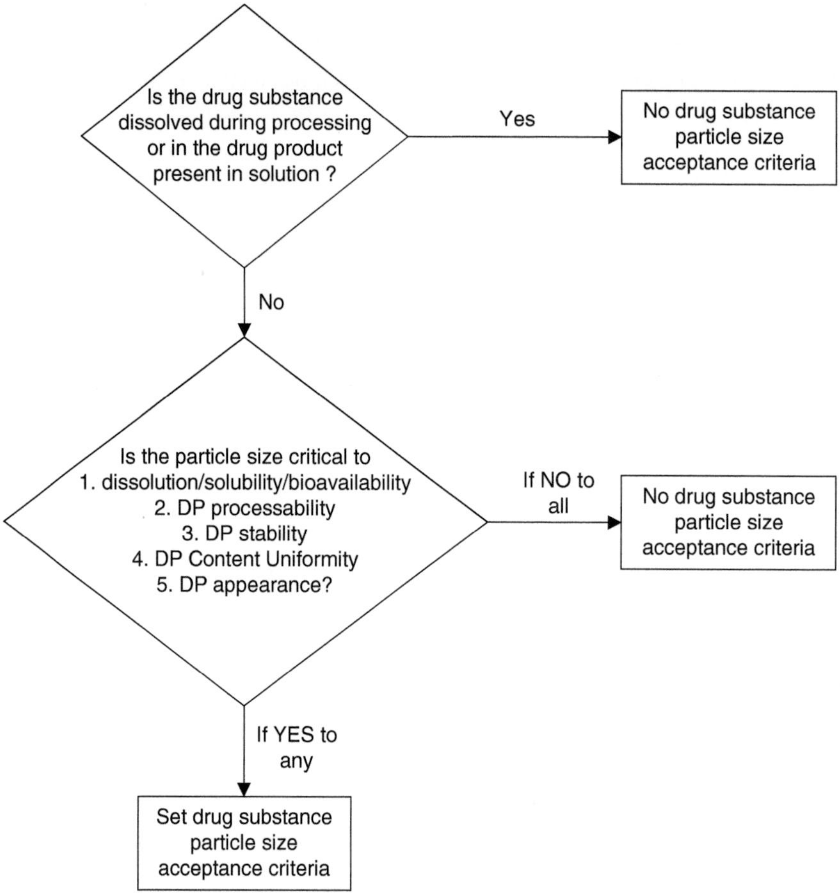

Figure 6.5 Adapted and modified decision tree from [157] to set acceptance criteria for drug substance particle size. Source: ICH [158].

drug applications (ANDAs). The following paragraph summarizes a few regulatory requirements, but it should not be considered complete, as regulations can differ between different regions and countries.

Key aspects to be considered during pharmaceutical development are outlined in ICH Q8(R2) [2]. The guideline describes the suggested content for the 3.2.P.2. (Pharmaceutical Development) section of a regulatory submission to the European Medicines Agency (EMA), the Pharmaceuticals and Medical Device Agency in Japan and the FDA in the USA. As an example, physico-chemical parameters and biological properties should be mentioned in the drug substance chapter (3.2.S.), which can influence the performance of the drug product or its manufacturability. For inhaled products for instance the respirable fraction should be determined, and aerodynamic properties of the particles are considered to be a CQA.

In addition, EMA has clearly stated [158] that a list of physico-chemical and other relevant properties should be provided under general properties (3.2.S.1.3.) in a marketing authorization application (MAA). In particular, those properties should be mentioned that affect pharmacological efficacy and toxicological safety, such as solubility, acid dissociation constant (pK_a) and polymorphism. Information on the proposed commercial solid state form including the potential for forming polymorphs, their solubility and information on the particle size (distribution) need to be presented in the same section (3.2.S.3.1.), and should be linked to the in vivo performance of the finished drug product (3.2.P.2.1.). Furthermore, all steps, which have an impact on the solid state properties and homogeneity of the active substance, are considered critical. This is particularly true, if the active substance is used within a solid dosage form, since, as a consequence, dissolution and thereby BA might be affected. The control of the critical steps and intermediates need to be described in 3.2.S.2.4. Depending on the nature of the drug substance and the impact on the BA and quality of the product, specifications for the polymorphic form and particle size need to be provided in 3.2.S.4.1.

Due to the impact of the polymorphic form on quality, safety, and efficacy of the drug product, FDA has also provided nonbinding recommendations for ANDAs [156]. It is recommended that the applicant assesses the sameness when the drug substance exists in different polymorphic forms. Accordingly, each ANDA applicant is required to demonstrate that the drug product exhibits sufficient stability and it is bioequivalent to the reference listed drug. Interestingly, FDA does not require that the drug substance is present in the same polymorphic form as the drug substance in the reference listed drug. The likelihood of being successful in showing bioequivalence (BE) in vivo is linked to the BCS. For drugs with dissolution rate limited absorption and a large difference in the apparent solubility of various polymorphic forms, an impact on BA/BE is expected, whereas the difference in apparent solubility does not play a major role in case of permeability limited absorption. The same is true, if the apparent solubilities of the polymorphic forms are sufficiently high and dissolution is rapid.

Besides of BA/BE, the impact of different polymorphic forms on manufacturability and stability needs to be assessed. The solid state properties of the drug substance are in particular critical for direct compression, whereas the solid state properties of the active ingredient are often altered by the process during wet granulation and they are therefore less likely to affect the manufacture of the drug product. In addition, polymorphic forms can undergo phase conversion when exposed to humidity, temperature, or stress during the manufacturing process. Nonetheless, this is not of concern, when the conversion occurs consistently as part of a validated manufacturing process, and when BA/BE has been demonstrated. In addition, recommendations are provided for monitoring and controlling polymorphs in the drug substance or drug product [156].

List of Abbreviations

A	surface area
AH	anhydrate
API	active pharmaceutical ingredient
ANDA	abbreviated new drug application
APSD	aerodynamic particle size distribution
BA	bioavailability
BCS	biopharmaceutics classification system
BE	bioequivalence
C_b	concentration in the bulk solution
C_s	saturation concentration
CAB	cohesion–adhesion balance
CMA	critical material attribute
CPP	critical process parameter
CQA	critical quality attribute
CSP	crystal structure prediction
d	thickness (of the diffusion layer)
D	diffusion coefficient
DCS	developability classification system
DDU	delivered dose uniformity
DMSO	dimethyl sulfoxide
DoE	design of experiments
DP	drug product
DPI	dry powder inhaler
DS	drug substance
EiH	entry into human
EMA	European Medicines Agency
F	fluorine
FDA	Food and Drug Administration
HH	hemihydrate
HLB	hydrophilic–lypophilic balance
HTS	high throughput screening
ICH	International Conference on Harmonization (of technical requirements for registration of pharmaceuticals for human use)
LBDDS	lipid-based drug delivery systems
LCFS	lipid formulation classification system
$\log P$	partition coefficient
m	mass
M	molecular weight
MAA	marketing authorization application
MCC	microcrystalline cellulose
MCS	manufacturing classification system
MP	melting point
MST	materials science tetrahedron

NDA	new drug application
NIR	near infrared (spectroscopy)
NSH	nonstoichiometric hydrate
PASS	partially automated solubility screening
pCMA	potential critical material attribute
pCPP	potential critical process parameter
PEG	polyethylene glycol
PIT	process-induced transformations
pK_a	acid dissociation constant
pMDIs	pressurized metered dose inhalers
QRA	quality risk assessment
QbD	quality by design
QTPP	quality target product profile
r	particle radius
R	universal gas constant
RH	relative humidity
S	solubility
SEDDS	self-emulsifying drug delivery systems
SMCC	silicified microcrystalline cellulose
SMEDDS	self-microemulsifying drug delivery systems
SORESOS	miniaturized assay for solubility and residual solid screening
ssNMR	solid-state nuclear magnetic resonance (spectroscopy)
t	time
T	temperature
γ	interfacial tension
ρ	density

References

1 Kuentz, M., Holm, R., and Elder, D.P. (2016). Methodology of oral formulation selection in the pharmaceutical industry. *European Journal of Pharmaceutical Sciences* 87: 136–163.

2 ICH (2009). ICH harmonised tripartite guideline - pharmaceutical development Q8(R2).

3 Charoo, N.A., Shamsher, A.A.A., Zidan, A.S., and Rahman, Z. (2012). Quality by design approach for formulation development: a case study of dispersible tablets. *International Journal of Pharmaceutics* 423 (2): 167–178.

4 Pallagi, E., Karimi, K., Ambrus, R. et al. (2016). New aspects of developing a dry powder inhalation formulation applying the quality-by-design approach. *International Journal of Pharmaceutics* 511 (1): 151–160.

5 Sun, C.C. (2009). Materials science tetrahedron—a useful tool for pharmaceutical research and development. *Journal of Pharmaceutical Sciences* 98 (5): 1671–1687.

6 Huang, L.-F. and Tong, W.-Q. (2004). Impact of solid state properties on developability assessment of drug candidates. *Advanced Drug Delivery Reviews* 56 (3): 321–334.
7 Aaltonen, J., Allesø, M., Mirza, S. et al. (2009). Solid form screening – a review. *European Journal of Pharmaceutics and Biopharmaceutics* 71 (1): 23–37.
8 Cerreia Vioglio, P., Chierotti, M.R., and Gobetto, R. (2017). Pharmaceutical aspects of salt and cocrystal forms of APIs and characterization challenges. *Advanced Drug Delivery Reviews* 117: 86–110.
9 Chan, H.K. and Doelker, E. (1985). Polymorphic transformation of some drugs under compression. *Drug Development and Industrial Pharmacy* 11 (2–3): 315–332.
10 Morris, K.R., Griesser, U.J., Eckhardt, C.J., and Stowell, J.G. (2001). Theoretical approaches to physical transformations of active pharmaceutical ingredients during manufacturing processes. *Advanced Drug Delivery Reviews* 48 (1): 91–114.
11 Li, X., Zhi, F., and Hu, Y. (2007). Investigation of excipient and processing on solid phase transformation and dissolution of ciprofloxacin. *International Journal of Pharmaceutics* 328 (2): 177–182.
12 Airaksinen, S., Karjalainen, M., Kivikero, N. et al. (2005). Excipient selection can significantly affect solid-state phase transformation in formulation during wet granulation. *AAPS PharmSciTech* 6 (2): E311–E322.
13 Picker, K.M. (2004). "Soft tableting": a new concept to tablet pressure sensitive materials. *Pharmaceutical Development and Technology* 9 (1): 107–121.
14 Wong, M.W.Y. and Mitchell, A.G. (1992). Physicochemical characterization of a phase change produced during the wet granulation of chlorpromazine hydrochloride and its effects on tableting. *International Journal of Pharmaceutics* 88 (1): 261–273.
15 Zhang, G.G.Z., Law, D., Schmitt, E.A., and Qiu, Y. (2004). Phase transformation considerations during process development and manufacture of solid oral dosage forms. *Advanced Drug Delivery Reviews* 56 (3): 371–390.
16 Alsenz, J., Meister, E., and Haenel, E. (2007). Development of a partially automated solubility screening (PASS) assay for early drug development. *Journal of Pharmaceutical Sciences* 96 (7): 1748–1762.
17 Wyttenbach, N., Alsenz, J., and Grassmann, O. (2007). Miniaturized assay for solubility and residual solid screening (SORESOS) in early drug development. *Pharmaceutical Research* 24 (5): 888–898.
18 Alsenz, J. and Kansy, M. (2007). High throughput solubility measurement in drug discovery and development. *Advanced Drug Delivery Reviews* 59 (7): 546–567.
19 Alsenz, J. and Haenel, E. (2003). Development of a 7-day, 96-well Caco-2 permeability assay with high-throughput direct UV compound analysis. *Pharmaceutical Research* 20 (12): 1961–1969.
20 Alsenz, J., Haenel, E., Anedda, A. et al. (2016). Miniaturized intrinsic dissolution screening (MINDISS) assay for preformulation. *European Journal of Pharmaceutical Sciences* 87: 3–13.

21 Carlson, E., Cong, P., Chandler, W.H. Jr. et al. (2003). An integrated high throughput workflow for pre-formulations: polymorph and salt selection studies. *Pharma Chemica* 2: 10–15.

22 Wyttenbach, N., Sutter, B., and Hidber, P. (2011). High-throughput strategies and techniques for salt screening. In: *Handbook of Pharmaceutical Salts: Properties, Selection, and Use*, 2e. Chapter 8 (eds. P.H. Stahl and C.G. Wermuth). Wiley-VCH.

23 Bastin, R.J., Bowker, M.J., and Slater, B.J. (2000). Salt selection and optimisation procedures for pharmaceutical new chemical entities. *Organic Process Research & Development* 4 (5): 427–435.

24 Fernández Casares, A., Nap, W.M., Ten Figás, G. et al. (2015). An evaluation of salt screening methodologies. *Journal of Pharmacy and Pharmacology* 67 (6): 812–822.

25 Chan, H.C.S., Kendrick, J., Neumann, M.A., and Leusen, F.J.J. (2013). Towards *ab initio* screening of co-crystal formation through lattice energy calculations and crystal structure prediction of nicotinamide, isonicotinamide, picolinamide and paracetamol multi-component crystals. *CrystEngComm* 15 (19): 3799–3807.

26 Childs, S.L., Rodríguez-Hornedo, N., Reddy, L.S. et al. (2008). Screening strategies based on solubility and solution composition generate pharmaceutically acceptable cocrystals of carbamazepine. *CrystEngComm* 10 (7): 856–864.

27 Collman, B.M., Miller, J.M., Seadeek, C. et al. (2013). Comparison of a rational vs. high throughput approach for rapid salt screening and selection. *Drug Development and Industrial Pharmacy* 39 (1): 29–38.

28 He, Y., Orton, E., and Yang, D. (2018). The selection of a pharmaceutical salt—the effect of the acidity of the counterion on its solubility and potential biopharmaceutical performance. *Journal of Pharmaceutical Sciences* 107 (1): 419–425.

29 Karagianni, A., Malamatari, M., and Kachrimanis, K. (2018). Pharmaceutical cocrystals: new solid phase modification approaches for the formulation of APIs. *Pharmaceutics* 10 (1): 18.

30 Korn, C. and Balbach, S. (2014). Compound selection for development – is salt formation the ultimate answer? Experiences with an extended concept of the "100mg approach". *European Journal of Pharmaceutical Sciences* 57: 257–263.

31 Morissette, S.L., Almarsson, Ö., Peterson, M.L. et al. (2004). High-throughput crystallization: polymorphs, salts, co-crystals and solvates of pharmaceutical solids. *Advanced Drug Delivery Reviews* 56 (3): 275–300.

32 Musumeci, D., Hunter, C.A., Prohens, R. et al. (2011). Virtual cocrystal screening. *Chemical Science* 2 (5): 883–890.

33 Yamashita, H., Hirakura, Y., Yuda, M. et al. (2013). Detection of cocrystal formation based on binary phase diagrams using thermal analysis. *Pharmaceutical Research* 30 (1): 70–80.

34 Price, S.L. (2014). Predicting crystal structures of organic compounds. *Chemical Society Reviews* 43 (7): 2098–2111.

35 Price, S.L., Braun, D.E., and Reutzel-Edens, S.M. (2016). Can computed crystal energy landscapes help understand pharmaceutical solids? *Chemical Communications* 52 (44): 7065–7077.

36 Price, S.L. (2018). Control and prediction of the organic solid state: a challenge to theory and experiment. *Proceedings of the Royal Society A: Mathematical, Physical and Engineering Sciences* 474 (2217): 1–16.

37 Reilly, A.M., Cooper, R.I., Adjiman, C.S. et al. (2016). Report on the sixth blind test of organic crystal structure prediction methods. *Acta Crystallographica Section B* 72 (4): 439–459.

38 Censi, R. and Di Martino, P. (2015). Polymorph impact on the bioavailability and stability of poorly soluble drugs. *Molecules* 20 (10): 18759–18776.

39 Stahly, G.P. (2007). Diversity in single- and multiple-component crystals. The search for and prevalence of polymorphs and cocrystals. *Crystal Growth & Design* 7 (6): 1007–1026.

40 Tilbury, C.J., Chen, J., Mattei, A. et al. (2018). Combining theoretical and data-driven approaches to predict drug substance hydrate formation. *Crystal Growth & Design* 18 (1): 57–67.

41 Kuentz, M. (2008). Drug absorption modeling as a tool to define the strategy in clinical formulation development. *The AAPS Journal* 10 (3): 473–479.

42 Ayad, M.H. (2015). Rational formulation strategy from drug discovery profiling to human proof of concept. *Drug Delivery* 22 (6): 877–884.

43 Reagan-Shaw, S., Nihal, M., and Ahmad, N. (2008). Dose translation from animal to human studies revisited. *The FASEB Journal* 22 (3): 659–661.

44 Alsenz, J. (2012). The impact of solubility and dissolution assessment on formulation strategy and implications for oral drug disposition. In: *Encyclopedia of Drug Metabolism and Interactions* (ed. A.V. Lyubimov), 1–70. Hoboken, NJ: Wiley.

45 Amidon, G.L., Lennernäs, H., Shah, V.P., and Crison, J.R. (1995). A theoretical basis for a biopharmaceutical drug classification: the correlation of in vitro drug product dissolution and in vivo bioavailability. *Pharmaceutical Research* 12: 413–420.

46 Butler, J.M. and Dressman, J.B. (2010). The developability classification system: application of biopharmaceutics concepts to formulation development. *Journal of Pharmaceutical Sciences* 99 (12): 4940–4954.

47 Raijada, D., Cornett, C., and Rantanen, J. (2013). A high throughput platform for understanding the influence of excipients on physical and chemical stability. *International Journal of Pharmaceutics* 453 (1): 285–292.

48 Raijada, D., Müllertz, A., Cornett, C. et al. (2014). Miniaturized approach for excipient selection during the development of oral solid dosage form. *Journal of Pharmaceutical Sciences* 103 (3): 900–908.

49 Edman, P. (1994). Pharmaceutical formulations--suspensions and solutions. *Journal of Aerosol Medicine: The Official Journal of the International Society for Aerosols in Medicine* 7 (Suppl 1): S3–S6.

50 Kulshreshtha, A.K., Singh, O.N., and Wall, G.M. (2010). *Pharmaceutical Suspensions – From Formulation Development to Manufacturing*. New York: Springer-Verlag.

51 van der Vossen, A.C., van der Velde, I., Smeets, O.S.N.M. et al. (2016). Design and stability study of an oral solution of amlodipine besylate for pediatric patients. *European Journal of Pharmaceutical Sciences* 92: 220–223.

52 van der Vossen, A.C., van der Velde, I., Smeets, O.S.N.M. et al. (2017). Formulating a poorly water soluble drug into an oral solution suitable for paediatric patients; lorazepam as a model drug. *European Journal of Pharmaceutical Sciences* 100: 205–210.

53 Yalkowsky, S.H. (1999). *Solubility and Solubilization in Aqueous Media*. New York: Oxford University Press.

54 Bak, A., Gore, A., Yanez, E. et al. (2008). The co-crystal approach to improve the exposure of a water-insoluble compound: AMG 517 sorbic acid co-crystal characterization and pharmacokinetics. *Journal of Pharmaceutical Sciences* 97 (9): 3942–3956.

55 Thakral, N.K. and Kelly, R.C. (2017). Salt disproportionation: a material science perspective. *International Journal of Pharmaceutics* 520 (1): 228–240.

56 Brewster, M.E. and Loftsson, T. (2007). Cyclodextrins as pharmaceutical solubilizers. *Advanced Drug Delivery Reviews* 59: 645–666.

57 Lu, G.W., Hawley, M., Smith, M. et al. (2006). Characterization of a novel polymorphic form of celecoxib. *Journal of Pharmaceutical Sciences* 95 (2): 305–317.

58 Möschwitzer, J.P. (2013). Drug nanocrystals in the commercial pharmaceutical development process. *International Journal of Pharmaceutics* 453 (1): 142–156.

59 Kesisoglou, F., Panmai, S., and Wu, Y. (2007). Nanosizing – oral formulation development and biopharmaceutical evaluation. *Advanced Drug Delivery Reviews* 59 (7): 631–644.

60 Yagnesh, T.N.S. and Kumar Rada, S. (2016). Pharmaceutical suspensions: patient compliance oral dosage forms. *World Journal of Pharmacy and Pharmaceutical Sciences* 5 (12): 1471–1537.

61 Gift, A.D., Hettenbaugh, J.A., Quandahl, R.A., and Mapes, M. (2018). Inhibition of the solid state transformation of carbamazepine in aqueous solution: impact of polymeric properties. *Pharmaceutical Development and Technology* 23 (9): 933–940.

62 Morri, M., Castellano, P., Leonardi, D., and Vignaduzzo, S. (2018). First development, optimization, and stability control of a pediatric oral atenolol formulation. *AAPS PharmSciTech* 19 (4): 1781–1788.

63 Bravo, P., Bertin, L., Pinon, A. et al. (2019). Development and stability of an oral suspension of procarbazine in pediatrics. *Journal of Drug Delivery Science and Technology* 49: 107–112.

64 Friciu, M., Roullin, V.G., and Leclair, G. (2017). Stability of gabapentin in extemporaneously compounded oral suspensions. *PLoS One* 12 (4): e0175208.

65 Vrignaud, S., Briot, T., Launay, A. et al. (2015). Design and stability study of a paediatric oral solution of methotrexate 2mg/ml. *International Journal of Pharmaceutics* 487 (1): 270–273.

66 Aitipamula, S., Wong, A.B.H., and Kanaujia, P. (2018). Evaluating suspension formulations of theophylline cocrystals with artificial sweeteners. *Journal of Pharmaceutical Sciences* 107 (2): 604–611.

67 Zietsman, S., Kilian, G., Worthington, M., and Stubbs, C. (2007). Formulation development and stability studies of aqueous metronidazole benzoate suspensions containing various suspending agents. *Drug Development and Industrial Pharmacy* 33 (2): 191–197.

68 Sheth, P., Sandell, D., Conti, D.S. et al. (2017). Influence of formulation factors on the aerosol performance of suspension and solution metered dose inhalers: a systematic approach. *The AAPS Journal* 19 (5): 1396–1410.

69 Ticehurst, M.D. and Marziano, I. (2015). Integration of active pharmaceutical ingredient solid form selection and particle engineering into drug product design. *Journal of Pharmacy and Pharmacology* 67 (6): 782–802.

70 Weda, M., Zanen, P., AHD, B. et al. (2008). The therapeutic index of locally acting inhaled drugs as a function of their fine particle mass and particle size distribution – a literature review. *Current Drug Delivery* 5 (2): 142–147.

71 Allen, A., Bareille, P.J., and Rousell, V.M. (2013). Fluticasone furoate, a novel inhaled corticosteroid, demonstrates prolonged lung absorption kinetics in man compared with inhaled fluticasone propionate. *Clinical Pharmacokinetics* 52 (1): 37–42.

72 Adi, S., Adi, H., Chan, H.-K. et al. (2013). Effects of mechanical impaction on aerosol performance of particles with different surface roughness. *Powder Technology* 236: 164–170.

73 Chan, H.-K. and Gonda, I. (1995). Physicochemical characterization of a new respirable form of nedocromil. *Journal of Pharmaceutical Sciences* 84 (6): 692–696.

74 Hooton, J.C., Jones, M.D., Harris, H. et al. (2008). The influence of crystal habit on the prediction of dry powder inhalation formulation performance using the cohesive–adhesive force balance approach. *Drug Development and Industrial Pharmacy* 34 (9): 974–983.

75 Leane, M., Pitt, K., and Reynolds, G. (2015). A proposal for a drug product manufacturing classification system (MCS) for oral solid dosage forms. *Pharmaceutical Development and Technology* 20 (1): 12–21.

76 Leane, M., Pitt, K., Reynolds, G.K. et al. (2018). Manufacturing classification system in the real world: factors influencing manufacturing process choices for filed commercial oral solid dosage formulations, case studies from industry and considerations for continuous processing. *Pharmaceutical Development and Technology* 23 (10): 964–977.

77 Wenzel, T., Stillhart, C., Kleinebudde, P., and Szepes, A. (2017). Influence of drug load on dissolution behavior of tablets containing a poorly water-soluble drug: estimation of the percolation threshold. *Drug Development and Industrial Pharmacy* 43 (8): 1265–1275.

78 Ahmed, H. and Shah, N. (2000). Formulation of low dose medicines – theory and practice. *American Pharmaceutical Review* 3 (3): 1–5.

79 EMA/CPMP (2001). Note for guidance on process validation. CPMP/QWP/848/96.

80 Desai, D., Li, D., Harianawala, A. et al. (2013). Solubilization of entecavir by povidone to overcome content uniformity challenges for low-dose tablet formulations. *Pharmaceutical Development and Technology* 18 (6): 1305–1313.

81 Rohrs, B.R., Amidon, G.E., Meury, R.H. et al. (2006). Particle size limits to meet USP content uniformity criteria for tablets and capsules. *Journal of Pharmaceutical Sciences* 95 (5): 1049–1059.

82 Zhang, Y. and Johnson, K.C. (1997). Effect of drug particle size on content uniformity of low-dose solid dosage forms. *International Journal of Pharmaceutics* 154 (2): 179–183.

83 Mirza, S., Miroshnyk, I., Rantanen, J. et al. (2007). Solid-state properties and relationship between anhydrate and monohydrate of baclofen. *Journal of Pharmaceutical Sciences* 96 (9): 2399–2408.

84 Kirchmeyer, W., Wyttenbach, N., Alsenz, J., and Kuentz, M. (2015). Influence of excipients on solvent-mediated hydrate formation of piroxicam studied by dynamic imaging and fractal analysis. *Crystal Growth & Design* 15 (10): 5002–5010.

85 Wyttenbach, N., Birringer, C., Alsenz, J., and Kuentz, M. (2005). Drug-excipient compatibility testing using a high-throughput approach and statistical design. *Pharmaceutical Development and Technology* 10 (4): 499–505.

86 Kirchmeyer, W., Grassmann, O., Wyttenbach, N. et al. (2016). Miniaturized X-ray powder diffraction assay (MixRay) for quantitative kinetic analysis of solvent-mediated phase transformations in pharmaceutics. *Journal of Pharmaceutical and Biomedical Analysis* 131: 195–201.

87 Wikström, H., Carroll, W.J., and Taylor, L.S. (2008). Manipulating theophylline monohydrate formation during high-shear wet granulation through improved understanding of the role of pharmaceutical excipients. *Pharmaceutical Research* 25 (4): 923–935.

88 Airaksinen, S., Luukkonen, P., Jørgensen, A. et al. (2003). Effects of excipients on hydrate formation in wet masses containing theophylline. *Journal of Pharmaceutical Sciences* 92 (3): 516–528.

89 Jørgensen, A.C., Airaksinen, S., Karjalainen, M. et al. (2004). Role of excipients in hydrate formation kinetics of theophylline in wet masses studied by near-infrared spectroscopy. *European Journal of Pharmaceutical Sciences* 23 (1): 99–104.

90 Tong, H.H.Y., Chow, A.S.F., Chan, H.M. et al. (2010). Process-induced phase transformation of berberine chloride hydrates. *Journal of Pharmaceutical Sciences* 99 (4): 1942–1954.

91 Paisana, M.C., Wahl, M.A., and Pinto, J.F. (2017). An insight into the impact of polymers on the hydrate conversion of olanzapine form I in aqueous suspensions. *Journal of Pharmaceutical Sciences* 106 (7): 1786–1794.

92 Chakravarty, P., Govindarajan, R., and Suryanarayanan, R. (2010). Investigation of solution and vapor phase mediated phase transformation in thiamine hydrochloride. *Journal of Pharmaceutical Sciences* 99 (9): 3941–3952.

93 Gift, A.D., Luner, P.E., Luedeman, L., and Taylor, L.S. (2009). Manipulating hydrate formation during high shear wet granulation using polymeric excipients. *Journal of Pharmaceutical Sciences* 98 (12): 4670–4683.
94 Thakral, S., Garcia-Barriocanal, J., and Thakral, N.K. (2019). Effect of processing conditions and excipients on dehydration kinetics of sodium naproxen hydrate in formulation. *International Journal of Pharmaceutics* 557: 221–228.
95 Di Martino, P., Malaj, L., Censi, R., and Martelli, S. (2008). Physico-chemical and technological properties of sodium naproxen granules prepared in a high-shear mixer-granulator. *Journal of Pharmaceutical Sciences* 97 (12): 5263–5273.
96 Wikström, H., Marsac, P.J., and Taylor, L.S. (2005). In-line monitoring of hydrate formation during wet granulation using Raman spectroscopy. *Journal of Pharmaceutical Sciences* 94 (1): 209–219.
97 Juban, A., Briançon, S., and Puel, F. (2016). Processing-induced-transformations (PITs) during direct compression: impact of tablet composition and compression load on phase transition of caffeine. *International Journal of Pharmaceutics* 501 (1): 253–264.
98 Mah, P.T., Novakovic, D., Saarinen, J. et al. (2017). Elucidation of compression-induced surface crystallization in amorphous tablets using sum frequency generation (SFG) microscopy. *Pharmaceutical Research* 34 (5): 957–970.
99 Huang, C., Klinzing, G., Procopio, A. et al. (2019). Understanding compression-induced amorphization of crystalline posaconazole. *Molecular Pharmaceutics* 16 (2): 825–833.
100 Schmidt, A.G., Wartewig, S., and Picker, K.M. (2003). Potential of carrageenans to protect drugs from polymorphic transformation. *European Journal of Pharmaceutics and Biopharmaceutics* 56 (1): 101–110.
101 Schmidt, A.G., Wartewig, S., and Picker, K.M. (2004). Polyethylene oxides: protection potential against polymorphic transitions of drugs? *Journal of Raman Spectroscopy* 35 (5): 360–367.
102 Feeney, O.M., Crum, M.F., McEvoy, C.L. et al. (2016). 50 years of oral lipid-based formulations: provenance, progress and future perspectives. *Advanced Drug Delivery Reviews* 101: 167–194.
103 Savla, R., Browne, J., Plassat, V. et al. (2017). Review and analysis of FDA approved drugs using lipid-based formulations. *Drug Development and Industrial Pharmacy* 43 (11): 1743–1758.
104 Holm, R. (2019). Bridging the gaps between academic research and industrial product developments of lipid-based formulations. *Advanced Drug Delivery Reviews* 142: 118–127.
105 Pouton, C.W. (2006). Formulation of poorly water-soluble drugs for oral administration: physicochemical and physiological issues and the lipid formulation classification system. *European Journal of Pharmaceutical Sciences* 29 (3): 278–287.
106 Kuentz, M. (2012). Lipid-based formulations for oral delivery of lipophilic drugs. *Drug Discovery Today: Technologies* 9 (2): e97–e104.

107 Mu, H., Holm, R., and Müllertz, A. (2013). Lipid-based formulations for oral administration of poorly water-soluble drugs. *International Journal of Pharmaceutics* 453 (1): 215–224.

108 Porter, C.J.H., Trevaskis, N.L., and Charman, W.N. (2007). Lipids and lipid-based formulations: optimizing the oral delivery of lipophilic drugs. *Nature Reviews Drug Discovery* 6: 231.

109 Williams, H.D., Trevaskis, N.L., Charman, S.A. et al. (2013). Strategies to address low drug solubility in discovery and development. *Pharmacological Reviews* 65 (1): 315–499.

110 Stillhart, C., Dürr, D., and Kuentz, M. (2014). Toward an improved understanding of the precipitation behavior of weakly basic drugs from oral lipid-based formulations. *Journal of Pharmaceutical Sciences* 103 (4): 1194–1203.

111 Stillhart, C. and Kuentz, M. (2016). Trends in the assessment of drug supersaturation and precipitation in vitro using lipid-based delivery systems. *Journal of Pharmaceutical Sciences* 105 (9): 2468–2476.

112 Williams, H.D., Ford, L., Lim, S. et al. (2018). Transformation of biopharmaceutical classification system class I and III drugs into ionic liquids and lipophilic salts for enhanced developability using lipid formulations. *Journal of Pharmaceutical Sciences* 107 (1): 203–216.

113 Jain, N. and Yalkowsky, S.H. (2001). Estimation of the aqueous solubility I: application to organic nonelectrolytes. *Journal of Pharmaceutical Sciences* 90 (2): 234–252.

114 Göke, K., Lorenz, T., Repanas, A. et al. (2018). Novel strategies for the formulation and processing of poorly water-soluble drugs. *European Journal of Pharmaceutics and Biopharmaceutics* 126: 40–56.

115 Ditzinger, F., Price, D.J., Ilie, A.-R. et al. (2019). Lipophilicity and hydrophobicity considerations in bio-enabling oral formulations approaches – a PEARRL review. *Journal of Pharmacy and Pharmacology* 71 (4): 464–482.

116 Alskär, L.C., Porter, C.J.H., and Bergström, C.A.S. (2016). Tools for early prediction of drug loading in lipid-based formulations. *Molecular Pharmaceutics* 13 (1): 251–261.

117 Chen, X.-Q., Gudmundsson, O.S., and Hageman, M.J. (2012). Application of lipid-based formulations in drug discovery. *Journal of Medicinal Chemistry* 55 (18): 7945–7956.

118 Beaulieu, P.L., De Marte, J., Garneau, M. et al. (2015). A prodrug strategy for the oral delivery of a poorly soluble HCV NS5B thumb pocket 1 polymerase inhibitor using self-emulsifying drug delivery systems (SEDDS). *Bioorganic & Medicinal Chemistry Letters* 25 (2): 210–215.

119 Morgen, M., Saxena, A., Chen, X.-Q. et al. (2017). Lipophilic salts of poorly soluble compounds to enable high-dose lipidic SEDDS formulations in drug discovery. *European Journal of Pharmaceutics and Biopharmaceutics* 117: 212–223.

120 Sahbaz, Y., Williams, H.D., Nguyen, T.-H. et al. (2015). Transformation of poorly water-soluble drugs into lipophilic ionic liquids enhances oral drug exposure from lipid based formulations. *Molecular Pharmaceutics* 12 (6): 1980–1991.

121 Page, S., Maurer, R., and Wyttenbach, N. (2016). Structured development approach for amorphous systems. In: *Formulating Poorly Water Soluble Drugs*. Chapter 8 (eds. R.O. Williams III, A.B. Watts and D. Miller). Springer International Publishing.

122 Kiekens, F., Eelen, S., Verheyden, L. et al. (2012). Use of ordered mesoporous silica to enhance the oral bioavailability of ezetimibe in dogs. *Journal of Pharmaceutical Sciences* 101 (3): 1136–1144.

123 Shi, Q., Moinuddin, S.M., and Cai, T. (2019). Advances in coamorphous drug delivery systems. *Acta Pharmaceutica Sinica B* 9 (1): 19–35.

124 Aroso, I.M., Silva, J.C., Mano, F. et al. (2016). Dissolution enhancement of active pharmaceutical ingredients by therapeutic deep eutectic systems. *European Journal of Pharmaceutics and Biopharmaceutics* 98: 57–66.

125 Shah, N., Sandhu, H., Phuapradit, W. et al. (2012). Development of novel microprecipitated bulk powder (MBP) technology for manufacturing stable amorphous formulations of poorly soluble drugs. *International Journal of Pharmaceutics* 438: 53–60.

126 Verreck, G., Brewster, M.E., and Assche, I.V. (2012). The use of supercritical fluid technology to broaden the applicability of hot melt extrusion for drug delivery applications. *Bulletin Technique Gattefossé* 105: 28–42.

127 Sheth, P. and Sandhu, H. (2014). Amorphous solid dispersion using supercritical fluid technology. In: *Amorphous Solid Dispersions: Theory and Practice* (eds. N. Shah, H. Sandhu, D.S. Choi, et al.), 579–591. New York: Springer New York.

128 Nagy, Z.K., Balogh, A., Démuth, B. et al. (2015). High speed electrospinning for scaled-up production of amorphous solid dispersion of itraconazole. *International Journal of Pharmaceutics* 480 (1): 137–142.

129 Ellenberger, D.J., Miller, D.A., and Williams, R.O. (2018). Expanding the application and formulation space of amorphous solid dispersions with KinetiSol®: a review. *AAPS PharmSciTech* 19 (5): 1933–1956.

130 Singh, A., Bharati, A., Frederiks, P. et al. (2016). Effect of compression on the molecular arrangement of itraconazole–soluplus solid dispersions: induction of liquid crystals or exacerbation of phase separation? *Molecular Pharmaceutics* 13 (6): 1879–1893.

131 Singh, A., Van Humbeeck, J., and van den Mooter, G. (2014). A new twist in the old story-can compression induce mixing of phase separated solid dispersions? A case study of spray-dried miconazole-PVP VA64 solid dispersions. *Pharmaceutical Research* 31 (11): 3191–3200.

132 Singh, A., De Bisschop, C., Schut, H. et al. (2015). Compression effects on the phase behaviour of miconazole-poly (1-vinylpyrrolidone-co-vinyl acetate) solid dispersions—role of pressure, dwell time, and preparation method. *Journal of Pharmaceutical Sciences* 104 (10): 3366–3376.

133 Ayenew, Z., Paudel, A., and van den Mooter, G. (2012). Can compression induce demixing in amorphous solid dispersions? A case study of naproxen-PVP K25. *European Journal of Pharmaceutics and Biopharmaceutics* 81: 207–213.

134 Ayenew, Z., Paudel, A., Rombaut, P., and van den Mooter, G. (2012). Effect of compression on non-isothermal crystallization behaviour of amorphous indomethacin. *Pharmaceutical Research* 29: 2489–2498.

135 Worku, Z., Aarts, J., and van den Mooter, G. (2013). Influence of compression forces on the physical and structural sttability of solid dispersions.

136 Newman, A. (2015). *Pharmaceutical Amorphous Solid Dispersions*, 296. Hoboken, NJ: Wiley.

137 Shah, N., Sandhu, H., Soon Choi, D. et al. (2014). *Amorphous Solid Dispersions: Theory and Practice*. New York: Springer.

138 Chaudhury, A., Wu, H., Khan, M., and Ramachandran, R. (2014). A mechanistic population balance model for granulation processes: effect of process and formulation parameters. *Chemical Engineering Science* 107: 76–92.

139 Kayrak-Talay, D., Dale, S., Wassgren, C., and Litster, J. (2013). Quality by design for wet granulation in pharmaceutical processing: assessing models for a priori design and scaling. *Powder Technology* 240: 7–18.

140 Matji, A., Donato, N., Gagol, A. et al. (2019). Predicting the critical quality attributes of ibuprofen tablets via modelling of process parameters for roller compaction and tabletting. *International Journal of Pharmaceutics* 565: 209–218.

141 Mazor, A., Orefice, L., Michrafy, A. et al. (2018). A combined DEM & FEM approach for modelling roll compaction process. *Powder Technology* 337: 3–16.

142 Reynolds, G.K., Campbell, J.I., and Roberts, R.J. (2017). A compressibility based model for predicting the tensile strength of directly compressed pharmaceutical powder mixtures. *International Journal of Pharmaceutics* 531 (1): 215–224.

143 Van Hauwermeiren, D., Verstraeten, M., Doshi, P. et al. (2019). On the modelling of granule size distributions in twin-screw wet granulation: calibration of a novel compartmental population balance model. *Powder Technology* 341: 116–125.

144 Pirttimäki, J., Laine, E., Ketolainen, J., and Paronen, P. (1993). Effects of grinding and compression on crystal structure of anhydrous caffeine. *International Journal of Pharmaceutics* 95 (1): 93–99.

145 EMA/CPMP (1998). Note for guidance on development pharmaceutics. CPMP/QWP/155/96.

146 Chakravarty, P., Suryanarayanan, R., and Govindarajan, R. (2012). Phase transformation in thiamine hydrochloride tablets: influence on tablet microstructure, physical properties, and performance. *Journal of Pharmaceutical Sciences* 101 (4): 1410–1422.

147 Koradia, V., Fontelonga de Lemos, A.F., Allesø, M. et al. (2011). Phase transformations of amlodipine besylate solid forms. *Journal of Pharmaceutical Sciences* 100 (7): 2896–2910.

148 Reddy, J.P., Jones, J.W., Wray, P.S. et al. (2018). Monitoring of multiple solvent induced form changes during high shear wet granulation and drying processes using online Raman spectroscopy. *International Journal of Pharmaceutics* 541 (1): 253–260.

149 Jørgensen, A., Rantanen, J., Karjalainen, M. et al. (2002). Hydrate formation during wet granulation studied by spectroscopic methods and multivariate analysis. *Pharmaceutical Research* 19 (9): 1285–1291.

150 Jørgensen, A.C., Luukkonen, P., Rantanen, J. et al. (2004). Comparison of torque measurements and near-infrared spectroscopy in characterization of a wet granulation process. *Journal of Pharmaceutical Sciences* 93 (9): 2232–2243.

151 Hédoux, A., Guinet, Y., Paccou, L. et al. (2013). Polymorphic transformation of anhydrous caffeine upon grinding and hydrostatic pressurizing analyzed by low-frequency Raman spectroscopy. *Journal of Pharmaceutical Sciences* 102 (1): 162–170.

152 Thakral, N.K., Thakral, S., Stephenson, G.A. et al. (2019). Compression-induced polymorphic transformation in tablets: role of shear stress and development of mitigation strategies. *Journal of Pharmaceutical Sciences* 108 (1): 476–484.

153 Koivisto, M., Heinänen, P., Tanninen, V., and Lehto, V.-P. (2006). Depth profiling of compression-induced disorders and polymorphic transition on tablet surfaces with grazing incidence X-ray diffraction. *Pharmaceutical Research* 23: 813–820.

154 Juban, A., Briancon, S., and Puel, F. (2016). Processing-induced-transformations (PITs) during direct compression: impact of compression speeds on phase transition of caffeine. *Drug Development and Industrial Pharmacy* 42 (11): 1857–1864.

155 ICH (2008). ICH harmonised tripartite guideline - pharmaceutical quality system Q10.

156 FDA (2007). Guidance for industry. ANDAs: pharmaceutical solid polymorphism.

157 ICH (1999). ICH harmonised tripartite guideline - specifications: test procedures and acceptance criteria for new drug substance and new drug products: chemical substances Q6A.

158 EMA/CHMP (2016). Guideline on the chemistry of active substances. EMA/454576/2016.

7

Workflow Management

Christian Große

Grünenthal GmbH, Chemical Development, Zieglerstr. 6, 52099 Aachen, Germany

7.1 Motivation

In the pharmaceutical industry, screening for polymorphs, salts, and cocrystals as well as the development of a crystallization process of an active ingredient (active pharmaceutical ingredient [API]) solid form is part of the drug development process. Thereby, mostly crystallization is a crucial step in the production of a solid compound. Therefore, such development is driven forward in interdisciplinary teams using many workflows, and thus, many interfaces are created. This chapter is about how development projects enter the solid-state group and how the development work can be organized. In this context, the topic of workflows and workflow management plays a key role. Since, from a personal perspective, not yet adequately published and accordingly its importance has not been acknowledged, the intention is to shed more light on the topics of workflow management [1, 2]. This chapter intends to serve as a basis for discussion and does not claim to have a full-scope view of all processes in the field of solid-state development. Heading a solid-state laboratory in the pharmaceutical industry for several years is a background allowing for, however subjective, presentation of some experiences in pharmaceutical research. The intention is to highlight crucial aspects and sensitize the topic of workflow management in connection with solid-state development. Furthermore, giving suggestions to rethink established workflows and encourage to allow new workflows in order to make development more efficient by implementing automated processes.

7.2 Workflow Management

Research in the pharmaceutical industry takes place in a highly regulated environment. Projects are often carried out under high entrepreneurial risk and time pressure. Frequently, repetitive processes are involved. It is therefore important to ensure the efficient use of resources. The goal should be to consider individual processes and optimize workflows. Automation, sequencing, and parallelization of workflows can help to overcome this challenge.

Solid State Development and Processing of Pharmaceutical Molecules: Salts, Cocrystals, and Polymorphism, First Edition. Edited by Michael Gruss.

A workflow is the sequence of individual actions within a business process, while a business process represents a higher-level process. Typical examples of business processes are the procurement process or the recruiting process. The workflow, in turn, is a subordinate subprocess at the operational level. Such a workflow is divided into a beginning, a process flow, and a defined end. It is important to plan the workflows that are run through effectively, to logically link them and to coordinate them with each other. Individual activities often depend on each other and build on each other. Thus, a process can only be started after completion of another process.

In the context of Workflow Management, it is therefore necessary to examine how processes are related and which prerequisites must be fulfilled in order to be able to carry them out. By skillfully organizing processes with organizational or technical solutions, processes can be coordinated with each other and bottlenecks can be prevented. A workflow management system can also help to support the optimization of workflows. By reorganizing existing workflows, the goal can be achieved to increase reliability and quality and avoid confusion. This is an important aspect especially in an environment with many operators in different laboratories, even beyond department boundaries. Thus, the optimization of workflows helps to prevent redundancies in the correct distribution of tasks and information and thus shortens operation times. Workflow management makes workflows and decision processes transparent and thus more comprehensible. Thereby documentation is an important aspect. Uniform processes also serve to increase quality and information availability and prevent media disruptions. All these requirements can lead to a reduction in process costs and processing times. At the same time, workflows should be prevented from being perceived as a too rigid framework by operators. Therefore, operators must always be involved in workflow management to increase their acceptance. Also the handling of unforeseen or rare events should be considered.

The following chapter deals with workflows in a solid-state laboratory used for the development process of an API. Starting with an analysis of stakeholders, the integration and organization in research projects are presented. A further emphasis is on the description of specific workflows in a solid-state laboratory. Finally, support processes, i.e. demonstrate to what extend a solid-state laboratory can also support other functions.

7.3 Organization of Solid-State Development by Project Management

7.3.1 Stakeholders

When considering workflows in the field of solid-state development, you have to clarify which stakeholders are involved in this part of pharmaceutical development. There are interfaces with many different functions in a research-based pharmaceutical company.

For projects at an early stage of development, these are mainly

- drug development,
- medical chemistry,
- synthesis development,
- analytical development,
- preformulation, and
- formulation development.

For more advanced projects and marketed products, in addition the following departments get involved:

- production or manufacturing,
- quality assurance (QA),
- intellectual property (IP), and
- regulatory affairs.

7.3.2 CMC Project Management

With handing over development candidates from the Research Team and Drug Discovery to the Chemistry Manufacturing and Control (CMC) organization, the development of the solid form of a new drug candidate starts. All development activities are managed by the CMC team. This team is led by a project manager, who as domain expert reports to a core development team. The tasks of the CMC team include coordination of activities for the supply of active ingredients and formulations for all phases of development, including, among others, the development of a scalable drug synthesis process, the development of suitable analytical methods, the elucidation of the impurity profile, and the development of suitable formulations. For the development phase from candidate selection (CS) to a decision point (DP), a major goal of the CMC team is the provision of material for toxicological studies conducted both under non-GMP and/or GMP (good manufacturing practice) control. Many aspects of drug development focus on satisfying the regulatory requirements of drug licensing authorities. At this point, reference is made to the ICH Q3C [3], ICH Q6A [4], and ICH Q11 [5] guidelines, which guide the development of a new chemical entity (NCE) with respect to the solid form. In addition to project costs, existing capacities and available know-how have to be taken into account. The team decides whether individual development packages are carried out with internal resources or externally with contract research organizations (CROs) or contract development and manufacturing organizations (CDMOs). Basis of a campaign for the development of the solid form of an NCE is a project plan. In close consultation with the members of the CMC team, planning of solid-state activities takes place. The initiation phase is the collection of all known solid-state data related to an active ingredient. Early solubility data provide information on whether the free form of the API is sufficiently soluble for the desired dosage form or whether it is necessary to change the form toward salts or cocrystals. The decision to begin with this investigation in depth is driven by the CMC team. Furthermore, in a risk-based approach, milestones and times for decisions are defined. Among others, there should be early awareness about possible solid forms of an active substance. At the same time, the scope of screening activities must not be too extensive. So, the study of polymorphism and the crystallization of salts and cocrystals must be done within a well-defined framework. In an early project status, the development work should be focused on the most promising activities, provided no project delaying results are obtained. Therefore, it is necessary to limit the crystallization studies of salts and cocrystals to a first selection of salt and cocrystallizing agents. The solid-form development work is the basis for the synthesis and crystallization of the final stage, during scale-up from the synthesis development laboratories to kg-scale laboratory and finally to pilot plant. Thus, it makes sense if the development of the solid form of the API and the development of its synthesis take place in parallel. The final step of drug substance synthesis is usually a crystallization step. It determines the properties of the final API used in further studies. Here both strands of development, synthesis and crystallization, come together. Outcomes of the crystallization studies and requirements for the solid form are routinely discussed in the CMC team. Decisive topics are, for instance, the indication, the administration route, and known data on the solubility of the active substance. Nevertheless, the requirements for the solid API form are often rather standardized and not project-specific: examining the free form of an NCE for polymorphic forms; identifying a stable, reproducible solid form, high yield, and purity with a manageable impurity profile; isometric particles; demonstrating options for salts and cocrystals. However, every compound is different and behaves differently. Therefore, every project has its own challenges.

7.3.3 Substance Requirement Plan

For better control of the required amount of material, a substance requirement plan should be created. In addition to the required substance quantities, the plan also contains

- a list of requisitioners,
- the date for the requirement,
- the required quality (non-GMP/GMP material), and
- the intended use.

Such an overview allows precise control and monitoring of synthesis campaigns for material supply and the delivery of material. Part of such a plan is therefore also the material supply for the solid-state laboratory delivered by the synthesis laboratories. If required, however, it is also possible to specify the supply with a specific polymorph, salt, or cocrystal form by the solid-state laboratory in coordination with the requesters. Responsibilities for crystallizing the demanded solid form depends on the individual organization; it might be synthesis or crystallization labs. Expertise, availabilities, and capabilities determine manufacturing.

7.3.4 Pre-CMC Data

In the very early stages of development, new compounds become most often purified by chromatographic methods and crystallization plays a minor role here. In addition, new APIs are typically used as solutions in early-stage investigations. Consequently, the solid state is usually not considered closer. Nevertheless, it may be useful for a full picture of the API's solid-state history to gain knowledge of the polymorphic form of the earliest batches as well. Whether X-ray powder diffraction (XRPD), Raman, or differential scanning calorimetry (DSC) data are used for characterization at this point is not crucial. These data can be a good starting point for later CMC activities. Therefore, they should also be available as raw data for further development and as part of a development candidate's patency package. In this way, they also contribute to the landscape of polymorphic forms.

7.4 Workflows in the Environment of the Crystallization Laboratory

7.4.1 Micro-Project Management

In order to control CMC activities, it is advisable to set up a project plan containing development milestones and dependencies on other activities in the project. This project plan is subject to a regular review cycle and will be adapted as necessary in consultation with the stakeholders. For the development of the solid form, specific work packages have to be planned in the laboratory. On the one hand, it must be ensured that the right work is carried out at the right time: micro-project management is the method of choice here. Most of the work is done for several different projects at the same time. However, only in the multi-project analysis of the laboratory activities one can plan resources. Downtime of devices or holiday periods must be taken into account. With the help of a generic subproject plan (Figure 7.1), the development process can be divided into individual processes. An adaptation takes

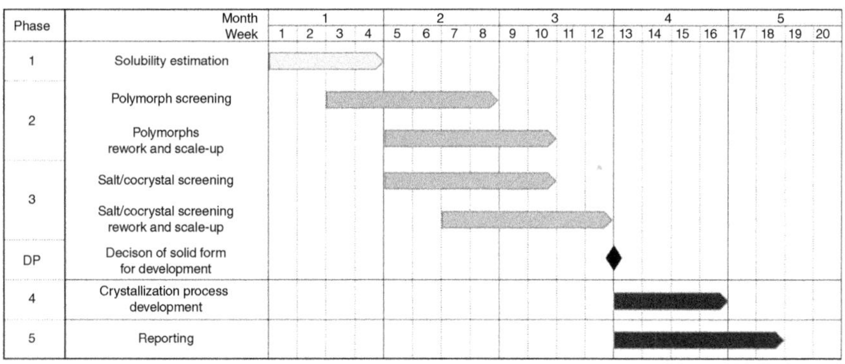

Figure 7.1 Generic subproject plan for controlling early solid-state development for a development candidate. A small team of two operators can handle a standard solid-state project in about 16 weeks. DP, decision point.

place depending on the project-specific requirements. This allows individual phases to be shorter or even longer if necessary.

The early development of an NCE in the crystallization laboratory can be divided into several phases. For a team of two operators, it is possible to develop the solid form of a moderately complex project up to the first milestone in about three months. This milestone is the decision on solid form for further development. Until this DP, all screening activities should be completed starting with the solubility estimation, the investigations on polymorphic forms and their scale-up as well as studies on salts and cocrystals, including their solvates and polymorphic forms. From the data obtained, an estimate of the landscape of polymorphic forms and options for salts and cocrystals can be made. Thus, taking into account the project-specific requirements, the solid form to run the further development phases can be determined. Ultimate goal is typically identification of the most stable solid form of the API. A later change of the form is in principle possible, however, causes additional activities and resources in the development and usually delays the project. If the most stable form could not be identified, a sufficiently chemically and physically stable form must be chosen based on a risk assessment. The CMC team collects all solid-state and related data to prepare the decision in the core development project team (CDPT). The first development activities in the crystallization laboratory are the initial investigation of the API, i.e. determination of particular properties such as initial solid state analytics of the received API batch, calculation of API's physical properties and estimation of solubility. In this phase, basic data on the active ingredient are collected and suitable solvents are selected for the following phases. Performing this initial phase needs about four weeks and two operators. The second stage of development includes the investigation and development of the free, i.e. neutral, form of a development candidate. Here polymorphs, solvates, and hydrates are identified and extensively characterized. Screening and scale-up takes approximately six weeks with two operators. By screening for polymorphic forms of the API, reference data is generated. Only with this reference data obtained in phase 2 is it meaningful to begin the investigation and development of salts and cocrystals. Identification of salts and cocrystals without knowledge of the polymorphism of the free form is extremely difficult. The third phase involves the identification of new salts and cocrystals. During scaling up appropriate forms, the analytical data set can be completed. Depending on the scope of the activities, it takes about six weeks for two operators to investigate salts and cocrystals. The screening phases end with the decision of the solid form for further development. Phase 4 serves to develop the crystallization process and is completed after about four weeks with a first laboratory protocol for the crystallization of the API. Finally, the data of solid-state development are summarized into development reports. This fifth phase can be estimated to take about six weeks. It can be started at the same time as Phase 4.

7.4.2 Dependencies

Process development and synthesis development are mutually dependent on each other. On the one hand, the solid-state development depends on material supply at the start of the campaign. Synthesis development provides most of this material from first batches. In individual cases, however, material might be taken from external suppliers. On the other hand, the synthesis development can only complete its synthesis development with the knowledge of the form to be developed and the provision of a crystallization process delivering this particular crystal form.

7.4.3 Material Flow

To process the experiments in the laboratory area, substances have to be exchanged between laboratories. There is a material input into the lab as well as a material output from the lab. For the transfer of substances, it is advisable to use a transfer document and to provide a substance data sheet. This allows tracking of when material was submitted or accepted. At least, the following information should be documented:

- compound name,
- sample and batch code,
- recipient,
- supplier,
- quality,
- quantity,
- date/time of submission, and
- safety information

In order to keep the efforts manageable, analytical samples should be handled differently. It is recommended to establish a safe and reliable, yet simplified, material and data transfer routine.

7.4.4 Designations and Code Assignment

Assignment of meaningful identifiers for samples, streams, and lab documents is an essential part of a functioning workflow in a well-organized laboratory. How to ensure that the exact history of a sample is linked to an analysis result? It is important to ensure the follow-up of samples and the correct assignment of analytical data to the corresponding experiments.

Basically, speaking and nonspeaking codes can be distinguished. As an example, a nonspeaking code can be some type of identifier, such as "12 345" or "ABCDE-12-34#5." A speaking code would be e.g. "2nd reaction of compound A with catalyst B in solvent X by operator Y for project Z".

On the one hand, talking codes apparently make it easy to assign samples to projects, labs, operators, and finally experiment parameters. On the other hand, such codes tend to be long and the system is not suited to address all available parameters or metadata, at all. In addition, such types of codes are not well standardizable and machine readable. Completely nonspeaking codes make communication in the lab and fast on-sight identification of samples difficult. Usually, they can be resolved only with the help of a sample management system [6] such as an electronic laboratory notebook (ELN) or a laboratory management information system (LIMS). To allow this they can be short and easily machine readable. In addition to pure text-based codes, often barcodes, 2D-codes, or radio-frequency identification (RFID) chips are in use on labels. Transient form of code allowing on-sight identification and data processing is recommended.

Depending on the intended use, different types of codes can be distinguished, such as

- project codes,
- substance codes,
- laboratory codes,
- experiment codes,
- parallel experiment codes,
- sample codes, and
- analytical codes.

Project codes should be centrally managed and commissioned through project teams. For the assignment of unambiguous one-to-one mapping substance codes, one should use a compound management system. Here you can differentiate between global and local substance codes. **Global substance codes** are intended for use in the whole company. They are useful for internal and external communication on business level. For daily use in different laboratory areas, it may be useful to use **local substance codes**. Polymorphs, salts, and cocrystals may also have assigned to their own substance codes. Therefore, the recommendation is to avoid a composite code made with API code and salt name. It might carry thread of confusion. This does not refer to and should not be mixed up with the formal INN (international nonproprietary name), because "An INN is usually designated for the active part of the molecule only, to avoid the multiplication of entries in cases where several salts, esters, etc. are actually used. In such cases, the user of the INN has to create a modified INN (INNM) himself; mepyramine maleate (a salt of mepyramine with maleic acid) is an example of an INNM" [7].

Laboratory codes are used to easily assign samples, batches, and documents to a laboratory area. As an example: for clarity, a substance code is combined with a prefix for laboratory assignment. The **experiment code** is a one-to-one identifier for an experiment. Each protocol has its own experiment code. Extension to an experimental code stem allows the designation of parallel experiments within a document. As **parallel experiment codes,** either enumeration, such as 1, 2, …, 96, etc. or the use of matrix identifiers, such as A1, B2, …, H12, etc. can be used.

Within an experiment, samples may be taken at different times, e.g. after filtration a sample can be taken from the filter cake and the mother liquor. For this purpose, a separate **sample code** should be assigned to each sample. This code uniquely references to an existing product of a synthesis or a downstream operation. Each further treatment of this material generates a new product that might have its own product code. Regardless of whether a new product code is assigned or not, the sample code should be uniquely assigned accordingly.

The final stage of the code assignment is the **analytic code**. Each sample can be examined with different analysis methods. For each of these types of analyses, a one-to-one analytic code should be assigned.

The codes explained earlier can be combined in cascades in the form: Substance Code – Experiment Code – Parallel Experiment Code – Sample Code – Analytical Code.

Generally recommended is a short speaking code that is standardized and well machine readable. You should refrain from combining codes with reaction or downstream conditions. These conditions should only be documented in the laboratory journal. A given analysis code references this sample.

7.4.5 Analytic Database System

In addition to the ELN [8], an analytical database (e.g. ACDLabs Spectrus [9]) is one of the central tools in solid-state development. This database system collects all analytical data on solid-state samples and can flexibly combine and present their analysis results. There are a number of innovative workflows around the analytic databases that make the handling and analysis of analytical data as well as the data flow between different software systems more efficient. An automated import, an automated evaluation, and an automated export of results to the ELN are implementable. A large amount of diverse raw analytical data is importable to the analytical database. To ensure reliable import, all file names may consistently be created having the respective analytic code in the filename and therefore refer uniquely to a sample documented in the ELN.

The automated import can be realized e.g. for nuclear magnetic resonance (NMR), high-performance liquid chromatography (HPLC), gas chromatography (GC), XRPD, DSC, thermogravimetric analysis (TGA), dynamic vapor sorption (DVS), infrared (IR), Raman, and image data files from various suppliers. An automated service has to be set up, which regularly checks an import folder in the filesystem for new data files. This import folder may be organized as part of a project-oriented folder structure. Thus, during import, the import directory can also be used as an information source for the names of projects and subprojects. Because many raw analytical data formats can be imported, it is recommended to differentiate raw data files via the file format (related to the file name extension) or via a separated identifier, such as "DSC," "TGA," or "DVS" in the filename. If new files are detected in the file system, a look-up routine in the ELN tries to match filename with the respective analytic code stored in the ELN. If the filename cannot be found in the ELN, the import service can parse the file name according to predefined rules. Once an entry is found, metadata about the experiment can be extracted from the ELN and transferred to the analytical database. These are, for example, substance code, experiment code, and parallel experiment code as well as the sample code or experiment parameters. During the import, all the various analytical data of a sample can automatically be combined into one record in the analytical database. This means that a record corresponds to a unique experiment sample, which makes it easy to search and display all solid-state (and further) analytics of a sample simultaneously.

Analytical data can be imported in two ways: as evaluated or not evaluated analysis. In case of the described system DSC, TGA, and NMR data are already evaluated by analytical operators before import. In contrast, because of the particular organization of the laboratory responsibilities, XRPD data is imported as raw data and evaluation takes place supported by the analytic database system. A machine learning (i.e. cluster) algorithm has been implemented, and during import, XRPD data is grouped to a cluster based on similarity. Each XRPD pattern is individually compared against all other XRPD data in the database. As long as similarity between the imported XRPD pattern and a known cluster is higher than a set threshold, the new XRPD becomes a member of this cluster. If no cluster could be found, a new cluster is created. The most representative XRPD for the cluster is marked as reference pattern. Besides many other parameters, cluster reference pattern is searchable. The cluster number is unique and consecutively numbered. Besides the cluster number, every cluster also gets a cluster name. After the import, new XRPD data are marked as "not supervised." An operator must confirm or correct the assignment to a cluster. This is done manually by visual comparison against known XRPD references. A parameter that can help the operator to identify pure phases and mixtures is the nearest-neighbor value. In a separate list, cluster numbers of the five best hits in similarity are shown. In addition to the cluster number, the cluster name can also be edited by the operator. Thus, names for polymorphs, solvates, hydrates, salts, and cocrystals are assigned. Basic functions for editing of analyses are available through the analytic database processor. Major functions are: similar analyses can be reanalyzed, superimposed, smoothed, zoomed, or pushed; peak tables can be generated; as well as labeling of certain signals is possible. Last, but not least, from the processor, an analysis of the whole database can be initiated to examine for the degree of similarity.

Although well implemented, there is much potential for improvement. Automated clustering of pure phases as well as of mixed phases usually works satisfactorily. However, the recognition of mixed phases as the sum of two or more pure phases is currently not addressed. Mixed phases must be identified manually. Machine learning, e.g. the cluster algorithm, could also be extended to other analysis methods. There is a lot of potential here for easier evaluation of analytic data and consequently a more efficient workflow. Also conceivable is an automated indexing of XRPD data and the categorization of images. Photographs of the analytical samples could be classified according to parameters such as crystallinity, crystal form, degree of agglomeration, or grain size and assigned to clusters. Eventually, these properties would be searchable, too.

If the analytical data are successfully imported and evaluated in the analytic database, they can be imported into the ELN. For this purpose, a look-up service database level is triggered by opening an ELN document. Based on the codes, an assignment of the analysis is done and data such as spectra or individual data point transferred to the ELN document.

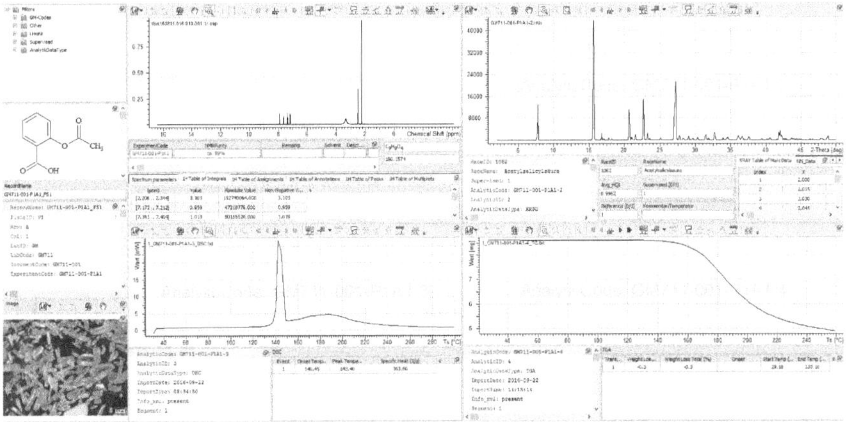

Figure 7.2 Solid-state screen form in the analytic database (ACDLabs) is shown. In a typical view for daily work, five different analytical results of a sample are shown: ^{1}H NMR, XRPD, DSC, TGA, and crystal image. Those data is stored into a single sample record. Source: Advanced Chemistry Development, Inc.

To visualize the analysis data in the user frontend of the analytic database system, several interfaces are available (Figure 7.2). Depending on configuration, either overviews of many analytical methods (e.g. NMR, XRPD, DSC, TGA, Image) or individual analysis methods can be presented. Selected XRPD pattern can be superimposed to check for individual differences within a cluster. The simple presentation of images of solids gives a good idea about particle properties. Combined view on TGA and DSC data gives a good overview about thermal events. Another big advantage of such a system is cross-project search for samples or results. Once imported, references of commercial compounds, reagents, or reactants can be used in many projects. All solid-state data from all development phases, regardless of operator or laboratory area, flows into the same system. Even complex searches and selections are feasible in such a database system.

7.4.6 Physical Sample Transfer

All samples for the solid-state investigation are accordingly labeled with the identifiers, i.e. analytic code and delivered straight to the solid-state analytics laboratory. The characterization lab becomes in addition informed by electronic transmission of measurement requests to a shared document.

7.4.7 Analytic Transfer Tool

For all analytical methods that are not solid-state analysis, a separate workflow has been established. In addition to samples for the NMR investigation, these include samples for chromatographic methods (HPLC, GC, liquid chromatography–mass spectrometry [LCMS]). Such analytical samples are registered in a separate software tool. With the help of the project code, the software searches for metadata such as project name and reaction scheme from the ELN system. Using an interactive input mask, analysis codes are created depending on the selected analysis method. A handover document contains the analysis code as human-readable text and as barcodes. Additional information, such as measurement parameters and explicit tasks can be specified for the analytic operators. Several analyses can be commissioned at the same time. The document is forwarded via e-mail to the particular laboratories. Appropriately, shared e-mail boxes are set up for this purpose.

Once the analysis order has been sent, the samples are transferred to the central office for analytical samples. From here, the samples are distributed to the analysis laboratories. When processing the analysis job, the sample name can be entered in the analysis software either manually or by scanner in the measurement software. Thus, a lean, automated process has been established, which also helps to avoid confusion and transmission errors. After evaluation of the analysis, documentation and report are carried out using established processes and are imported into a LIMS and the analytic database.

7.4.8 Analytical Processes – Timely Measurement

A timely measurement of analytical samples is a high demand on the analysis. In the development of the solid form of an API, much effort is spent on the screening and synthesis of polymorphs, salts, and cocrystals. A wide variety of synthetic methods, solvents, and parameters are applied. Obtained solids can be dried specifically or deliberately not dried. Thus, a large amount of samples is generated that must be analyzed.

Two aspects are important: on the one hand, the time span between sampling and generation of the analytical measurement data has to be considered. On the other hand, the measurement start times by using many different methods are sometimes far apart. In the waiting time for the analytical measurement, metastable forms may have already transformed into more stable solid forms. At the same time, samples of anhydrate may have transformed to solvates in residual moisture, or the other way round in dry atmosphere.

In order to keep the time between sampling and analysis as short as possible, samples should always be freshly analyzed. This means that after a sample has been generated, it is documented in the laboratory journal, the analytical methods are selected, and the analysis codes are generated. Sample is filled directly into labeled sample containers. The label for a sample container can be fixed as a sticker from a label printer. Such a label has distinct advantages over marking with pens: it is abrasion-resistant, solvent-resistant, and easy to read, for an operator and/or a scanner. For each analysis, a separate label with analytic code and analytical method should be pasted onto the analysis samples. Thus, the sample name does not have to be entered manually in the logbooks of the measuring instruments. It can be glued in and transmission errors or mistakes are avoided. At this point, it should be mentioned that it may be advantageous to prepare subsets of the sample for different analytical laboratories. It speeds up the sample processing because all laboratories can start at the same time and get comparable samples. If only one sample vessel circulates through all the analytical laboratories, this is not guaranteed. This is particularly critical, for example, for the measurement of particle sizes, as samples can segregate rapidly in terms of grain size. Depending on the number of analytical methods commissioned in a laboratory, less than 100 mg is generally required. A 2 ml HPLC vial with a wide neck has been found to be a very suitable vessel for the analytical samples. It offers sufficient capacity even with voluminous solids, and the wide neck makes filling and removal of material with a spatula very easy.

In many solid-state analyses, the measurement and evaluation method can be standardized. These include XRPD, Raman, DSC, TGA, and DVS. If there are no special questions, a fast measurement with limited angular range can be used for the measurement of the XRPD. The same applies to thermoanalysis and DVS. First measurements can be carried out quickly in standard settings. Only with conspicuous measurement results, the measurement should be adjusted. Interfaces, e.g. additional communication, between synthesis and analysis department are thus standardized and reduced to a minimum, analysis times are reduced, standardized reports can be used, and the comparability of the measurements is increased.

Already touched is another aspect: it must be taken into account that not all commissioned solid-state analyses are measured at the same time. The reasons for this can be manifold: different availability of the measuring instruments, duration of the measurement, use of autosamplers, time required for sample preparation, downtimes, and maintenance times. Most importantly, different measurement methods are usually processed in different laboratories by different lab teams. Here, the timely registration of the samples can help. This allows analysis times to be better planned and coordinated. Through regular joint meeting with the analysis team operators can be sensitized to the topic. At least you can get information about the exact measuring time by tracing the exact measuring time. In the case of discrepancies in the measurement results, differences in the measurement times could be reconstructed.

7.4.9 Sample Storage Processes

The concept of sample storage is difficult to grasp. Among others, the following questions arise on this topic, which can help to approach the topic:

- Is storage a process step?
- When can one speak of storage?
- Can depositing a sample in the lab overnight until analyzed the next day already be considered as a storage process?
- Should storage always be associated with the control of storage conditions?
- How to deal with reanalysis after a long period?
- When ends the drying period and when starts storage?

Storage of samples might be a problem. In the context of solid-state development, it may happen that batches are produced that are not stable after prolonged storage and recrystallize into other solid forms. All processes that expose a substance include a storage step that must be documented and evaluated. Through a regular reanalysis of API batches, including the batch used for solid-state development, consistent quality can be controlled. Changes can give indications of bearing instabilities of its solid form. Does the polymorph change? Is there a change to a new solid form or a mixture of forms? Are salts or cocrystals chemically and physically stable? If storage over a long period of time is intended, the temperature and humidity should be kept constant, if necessary controlled. A dry, dark, and cool environment is an advantage. Possible transformation processes are slowed down but not excluded. And in contrast, phase transformation might also be initiated by cold conditions. Beyond this, there is the ICH Q1A (R2) guidance [10] for controlled stability testing.

7.4.10 Documentation

The most complex processes in the laboratory include the documentary processes. This usually raises the question of what data and information should be documented and reported. On the one hand, a development phase often involves a large number of many individual but also parallel experiments. On the other hand, large amounts of raw data are collected in the various laboratories. In addition to process observations, measurement data, and individual analysis, these include above all images, film recordings, and signals from process probes. How should this large amount of data be meaningfully documented?

For reporting the results, one can clearly differentiate between documentation of raw data and reporting of results. A laboratory journal is used to log the experiments carried out in the laboratory. The times of the handwritten protocols are over. State of the art is the use of electronic laboratory journals, which is implemented in the described workflow as well. In addition to the operators of the solid-state group, the groups of synthesis development and analytical development also access the ELN and analytic database system. The aim is clearly a common data basis based on the protocols for the development of synthesis and crystallization. The ELN is a database system having functionalities such as structures drawing, chemical calculations, documentation, handling of analytical data, and import of external documents. In the ELN with the experiment planning with objective, the documentation of weighing results, used devices, observations, yield calculations, analytics, and finally documenting, a conclusion takes place. The ELN uses the data of the compound management system to assign experiment codes for new documents. The name of the operator, exact dates, and times as well as electronic signatures of the participants are obligatory and can be automatically logged in electronic systems. The protocol is the basis of legal documentation and supports ensuring IP for inventions. There is no room for speculation, assumptions, and blanket statements here. Well-defined templates, specifically targeted to map workflows, are starting point for the documentation. Implemented custom interfaces with predefined fields for administration of reactants, batch calculation, yield calculation, and analytic data simplify daily work. So, the system guides the operator through the documentation of the experiment by the sequence of tabs in the graphical user interface. An additional section for documentation in free text gives the user maximum flexibility for the detailed description of the work processes, observations, and results. Images, tables, and texts can also be imported from external sources and edited further here.

All these functionalities are collected in a document template. An automatic import of measurement results is realized for the analytic data tab as part of the workflow. To use the service, all analytic samples must be registered in a separate analytic section. Based on the analytic code, each individual analysis with type of analysis, sample code, and timestamp for delivery and receipt can be identified. When opening an ELN document, a look-up service compares the registered analytic samples with the results in the analytic database. On the one hand, it is possible to check whether the results of analyses are already available and at the same time to ensure that the analysis results are up-to-date. For mainly needed data, such as XRPD, DSC, TGA such a transfer service is implemented, i.e. machine-readable data is imported and readily transferred and available in sortable tables. For other types of analytics, the import is done manually.

The second form of documentation includes research reports, invention disclosures, and presentations. It is recommended to split research reports and write smaller phase-dependent reports. These are put together quickly and complete a development phase thematically. A solubility data report can be used to complete the solvent selection and synthesis support section (e.g. for cleaning purpose). All screening replication and scale-up experiments can be summarized in a report on polymorphic forms. Not only the pure number of polymorphs is documented, but also preparation pathways and representative analytic data can be found here. The report about API salts and cocrystals may be combined with the report on the polymorphic forms or be available as a single report. In terms of content, reports should refrain from a pure enumeration of raw data. A report thematically summarizes experiments, interprets the results, sets the path to gain insights, and draws conclusions for further experiments. Also useful is a list of process-relevant parameters in tabular form. The documentation of process data in movies is a real problem. While electronic presentations can be linked to movies, this is not possible in text documents. Here, the course of a crystallization can be represented by means of selecting representative images and adding them to the report. Furthermore, the following should be considered:

- What information does the customer need?
- Does the report have to be adapted to the course of the project or objective?

Corresponding templates for documents can be prepared. In an optimized workflow, data can be extracted from the ELN, depending on the requirements of the report. By specific searches for parameters e.g. synthesis, downstream processes, or analysis, a document-based compilation of results reduces the high proportion of manual work required to produce a report.

7.4.11 Review Process for ELN Documents

7.4.11.1 Document Status

All ELN documents go through several stages in a review cycle. The first stage is the “In Progress” phase, which begins with the creation of the document. Until the processing of the documentation is complete, only the author can add, modify,

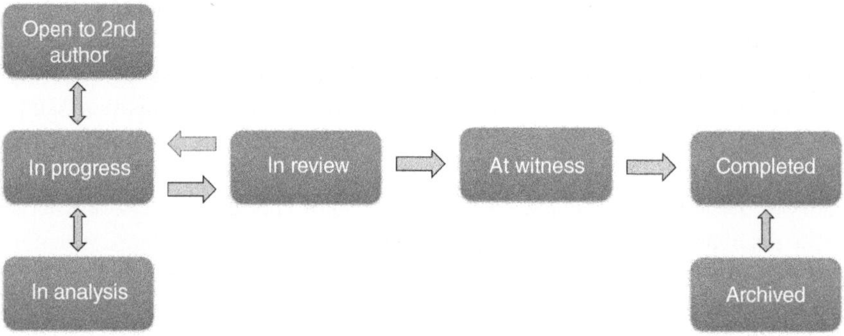

Figure 7.3 Example for ELN document status workflow.

and delete data in this document. Other ELN users only have read access to the document. If the document is in the "Open to second author" status, other editors can also be authorized to edit the document. Useful is this function for substitution of primary workers during holiday and sickness times or desired teamwork projects. Of course, an author can revoke allowance for teamwork. If an ELN document is almost finished and there are only some missing analytical data left, it can be set to the status "In Analysis." The document remains fully editable for the author. After completing the documentation, the document returns to the "In Progress" status.

The review cycle (Figure 7.3) is organized in such a way that two additional witnesses have to sign the document electronically. In the first step, a document in status "In Progress" is promoted to "In Review" and thereby automatically transferred to the inbox of the laboratory management. Here a thorough examination, i.e. review, of the content of the document takes place. When the review is complete, the document is set into the "At Witness" state. If the reviewer requests additions or changes, the document is reset to "In Progress" in order to allow amendment. All potential witnesses can see the document in this status. Based on a witness list within the department, the witnesses are assigned to a project. The witness should not work in the same project in order to preserve his independence. After the document has been checked and signed electronically, the document is set into the "Completed" state, thus closing the document. The last document status is "Archived," which is immediately set after an automated export of the document as portable document format (PDF) file to the file system. The review cycle is now complete. If it is necessary to edit a completed and archived document again, a document can be moved from the status "Archived" to the status "In Progress" and the review cycle starts again. All status changes are logged by an audit trail.

7.4.11.2 Manual ELN Review Process

Before establishing the process of electronic reviewing and witnessing ELN documents, there was a media break. The electronic documents in the ELN systems were printed to paper as short reports with specially implemented templates. These prints were glued into hardback paper lab books for archival purposes. Both the editors, laboratory supervisors, and one other witness had the task of signing the paper document for legal certainty: an unpopular and time-consuming job. The experiment documentation assembled into books was transferred to the archive via an archiving process and stored. If required, laboratory journals could also be retrieved back from the archive. Therefore, it must have been known which lab book would have been required. From the perspective of workflow management, these processes were laborious and offered much potential for improvement. Electronic systems show clear advantages here. They are fully available at all times and searchable by many aspects.

7.4.11.3 Archive Process

The subject of archiving concerns the development documents, data from the ELN, and raw data from the laboratory. Development documents can be stored as signed PDF files in project folders or project e-rooms. For approval-relevant documents, a document management system is recommend. An automated workflow has been established for the archiving of ELN data. Documents that have gone through the entire review cycle are automatically exported as PDF documents from the ELN system to an e-archive. This archive is a specific folder in the file system and saves the data for min. 30 years. Data security, data integrity, and readability over the entire lifetime of the documents are ensured by the central IT department. Raw data from laboratory automation systems can be archived via local data media or, even better, global information technology (IT) solutions. One should be aware that the access to raw data via corresponding measurement software must also be ensured for future retrieval. Relevant analytic samples should therefore be measured by the central analytic department. In that case, an established and validated LIMS system guarantees adequate archiving.

7.4.12 Communication with CROs

If capacities or know-how is not available in-house, projects or subprojects can be realized in cooperation with external partners. For the assignment, several processes have to be run through: supplier selection, budget processes, and contracting process. The content is provided in the form of clear communication of requirements and expectations. These include phone and conference calls, presentations, and requirement documents. On a regular basis, the current status of the project and its progress should be discussed in conference calls. In addition, written status reports are recommended in short intervals (two to four weeks), after completing a project phase or project milestone. When communicating with CROs and CDMOs, it should be borne in mind that, as a rule, raw data and protocols cannot be used to the same extent as internally processed projects. Usually only results in the form of a report as PDF or Word document are received. The joint use of an electronic laboratory journal would be desirable. However, it is possible to establish the same quality standards and to create a much larger data background to address questions in a later project phase as well. Online access or cloud solutions can help here. As far as the diversity of raw data formats and measurement methods is concerned, the exchange of data with external partners can be a challenge. Software solutions are not available internally for all raw data from measuring instruments. Therefore, it is advisable to agree on common data formats for the exchange. Furthermore, for the measuring methods used externally, an exact description should be available. Last, but not least, it should be considered whether externally generated measurement data should be included in the internal analytic database. This makes the data easily accessible via internal workflows and fits seamlessly into the suite of internal analytic data.

7.4.13 Fundamental Lab Processes

Workflows for the maintenance of the laboratory are the basis of all experimental work. They shall be mentioned here only for the sake of completeness. Basic requirements for the operation of a crystallization laboratory in the field of pharmaceutical development is a technically functioning laboratory. These include the infrastructure with electricity, water, technical gases, ventilation of the laboratories, but also the connection to the IT infrastructure and procedures for the supply and disposal of chemicals, ordering of equipment and consumables, i.e. supply chain and waste management. The use of the laboratory and ancillary rooms must be regulated in guidelines and standard operating procedures (SOP). A strict separation of laboratory area and evaluation or office area is urgently recommended. Here is the structural separation of advantage. Risk analysis documents and explosion protection documents are required for work processes and devices. Furnishings that serve the work safety must be regularly maintained. To ensure high quality of the experiments, it is recommended to regularly check, calibrate, and, if necessary, repair laboratory equipment. Software for operating reactors, measuring devices, and other laboratory equipment should be updated regularly. This includes operating systems, firmware, and device drivers. Although the crystallization laboratory is usually a non-GMP area, maintenance, calibration, and repair of equipment should be documented. For repetitive work processes, best practice documents and templates can be made. This allows quality standards for workflows to be defined. All employees must be educated and regularly trained according to their field of activity. Occupational safety measures and training should also be documented. For the incorporation of new employees into the laboratory processes, appropriate documents with workflow descriptions are recommended. As part of further education and further qualification, employees can become more familiar with the topic of solid-state development and develop into solid-state experts. Internal trainings by experts, specialist conferences, workshops, and white papers should be used here.

7.5 Processes in the Solid-State Lab

7.5.1 Initial Testing

Roughly speaking, the process of solid-state development can be subdivided into the following phases:

- initial analysis,
- screening phase,

- rework,
- scale-up phase, and
- development of a crystallization process.

In principle, all investigations should begin with the initial analysis of the substance batch that was provided for the studies. This is an important means of QA. A batch of the active substance should be handed over by means of a transfer document and substance data sheet. It may be necessary to train employees on how to handle the API at this point. Sources for this material may be

- internal substance management,
- direct delivery from medicinal chemistry laboratories and synthesis laboratories, or
- external sources.

The minimum API for polymorphism studies can be estimated to be between 2 and 10 g of material. The number of feasible experiments correlates directly with the available amount of substance. The content of the obtained batch should be as high as possible. Low-content material can be purified by repeated recrystallization in various solvents. With the data of the transfer document, the compound can be registered in the compound management system if it is not yet registered there. Each compound receives its individual laboratory code. The project folders in the IT systems are also created at this stage:

- project folders in the file structure,
- project folders in the ELN, and
- import folders for the automated import of the analysis data.

Only now first experiments can be created in the ELN using the API's laboratory code. In the ELN, document quantity, batch designation, and source of the material are entered and the input analysis is marked as such. The choice of analysis methods for samples can be standardized. These include ^{1}H NMR, XRPD, DSC, TGA, and an imaging. A proposal is that these five methods should form the basis of all sample analyses. The analytic code of these studies can be easily standardized. This makes it easier for the processor to assign the analyses. If both HPLC purity and content method are already available for the API, both data should be determined also. The material requirements for all these analyses should not amount to more than 100 mg in total. The time required is only a few days. After evaluation of the analytical data and documentation in the ELN, the sample is well characterized and can be used as starting material for all further investigation activities. Should the starting material prove to be representative of a particular polymorph at a later date, further analysis such as DVS or XRPD at variable temperatures or variable humidities can also be commissioned.

Initial analysis is completed by the compilation of the batch history. It is beneficial for the development of the drug substance with respect to the solid form when at least XRPD or Raman data from the earliest development batches were measured. If there are already several API batches with solid-state analysis, you get a first impression of solid forms of the API. Earlier batches can be of good crystallinity but also amorphous. First references for XRPD or Raman can be derived. A review on the synthesis method with focus on synthesis route and solvents can also provide information on residual solvents or impurities. If the API is already well crystalline, the last process solvent used should later also be used in screening experiments.

7.5.2 Solubility Estimation

The starting point for screening experiments is solubility data in organic solvents and aqueous mixtures. If such data is available, one should check if for all solvents of interest data in the required quality are available. The selection of appropriate solvents for the study of solid forms has already been reported [11–13]. If no solubility data is available, they must be generated. By way of example, the determination of saturation solubility by HPLC and the estimation of solubility by titration may be mentioned here. The advantage of the HPLC method is the higher accuracy of the data and the high degree of automation. However, if no HPLC content method is available, which is often the case at an early development stage, the titration method can be used. Solvent will be added to a solid sample until a complete solution is obtained. The titration method can be carried out with very small amounts of sample. With less than 10 mg per solvent, solubility can be estimated. In order to avoid having to rethink the question of solvent selection every time, selected solvents, and solvent mixtures can be divided into predefined groups. Here, aspects of the solvent class, usability in the synthesis, or green chemistry aspects can play a role. Mixtures of organic solvents with water and organic solvent mixtures should always be taken into account. Thus, depending on the application, simply one or more groups of solvents can be selected, which also greatly simplifies the workflow in the laboratory. For example, predefined documents can be used for each corresponding group. These individual documents can be merged into one master document. In addition, a fixed, standardized sequence of solvents within a group of selected solvents offers more advantages. This makes the evaluation of syntheses and analyses much easier, because it is simpler for operators to assign results to reaction conditions. For solubility estimation, the temperature of the solvent can be easily varied. One solubility estimation at room temperature and one at 50 °C, respectively, give a good impression of the solubility behavior of the API. Such a routine assists to quickly classify solvents. Depending on available material and time resources, the choice of solvent groups can be varied. It should be remembered, however, that selecting all solvents at two temperatures can quickly give rise to large numbers of samples. As a good compromise, one can start with the solvent group of the most commonly used solvents at two temperatures. If no usable solvent has been found yet, the selection can be expanded as desired. Based on the results of the solubility estimation, the following screening experiments are planned. A first selection of suitable solvents can be made and the crystallization methods for the screening phase can be determined.

7.5.3 Manual Screening

After estimating solubilities, the resulting solutions and suspensions can be easily used for further screening experiments [14–17]. From the point of view of the workflow management, it is an ideal situation, since no samples need to be rebuilt.

It is particularly easy to reuse the samples for screening experiments by evaporation. This can be done actively with compressed air (i.e. fast evaporation) but also passively (i.e. slow evaporation). In case of the passive process, it more likely to crystallize stable solid forms. Since up to two samples per solvent are available, two different temperatures can also be selected for the evaporation. If one chooses a low as well as a high temperature for evaporation, it can be screened for different solid forms over a wide temperature range. One may thus crystallize in addition to stabilizing also metastable forms and solvates. For this purpose, the reactors are heated in heating blocks and overflowed with compressed air. By doing so, one should finally gain oily residues or crystallize solids. The resulting solids are examined for their polymorphic forms. The data obtained form the basis for further screening experiments. Most often, they represent the first referential data for polymorphic forms.

Further screening experiments can, for instance, be carried out as crystallization by cooling, crystallization by antisolvent, and as slurry experiments. Thermoshakers can be used to manually carry out slurry or interconversion experiments at different temperatures. Evaporation can also be realized in a heated blocks by means of an exhaust head. Software-controlled multireactor systems [18, 19] are better suited for exact temperature control for fast, slow, or and repeated impressing of temperature profiles. To observe crystallization process, turbidity probes or cameras can be installed in addition. After the end of the screening experiments, samples are prepared, the analysis, typically just XRPD, commissioned and documented, as described.

Subsequently, the results of the screening experiments are compiled. Particular attention is paid to obtaining an overview of the resulting solid forms. This can be done in a spreadsheet, preferably in the ELN or another database system. For better visualization, different solid forms (e.g. cluster of same XRPD pattern) can be highlighted in colors. The question of whether these are ansolvates, hydrates, or solvates can usually not be answered with certainty in this phase yet. Once a few experiments have been carried out, it is often impossible to distinguish between pure phases and mixtures of phases. Only with an increasing number of data, mixtures can be identified as the sum of two (or more) known solid forms. X-ray single-crystal structure analysis would be supportive in identification, albeit consuming more resources.

Furthermore, screening experiments often do not provide enough material for all standard analytic methods (see Section 7.4.8) or for further analyses such as DVS, variable temperature X-ray powder diffraction (vT-XRPD), or to determine solubilities and stabilities. For this purpose, the polymorphs, salts, and cocrystals must be produced in larger quantities.

7.5.4 High-Throughput Screening

High-throughput (HT) screening platforms [20–24] have been used for many years in the elucidation of polymorphism, the screening for salts and cocrystals but also compatibility studies of the active ingredient with excipients. If the focus is on carrying out a large number of experiments, the use of this HT screening platform is an obvious choice. Already during the setup of the crystallization laboratory in the mid-2000s, an HT platform [25] was implemented for the handling of 96-well multiwellplates.

As already reported, all experiments in the crystallization laboratory are planned in the ELN. Here all calculations for the reaction with all starting materials are carried out, reaction procedure and reaction conditions are described. For the 96-well multiwellplates, the workflow was adopted. Since the programming of the HT platform could not take place from the ELN, the reaction planning was carried out again with a second software, called Library Studio [26]: a circumstance that made the use of the robot even more difficult. Chemicals selected from a built-in database can be assigned to wells individually or with volume or mass gradients, while subsequent dosing of reagents and reaction steps, such as temperature control by heating and cooling, shaking as well as filtration steps and withdrawals from the reactors can be programmed.

After export of the recipe to an Excel sheet and import into the robot, a 96-well multiwellplate is prepared accordingly by means of liquid and solid dispensing.

Often transfer of the API and reactants is described by transferring a solution of the compounds. Unfortunately, the used solvent must then be removed later to get a dry solid for further experimentation. Otherwise, residual solvent might affect the screening results. To avoid this circumstance, solids were dosed. However, solid dispensing required prior preparation of the solids by sieving. Charging and running the HT robot was useful only if larger amounts of substance (about 1 g per plate) and thus a sufficient number of experiments were objected. Consequently, in order to use the platform for a polymorphism and salt study, at least 5 g of API is needed. If one considers the more elaborate programming, such a platform will only achieve considerable time savings compared to manual work when using 2–3 multiwellplates with approximately 200–300 individual experiments.

In order to maximize the capacity of the laboratory, it has proven to be advantageous to prepare the platform during the day and equip it with all chemicals by the afternoon. Since the pipetting of solid compounds can only be done with an open multiwellplate afterward, the plate must also be lidded manually. All manual steps had therefore to be completed at the end of the working day. The automated synthesis platform is run during the night, which significantly increased the productivity of the crystallization laboratory.

After the synthesis has been completed, the 96-well multiwellplate is opened. As a routine procedure, evaporation of the synthesis solvents as final downstream step has been proven meaningful. Documentation of the dosing and synthesis sequence along with values and parameters was performed by the robot control software. All data can be exported as an Excel file. An automated workflow was not established, but the data was manually copied into the particular ELN document. Further calculations on transferred masses and yields must be carried out separately in the ELN. Nevertheless, it is desirable to enable an import into individual database fields of the ELN in order to optimize the workflow further.

A separate procedure was established for the analysis of the 96-well microplates by XRPD. After working up the samples, the 96-well microplate was sealed with a transparent foil. This way, all movable stamps in each reactor can be carefully pushed through the synthesis block until the foil was reached and ensured an even surface for all samples. Consequently, the quality of the XRPD pattern was sufficiently high because the XRPD could be measured through the foil in reflection geometry. In addition, one of the wells could be filled with silicon for reference measurement. Despite all this, it could not be avoided that, depending on measuring time per well, some hours can pass between the first and last measurement on the plate. The XRPD measurement was not optimized for high resolution but for fast throughput. Short measuring times reduce the waiting times of each sample to a minimum. A cooling of the samples was not implemented. Here is potential for optimization.

In addition to pure data collection, the analysis of the analytical data is often the bottleneck. For reasons of efficiency, the high-throughput X-ray powder diffraction (HT-XRPD) data was not baseline corrected. The measurement data was imported into the analytic database without any further processing and passes through the established evaluation process, as described in Section 7.4.5 with a final approval by an engineer.

7.5.5 Processes for Replica Experiments and Scale-Up of Solid Forms

The workflows for the replica experiments and scale-up phase are based on the processes of the screening phase and are essential input for the development of a crystallization process. Single screening experiments are exactly repeated and gradually increased in scale. It has to be taken into consideration that the material requirements for the experiments are constantly increasing and the available amount of API has to be kept in mind. If the experiments are successful and larger quantities of certain solids (polymorphs, salts, cocrystals) are available, more analysis can be done. In addition to standard analysis, sufficient material should now be available for e.g. DVS, hot-stage microscopy, vT-XRPD, or single-crystal structure analysis. Material can be dried specifically and interconversion experiments using various solid forms can be performed. Initial studies on stabilities when stored under moist and dry conditions are feasible. Furthermore, a wide variety of solubility testing, including pH-dependent, can begin. At the end of the replica experiments and scale-up phase, one should have a good overview of the landscape of the polymorphic form. Also, basic options for salts and cocrystals are already known and studied. The relationships between the individual forms, conversions, and a detailed analytical characterization should be almost complete. All these data support the decision on the solid form for further development. In coordination with the project-specific requirements for the solid form, the results are discussed and a decision prepared in the CMC team.

7.6 Development of Crystallization Processes

The development of crystallization processes is carried out depending on the development phase. A risk-based approach is beneficial to scope the experiments, because the requirements for crystallization processes in an early development phase differ from those of late-phase processes.

The workflows addressing development of crystallization processes are not fundamentally different from those of the screening and reproduction scale-up phase. In larger scales, i.e. 100–1000 ml, screening tests are carried out to determine process-relevant parameters. The use of design of experiment (DoE) methods is recommended here. With regard to the workflow, hardly any automated processes exist. The lab journal is not set up to communicate with automated laboratory equipment and the software to plan DoE experiments runs without interfaces to the ELN, too. An import and export functionallity of experimental setup in DoE softwares such as Modde, Design-Expert, Unscrambler (e.g. Modde [27], Design-Expert [28], Unscrambler [29]) are usually only rudimentary and limited to Excel formats. In order to be able to use the ELN as a central control system in the laboratory, it is necessary to work on functional solutions. It would make sense to implement the complete planning of an experiment in the ELN. If this is not possible, an external design plan should be imported into the ELN so that the difference to a native ELN documentation is minimal. A standardized exchange format is desirable here.

At this stage of crystallization optimization, process analytical techniques (PAT), i.e. in-process probes, such as temperature, pH, particle sizing, or IR probes, are often introduced into the reactors to enhance and speed up process understanding. The documentation of the results can be a challenge. Screenshot from lab device control software is one option to document evaluations in the ELN. The raw data remains on the instrument and is available for further analysis at a later point in time and from a different point of interest. Alternatively, electronic reports generated by the control software might also be imported into the ELN. Once all measurement data have been recorded and all ELN protocols finalized, the results are summarized and prepared for the customer. Finally, the development is completed by a development report and a laboratory specification.

7.7 Support Processes

In addition to elucidating the solid form of an active ingredient, a crystallization laboratory can also act as service supplier for other development units in course of a project. Support processes in a crystallization lab can support, e.g.

- resolution of enantiomers or diastereomers,
- scale-up,
- tech transfer,
- targeted crystallization of intermediates of synthesis,
- deliver solubility and solid-state data for intermediates,
- provision of reference material for
  - preformulation,

- o formulation purposes, and
- o stability testing.

The following sections may exemplify some details.

7.7.1 Route Scouting Process

The established solubility estimation process can also be used to determine data for reactants, intermediates, and impurities occurring in drug synthesis. Thus, synthesis development is actively supported. Solvents exhibiting good or high solubility may serve as extractants or for synthesis purposes to raise space–time yield. To deplete impurities and intermediates, solvents with low solubility are beneficial. Samples of API, reactants, intermediates or impurities can be dissolved in solvent, respectively to get dried again by evaporation (even fast or slow). Analytic data from received solids such as XRPD or spectroscopic data reveal indications of complex polymorphism or the presence of solvates, which both could lead to problems in the following synthesis. Solvent residues can accumulate and disrupt the synthesis in the long term or be contained as an impurity in the final product.

7.7.2 Crystallization of Impurities and Intermediates

The crystallization laboratory can also provide support for the separation of intermediates through specific crystallization techniques as well as for the controlled crystallization of intermediates and impurities. Established processes and expertise for the development of crystallization processes are contributed for this purposes. Depending on the amount of substance available, screening approaches can be carried out or crystallization is performed on a larger laboratory scale. In addition to the separation of intermediates as a free form, a separation can take place by transformation to salts or cocrystals. Screening experiments on intermediate salts quickly provide information about general feasibility. The use of PAT adds deeper insights and details to speed up crystallization development. In addition to the development of process procedures, the crystallization laboratory can also realize small supplies of material. This can be both provision of material for biological or preformulation testing and reference material.

7.7.3 Downstream Processes

In addition to the investigation of crystallization processes, the following tasks with respect to the behavior and properties of solids can be considered and reported to the corresponding working groups:

- isolation, i.e. filtration or centrifugation,
- drying,
- storage under controlled temperature and humidity conditions,
- mechanical process steps, such as milling or de-agglomeration.

Receiving groups for the results are above all synthesis development, preformulation, and formulation development.

7.7.4 Scale-Up and Technology Transfer Process

As soon a first crystallization process of an active ingredient has been worked out by the solid-state group, it can be handed over (or returned) to the synthesis laboratories. The requirements of such a process are:

- control of proper polymorphic form,
- control of morphology and particle size distribution,
- reproducibility,
- scalability,
- cost efficiency by high yield and purity.

The depletion of impurities is particularly important as the last step in drug synthesis. Intimate coordination within the CMC team is required when dealing with impurities profiles and the depletion to their maximum tolerable content. As workflows, a shared list of all the impurities has proven helpful. Rapid sharing of new insights into structures and detected toxicities is possible by such means. If necessary, the crystallization process must be adapted to fulfill the requirements.

Technology transfer of the crystallization process takes place in two ways. On the one hand, the development of the process is described in a development report. On the other hand, besides delivering a laboratory specification for the crystallization process, a compound data sheet is prepared. The laboratory specification is the basis for the further development of the crystallization process of the API. After technology transfer, the crystallization, as final step of API synthesis, is performed through the synthesis laboratory or pilot plant; however, actively and continuously supported by the crystallization laboratory.

The compound data sheet is a list of substance-specific properties and includes contributions from the crystallization laboratory such as

- chemical structure,
- crystal form,
- morphology,
- analytical data,
- solubility data for synthesis and cleaning purposes.

7.7.5 Analytical Development

Part of the development of an API is also the study of stability to light, heat, humidity, acidic and basic conditions, as well as reduction and oxidizing agents. Other aspects are solubilities in buffered solutions and dissolution data. Polymorphs and especially salts and cocrystals of an API may differ dramatically in these characteristics. The crystallization laboratory can provide reference materials for specific solid forms. In coordination with the CMC team, these analytical testings are carried out. The results contribute to the decision on polymorphic form for further development.

7.7.6 Preformulation

The development of the active ingredient is also undergoing preformulation. Task is the development of all preclinical formulations, e.g. development of formulations for toxicology studies. Preformulations could be basis for the subsequent development of the final API formulation. Certainly, data and knowledge of solid-state development are relevant. Focus is on identifying the solid form of an API, considered for further development, as early as possible. This minimizes the risk of later change of the solid form. In most cases, preformulations are more simple formulations such as solutions or suspensions. In the case of solutions, the solid form should completely dissolve and not form sparingly soluble solvates. Since they are mostly aqueous media, from the point of view of preformulation, the ability of the API to form hydrates is of particular interest. To clarify the dissolution behavior of certain solid forms, salts, and cocrystals, the crystallization laboratory can provide reference material. Depending on the indication, the API should dissolve rapidly or slowly in the medium. By early cross-functional consultation, the development of the solid can be concentrated on fast or slow-release solids. In the case of suspension, the particle shape can also be a decisive property. Therefore, isometric solids are preferable. Active ingredients that are present as thin needles can break easily and tend to agglomerate. This can adversely affect the uniform distribution of the active ingredient. As part of the workflow, an early exchange of development data in the form of a presentation on the development status and also with development reports about the CMC team is recommended.

7.7.7 Formulation

At a later stage of development, the CMC team establishes an interface with formulation development. In contrast to preformulation development, which prefers liquid or semiliquid formulations, solid formulations are more preferred as pharmaceutical dosage forms. The contribution of the crystallization laboratory for this purpose is mainly data on the drying behavior, behavior under mechanical stress during grinding, or the degree of agglomeration of the active ingredient. In addition to the particle properties of the active ingredient, the particles of the formulation matrix are also taken into account for the development of the dosage form. The crystallization laboratory can support this development by providing particle size data and particle size distribution data. To counteract segregation within a solid formulation, the particles should be as similar in size and shape as possible. Therefore, data from process probes [19] such as focused beam reflectance measurement (FBRM) or real-time microscopy provide insights into API properties and changes during and at the end of the crystallization and formulation process. Furthermore, data about particles, their size, appearance, and degree of agglomeration is accessible. Thus, early on, this data serves to estimate the risk of how drug substance and excipients react to process steps such as compaction or granulation and to a warm and humid environment.

7.8 Conclusion

Establishment and further development of the workflows in the crystallization laboratory have been continuously advanced over the past 20 years. This was done in close collaboration with our colleagues in synthesis development and analytical development. The focus was on interface management and its consistent improvement. Central tools applied are the compound management system, the ELN, the analytic database, and the LIMS system. Coordinated, automated data streams between these systems and the automated laboratory devices were in the focus of technical and IT development activities. As described, experiment procedures till sample preparation can be conducted automatically, data transferred electronically, analyzed by experts who are supported by machine learning tools, and the results can be documented again in the ELN. The aim was always to define workflows and processes as precisely as possible in order to ensure a high quality of work, to simplify work, to increase the sample throughput, and to relieve employees of unnecessary work at the same time. During the development time, many colleagues contributed to the overall project "Workflow Management Solid State Development" with different emphases: whether through project conception, content development, or implementation in software solutions. Thank you very much for this!

As the essential components of the established workflows, I would like to emphasize again the following points

- The use of standardized, uniform identifiers ranging from laboratory codes, document codes to analysis codes is essential. A compound management system is the important component here.
- A central electronic laboratory journal for as many laboratory areas as possible with access to the compound management system should be the state of the art today.
- A major relief is working with templates for various processes. They execute workflows and give structure.
- IT systems connected via automated electronic workflows sustainably relieve the employees and increase the efficiency of the laboratory and of interfaces. As far as possible, global IT solutions should be used. In-house IT developments or isolated IT solutions rather prevent effective information exchange in the company and are often expensive to buy and to maintain.
- In addition to the LIMS system established in analytic department, the use of an analytic database with access to as many analytic data as possible is an advantage for development projects. Not the mere documentation of all analyses but an adapted, depending on the objective, alternative representation of analytical data is the benefit of such a system. Different individual samples can be brought into new contexts, thereby the record bundles all analytical data points of a sample.

List of Abbreviations

API	active pharmaceutical ingredient
CDPT	core development project team
CMC	chemistry manufacturing and control
DoE	design of experiment
DSC	differential scanning calorimetry
DVS	dynamic vapor sorption
ELN	electronic laboratory notebook
GC	gas chromatography
GMP	good manufacturing practice
HPLC	high-performance liquid chromatography
HT	high-throughput
HT-XRPD	high-throughput X-ray powder diffraction
IR	infrared (spectroscopy)
IT	information technology
LIMS	laboratory management information system
NMR	nuclear magnetic resonance (spectroscopy)
non-GMP	area outside the good manufacturing practice system
PDF	portable document format
TGA	thermogravimetric analysis
vT-XRPD	variable temperature X-ray powder diffraction
XRPD	X-ray powder diffraction

References

1 Cheng, Y., Lam, K., Ng, K., and Wibowo, C. (2009). Workflow for managing impurities in an integrated crystallization process. *AIChE Journal* 56: 633–649.
2 Wibowo, C. and Ng, K.M. (2002). Workflow for process synthesis and development: crystallization and solids processing. *Industrial and Engineering Chemistry Research* 41: 3839–3848.
3 EMA (2019). ICH guideline Q3C (R6) on impurities: guideline for residual solvents (step 5) (EMA/CHMP/ICH/82260/2006) (accessed 11 November 2019).
4 EMA (2000). ICH guideline Q6A specifications: test procedures and acceptance criteria for new drug substances and new drug products: chemical substances (step 5) (CPMP/ICH/367/96) (accessed 11 November 2019).
5 EMA (2012). ICH guideline Q11: development and manufacture of drug substances (chemical entities and biotechnological/biological entities) (step 4) (CHMP/ICH/425213/2011) (accessed 20 November 2019).
6 Martin, E., Monge, A., Duret, J.-A. et al. (2012). Building an R&D chemical registration system. *Journal of Cheminformatics* 4: 11.

7 WHO Guidance on INN. https://www.who.int/medicines/services/inn/innquidance/en (accessed 17 November 2019).

8 Dassault Systèmes BIOVIA Electronic laboratory notebooks. https://www.3dsbiovia.com/products/unified-lab-management/biovia-electronic-lab-notebooks

9 ACD (2019). *Spectrus Platform*. Totonto, ON, www.acdlabs.com: Advanced Chemistry Development, Inc.

10 EMA (2003). ICH Q1A (R2) stability testing of new drug substances and drug products (CPMP/ICH/2736/99) (accessed 17 November 2019).

11 Qiu, J. and Albrecht, J. (2018). Solubility correlations of common organic solvents. *Organic Process Research and Development* 22: 829–835.

12 Allesø, M., van den Berg, F., Cornett, C. et al. (2008). Solvent diversity in polymorph screening. *Journal of Pharmaceutical Sciences* 97: 2145–2159.

13 Piccione, P.M., Baumeister, J., Salvesen, T. et al. (2019). Solvent selection methods and tool. *Organic Process Research and Development* 23: 998–1016.

14 Anderton, C. (2007). A valuable technique for polymorph screening. *American Pharmaceutical Review* 10: 34–40.

15 Balbach, S. and Korn, C. (2004). Pharmaceutical evaluation of early development candidates 'the 100 mg-approach'. *International Journal of Pharmaceutics* 275: 1–12.

16 Kumar, L., Amin, A., and Bansal, A.K. (2007). An overview of automated systems relevant in pharmaceutical salt screening. *Drug Discovery Today* 12: 1046–1053.

17 Fernández Casares, A., Nap, W.M., Ten Figás, G. et al. (2015). An evaluation of salt screening methodologies. *The Journal of Pharmacy and Pharmacology* 67: 812–822.

18 Technobis Crystallization Systems https://www.crystallizationsystems.com/home

19 Automatische reaktoren und in-situ-analyse. https://www.mt.com/de/de/home/products/L1_AutochemProducts.html

20 Detoisien, T., Forite, M., Taulelle, P. et al. (2009). A rapid method for screening crystallization conditions and phases of an active pharmaceutical ingredient. *Organic Process Research and Development* 13: 1338–1342.

21 Florence, A.J. (2018). Approaches to high-throughput physical form screening and discovery. *Polymorphism in Pharmaceutical Solids*: 139–184. https://www.taylorfrancis.com. https://doi.org/10.3109/9781420073225-8.

22 Morissette, S.L., Almarsson, O., Peterson, M.L. et al. (2004). High-throughput crystallization: polymorphs, salts, cocrystals and solvates of pharmaceutical solids. *Advanced Drug Delivery Reviews* 56: 275–300.

23 Hilfiker, R., Berghausen, J., Blatter, F. et al. (2003). Polymorphism - integrated approach from high-throughput screening to crystallization optimization. *Journal of Thermal Analysis and Calorimetry* 73: 429–440.

24 Guo, J., Elzinga, P.A., Hageman, M.J., and Herron, J.N. (2008). Rapid throughput screening of apparent KSP values for weakly basic drugs using 96-well format. *Journal of Pharmaceutical Sciences* 97: 2080–2090.

25 Zinsser Crissy. http://www.zinsserna.com/crissy.htm (accessed 27 November 2019).
26 Library studio. https://www.unchainedlabs.com/lea (accessed 11 November 2019).
27 MODDE Umesoft. http://www.umesoft.de/modde (accessed 27 November 2019).
28 Stat-Ease Inc. Design-expert. https://www.statease.com/software/design-expert
29 Camo Analytics Design of experiments. https://www.camo.com/doe

8

Digitalization in Laboratories of the Pharmaceutical Industry

Tanja S. Picker

adesso SE, Competence Center Life Sciences, Jülicher Str. 116, 52070 Aachen, Germany

8.1 Introduction

In laboratories, information technology (IT) systems and data are omnipresent and effective data management is crucial for success.

As in all industries, the term 4.0 is rising in the pharmaceutical world as well. Pharma 4.0 is defined as the pharmaceutical production based on Industry 4.0, whereas Industry 4.0 describes the digitalization as fourth industrial revolution [1]. The vision of the derived Laboratory 4.0 or shorter "smart lab" means

- visibility (what happens?),
- transparency (why does it happen?),
- prognosis (what might happen in future and finally
- adaptability (automated processes, able to be self-optimizing).

As essential prerequisite for the realization of Pharma and Laboratory 4.0, a valid and reliable set of data is needed [2]. The findable, accessible, interoperable, and reusable (FAIR) data principles, first published in 2016 by a diverse group of more than 50 participants out of academia, industry, funding agencies, and scholarly publishers, describe the requirements of sustainable data management [3].

Looking into current processes in laboratories, including the use of universal series bus (USB) sticks or even paper notices as interfaces between electronic devices and laboratory IT solutions, the industry seems to be far away from Pharma 4.0, but need to fulfill the transition to full connectivity of all items first.

IT solutions are often seen as cost center, but should be considered as an opportunity to enhance quality and efficiency in research and development (R&D) as well as in manufacturing.

However, a look at real-world laboratories demonstrates outdated ineffective manual processes or disconnected systems.

Solid State Development and Processing of Pharmaceutical Molecules: Salts, Cocrystals, and Polymorphism, First Edition. Edited by Michael Gruss.

Analysis of the workflows of major companies ([Report] Everything you need to build a Smart Lab [4]) identified five key processes to improve:

- Data collection
- Availability of data
- Automated data flow, reduced manual steps
- Centralized Master Data Management
- Integrated migration management (following the data flows)

The authors discussed motivation and impediments of further digitalization in laboratories and suggested best practice processes for this transition.

Although some general principles of digitalization are included, this chapter focuses on IT solutions that are used in pharmaceutical development, however typical, but not only applicable, for laboratories that investigate, develop, and optimize solid state properties and related processes. Some of the common processes in laboratories are listed in Table 8.1.

8.2 Motivation of Digitalization in the Laboratory

The Gartner IT Glossary [5] defines digitalization as "the use of digital technologies to change a business model and provide new revenue and value-producing opportunities; it is the process of moving to a digital business."

The most obvious reason of any digitalization effort is the expectation of a higher work efficiency and a respective cost saving.

However, there are much more reasons and new opportunities connected to the recent trend of digitalization and the ongoing discussion of the smart lab.

8.2.1 Expectations of the Staff

In Germany, the lack of specialists is one of the main upcoming risks for workplaces with considerable degree of manual work like laboratories in the pharmaceutical industry, including solid-state characterization and processing labs.

For recruitment and long-term binding of the technicians and academic personnel, employers, and laboratory work should meet the expectations of the upcoming generation of employees. The "digital natives," leaving apprenticeship and universities today, will not expect to spend time in unnecessary manual processes like handwriting and copying values from an analytics printout into different IT system, like an electronic lab notebook (ELN) or laboratory information management system (LIMS). Pallast et al. [6] postulate "Good scientific practice requires easy access and safe data recording and storing." In the world of "Plug-and-Play," devices without automated and continuous data flow are no longer acceptable. However, the diversity of people working in laboratories and manufacturing should be considered. Those not having affinity to IT systems may be more reluctant but can be easily convinced if the electronic systems serves by introducing an advantage.

Table 8.1 Solid-state laboratory processes, supported by IT solutions.

Tasks	Explanation	IT systems, examples
Search for initial information	• Define objective • Search for available information in in-house and external libraries, databases • Ideation	• Electronic laboratory notebook (ELN) • SciFinder, Scopus, Espacenet, online search tools (Bing, Google [Scholar], Cambridge crystallographic database, crystallography open database, ELN, books, and journals [electronic, websites of publishers]) • Mind mapping, video conferencing
Set-up experiments	• Purchase chemicals, glass ware, equipment • Educate and train staff • Codify, i.e. name experiments and samples	• Web shops, catalogs • eLearning, videos, supplier demo • Registration system, label printer
Design	• Calculate system properties (e.g. pK_A, solubility, crystal structures) • Define range and details for experiments • Estimate and prospect progress (e.g. mixing, heat transfer)	• In silico prediction tools • Design of experiments (DoE) • Simulation tools (e.g. computational fluid dynamics [CFD], process modeling)
Execute/control experiments	• Program electronic equipment (heating/cooling, dosing, stirring), define profiles • Sampling and process analytics (timing, actions, measures, e.g. temperature, pH, pressure, particle size [distribution], concentration)	• (Semi)automated lab reactors • Process analytical technologies (PAT) (sensors, probes)
Document process recipes	• Planned parameters and methods vs. • Executed parameters and methods • Data integrity	• Lab devices, electronic lab notebook (ELN), lab execution systems (LES), lab information management system (LIMS)
Document chemical analytics (observations, results	• Data retrieved from sampling and process analytics (at-, in-, off-line) • Ensure data integrity (ALCOA+)	• Lab devices, electronic lab notebook (ELN), lab execution systems (LES), lab information management system (LIMS) • Scientific data management systems (SDMS)
Process and analyze the data	• Data tidying (if necessary) • Understand the data and relations • Transform to knowledge • Derive conclusions (result or set up further experiments)	• Self-made macros, processing • Machine Learning (statistics, uni-, multivariate analysis, clustering) • Visualization (charts, video, animation) • Automated processing • Artificial intelligence (AI) (i.e. self-training/self-improving algorithms)

(Continued)

Table 8.1 (Continued)

Tasks	Explanation	IT systems, examples
Reporting	• Summarize, communicate, and explain results • Consider different type of addressees o Level of detail, e.g. operators, supervisors, project team (CMC, core team), management, customers o Functional experts, e.g. synthesis, solid state, analytic, formulation, marketing, regulatory, patents o Purposes, e.g. internal vs. external publication, functional expert vs. layman	• Reporting tools • Text processor, presentation tools, Wikis, corporate databases • Automated processing o Aggregation o Visualization
Distribute data	• Submission • Further analysis	• Corporate data collections • Data mining (big data, smart data) • Artificial intelligence (AI) • Frameworks

8.2.2 Increasing Throughput

As in many other industries, the faster time-to-market expectation of the management leads to higher workload and rising complexity in the laboratory [7]. To manage the upcoming tasks, smart processes without manual processes and system discontinuity are needed to decrease the effort per experiment (throughput time) and increase the sample throughput or give time for creative work on new experiments.

8.2.3 Repeatability

In 2016, more than 1.500 scientists answered a survey about their attempts to repeat successful experiments.

Among the participants, the group of chemists ($n = 708$) resulted with the worst rate: more than 80% of them failed at least once in reproducing the experiments of someone else, more than 60% failed even in reproducing experiments they had done personally [8].

Having in mind the common practice to report the results, but keep the raw data on the devices, the repeatability might be improved by including this information in the data analysis in order to share them.

8.2.4 Enhanced Requirements on Data Integrity

It is expected that the regulatory requirements and regulations will be enforced in earlier steps of the pharmaceutical development in order to ensure high quality of results and reduce risks for the patients. With respect to laboratory data, the regulatory requirement of data integrity is asked and audited by the inspectors with increasing frequency, visible in the number of Food and Drug Administration (FDA) warning letters [9]:

- *2008–2013:* less than 10 letters per year,
- *2016–2018:* more than 40 letters per year.

The Medicines and Healthcare Products Regulatory Agency of Great Britain defines data integrity as "the degree to which data are complete, consistent, accurate, trustworthy, reliable" (Medicines and Healthcare Products Regulatory Agency of Great Britain [MHRA] [10], p. 9). In the most recent document of the Pharmaceutical Inspection Convention (3rd Draft from 30 November 2018), the requirements on data are described with the ALCOA + principles:

- A – Attributable
- L – Legible
- C – Contemporaneous
- O – Original
- A – Accurate
- + – Complete, Consistent, Enduring, and Available [11]

The compliance to the requirements, defined by the authorities, seems to be impossible without further digitalization. As an example, in 2018, the FDA answered the question:

"Is it acceptable to only save the final results from reprocessed laboratory chromatography?" with "No" (U.S. Department of Health and Human Services, Food and Drug Administration [12], p. 11).

8.2.5 Centralized Archiving

The long-term storage of raw data without undocumented update (archiving) is a regulatory and scientific requirement. Multiple individual archiving procedures for any device and IT solution cause high effort for the implementation and maintenance – including check of completeness and periodical restore tests of each single archiving process. In an environment with an integrated data flow and centralized storage of all relevant data, the implementation of a more automated and monitored central archiving solution is worth the higher initial investment with less operation effort and lower risks of data loss.

Note: It is recommended to implement a validated preprocessing of the raw data and store the preprocessed data in a standardized data format in addition to the raw data. This enables later analysis also in case the initial solution or device with specific software is no longer available (see Section 8.4.3).

8.2.6 Ad Hoc Analysis

Electronic data that are easily accessible for laboratory staff, lab heads, project leads, and clients enable those who are interested in a deeper or more detailed analysis of the data to do so. Different people share different backgrounds and therefore provide additional perspectives to look at laboratory data, like synthesis procedures, sampling protocols, and analytical results. They may like to interpret or process data by different representations or even means. Shared access is simplified if data are stored in a standardized format. Proprietary data formats, just dedicated to be processed by particular lab devices, are not recommended if data shall be explorable by a variety of functional users.

Another advantage of standardized and centralized access routines is that cumbersome familiarization with specific software is not needed. Functional experts may use the software tools perfectly suited to their needs and fields of application. Often experts apply tools having advanced functionality for statistical analysis and reporting.

If the data sharing workflow is well implemented and optimized, data are available for distribution to functional experts shortly after recording. In addition, release of the data by those who acquired them is possible in the electronic system.

From a solid-state perspective, the advantages are manifold. Underlying experimental conditions, like temperatures and solvents, can be looked up. Analytical results can be collected over a series of batches and various analytical techniques. Results that deviate from known solid forms may readily be visible by striking data points (e.g. melting point, mass loss, residual solvent). Furthermore, overlays or arrangement of graphs is possible, e.g. to compare crystal images, X-ray diffractograms, thermal analysis curves, or heating and cooling profiles of crystallization profiles.

However, sharing data requires responsible handling and interpretation. Conclusions without sufficient information about provenience of data and missing knowledge about experimental conditions, objectives, and circumstances (meta data) may lead to wrong impressions. Furthermore, a lack of sufficient expertise of understanding the data or a lack of statistical knowledge may lead to misinterpretation of the results. It is essential and must be assured that prior to concluding decisions, the functional experts who originated and processed the data must be contacted for affirmation or verification.

8.2.7 The Value of Data

The main argument for increasing digitalization is the potential value of the collected data.

Most big players in the pharmaceutical industry are going through the mind change from the individually owned data toward the new paradigm of the joined data lake with multiple reuse of data. In the recently founded project [13] concerning the implementation of the FAIR guiding principles for the data management, six of the top 10 pharmaceutical companies are included as founders or partners. As in any fundamental changes, some walls need to be broken down. Traditionally, each company unit manages the data in its own kingdom, fighting against access from other departments. Justification for limited access rights comprises additional internal efforts or missing external knowledge about the metadata. In addition to technical issues and lack of standardization (see also Section 8.4), the organizational hurdles are enormous. New responsibilities and processes need to be established:

- The responsibility of the data creator to store the data following the FAIR principles and
- The responsibility of the Data User to take care of the correct use and communication with the creator.

As an example for the further use of data may serve upcoming techniques described as artificial intelligence (AI). FAIR data are crucial for AI and machine learning which benefits from large, harmonized data sets for better predictions [14]. Digitalization in laboratory processes will help to collect structured data with high information density.

8.3 Categories of Laboratory IT Systems

A hierarchical overview suits to categorize the innumerable systems that are connected with IT in a laboratory environment (Figure 8.1). The pyramid is based on the number of different solutions, commonly available in a company with at least one laboratory.

The following description of solutions does not claim completeness, but focuses on the kind of solutions that are considered as important in the discussion of digitalization in the laboratory.

8.3.1 Devices

The base level of the pyramid, with the highest number of entities, comprises all devices that are able to receive or produce data. This group starts with quite simple

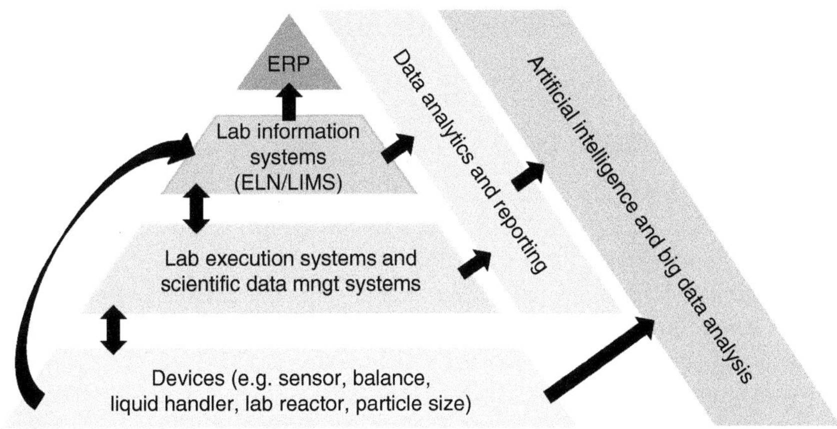

Figure 8.1 Hierarchical categories of laboratory information systems. ERP, enterprise resource planning; ELN, electronic lab notebook; LIMS, lab information management system. Source: adesso SE, internal.

sensors for temperature or humidity, creating two-dimensional data (timestamp and measured value) and includes also very complex instruments, replacing a complete laboratory work bench with repetitive tasks otherwise performed by a lab technician. Structured, often individually defined receipts must be provided to execute/control lab reactors or lab-on-a-chip systems. The results returned may consist of several files or data signals including the executed receipt and complex measurement results such as temperature curves or series of spectroscopic data. In addition to the required data formats for import and export, possible adapters and interface of the devices should be considered (see Section 8.4).

8.3.2 Lab Execution Systems (LES) and Scientific Data Management Systems (SDMS)

The second bottom layer consists typically of selected IT solutions for the controlling of complex devices (experiment processing) on the one hand (lab execution systems [LES]) and the collection of the recorded data on the other hand (scientific data management systems [SDMS]).

Often, these solutions are created for the use for devices of one brand only (manufacturer-bounded systems). While admitting that specialized systems are optimized for the connection to the devices of the same manufacturer, one should not forget that the devices probably are not fitted for the individual purposes of the particular laboratory. Furthermore, requirements of a laboratory may change over time. Time will show whether the strategy of binding the customers to only a single manufacturer will survive. Liberated consumers usually select the best of breed in all hierarchical levels of devices and IT solutions and expect universal connectivity. Multimanufacturer and multibrand solutions appear more future orientated and will be quite likely the systems of choice if flexible implementation and operation is sought. The impression is that systems become more and more designed to cooperate with third-party solutions. Although controlling (sending appropriate data) might be challenging, if not based on standards, but retrieving data is typically enabled by implementation of data converters.

8.3.3 Lab Data Systems

The systems in the second top layer of the pyramid, involved in nearly each action of the laboratory, can be divided into the sample-based LIMSs and the experiment-based ELNs. LIMSs are mainly used in analytic laboratories with limited number of different experiments, performed with a high number of samples. In contrast, the requirements of preparative working laboratories with a high number of different synthetic methods, performed with a manageable number of different analytical samples, are better fulfilled by ELN solutions. For both kinds of lab data systems, a wide selection of solutions is available, some of them very specialized for a narrow field of purposes. Others are designed for a wide variety of processes, even blurring the boundary between LIMS and ELN, containing also functions of data collection as the already mentioned SDMS (see Section 8.3.2). Again, systems are developing rapidly. New solutions become introduced, some become obsolete. To get an impression and comparison of various systems currently on the market, a search for websites providing relevant information is advised (e.g. https://www.gurdon.cam.ac.uk/institute-life/computing/elnguidance, accessed 4 November 2019). For software evaluation or implementation projects it is recommended to contact experienced internal or external consulting services to avoid pitfalls and make benefit of experiences and best practices.

8.3.4 Enterprise Resource Planning (ERP)

Most of the companies in the pharmaceutical, chemical, or nutrition industry use a single, centralized system for the production planning and recording of relevant economic data.

These systems can also be used for the management of centralized master data or host the chemistry warehouse or stock and might be relevant for the data management of laboratories, too. However, the integration of manufacturing sites is more common.

Another reason for laboratories to implement an interface to the enterprise resource planning (ERP) system might be the compliant storage and handling of quality-related data such as batch recording. In case of any connection between the output of the laboratory and patients or any influence on information, used for the submission, the relevant guidance documents need to be respected (good manufacturing practice [GMP] [15] and good clinical practice [GCP] [16]).

The ERP solutions such as SAP® are equipped with standardized interfaces (very often file based). As they are serving and consuming data from many other IT systems, the interfaces are very inflexible with regard to adaptions of the interfaces. The saying "never change a running system" becomes sadly true because adaptation for the sake of innovation and modernization is often postponed or canceled due to the complexity of interdependent systems and high effort of documentation in validated environments. The burden is often on the users whom have to continue working with suboptimal solutions.

8.3.5 Further Use of Data

Solutions with access to several levels and integration functionality are placed outside of the hierarchy. To gain the full use, it is recommendable to establish technical and organizational access to several levels (see also Sections 8.2.6 and 8.2.7).

8.3.5.1 Data Analysis and Reporting

To enable ad hoc analysis and periodically reporting, specialized solutions are in use. Although the devices and the solutions, described in the previous sections, also allow basic analysis of the data, it is recommendable, to implement additional solutions for these tasks.

The solutions are able to integrate several data sources and apply the provided methods and reports to the results of several devices and systems.

Open source solutions such as R [17] or KNIME [18] allow the use of state-of-the-art methods for statistical methods, graphic representation, or data manipulation up to data science workflows supported by easy-to-use graphical user interfaces (GUIs). Wizards for the guided creation of analysis, and numerous examples that can be copied and adapted for the individual purpose.

For single or periodical reporting in predefined formats, also open source solutions are available, e.g. Jasper Reports [19] or the already mentioned KNIME [18]. In contrast to the individual or ad hoc data analysis, the reports inform stakeholder not directly involved with layouts and formats, predefined by the reporting user.

8.3.5.2 Big Data Analytics and Artificial Intelligence

In addition to the ad hoc analysis, the more comprehensive Data Analytics uses systematical approaches to gain new knowledge from the existing data. While solutions like Rapid Miner [20] or Tibco Spotfire [21] are often used in research departments, it might be advantageous to evaluate also software from major companies. If products are already in use, e.g. SAS Statistic solutions for clinical data claiming, SAP as ERP system or Microsoft for the business software, the implementation and maintenance of the respective analytics software like SAS Analytics platforms [22], SAP Business Intelligence [23], or Microsoft Big Data and Advanced Analytics Solutions [24] from these providers is often easier to realize with the existing IT staff.

The borderline between the advanced data analytics and AI is often flowing and depends on the intention. Volker Gruhn, founder of adesso, "consider(s) AI systems to be systems that can automatically or independently take decisions and respond to input such as images, the written word or even spoken language" [25]. An important characteristic of AI is the ability of self-optimization, meaning the continuous improvement of the results based on the delivered data. AI is seen as current hype technology, and more and more frameworks are coming up to market.

In pharmaceutical laboratories, the additional services, provided by data analytics and AI, can be used to optimize processes and products, e.g. by reducing waiting times caused by device failures by predictive maintenance and or by reducing the out-of-spec product by continuous monitoring of analysis results (shift detection).

8.4 System Interfaces for Data Exchange

One of the most mentioned hurdles toward the digital transformation of laboratories is the missing standardization of interfaces [26].

On the different levels needed for the data exchange (adapters, communication standards, and data formats), several standards are available, and a final decision in the laboratory world is not yet taken. In the following section, commonly used technologies and standards of the three levels of building the interfaces between Laboratory IT Systems are discussed.

8.4.1 Adapters

The physical gateway to the systems is given by adapters, enabling laboratory instruments to communicate with each other on the same or different hierarchical level. Typically, these interfaces are defined by the manufacturing party. Technical reasons might be the foundation of supplier's choice. Consumers and clients generally do not have a big influence on the choice made but must consider the available adapters during the selection of laboratory devices.

Although several tools for transformation are available for each of the adapters, additional effort and version dependency imply complexity during the digitalization efforts.

8.4.1.1 Serial Port (RS232)

The historic recommended standard 232 (RS232) interface was introduced in the 1960s and is still found very often. Although most modern personal computer do not include a serial port, knowledge about the RS232 adapter is needed due to the existing laboratory devices. The low costs and the simple implementation are advantages, while the limitation of the cable length with regard to the baud rate and the issues and probably needed reimplantation caused by firmware updates are known disadvantages.

8.4.1.2 Universal Series Bus (USB)

USB-slots are the de facto standard on most electronic devices and exist currently in four main generations. The rising baud rates and the fast and easy implementation are often the reason for the choice of USB connections. During the setup, the form of connection should be considered: Whereas a computer usually provides several USB ports, the USB connection of the device is mostly set up to one computer, implementing a dependency from this computer and its performance.

8.4.1.3 Ethernet

Ethernet adapters enable a direct connection into the laboratory network. Standard in nearly all IT systems, only minor configuration of the Network Interface Card effort, is needed for the use. A common local area network (LAN) in the laboratory offers multiple possibilities to share data including centralized backup and archiving of raw data, but costs additional effort and hardware for the initial setup.

8.4.1.4 Cable Less Connections

In laboratories with regular relocation of the devices, cables imply an additional effort and a higher risk. Cable less adapters like WiFi (access to wireless LAN) or Bluetooth give a higher flexibility, but require additional security functions like encryption or checksums. Whereas Bluetooth is traditionally used for one-to-one connections, existing frameworks also allow the setup of a network with several participants providing acceptable baud rates [27].

8.4.2 Communication Medium and Protocols

In addition to the access points, the exchange of information needs a medium and detailed rules. Initially, each communication has been defined individually between a data source and receptor. Limiting the different ways of communication in a laboratory and using of international standards facilitate the data exchange and reduce the efforts for implementation and maintenance.

8.4.2.1 File-Based Communication

The easiest way for the exchange of data seems to be the storage and retrieval of files in a shared storage location. Although the initial effort is quite low, there are some disadvantages to be considered. The access rights need to be set wisely, to ensure the data integrity. Permanent monitoring of the shared location allows a timely use of the delivered files, while simultaneous access must be prevented, and intermediate updates and errors need to be handled.

8.4.2.2 ANSI/ISA-88 Batch Control (S-88)

Guiding the design and specifications of batch control systems is the intention of the American National Standards Institute (ANSI)/International Society of Automation (ISA)-88 also known as S-88 standard. A standardized data structure shall simplify programming, configuration, and communication between system devices. S-88 covers the physical model (i.e. locations and devices) as well as the functional or procedural model (i.e. recipe and unit operations) [28].

8.4.2.3 Open Platform Communications Unified Architecture (OPC UA)

OPC UA stands for open platform communications unified architecture [29] and is widely used in the process industry especially in internet of things (IoT)-related projects.

The standard is independent from the software platform and includes features relevant for data integrity such as access control, authentication, and encryption.

Due to its high complexity, most users do not implement the standard completely, but only with the individually needed features. In addition, the implementation is often not transferable between the suppliers of various devices.

OPC UA is optimized for the connection of multiple data sources with simple data structures to a central system, but not for the typical setup in a laboratory with defined source systems having very individual and sophisticated data structures, created by specialized devices.

8.4.2.4 Standards in Lab Automation (SiLA)

Standards in lab automation (SiLA) is the most specialized standard in the topic of laboratory informatics and is available in Version 2. The SiLA Consortium is a nonprofit organization, the board of directors is joined by members of staff from several pharmaceutical companies such as F. Hoffmann-La Roche and Novartis as well from IT companies and research organizations [30].

Version 2 implements in addition to the existing communication standard a communication protocol between microservices based on established web service technology.

The communication is established directly between server (e.g. a device) and client (e.g. a lab data system), whereby the roles can be exchanged due to the kind of communication. In a typical scenario, the device informs about its features, including the accepted commands and the provided data structures. The lab data system discovers the available devices with the offered services and is able to identify and connect them [31]. So SiLA can be used for the controlling of devices as well for the collection of the produced data in an interactive exchange.

As open source initiative, the standard definition, and existing drivers and implementations are available in the SiLA 2 repository.

8.4.3 Data Formats

Devices produce data in individual file formats, optimized for the purpose of the device. To implement the FAIR data principles with emphasis on the interoperability and the reusability (additional use of the data with reduced additional transformation effort), a central data format for the laboratory and even better on company level is recommended.

8.4.3.1 Common Data Formats (e.g. TXT, XML, JSON)

Whereas the text (TXT) format is familiar for nearly each user of a computer, there are much more easy, human readable formats used for the data exchange.

As examples can be mentioned, the extensible markup language (XML) widely used in internet communication [32] or the javascript object notation (JSON), a lightweight data format for the exchange of information between Client and Server [33].

Professional laboratory solutions mostly provides converters from the common file types such as portable document format (PDF), TXT, and XML. On a technical point of view, the import is easily possible, but the very individual structure of the files implies considerable effort and high risks of errors and missing relevant data.

8.4.3.2 Analytical Information Markup Language (AnIML)

The open data format analytical information markup language (AnIML) was defined for storage and transfer of structured data used in analytical chemistry and biology. The development has been started as subcommittee of American Society for Testing and Materials (ASTM) International with participation of users, suppliers, academics, and governmental representatives [34]. It is based on the common xml format with specialized xml tags for information concerning sample, analytic data, methodology, instrument, events, and even digital signatures. AnIML files may contain the multidimensional results as well as a detailed description of the processes (experiment steps).

The published standard includes the core schema with the allowed tags for the analytical data and the technique schema with the rules for the storage of an analytical technique. In addition, technical definitions for several technologies such as ultraviolet (UV)/visible (Vis) or chromatography methods are already available and can be directly used or extended for the individual implementation [35].

The standard can be used for any kind of devices, independent from the supplier. Adapter for several devices and IT solutions are already available, due to the straightforward and published structure, missing interfaces can be created at low cost or even internally.

The human readable form, the concept of audit trail for the recording of any updates as well as digital signatures and numerous open source tools for the handling of xml data, suits AnIML for the long-term archiving without additional license cost.

The definition groups of SiLA and AnIML are working together, supporting the cooperation of both technologies. AniML files can be easily transferred using SiLA as communication standard [36].

In addition to the available sources on the official home page (https://www.animl.org, accessed 27 November 2019), commercial tools such as the Seahorse suite and support are available [37]. According to the company's information, currently, adapters for about 150 devices are available, and the AnIML Viewer is implemented in several lab data systems [38].

8.4.3.3 Allotrope Data Format (ADF)

The allotrope data format (ADF) is developed by the Allotrope Foundation, an international consortium of pharmaceutical, biopharmaceutical, and other research-intensive industries (e.g. Bayer, gsk, Pfizer).

In addition to the foundation, a partner network has been established with supplier of devices and IT service providers. The access to the source code and the full documentation as well as the development of software for commercial use including the standard is restricted to members and associated with costs, whereas the access to the specification and the use of the software is allowed to everyone [39].

ADF as data standard is a data container containing of a data description, one or several data cubes and optional data packages. It is based on the hierarchical data format (HDF), developed for the management of large and complex data [40].

Due to the underlying technology, the full-text search in the files is performant, and a compliant long-term archiving as already approved by the authorities can be established [41].

In contrast, the binary format of the files is not human readable and an additional application as the ADF Explorer or any other HDF5 tool is needed.

The current focus lies on the development of the taxonomies and ontologies to ensure controlled vocabulary and relationships for the metadata.

Concerning a private and informal investigation among several players in the laboratory environment, specialized standards (such as AnIML and ADF) are widely but not completely known. The trend to open source products even in regulated areas seems to support AnIML as the open standard, whereas the prominent support by big pharmaceutical and device companies may influence decisions toward ADF.

8.5 Implementation of IT Solutions

Each implementation starts with the selection of the project and the general project setup. Chemists, working in the area of solid state development, are expected to be open minded and creative. This is visible also in the way, how they set up their working environment. In order to do the right project in the best way, some preliminary considerations are recommended.

8.5.1 Identification of Digital Gaps in the Lab Processes

In order to find the best topic for the next, i.e. the most needed, digitalization project, companies follow different strategies. Staying informed about the "Best Practice" by visiting the respective web sites or conferences, reading news letters or (IT) journals, and discuss with the laboratory staff is basic. More structured approaches are found in the area of human–computer interaction (Usability Engineering) and Software Technology as described in the following examples.

8.5.1.1 Contextual Inquiry

The Pistoia Alliance, a nonprofit organization for the support of innovation in life sciences (www.pistoiaalliance.org), established in 2017 the "*User Experience (UX) for Life Sciences (LS) project to enable business to adopt UX principles and methods as they develop scientific software*" (User Experience for Life Sciences – Pistoia Alliance [42]).

The community, comprising UX practitioners in life sciences, worked out a toolkit of UX methods and case studies. The method of contextual inquiry, i.e. the observation of users, accompanied by detailed questions, is particularly suitable for the identification and understanding of the most cumbersome and error-prone data processes in the environment of the key personal in laboratory.

Not really surprising are the results, documented in a case study [43]:

> "Also, scientists found it very difficult to manage data scattered between paper, desktop computers, and labs, and redundant data entry. These difficulties increased error rate in data, made scientists inefficient, and contributed to a generally unsatisfying experience."

Based on the observation of a representative selection of users, this method allows the selection of worthwhile areas for improved support by IT solutions.

8.5.1.2 Interaction Room

In order to align interdisciplinary teams and enhance the communication between IT experts and subject matter specialists, a workshop in the interaction room (IR) could be helpful. This method during an IT project, or preferable before the start, serves as organized communication platform. In a real space, the stakeholders meet for a defined time and follow a moderated process [44].

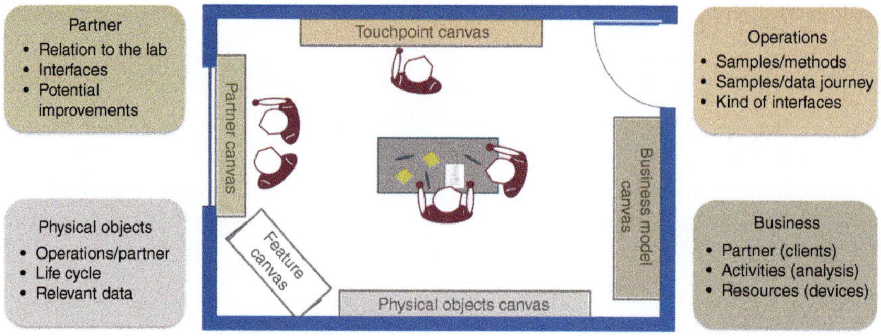

Figure 8.2 Interaction room digital, a well-proven approach in digitalization initiatives. Source: adesso SE, internal.

For digitalization initiatives the standard setting of the IR has been adapted to the "IR:digital" (Figure 8.2). In case of seeking solutions for a laboratory environment, the **business model** including internal and external clients, the activities in form of available services, methods, and experiments, as well as the resources, meaning the number and quality of devices and lab technicians will be discussed first. Looking at typical business processes, the interactions or **touchpoints** of the customers and the samples are examined: typically, a virtual "sample journey" and the recording of the actions and data consuming or producing steps of a sample during the time in the laboratory are included. On the **partner canvas**, selected customers and service or material providers, such as the project team, requesting further investigations or the central chemical storage are documented. What kind of experiences do they gain during the previous discovered scenarios? Which interfaces are used? In the final dimension, the states and state transition of the **physical objects** are observed and a typical life cycle of the samples or the data is described. Any possible enhancement, detected during the workshop will be captured on the **feature canvas**. It is not expected to define a complete list of possible projects and requirements during a single IR:digital workshop. But the participants will receive a common vision of the respective target and agree on the priorities and the next steps to be followed.

8.5.2 Implementation Approach

Computer system validation (CSV), as described in the guidance of good … practice (GxP), is usually not deemed required in early R&D solid-state laboratories with duties like polymorphism studies, salt screening, or pre-formulations tasks. However, these regulations might apply in solid-state characterization labs. With respect to later development stages crystallization or formulation scale-up labs, pilot plants or manufacturing facilities could be affected.

Regardless of the regulatory obligation, it is worth to consider in all software implementation projects if following the principles of the validation process is advised (Figure 8.3). The best practices, described in good automated manufacturing practice (GAMP 5), claim to be a "a cost effective framework of good practice to

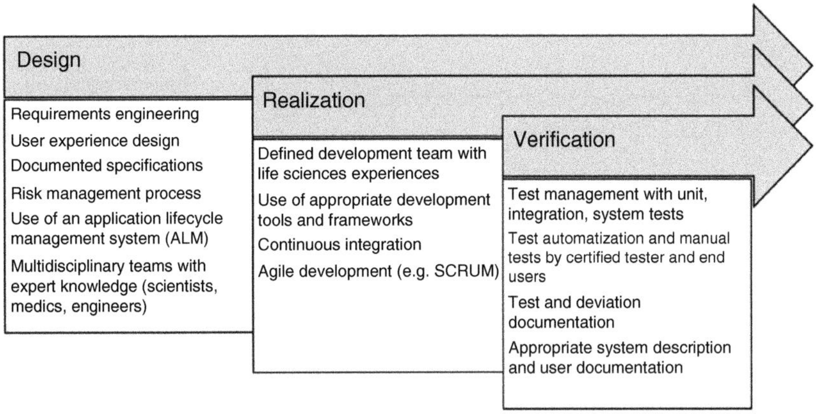

Figure 8.3 Recommended software development process based on general GAMP principles. Source: adesso SE, internal.

ensure that computerized systems are fit for intended use (…). The framework aims to safeguard patient safety, product quality, and data integrity, while also delivering business benefit" [45].

In the regulated area, each step and activity must be documented in a valid form (mostly printed on paper and signed documents). Creating documents during the activity costs additional efforts and should be evaluated carefully. However, having a written consent on each activity brings the great benefit of having the possibility to go back after a time and read about the reason for a particular decision.

In this context, it might be helpful to use a specialized solution for the application lifecycle management (ALM). These solutions enable the structured and linked documentation of all development-related artifacts like requirements, specifications, risks, codes, and tests with enhanced possibilities of completeness checks, reporting, and change management of the single items. More details can be found on the websites of the respective suppliers such as Intland, Polarion, or Medsoto.

8.5.2.1 Design

Design activities start with the project and should be ongoing during the whole software development process. The former way to close the design phase before the realization comprises the danger to create solutions, not adapted to the user needs. Recent agile development processes allow the adaption of the design based on the experiences made during the realization (see Section 8.5.2.2).

Requirements Understanding the real user need is essential for the success of an IT solution. The International Requirements Engineering Board (IREP) was founded in 2006 as a nonprofit organization and intends to enhance the requirement engineering by sharing the knowledge about practical methods [46].

The requirement engineering process consists of four main activities [47]:

- Identification
- Documentation

- Review and agreement (coordination)
- Management

For each of the activities, intensive collaboration of the stakeholder in interviews, questionnaires, workshops, or observation is needed.

While giving several details, already reaching the level of specifications (e.g. the name and order of GUI elements, i.e. graphical user interface, the appearance on the screen), several requirement documents are lacking in a clear demarcation between the system, the system context, and the not relevant part of the environment.

Particular attention is required for the implementation of interfaces between solutions and data management to fulfill the FAIR data principles (see Section 8.2.7).

Usability In comparison to tools for the mass-market tools, e.g. smartphones, the usability of scientific IT systems is often very poor. Scientists are expected to be able to work with complex expert systems, so why should the implementation project waste time for usability analysis and enhancements?

The answer is simple: because the return of this investment is highly probable. The effort, spent into the design process, will be paid back by higher reliability (and therefore quality of the results) and in addition efficiency in using the system.

The user experience and life sciences (UXLS) toolkit, prepared by the Pistoia Alliance (see Section 8.5.1.1), contains several UX methods adapted to the special needs in life sciences and is complemented by case studies of the life sciences area [48].

Specification The amount of details, to be specified before the realization starts, depends on the complexity and the overall risk of the intended IT solution.

In any case, the solution architecture and the high-level workflow and screen flow should be defined. This includes also the selection of the best-suited method and technology for the realization of the interfaces (see Section 8.4).

Risk Management Risk management is done in every software development project, but in most cases, the identification and assessment of risks are performed implicitly by the developer without communication and documentation. Following this common procedure, many erroneous decisions are not identified, and many possible opportunities are not taken.

A managed risk assessment workshop with a multidisciplinary team built by developer, architect, test specialist, end user, and project manager analyzes the defined specifications and describe possible malfunctions or operating errors. The risks are characterized by

- severity (possible harm),
- probability (expected number of occurrence), and
- detectability (possibility of correction during the defined process).

For each risk mitigation actions should be defined. The mitigation actions may result in new or adapted requirements and specifications, or describe needed user training or additional hints in the user manual.

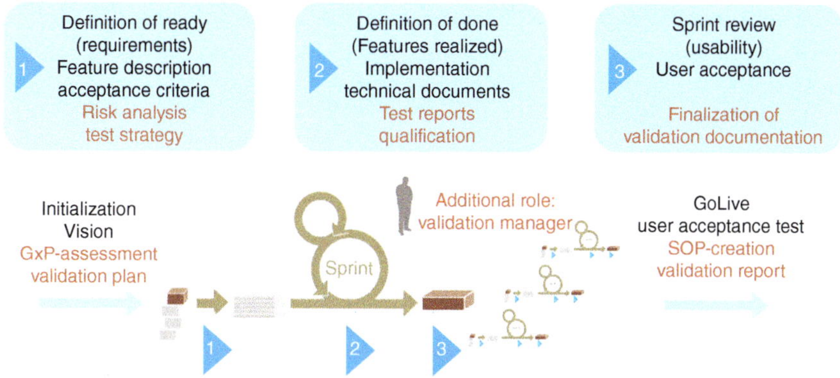

Figure 8.4 SCRUM process. Content in red is needed in regulated areas. Source: adesso SE, internal.

8.5.2.2 Realization

The realization of IT solutions is usually a process with low impact on the laboratory itself.

Based on the experiences of the delivery of inappropriate IT solutions after long realization periods, an agile approach with intermediate deliveries such as SCRUM is recommended.

This lightweight framework implements the principles of transparency, inspection, and adaption in a set of roles, events, and artifacts. The development is split into defined sections (called sprints). The team collects the features, to be realized in the next sprint out of a prioritized list of requirements (product backlog), and commits itself to the delivery of an executable version of the software after the sprint (increment) [49].

After each delivery, the group of key user is invited to review the current solution and gives feedback in order to update and reprioritize the remaining features in the product backlog. Due to the fact, that each implement is executable, the development process can be stopped if the end-users agree on the realized features.

All activities are performed “time boxed”, meaning that the result is to be delivered after the defined time. This principle prevents the team from the creation of the optimized solution, but forces to concentrate on the essential topics first – during development as well as during test and discussion by end users (Figure 8.4).

8.5.2.3 Verification

In addition to the time boxed review of the intermediate deliveries, a structured and multistage test process is performed in the verification phase. This includes not only the solution internal unit tests but also the integration of the solution into the data management process and the test of all interfaces between the single IT solutions.

Especially in case of an incremental realization with the continuous integration of realized new features as described in the SCRUM process, the automation of test cases and the periodically execution of regression tests is highly recommended.

The verification includes also finalization of the technical documentation in a system description and a usable and appropriate user documentation containing also the mitigation actions, defined during the risk management process (see section titled "Risk Management").

8.5.2.4 Rollout

The start of the productive use in the laboratory environment impacts the whole laboratory staff. Workshops and trainings regarding the adapted processes and the handling of the IT solution have to be performed.

The amount of user-related rollout activities depends on the individual IT affinity and the participation in the previous implementation process. Having natural scientists in the group of users, the creativity to adapt given IT structures toward individual redefinition should not be underestimated. In regulated areas, specific standard operational procedures (SOPs) need to be created. This is recommended in not regulated laboratories as well, in order to agree on the same way of usage and make the best use of the system.

8.6 Conclusion

Leveraging the transfer of data in the laboratory is one of the main tasks on the way to Pharma 4.0. Implementation of appropriate standards and IT solutions and replacement of manual processes will help to reduce risks of transmission errors and free up resources for innovative research. Observation of already available technology and use of pragmatic implementation processes support the laboratory to take the best decisions and limit the efforts on this journey.

List of Abbreviations

AI	artificial intelligence
ADF	allotrope data format
ALCOA	attributable, legible, contemporaneous, original, accurate
ALM	application lifecycle management
AnIML	analytical information markup language
ASCII	American Standard Code for Information Interchange
CSV	computer system validation
DSC	differential scan
ELN	electronic lab notebook
ERP	enterprise resource planning
FAIR	findable, accessible, interoperable, and reusable
GAMP	good automated manufacturing practice
GCP	good clinical practice
GMP	good manufacturing practice
GUI	graphical user interface

GxP	good ... practice
HDF	hierarchical data format
IR	interaction room
IREP	International Requirements Engineering Board
IT	information technology
IoT	internet of things
JSON	javascript object notation
LAN	local area network
LIMS	lab information management system
LS	life sciences
OPC	open platform communications
UA	unified architecture
PDF	portable document format
R&D	research and development
SiLA	Standards in Lab Automation
TSV	tab separated value
USB	universal series bus
UX	user experience
UXLS	user experience and life sciences
XML	extended markup language
XRPD	X-ray powder diffraction

References

1 Wölbeling, C. (2019). The ISPE Pharma 4.0 operating model. Basis for the ICH holistic control strategy – an introduction. ISPE Webinars. ISPE. (zuletzt geprüft am 18 October 2019).

2 Schmitz, S (2019). Erfolgreiche Umsetzung der digitalen Transformation: VBU | Webinar. http://v-b-u.org/VBU_Webinare/Webinar_+Erfolgreiche+Umsetzung+der+digitalen+Transformation-p-3872.html (accessed 06 June 2019).

3 Wilkinson, M.D., Dumontier, M., Aalbersberg, I. et al. (2016). The FAIR guiding principles for scientific data management and stewardship. *Scientific Data* 3: 160018. https://doi.org/10.1038/sdata.2016.18.

4 [Report] Everything you need to build a Smart Lab (2019). Online verfügbar unter. https://www.pharma-iq.com/informatics/reports/report-building-a-smart-lab (zuletzt geprüft am 12 June 2019).

5 Gartner IT. Glossary: digitalization. https://www.gartner.com/en/information-technology/glossary/digitalization (zuletzt geprüft am 6 November 2019).

6 Pallast, N., Wieters, F., Nill, M. et al. (2018). Cloud-based relational database for multimodal animal data. *Database: The Journal of Biological Databases and Curation* 2018: bay124. https://doi.org/10.1093/database/bay124.

7 Analytica/Messe München (2018). Trend report – future lab/analytica 2018. Digital transformation: towards the laboratory of the future. https://www.analytica.de/media/website/files/pdf/trend-reports/analytica-trendbericht-labor-der-

zukunft.pdf (zuletzt aktualisiert am 26 February 2018, zuletzt geprüft am 20 September 2019).

8 Baker, M. (2016). 1,500 scientists lift the lid on reproducibility. *Nature News* 533 (7604): 452. https://doi.org/10.1038/533452a.

9 Unger, B. (2019). An analysis of 2018 FDA warning letters citing data integrity failures. Hg. v. Pharmaceutical online. https://www.pharmaceuticalonline.com/doc/an-analysis-of-fda-warning-letters-citing-data-integrity-failures-0001 (zuletzt aktualisiert am 12 June 2019, zuletzt geprüft am 4 November 2019).

10 MHRA (2018). 'GXP' data integrity guidance and definitions. (zuletzt aktualisiert am March 2018, zuletzt geprüft am 4 November 2019).

11 PIC/S (2018). Guidance on data integrity. Good practices for data management and integrity in regulated GMP/GDP environments (PIC/S guidance). (zuletzt geprüft am 4 November 2019).

12 U.S. Department of Health and Human Services, Food and Drug Administration (Hg.) (2018). Data integrity and compliance with drug CGMP. Questions and answers guidance for industry. https://www.fda.gov/regulatory-information/search-fda-guidance-documents/data-integrity-and-compliance-drug-cgmp-questions-and-answers-guidance-industry (zuletzt aktualisiert am December 2018, zuletzt geprüft am 4 November 2019).

13 Pistoia Alliance (2019a). FAIR implementation. Supported by a FAIR Toolkit for Life Science industry. https://www.pistoiaalliance.org/projects/current-projects/fair-implementation (zuletzt aktualisiert am 6 September 2019, zuletzt geprüft am 4 November 2019).

14 Lorica, B. and Loukides, M. (2019). How new tools in data and AI are being used in health care and medicine. https://www.oreilly.com/radar/how-new-tools-in-data-and-ai-are-being-used-in-health-care-and-medicine (zuletzt aktualisiert am 25 September 2019, zuletzt geprüft am 25 September 2019).

15 European Commission (2019). EU guidelines to good manufacturing practice (GMP) medicinal products for human and veterinary use. https://ec.europa.eu/health/documents/eudralex/vol-4_en (zuletzt aktualisiert am 21 November 2019, zuletzt geprüft am 21 November 2019).

16 European Medicines Agency (EMA) (2018). Good clinical practice (GCP). https://www.ema.europa.eu/en/human-regulatory/research-development/compliance/good-clinical-practice (zuletzt aktualisiert am 21 November 2019, zuletzt geprüft am 21 November 2019).

17 R (2019). The R project for statistical computing. https://www.r-project.org (zuletzt aktualisiert am 13 November 2019, zuletzt geprüft am 21 November 2019).

18 KNIME. KNIME analytics platform. https://www.knime.com/knime-analytics-platform (zuletzt geprüft am 21 November 2019).

19 JasperReports® Library (2019). Jaspersoft® Community. https://community.jaspersoft.com/project/jasperreports-library (zuletzt aktualisiert am 21 November 2019, zuletzt geprüft am 21 November 2019).

20 RapidMiner. Lightning fast data science platform. https://rapidminer.com (zuletzt geprüft am 27 November 2019).

21 Tibco. Spotfire. https://www.tibco.com/products/tibco-spotfire (zuletzt geprüft am 27 November 2019).

22 SAS. Analytic software & solutions. www.sas.com (zuletzt geprüft am 21 November 2019).

23 SAP. Business intelligence (BI) & business analytics. https://www.sap.com/uk/products/analytics.html (zuletzt geprüft am 21 November 2019).

24 Microsoft. Big data and advanced analytics solutions. https://azure.microsoft.com/en-us/solutions/big-data (zuletzt geprüft am 21 November 2019).

25 adesso AG (2019). Facts about AI. https://ki.adesso.de/en/facts (zuletzt aktualisiert am 19 August 2019, zuletzt geprüft am 21 November 2019).

26 Spectaris (2019). Trendreport 2019. Analysen- bio- und labortechnik. Märkte, entwicklungen, potenziale. https://www.spectaris.de/fileadmin/Content/Analysen-Bio-und-Labortechnik/Zahlen-Fakten-Publikationen/Trendreport_ABL_2019.pdf (zuletzt geprüft am 20 September 2019).

27 Bluetooth SIG (2019). Bluetooth technology website. https://www.bluetooth.com/bluetooth-technology (zuletzt aktualisiert am 9 October 2019, zuletzt geprüft am 24 November 2019).

28 Johnson, C. S88 white paper – beginners.doc. Hg. v. ResearchGate. (zuletzt geprüft am 3 October 2019).

29 OPC Foundation. The industrial interoperability standard. https://opcfoundation.org (zuletzt geprüft am 25 November 2019).

30 SiLA. Standardization in Lab Automation. https://sila-standard.com (zuletzt geprüft am 25 November 2019).

31 Croissant, M. (2019). SiLA 2 hands on : bringing automation to the laboratory. https://medium.com/@matthieu.croissant/sila-2-hands-on-bringing-automation-to-the-laboratory-dacc12df7152 (zuletzt geprüft am 25 November 2019).

32 W3C (2019). XML technology. https://www.w3.org/standards/xml/ (zuletzt aktualisiert am 21 January 2019, zuletzt geprüft am 25 November 2019).

33 ecma International (2017). The JSON data interchange snytax. http://www.ecma-international.org/publications/files/ECMA-ST/ECMA-404.pdf (zuletzt aktualisiert am 1 December 2017, zuletzt geprüft am 25 November 2019).

34 Schaefer, B. (2018). Datenaustausch im labor der zukunft. https://www.git-labor.de (zuletzt aktualisiert am 9 October 2018, zuletzt geprüft am 24 September 2019).

35 AnIML Working Group. AnIML. https://www.animl.org (zuletzt geprüft am 27 November 2019).

36 Jüchli, D. (2019). Leitfaden zum einsatz von standards im automatisierten labor der zukunft. LIMS-Forum 2019. Dr. Klinkner & Partner. Jülich.

37 BSSN Software. Seahorse scientific workbench. https://www.bssn-software.com/products/seahorse-scientific-workbench (zuletzt geprüft am 28 November 2019).

38 Schäfer, B. (2019). Unleash the power of lab data. LIMS-Forum 2019. Dr. Klinkner & Partner. Jülich.

39 Allotrope Foundation. The allotrope framework. https://www.allotrope.org/allotrope-framework (zuletzt geprüft am 24 November 2019).

40 The HDF Group. HDF support portal. https://portal.hdfgroup.org/display/support (zuletzt geprüft am 25 November 2019).
41 Walter, P. (2018). Rohdatenarchivierung mittels ADF. LIMS-Forum 2018. Dr. Klinkner & Partner. Jülich.
42 Pistoia Alliance (2017). User experience for life sciences. https://www.pistoiaalliance.org/projects/user-experience-for-life-sciences (zuletzt geprüft am 1 November 2019).
43 Pistoia Alliance (2019b). Seeing scientists work. https://uxls.org/case-studies/contextual-inquiry (zuletzt aktualisiert am 1 January 2019, zuletzt geprüft am 1 November 2019).
44 adesso AG (Hg.). Creating space for developments - software projects in the interaction room (New School of IT) (zuletzt geprüft am 1 November 2019).
45 GAMP 5 (2008). *A Risk-Based Approach to Compliant GxP Computerized Systems*. Tampa, FL: ISPE Headquarters.
46 IREP Board. IREB in depth - about IREB - IREB – International Requirements Engineering Board. https://www.ireb.org/en/about/ireb, (zuletzt geprüft am 1 November 2019).
47 Pohl, K. and Rupp, C. (2015). *Basiswissen Requirements Engineering*. Aus- und Weiterbildung nach IREB-Standard zum Certified Professional for Requirements Engineering: Foundation Level Nach IREB-Standard. 4., Überarbeitete Auflage. Heidelberg: dpunkt https://ebookcentral.proquest.com/lib/subhh/detail.action?docID=2029882.
48 Pistoia Alliance (2019c). User experience for life sciences (UXLS). A UX Toolkit for the Life Sciences Community, by the Life Sciences Community. https://uxls.org (zuletzt aktualisiert am 1 January 2019, zuletzt geprüft am 20 September 2019).
49 Schwaber, K. (2017). The scrum guide. The definitive guide to scrum: the rules of the game. Unter mitarbeit von Jeff Sutherland. Hg. v. http://Scrum.org (zuletzt geprüft am 1 November 2019).

9.1

Polymorphs and Patents – the US Perspective

Kristi McIntyre

Bausch Health US, LLC, 400 Somerset Corp. Blvd., Bridgewater, NJ 08807, USA

9.1.1 Introduction

Patents have the potential to be extremely valuable – but it is important to understand some basics about what patents can and cannot do in order to understand how they may benefit you or your company. It is important to remember that legal landscape for intellectual property rights varies from country to country and invention to invention. What is patentable in one country may not be patentable in another, and just because a certain invention related to a polymorph is protectable does not mean all discoveries related to all polymorphs are protectable.

This chapter will provide a basic overview of the patent system and review certain examples of outcomes from the United States litigations involving crystal form and salt patents [1].

9.1.2 What Is a Patent?

A patent is a legal document, providing a set of exclusive rights granted for an invention for a limited period of time. In short, the patent system seeks to foster innovation by granting a limited duration monopoly in return for the public disclosure of an invention – especially assuming that some period of exclusivity is needed to induce an optimum level of innovation and public disclosure.[1]

The rules and requirements for patents differ from country to country, but there are some common threads. Generally, a patent is intended to include a full disclosure of an invention and describes how to make and use the invention. It typically consists of several predefined parts, for example, the title, introduction, background, short description of the invention/figures, detailed description, claims, and drawings.

Perhaps the most important part of a patent is its *claims*, which define the scope of patent protection, serving as a "fence" to let the public know where property rights begin and end. Claims, ideally, should cover *what you practice* (i.e. your

Solid State Development and Processing of Pharmaceutical Molecules: Salts, Cocrystals, and Polymorphism, First Edition. Edited by Michael Gruss.

product, your method, your process) as well as *what you want to stop competitors from practicing* [2].

Importantly, a patent provides a *right to exclude*. That is, in principle, a patent owner has the right to exclude others from making, using, distributing, importing, or selling his invention without his consent. But it is imperative to understand that a patent does not provide an affirmative right to make and use your own invention. In fact, there may be other patents out there that block you from doing so.

9.1.3 How Are Patents Obtained?

Patents are procured through a process called *patent prosecution*. Patent prosecution involves drafting patent applications, filing relevant documents with a country's patent office, and essentially negotiating with the patent office in order to obtain patent protection. While it may be possible for inventors to prosecute their own patents in worldwide patent offices, it is generally advisable to utilize a patent agent or patent attorney who is familiar with domestic and international patent prosecution procedures [3].

The application and associated documents will be reviewed by a patent examiner, who evaluates an application and can (and often will) issue rejections, which many times involve formalities and many times involve substantive rejections based on the national patentability requirements. Often, responses to such rejections include both a substantive response and amendments to the proposed claims.

It is important to understand that a patent is not infallible after it is granted. For example, in the United States, issued patents can be challenged through various *post-grant procedures* before the United States patent office, all with their own nuances, or during *litigation* in one of many United States district courts or at the United States International Trade Commission.

9.1.4 United States Patent Law

In the United States, there are many requirements that must be met to obtain a utility patent[2] – the first being that the claims constitute something called *patentable subject matter*. United States patent law identifies four categories of inventions that constitute patentable subject matter: processes, machines, articles of manufacture, and compositions of matter.[3, 4] Separate and apart from constituting patentable subject matter, an invention must also be *new (novel)*, *useful*, and *nonobvious*.[5, 6] While these terms have nuanced legal meanings, at a high level, this means that an invention must not have already been disclosed in the prior art[7]; must have some utility; and must be sufficiently different from what has been used or described before that it may be said to be nonobvious to a person of ordinary skill in the art.[8]

As of the time of writing (August 2019), the Federal Circuit – which is the court of appeals that hears patent cases in the United States – has weighed in on a handful of polymorph and salt cases. Below we describe select cases that may be informative regarding polymorph and salt patents. The intention is to describe some of the challenges raised during litigation and the legal outcome, not to provide legal advice. United States litigations generally take years and involve large volumes of documents, briefs, and fact and expert testimony – the summaries and simplifications presented herein are intended to provide general background information solely for educational purposes.[9] Each case has its own specific facts, each patent is different, and the materials herein may not be relevant to any particular situation – the appropriate solution in any case will vary, and the materials presented herein cannot take the place of individualized legal advice. The information here is presented based on the public records from the below mentioned cases and the court decisions associated with those cases. The referenced standards and litigations are a sampling of cases. Additional cases exist, both at the Federal Circuit and district court level, but the cases below touch on a variety of issues that may be interesting to a scientist.

9.1.4.1 Tapentadol Hydrochloride

Quick summary: Tapentadol hydrochloride Form A is not obvious over the prior art, including art discussing polymorph screening, and has utility.

Tapentadol hydrochloride is the active ingredient in commercial products branded Nucynta® (US) and Palexia® (European Union [EU] and rest of world [ROW]). Certain patents covering Nucynta products were challenged during litigation in the United States, including the tapentadol hydrochloride polymorph patent (US Patent No. 7,994,364 ("the '364 patent")) being challenged for, among other things, obviousness and lack of utility. After several weeks trial and subsequent appeal, the Federal Circuit determined that the claimed Form A was not obvious over the prior art and that Form A was shown to have sufficient utility [4]. Below a brief overview of the evidence and arguments is presented on obviousness and utility.

9.1.4.1.1 Tapentadol Hydrochloride Form A Held Not Obvious

The active pharmaceutical ingredients (API) tapentadol hydrochloride, used to treat pain, was discovered and patented by Grünenthal in the mid-1990s. US Patent No. 6,248,737 ("the '737 patent") discloses tapentadol hydrochloride and sets forth a method of synthesizing it in Example 25. The '737 patent indicates that tapentadol hydrochloride was crystallized, but it does not describe the resulting crystal structure or discuss whether tapentadol hydrochloride is polymorphic. Later investigation would show that faithful reproduction of Example 25 results in what is now known as **Form B** of tapentadol hydrochloride.

Years after the discovery of the compound, scientists at Grünenthal discovered that tapentadol hydrochloride was polymorphic and eventually identified and characterized two forms, later called Form A and Form B. Forms A and B are enantiotropic polymorphs, with Form A being the thermodynamically more stable form at room temperature. The discovery of **Form A** led to the filing of a patent application, which resulted in the issuance of the '364 patent.

At the time of the application of the ’364 patent, several things were known in the art: the structure of tapentadol hydrochloride; the synthesis method of Example 25 for manufacturing tapentadol hydrochloride; the concept of polymorph screening; and Food and Drug Administration (FDA) guidance concerning polymorph screening. On this basis, the patent challengers argued, among other things, that Form A of tapentadol hydrochloride was rendered obvious by virtue of the disclosure of Example 25 of the ’737 patent, plus general information regarding polymorph screening – in particular, the article *Pharmaceutical Solids: A Strategic Approach to Regulatory Considerations* published in Pharmaceutical Research [5].

The United States Court of Appeals for the Federal Circuit explained that because the record indicates that there was (i) no known or expected polymorphism of tapentadol hydrochloride (not all compounds are polymorphic); (ii) no evidence that the synthesis of Example 25 results in any Form A; and (iii) no guidance as to what particular solvents, temperatures, agitation rates, etc., were likely to result in Form A, the patent challengers failed to prove that there would have been a reasonable expectation that a polymorph screen of Form B (which is what results from Example 25 of the ’737 patent) would result in Form A. The court explained that the lack of knowledge in the field shows that there was “little to no basis” to expect a probability of success in producing the claimed Form A.

9.1.4.1.2 Tapentadol Hydrochloride Form A Was Found to Have Utility

The ’364 patent states that “Crystalline Form A ... has the same pharmacological activity as Form B but is more stable under ambient conditions. It can be advantageously used as [an] active ingredient in pharmaceutical compositions.”[10] The patent challenger contended that the ’364 patent lacks utility because the ’364 patent’s disclosure purportedly is vague; fails to provide benefit; and it does not demonstrate that Form A is superior to Form B. The Federal Circuit explained that the ’364 patent’s disclosure concretely discloses the practical benefit of Form A as an analgesic and, further, explained that the patentee need not prove that Form A has superior stability over Form B for purposes of determining utility.

9.1.4.2 Paroxetine Hydrochloride Hemihydrate

> *Quick summary: Paroxetine hydrochloride hemihydrate patent infringed by tablets using paroxetine hydrochloride anhydrate, under a theory of inevitable conversion, but hemihydrate patent invalid for inherent anticipation.*

Paroxetine hydrochloride (PHC) hemihydrate is the active ingredient in the commercial product Paxil®, an antidepressant. A patent covering Paxil was challenged during litigation in the United States, including the PHC hemihydrate crystal form patent (US Patent No. 4,721,723 (“the ’723 patent”)) being challenged for, among other things, anticipation and indefiniteness, and the generic challenger also alleged that it did not infringe the patent. After a trial and subsequent appeal, the Federal Circuit determined that the claimed crystalline PHC hemihydrate was infringed by Apotex and that the patent claim was definite, but that the patent claim was invalid as inherently anticipated by the prior art [6]. In particular, in an *en banc* decision, the Federal Circuit determined that the prior art did inherently disclose crystalline PHC hemihydrate, despite not discussing it, and rejected the patentee's "disappearing polymorph" theory. A brief overview of the evidence and arguments is presented later.

9.1.4.2.1 PHC Hemihydrate History

In the late 1970s, scientists at Ferrosan developed and patented a new class of compounds known as paroxetine. Ferrosan's scientists later developed crystalline PHC – crystalline PHC **anhydrate**. Ferrosan eventually licensed its patent and other PHC technology to SmithKline. While trying to improve PHC production, SmithKline synthesized crystalline PHC hemihydrate – different than the original PHC anhydrate. SmithKline eventually patented crystalline PHC hemihydrate, claiming "crystalline PHC hemihydrate." As mentioned earlier, PHC hemihydrate became the active ingredient in Paxil.

Apotex, seeking to market a generic PHC product, sought approval for a product with PHC **anhydrate** identified as the active ingredient. Apotex notified SmithKline of its intention to market its product, and SmithKline initiated a lawsuit, asserting that Apotex would infringe the '723 patent because its PHC anhydrate tablets would necessarily contain, by a conversion process, at least trace amounts of PHC hemihydrate.

In arguing that Apotex's PHC anhydrate tablets necessarily contain PHC hemihydrate, SmithKline offered expert testimony about "seeding" and "disappearing polymorphs" to explain why PHC anhydrate would necessarily convert to the hemihydrate. SmithKline explained that although Ferrosan may have originally created crystalline PHC anhydrate with the prior art method, **after** the environment became seeded with SmithKline's crystalline PHC hemihydrate, PHC anhydrate inevitably converts to PHC hemihydrate upon its inevitable contact with seeds of the hemihydrate. Essentially, after PHC hemihydrate came into existence, creation of PHC anhydrate became difficult or impossible, meaning the old polymorph (the anhydrate) was a "disappearing polymorph." SmithKline argued that the PHC anhydrate used by Apotex would necessarily convert to the hemihydrate due to seeding, but the prior art did **not** necessarily disclose the hemihydrate.

9.1.4.2.2 Meaning of "Crystalline Paroxetine Hydrochloride Hemihydrate"

At the district court, there were various arguments about the scope of the claim "crystalline PHC hemihydrate," including arguments that it should include only commercially significant quantities of crystalline PHC hemihydrate, such that tablets with undetectable crystalline PHC hemihydrate would not infringe. The Federal Circuit rejected this argument, calling the claim "plain on its face" and "covering a definite chemical structure." It explained that it would be difficult to imagine a more clear and definite claim and that a skilled artisan would readily understand the bounds of the invention.

9.1.4.2.3 PHC Hemihydrate: Infringed, But Invalid for Anticipation

The Federal Circuit has stated *that which would literally infringe if later in time anticipates if earlier in time.* With that in mind, that Federal Circuit discussed infringement and anticipation of the '723 patent claim hand in hand.

SmithKline argued that while Apotex's practice of the prior art infringed its PHC hemihydrate patent, the prior art did not anticipate its PHC hemihydrate patent. In support of this, SmithKline cited the "disappearing polymorph" theory. It argued that following the prior art **prior to** its discovery of PHC hemihydrate would **not** have resulted in production of the hemihydrate, while, on the other hand, Apotex's practice of the prior art **after** the discovery of PHC hemihydrate **did** result in production of the hemihydrate.

The Federal Circuit agreed with only part of SmithKline's position, however. The court explained that the record – and in particular the discussion of inevitable conversion from the anhydrate to the hemihydrate – supported a finding of infringement of the claim by Apotex's PHC anhydrate product. But the court's view differed from SmithKline's argument when it came to anticipation, concluding that whether it was actually possible to make pure PHC anhydrate in the past was not relevant and that the prior art suffices as anticipatory if it discloses in an enabling manner the production of PHC hemihydrate. Because it found the prior art disclosed a method of manufacturing PHC anhydrate that naturally resulted in the production of the hemihydrate, it held that the claim "crystalline PHC hemihydrate" was invalid as anticipated by the prior art. In other words, it held that the claim was invalid because it was not legally new or novel.

9.1.4.3 Ranitidine Hydrochloride

Quick summary: Where following a prior art process results sometimes in Form 2 and other times in Form 1, Form 2 is not inherently anticipated.

Ranitidine hydrochloride (RHCl) is the active ingredient in the commercial product Zantac®, an antiulcer medication. A patent covering **Form 2** of RHCl (US Patent No. 4,521,431 ("the '431 patent")) was the subject of litigation in the United States, including a lengthy discussion of inherent anticipation. After a trial and subsequent appeal, the Federal Circuit determined that the '431 patent was valid and that the prior art process did **not** inherently disclose Form 2 [7]. Below a brief overview of the evidence and arguments is presented.

9.1.4.3.1 History of RHCl Form 2

Glaxo chemists discovered and patented RHCl in the 1970s. The patent disclosed one method of preparing RHCl, set forth in Example 32. Internally, Glaxo produced large quantities of RHCl, using procedures called Processes 3A and 3B, which were different from the Example 32 procedure. But until April 1980, the syntheses of Example 32, Process 3A, and Process 3B all resulted in identical RHCl.

In April 1980, Process 3B created a different type of RHCl – what would be called Form 2 of RHCl, with the old polymorph being referred to as Form 1. Glaxo eventually patented Form 2 of RHCl.

9.1.4.3.2 RHCl Form 2 Not Anticipated by Example 32

During litigation, Novopharm argued that RHCl Form 2 was anticipated by the disclosure of Example 32 of the prior art. In support of its argument, Novopharm's experts performed Example 32 thirteen times. Each time, they got Form 2, not Form 1. By contrast, Glaxo presented evidence that it had made Form 1 by Example 32, and Glaxo's expert likewise followed Example 32 and obtained Form 1. Based on this, the district court and the Federal Circuit determined that Form 2 was ***not*** anticipated by the prior art.

9.1.4.4 Cefdinir

> *Quick summary: Where a priority document describes multiple polymorphic forms and the U.S. application describes and claims only one form, the other polymorphic forms that were jettisoned between the priority document and U.S. application are dedicated to the public.*

Cefdinir is the active ingredient in the commercial product Omnicef®, an antibacterial. A patent covering Omnicef (US Patent No. 4,935,507 ("the '507 patent")) was the subject of litigation in the United States – in particular, the question of infringement of the '507 patent was the subject of litigation. After a summary judgment determination and subsequent appeal, the Federal Circuit determined that Lupin's generic product, which contained Crystal Form B, did not infringe the '507 patent, which was directed to Crystal Form A [8]. Below a brief overview of the evidence and arguments is presented.

Claim 1 of the '507 patent recites crystalline cefdinir showing peaks at seven diffraction angles. Although the claim does not specifically recite "Crystal Form A," the patent specification states that Crystal Form A "shows its distinguishing peaks" at the seven diffraction angles listed in claim 1 and does not mention any other crystal forms. In light of this, the district court concluded that "crystalline" in the claims means Crystal Form A as outlined in the specification.

In contrast to the '507 patent, which only mentioned Crystal Form A, the '507 patent claims priority to a foreign patent that claimed two crystalline forms of cefdinir: Crystal Form A and Crystal Form B. Both forms were defined by X-ray powder diffraction (XRPD) and infrared (IR) in the priority document. Despite claiming priority to the foreign application, the '507 patent did not mention Crystal Form B.

The patentee argued infringement of claim 1 by Crystal Form B under the doctrine of equivalents. The Federal Circuit concluded that use of Crystal Form B did **not** infringe the claim, specifically noting that the '507 patent **declined to claim** an embodiment expressly disclosed in its priority document, and therefore dedicated Crystal From B to the public, foreclosing the argument of infringement under the doctrine of equivalents.

9.1.4.5 Amlodipine Besylate

Quick summary: Patent claim directed to amlodipine besylate found obvious where "it was reasonable to expect that the combination of amlodipine and benzene sulphonate—one of the 53 anions disclosed the prior art—would likely result in amlodipine besylate because of the known acid strength, solubility, and other chemical characteristics of the benzene sulphonate" [9].

Amlodipine besylate is the active ingredient in the commercial product Norvasc®, which is approved for treating hypertension and chronic stable and vasospastic angina. A patent covering pharmaceutically acceptable salts of amlodipine (US Patent No. 4,879,303 ("the '303 patent")) was the subject of litigation in the United States. After a trial and subsequent appeal, the Federal Circuit determined that the district court erred in holding that the subject matter of the '303 patent would not have been obvious [10].

9.1.4.5.1 History of Amlodipine Besylate

Scientists at Pfizer invented amlodipine and discovered its antihypertensive and anti-ischemic properties prior to 1982. Pfizer filed a patent application on – and eventually obtained US Patent No. 4,572,909 ("the '909 patent") – covering amlodipine and pharmaceutically acceptable acid-addition salts, which are described as those formed from acids which form nontoxic acid addition salts containing pharmaceutically acceptable anions, such as hydrochloride, hydrobromide, sulfate, phosphate or acid phosphate, acetate, maleate, fumarate, lactate, tartrate, citrate, and gluconate salts. The '909 patent discloses the preferred salt as maleate.

During drug development, Pfizer identified a formulation for amlodipine maleate that produced excellent capsules, but had issues with tableting, including chemical instability and issues with the tablet blend. So Pfizer attempted to make other salts, selecting seven alternative anions based on differing structures and properties. While both amlodipine maleate and amlodipine besylate ultimately ended up in clinical trials, Pfizer discovered that the besylate salt showed much improved stability over the maleate salt in all cases and also showed superiority in processing characteristics. Eventually, Pfizer filed a patent application to amlodipine besylate, which issued as the '303 patent.

9.1.4.5.2 Amlodipine Besylate Found Obvious

At the district court, the claims of the '303 patent were found nonobvious over the '909 patent and various prior art, including the Berge article, which disclosed 53 anions, including besylate, used in FDA-approved, commercially marketed products. The Federal Circuit reversed this determination, instead finding that the claims of the '303 patent were, in fact, obvious. The Federal Circuit explained that out of the list of 53 anions in the Berge article, one of ordinary skill in the art would have considered besylate because of its known acid strength, solubility, and other known chemical characteristics. Despite Pfizer's argument (and the district court's previous finding) that there would have been no reasonable expectation of success in making amlodipine besylate, the Federal Circuit disagreed, explaining that a finding of obviousness cannot be avoided simply by a showing a degree of unpredictability in the art, so long as there was a reasonable probability of success. And while the district court had found that the besylate salt resulted in unexpected superior results, supporting a finding of nonobviousness, the Federal Circuit disagreed, explaining that "given the range of 53 anions disclosed by Berge, one skilled in the art would expect those anions to provide salts having a range of properties, some of which would be superior, and some of which would be inferior, to amlodipine maleate" [11]. On this basis, the Federal Circuit explained that the properties of the besylate salt were not unexpected, but even if they were, that the results could not overcome the strong case of obviousness.

Notably, the Federal Circuit has since stated that the holding in *Pfizer v. Apotex* was based on the particularized facts at issue in the case. For example, in *Grunenthal v. Actavis*, discussed above and finding a polymorphic form nonobvious, the court noted: "Under the 'particularized facts of [*Pfizer*],' such expectation was reasonable because a "person skilled in the arts" (POSA) would have narrowed the list of 53 anions 'to a few' due to *known* characteristics of the anions" [12].

9.1.4.6 Concluding Remarks

Whether or not your solid state or salt form is patentable in the United States or around the world is a fact-specific matter – a fact specific matter worth documenting and discussing with a team of qualified attorneys and expert scientists. But one theme remains clear: a valid US patent can be a very valuable asset.

Notes

1 There are many sources that speak generally about the purpose behind the patent system and substantiate the general points offered here, including: The World Intellectual Property Organization on Patents (https://www.wipo.int/patents/en/) or The United States Patent and Trademark Office's General Information Concerning Patents (https://www.uspto.gov/patents-getting-started/general-information-concerning-patents) (hereinafter "USPTO General Information") (accessed 28 August 2019).
2 This chapter will focus on utility patents and will not discuss design patents or plant patents, which are also available in the United States.
3 To date, while there has been much to be said about what constitutes patentable subject matter, the patentable subject matter requirement traditionally has not been a problem for inventions related to solid forms of man-made chemicals, and so the subject will not be addressed here.
4 *See generally* USPTO General Information at "What Can Be Patented."
5 The claims must also comply with certain other statutory requirements—for example, the written description, enablement, and definiteness requirements. These requirements will not be discussed in detail herein.

6 35 U.S.C. §§ 101–103, 112.

7 Prior art is a legally defined, nuanced concept. But, essentially (and with some exceptions), a reference (or public use) is prior art if it was publicly available before the effective filing date of the claimed invention.

8 *See generally* USPTO General Information at "Novelty And Non-Obviousness, Conditions For Obtaining A Patent."

9 Additional case summaries, from the perspective of the late polymorph expert Dr. Joel Bernstein, can be found in Chapter 10 of his book Polymorphism in Molecular Crystals (2002). Dr. Bernstein was a wealth of knowledge and is greatly missed.

10 '364 patent at col. 4, ll. 13–16.

References

1 Another resource for chemists interested in learning about patents is the American Chemical Society's (2019). What every chemist needs to know about patents. (4. https://www.acs.org/content/dam/acsorg/about/governance/committees/patents/what-every-chemist-knows-patents.pdf (accessed 25 September 2019).

2 See, e.g.: Cook, E. (2017). Patents: what to be aware of, IT News. https://www.finnegan.com/en/insights/patents-what-to-be-aware-of.html (accessed 28 August 2019) for a brief discussion regarding deciding what to patent and claim strategy, generally.

3 See: USPTO (2011). Advice: working with a patent practitioner. https://www.uspto.gov/inventors/independent/eye/201101/advicepracticioner.jsp (accessed 28 August 2019).

4 See generally Grunenthal GmbH v. Alkem Labs. Ltd., 919 F.3d 1333 (Fed. Cir. 2019).

5 Byrn, S., Pfeiffer, R., Ganey, M. et al. (1995). Pharmaceutical solids: a strategic approach to regulatory considerations. *Pharmaceutical Research* 12: 945–954.

6 See generally SmithKline Beecham Corp. v. Apotex Corp., 403 F.3d 1331 (Fed. Cir. 2005).

7 See generally Glaxo Inc. v. Novopharm Ltd., 52 F.3d 1043 (Fed. Cir. 1995).

8 See generally Abbott Labs. v. Sandoz, Inc., 566 F.3d 1282 (Fed. Cir. 2009).

9 See Grunenthal, 919 F.3d at 1343–44 (discussing the particulars of Pfizer v. Apotex, 480 F.3d 1348 (Fed. Cir. 2007).

10 See generally Pfizer, 480 F.3d 1348.

11 See generally Pfizer, 480 F.3d 1371.

12 See Grunenthal, 919 F.3d at 1343–44 (discussing Pfizer, 480 F.3d 1348).

9.2

Polymorphs and Patents – The EU Perspective

Oliver Brosch

European Patent and Trademark Attorneys, Kutzenberger Wolff & Partner, Waidmarkt 11, 50676 Cologne, Germany

9.2.1 European Patent Applications and European Patents

9.2.1.1 Introduction

In addition to the US perspective, this chapter will provide an outline of the procedure involved in applying for a European patent and in getting a European patent especially against the background of crystal forms. Still further, this chapter will provide a brief outlook to the Unitary Patent system the start of which is currently expected for the first half of 2022.

9.2.1.2 Summary of the Processing of Applications and Patents Before the European Patent Office (EPO)

The processing of a European application and of a European patent is carried out in a number of distinct steps which may be summarized as follows:

(i) the application is filled with the European Patent Office (EPO) or a competent national authority;
(ii) the Receiving Section examines the application to determine if a date of filing can be accorded to the application;
(iii) the Receiving Section carries out the formal examination of the application;
(iv) if the Receiving Section has established that the application complies with the formal requirements, the search division draws up an EESR (extended European search report), a copy of which is forwarded to the applicant;
(v) the application and the search report are published by the EPO either together or separately;
(vi) on receipt of a request for examination from the applicant, or, if the request for examination has been filed before the search report has been transmitted to the applicant, on confirmation by the applicant that it is desired to proceed further

Solid State Development and Processing of Pharmaceutical Molecules: Salts, Cocrystals, and Polymorphism, First Edition. Edited by Michael Gruss.

with the European patent application, the application is subjected to substantive examination and any necessary formal examination before a European patent is granted by the examining division;

(vii) provided the requirements of the European Patent Convention (EPC) are met, a European patent is granted for the states designated;

(viii) the specification of the European patent is published by the EPO;

(ix) within nine months from publication, any person may give notice of opposition to the European patent granted; after examining the opposition, the opposition division decides whether to reject the opposition, maintain the patent in amended form, or revoke the patent;

(x) the patent proprietor may request limitation or revocation of the granted European patent; the examining division will decide on this request;

(xi) if the European patent is amended, the EPO publishes a new specification of the European patent amended accordingly.

The patent owner's exclusive rights to an invention are defined by the patent claims. Thus, throughout the entire process of drafting and prosecuting a patent application for an invention, particular care must be exerted regarding the wording of the claims. In many cases, various set of patent claims are submitted to the EPO, referred to as main request and auxiliary requests. If a main request is not considered to be patentable, the auxiliary requests will then be considered by the Office in ascending order and – if the criteria of patentability are fulfilled – the claims will be granted. This approach allows for an efficient prosecution.

Any decision by the Receiving Section, an examining division, an opposition division, or the Legal Division which adversely affects a party is appealable and, thus, subjects to review before a board of appeal of the EPO.

Once granted, as of 2019, the European patent may be validated in one or more of the following Contracting States: Albania, Austria, Belgium, Bulgaria, Croatia, Cyprus, Czech Republic, Denmark, Estonia, Finland, France, Germany, Greece, Hungary, Iceland, Ireland, Italy, Latvia, Liechtenstein, Lithuania, Luxembourg, Malta, Monaco, Netherlands, Republic of North Macedonia, Norway, Poland, Portugal, Romania, Serbia, San Marino, Slovak Republic, Slovenia, Sweden, Switzerland, Turkey, and United Kingdom.

9.2.1.3 Economic Factors

Processing fees in the European patent grant procedure are staggered, so that at each stage of the procedure, the applicant has a further chance to decide, in the light of the previously completed stages, whether the interest in obtaining patent protection is still great enough to justify paying the next fee.

In particular, the separation between search and substantive examination enables the applicant to decide in the light of the European search report whether it is worth requesting substantive examination.

Taking into account the fees levied for the European grant procedure, costs for representation by a single agent, and the cost of conducting the proceedings in a single language, a European patent as a rule costs about as much as the prosecution of three or four national patents.

9.2.1.4 Unitary Patents

A Unitary Patent is a "European patent with unitary effect," which means a European patent granted by the EPO under the rules and procedures of the EPC to which, after grant, unitary effect is attributed for the territory of the participating Member States at its proprietor's request.

The Unitary Patent system builds on the EPC. This means that the pre-grant phase is exactly the same as for European patents. Applicants file a European patent application seeking the grant of a European patent for some or all of the EPC contracting states with the EPO. It examines their application in accordance with the EPC and, if all formal and substantive requirements for patentability are met, grants a European patent.

The EPO has been entrusted by the participating Member States with certain administrative tasks relating to Unitary Patents, in particular the administration of requests for a Unitary Patent (requests for unitary effect). Accordingly, once a European patent has been granted, a separate, post-grant procedure can be initiated at the EPO with a view to obtaining a Unitary Patent.

This procedure is less complex and significantly cheaper than the existing system of national validation according to item 1.2 above and so offers an attractive new alternative for proprietors of European patents.

The start of the Unitary Patent system is currently expected for the first half of 2022.

9.2.1.5 Protection of Polymorphs and Solid Forms in General

European patents are granted for inventions that are new, involve an inventive step, and are susceptible of industrial application. An invention can belong to any field of technology.

The current practice on the protection of solid forms, i.e. polymorphs, solvates, salts, and cocrystals of known drugs principally follow these general European rules for assessing patentability. If not specified differently, in the following text, the term polymorph(s) shall comprise the solid forms mentioned before.

First, the polymorph needs to be clearly defined. This requirement is typically satisfied by specifying a series of several values for an analytical parameter such as X-ray powder diffraction signals, Raman signals, infrared (IR) signals, solid nuclear magnetic resonance spectroscopy (NMR) signals, thermal events, and the like. It is principally possible to cover a mixture of polymorphs, i.e. there is no formal requirement of limiting the claims to a single polymorphic species.

Second, the polymorph needs to be enabled, i.e. it must be obtainable in a reproducible manner. This requirement is usually satisfied by providing at least one example relating to the manufacture of the polymorph.

Third, the polymorph needs to be novel. In this regard, it can be particularly relevant whether any synthesis protocol disclosed in a prior art reference inevitably provides a particular polymorph, even if the crystalline structure of said polymorph is not explicitly characterized in said reference. Provided that a faithful reproduction of the synthesis protocol inevitably yields the polymorph in question, it is inherent to the disclosure of said prior art reference and thus not novel. Mixtures of various polymorphs may also be relevant. A properly defined single specific crystalline form can be novel over a mixture of polymorphs, even if said mixture inter alia contains said specific crystalline form besides another crystalline form.

Fourth, the polymorph needs to involve in an inventive step. Most decisions of the Technical Boards of Appeal of the EPO are focused on this requirement of patentability. As polymorph screening has become routine work, however, many aspects of polymorph screening are nowadays within common general knowledge of the skilled person. Essentially for that reason, the assessment as to whether a given polymorph involves an inventive step has changed over the years more significantly than the assessment of the other patentability requirements.

9.2.1.6 Polymorph Screening

Polymorph screening has meanwhile become a regulatory requirement for marketing approval. Regulatory authorities have published guidance to assist applicants seeking for regulatory approval for drug substances that exist in polymorphic forms. The systematic investigation of drug polymorphism has become a routine practice. Standard procedures for polymorph screening are well known in the art. Therefore, when assessing inventive step, it must be taken into account that the person skilled in the art is familiar with routine polymorph screening by means of such standard procedures.

In the absence of any comparative experimental data, different polymorphs, solvates, cocrystals, or salts of one and the same drug are usually considered as being functionally equivalent alternatives having the same pharmacokinetic and pharmacodynamic properties. Further, it is typically expected that standard procedures for polymorph screening will reveal new polymorphs. The mere provision of a new polymorph as a result of performing standard procedures for polymorph screening is therefore not regarded as a significant technical contribution to the art.

In general, according to the established jurisdiction of the Technical Boards of Appeal of the EPO, the mere discovery of a (new) polymorph by means of standard procedures for polymorph screening does not require inventive skills and hence does not involve an inventive step.

This is set forth in the first headnote of the landmark decision T 777/08 of 24 May 2011 (EPO, T 0777/08 (Atorvastatin polymorphs/WARNER-LAMBERT) of 24.5.2011 [1]):

> "At the priority date of the patent in suit, the skilled person in the field of pharmaceutical drug development would have been aware of the fact that instances of polymorphism were commonplace in molecules of interest to the pharmaceutical industry, and have known it to be advisable to screen for

polymorphs early on in the drug development process. Moreover, he would be familiar with routine methods of screening. Consequently, in the absence of any technical prejudice and in the absence of any unexpected property, the mere provision of a crystalline form of a known pharmaceutically active compound cannot be regarded as involving an inventive step."

In order to render a polymorph inventive, especially the following aspects of polymorph screening can become relevant:

(i) it can be made plausible or demonstrated by comparative experimental data that the newly discovered polymorph has unexpected technical advantages, e.g. with regard to hygroscopicity, stability, bioavailability, and the like; or
(ii) there is a technical prejudice in the art which would have prevented a skilled person from taking the measures which have led to the discovery of the new polymorph; or
(iii) the newly discovered polymorph is not obtainable by routine procedures, but requires specific procedures which go far beyond of what is currently to be considered as routine procedures for polymorph screening.

The question of whether a given procedure is among the standard procedures for polymorph screening is of course subject to further progress in this technical field. This will be reflected by textbooks and review articles. Thus, over the next years, it can be foreseen that procedures for obtaining solid forms of chemical substances, which are presently considered rather innovative, will be among the standard procedure in the future. Currently, the most common methods to prepare polymorphs include solvent evaporation, slow cooling, solvent diffusion, vapor diffusion, and vacuum sublimation.

9.2.2 Decisions of Technical Boards of Appeal of the EPO

In the following, we summarize five representative decisions of Technical Boards of Appeal of the EPO illustrating that the assessment of patentability highly depends upon the specific circumstances of the individual case.

In general, decisions can be searched for in the EPO decision database (EPO, Search in the Board of Appeal decisions database [2]) as exemplified by a hitlist for the term "polymorph" Table 9.2.1.

9.2.2.1 Decision T 777/08 of 24 May 2011

This decision (EPO, T 0777/08 (Atorvastatin polymorphs/WARNER-LAMBERT) of 24.5.2011 [1]) has already been addressed in the previous Section 9.2.1.6.

9.2.2.2 Decision T 1555/12 Dated 29 April 2015

The granted claims of the patent in suit were drawn to anhydrous Aripiprazole crystals C having an X-ray powder diffractogram (XRPD) with eight characteristics

Table 9.2.1 Hitlist of search for "polymorph" in the EPO decision database (selected hits, as of 23 November 2019).

Decision no.	Decision title	Decision date	Application no.	Application title
T 1450/14	Mesylatsalz	24 July 2019	7115663	3-[(2-{[4-(Hexyloxycarbonylamino-imino-methyl)-phenylamino]methyl}-1-methyl-1*H*-benzimidazol-5-carbonyl)-pyridin-2-yl-amino]-propionsäure-ethylester Methans
T 1894/15	(No title)	7 May 2019	6760544	Crystalline solid forms of tigecycline and methods of preparing same
T 0246/15	Vardenafil enthaltende Arzneimittel	13 November 2018	3763695	Vardenafil hydrochloride trihydrate enthaltende arzneimittel und deren herstellungsverfahren
T 0264/16	Pitolisant	27 September 2018	6744466	Treatment of symptoms of Parkinson's disease with non-imidazole alkylamines histamine H3-receptor ligands
T 0662/15	Using of organic solvents in wet granulation of moxifloxacin	18 September 2018	8807939	Using of organic solvents in wet granulation of moxifloxacin
T 2397/12	Crystalline squalamine dilactate	12 March 2018	6751251	Polymorphic and amorphous salt forms of squalamine dilactate
T 0182/13	Moxifloxacin HCl polymorphs	25 August 2017	4791330	Polymorphs of 1-cyclopropyl-7-([S...
T 0758/12	Selecting psychotropic medications	4 July 2017	7840663	Methods for selecting medications
T 1936/13	Olanzapine	27 June 2017	6841451	Oral formulation of anhydrous olanzapine form I
T 0168/12	Lizenolid form III	14 March 2017	9163348	Crystalline form of linezolid
T 0565/15	Mycophenolate sodium salt	2 February 2017	4703158	Process for modifying drug crystal formation of mycophenolate sodium salt
T 2114/13	Stable polymorphic form of febuxostat	12 October 2016	8005934	Polymorph of 2-(3-cyano-4-isobutyloxyphenyl)-4-methyl-5-thiazolecarboxylic acid and method of producing the same

T 0455/13	Tolperisone	22 June 2016	8767427	Compositions of tolperisone
T 1872/14	Travoprost	21 April 2016	8100474	Fluprostenol isopropyl ester for use in the treatment of glaucoma and ocular hypertension
T 2399/12	(No title)	19 January 2016	8725456	Method of therapeutic administration of DHE to enable rapid relief of migraine while minimizing side effect profile
T 2397/11	Erlotinib hydrochloride polymorph	15 October 2015	4710031	Polymorph of {6,7-bis(2-methoxy-ethoxy)-quinazolin-4-yl}-(3E)
T 2566/11	(No title)	17 September 2015	2785879	Extended release compositions comprising as active compound venlafaxine hydrochloride
T 2597/11	(No title)	25 June 2015	3782368	Mastitis treatment with a combination of prednisolone and cephalosporin
T 0517/14	Ibandronate sodium, Form T	19 June 2015	8002626	Crystalline form of ibandronate sodium
T 2128/13	(No title)	18 June 2015	7810005	Methods of manufacturing and modifying taxane coatings for implantable medical devices
T 0205/14	Ibandronate sodium, Form QQ	18 June 2015	5791142	Crystalline form of ibandronate sodium and processes for preparation thereof
T 0177/13	Enteric solid oral dosage form of a bisphosphonate containing a …	11 June 2015	5735411	Enteric solid oral dosage form of a bisphosphonate containing a chelating agent
T 1657/11	(No title)	9 June 2015	4797879	Verfahren zur Herstellung einer festen oral applizierbaren pharmazeutischen Zusammensetzung mit 5-Chlor-*N*({(5-2-oxo-3-[4-(3-oxo-4-morpholinyl)-phenyl]-1,3...

(Continued)

Table 9.2.1 (Continued)

Decision no.	Decision title	Decision date	Application no.	Application title
T 0256/13	(No title)	2 June 2015	5795965	Solid pharmaceutical composition comprising donepezil hydrochloride
T 1555/12	Aripiprazole polymorph	29 April 2015	4002427	Low hygroscopic aripiprazole drug substance and processes for the preparation thereof
T 1651/11	Apremilast against psoriasis	17 March 2015	3721414	(+)-2-[1-(3-*ethoxy*-4-methoxyphenyl)-2-methylsulfonylethyl]-4-acetylaminoisoindoline-1,3-dione for use in treating psoriasis by oral administration
T 0094/11	(No title)	27 November 2014	2718949	Succinate salt of o-desmethyl-venlafaxine
T 2007/11	(No title)	7 October 2014	7809032	Polymorphic form of imatinib mesylate ethanol solvate and process for its preparation
T 0643/12	Lenvatinib mesylate polymorphs	18 June .2014	4807580	Crystal of salt of 4-(3-chloro-4-(cyclopropylaminocarbonyl)cyclopropylamino-carbonyl)-7-methoxy-6-methoxy...
T 0833/11	Fluticasone furoate	8 April 2014	1953272	6alpha, 9alpha-Difluoro-17alpha-(2-furanylcarbonyl)-oxy-11beta-hydroxy-16alpha.-methyl-3-oxo-androst-1,4-diene-17beta-carbothioic acid S-fluoromet...
T 0335/10	Sorbitol	22 November 2013	99403095	Sorbitol pulvérulent et son procédé de préparation
T 0598/12	Tablets comprising ethinylestradiol and drospirenone	5 November 2013	3017743	Pharmaceutical combination of ethinylestradiol and drospirenone for use as a contraceptive
T 1723/10	Amorphous forms	1 October 2013	5706666	Amorphous forms of risedronate monosodium
T 1772/09	(No title)	5 July 2013	2782507	Low hygroscopic aripiprazole drug substance and processes for the preparation thereof

T 1422/12	Tigecycline crystalline forms	11 April 2013	7776245	Tigecycline crystalline forms and processes for preparation thereof
T 1677/11	(−)-omeprazole Na	27 November 2012	108479	The sodium salt of the (−)-enantiomer of omeprazole
T 1760/11	(−)-Omeprazole Mg	13 November 2012	108480	Magnesium salt of the (−)-enantiomer of omeprazole and its use
T 1572/08	(No title)	16 October 2012	4737220	Immediate-release pharmaceutical dosage form comprising polymorphous tibolone
T 0097/11	Crystalline meropenem	30 May 2012	8101985	Crystalline carbapenem compound and produced method thereof
T 0075/11	(No title)	24 April 2012	2764081	Crystalline form of omeprazole
T 1285/11	Polymorphs	15 December 2011	8806952	Polymorphs of 3-(((4-tert-butyl-benzyl...
T 0777/08	Atorvastatin polymorphs	24 May 2011	1116338	Crystalline R-(R*, R*)$-2-(4-fluorophenyl)-beta, delta-dihydroxy-5-(1-methylethyl)-3-phenyl-4-(phenylamino)carbonyl -1*H*-pyrrole-1-heptanoic acid hemi calci...
T 0058/08	Combination particles	10 February 2011	1974368	Combination particles for the treatment of asthma
T 1753/06	Tert-butylamine du perindopril	6 May 2009	1954058	Forme cristalline alpha du sel de tert-butylamine du perindopril
T 0912/06	High purity tibolone	17 April 2009	99948994	High purity compound (7alpha, 17alpha)-17-hydroxy-7-methyl-19-nor-17-pregn-5(10)-en-20-yn-3-one
T 0392/06	Triazinylaminostilbene	22 October 2008	98810443	Triazinylaminostilbene compounds
T 0699/05	Using laser light to control crystal form	23 June 2008	913216	Method for using laser light to control crystal form
T 0550/04	Production of quinacridones	2 April 2008	94304782	Process for the production of 2,5-di(arylamino)-3,6-dihydroterephthalic acid dialkyl ester, and process for the production of quinacridone from said ester ...
T 1210/05	2-(2-Pyridylmethylsulfinyl-benzimidazole crystals	24 January 2008	97912445	Crystals of benzimidazole derivatives and their production

(Continued)

Table 9.2.1 (Continued)

Decision no.	Decision title	Decision date	Application no.	Application title
T 0323/05	Finasteride	9 August 2007	93203163	A process for the production of finasteride
T 1341/04	(No title)	10 May 2007	95306832	Process for accelerating the polymorphic transformation of edible fats using ultrasonication
T 0250/04	(No title)	8 February 2007	95306833	Process for retarding fat bloom in fat-based confectionery masses
T 0256/03	Aerosol formulation	7 November 2006	95901923	Flunisolide aerosol formulations
T 1246/05	Donepezil hydrochloride	25 July 2006	2005248	Polymorphs of donepezil hydrochloride and process for production
T 1066/03	Polymorphic atorvastatin	11 July 2006	96924553	Novel process for the production of amorphous [R-(R*,R*)]-2-(4-fluorophenyl)-beta,delta-dihydroxy-5-(1-methylethyl)-3-phenyl-4-[(phenylamino)carbonyl]1*H*-P...
T 0817/03	Kristalliner D-sorbit	28 June 2006	97935531	Verfahren zur herstellung von kristallinem D-sorbit
T 0605/02	Finasteride	27 September 2005	97201712	Polymorphic forms I and II of finasteride
T 0885/02	Paroxetine methanesulfonate	15 December 2004	99303151	Paroxetine methanesulfonate
T 0025/01	Finasteride	16 September 2003	93203163	A process for the production of finasteride
T 0455/98	Crystalline sugar alcohols	7 December 2001	90312752	Crystalline sugar alcohol containing uniformly dispersed particulate pharmaceutical compound
T 0226/98	Famotidine	7 February 2001	87306882	Morphologically homogenous forms of famotidine and processes for their preparation
T 0020/94	Amorphous TPM	4 November 1998	87201409	Tetrakis [3-(3,5-di-tert.buthyl-4-hydroxyphenyl) propionyl-oxymethyl] methane with amorphous structure, process for its preparation and its use as a stabilizer
T 0173/89	(No title)	29 August 1990	80303773	Improved gamma-sorbitol polymorph and uses thereof

Source: Adapted from EPO [2].

peaks, irrespective of whether this form was a single crystalline form (i.e. a single polymorph) or a mixture. Heating known crystals under the conditions described in a prior art reference yielded a crystalline material of Aripiprazole with XRPD peaks according to the granted claims. The Board did not consider it relevant whether or not the skilled person was aware of the fact that some of the identifiable peaks may belong to different polymorphic forms. Neither was it relevant whether or not it would have been possible for the skilled person to separate the different crystalline forms from one another. Novelty was thus denied (EPO, T 1555/12 (Aripiprazole polymorph/OTSUKA) of 29.4.2015 [3]).

The claims of auxiliary request 1 differed from the granted claims inter alia in that particular solid 13C-NMR peaks were recited. The Board held that in the absence of solid 13C NMR data on the prior art material, it could not be concluded with the required certainty that the crystalline form according to auxiliary request 1 was disclosed in the prior art. Novelty was thus acknowledged. With regard to inventive step, the Board emphasized that the claims of auxiliary request 1 were still not only limited to a thermally stable, highly pure crystalline form of Aripiprazole (i.e. a single polymorph) but also encompassed mixtures with other crystals which were thermally instable, in more or less any amount. The objective technical problem was regarded as the mere provision of a further crystalline form of Aripiprazole. According to the prior art, one crystalline form was prepared by heating another crystalline form, but under the heating conditions, the crystals were apparently not stable and transformed into a different crystalline form or forms (i.e. mixtures of polymorphic forms). Although the proprietor was the first to report this transformation, the Board was of the opinion that the same observation was within the normal skills of any person skilled in the art following the teaching of the prior art. "In view of regulatory requirements the skilled person would then routinely examine the conditions for this transformation and investigate the respectively obtained crystalline form(s). The mere provision of a further crystalline form, including mixtures, as the result of such routine investigations and routine experimentation does not require inventive skills." Thus, inventive step was denied.

The claims of auxiliary request 2 differed from the granted claims in that a reference to the full XRPD according to a published figure was added. In view of the additional XRPD peaks, novelty was not an issue. The complete XRPD characterized a single specific crystalline form, contrary to the claims of the main request and auxiliary request 1. In view of comparative experimental data on improved stability, the Board regarded the objective technical problem as the provision of a thermally stable crystalline form of Aripiprazole which could be obtained in high purity in a reliable manner. The claimed invention was considered to be based on the proprietor's realization that known crystals were not stable under the conditions at which they were prepared and were prone to transform into a different crystalline form. This transformation rendered it difficult to prepare the known crystals of constant quality in a reliable manner. Although any skilled person applying the teaching of the prior art would have realized that a problem existed in this respect, no indication as to the technical measures necessary to solve it could be found. Thus, inventive step was acknowledged.

9.2.2.3 Decision T 2114/13 Dated 12 October 2016

The claims of the patent in suit were drawn to a single specific crystalline form, namely form C of Febuxostat. The closest prior art reference neither mentioned polymorphs of Febuxostat nor their potential existence. It was shown by experimental tests that the crystalline material obtained according to said reference inherently was a mixture of crystalline form A and crystalline form C in an 80 : 20 ratio. Comparative conversion experiments (slurries in acetone) revealed that crystalline form C is more stable than crystalline form A. The Board considered these conversion experiments relevant, although other standard tests not using acetone were commonly applied for assessing long-term stability of drugs. According to the Board, polymorphic stability during formulation was also crucial for any drug manufacturer and solvents such as acetone were frequently used in formulation technology. The Board therefore considered the objective technical problem as the provision of a pharmaceutical composition of Febuxostat with improved quality and reliability, that is improved polymorphic stability, in particular during formulation. As none of the cited prior art references contained any guidance as to how a particular crystalline form with desirable properties could be obtained in a targeted manner, inventive step was acknowledged (EPO, T 2114/13 (Stable polymorphic form of febuxostat/TEIJIN) of 12.10.2016 [4]).

9.2.2.4 Decision T 2397/12 Dated 12 March 2018

The claims of the patent in suit were drawn to a crystalline form of Squalamine dilactate salt. The closest prior art reference neither mentioned a crystalline form of Squalamine nor that its lactate salt is a dilactate. The Board considered the objective technical problem as the provision of a stable form of Squalamine suitable for use in pharmacy. Two differences were identified with regard to the closest prior art. In addition to finding suitable conditions, especially solvents, for crystallization, the skilled person had to settle on a specific salt form of Squalamine. Since Squalamine dilactate was not disclosed in any prior art reference, it was not a compound that the skilled person would inevitably have selected when carrying out routine tests for finding suitable crystallization conditions. Consequently, the solution of the problem, i.e. the provision of crystalline Squalamine dilactate, could not have been found by the skilled person following routine approach. Thus, inventive step was acknowledged (EPO, T 2397/12 (Crystalline Squalamine Dilactate/GENERA) of 12.3.2018 [5]).

9.2.2.5 Decision T 246/15 Dated 13 November 2018

The claims of the patent in suit were drawn to a coated tablet obtainable by a method wherein Vardenafil hydrochloride was converted into the trihydrate form by contact with humidified gas until at least 90 mol.% of Vardenafil hydrochloride were present in trihydrate form. Pure Vardenafil hydrochloride trihydrate (100 mol.%) was known from the prior art, but not in the form of film coated tablets. The proprietor argued that each and every common process for the manufacture of coated tablets involved a drying step inevitably reducing the content of the trihydrate form below 90 mol.%. The Board held, however, that this was not true for dry compression coating by which the pure Vardenafil hydrochloride trihydrate known from the prior art could principally be formulated as coated tablet without any loss of crystal water. Consequently, the content of trihydrate form on the one hand and the coated tablet on the other hand was regarded as features independent of one another and only the latter feature was considered as a distinguishing feature over the closest prior art reference. The Board considered the objective technical problem as the provision of a dosage form suitable for Vardenafil hydrochloride trihydrate. As methods for the manufacture of coated tablets were within common general knowledge, inventive step was denied (EPO, T 0246/15 (Vardenafil enthaltende Arzneimittel/BAYER) of 13.11.2018 [6]).

9.2.3 Jurisdiction of the Federal Patent Court and the German Federal Supreme Court

Besides the above European case law, we would also like to comment on the jurisdiction of the German Federal Court and the German Federal Supreme Court on patentability of polymorphs. The number of decisions is limited, and in the following sections, we provide a summary of four representative decisions. As Table 9.2.2 indicates, a web-based search for the term "polymorph" in the database of the German Federal Supreme Court (Bundesgerichtshof, Aktuelle Entscheidungen des Bundesgerichtshofs [7]) can reveal decisions relevant for pharmaceutical solid forms since January 2000.

9.2.3.1 Decision "Kristallformen" German Federal Court

In January 1977, the Federal Patent Court of Germany (BPatGE) has decided that new crystalline forms are patentable. Subject matter of the decision was Cephaloridine and the conclusion that two newly prepared nonsolvated and nonhygroscopic forms were new. Although the chemical constitution of formerly known solvated forms was different because of differences in the exact chemical composition, difference was also acknowledged based on differences in the physical form, i.e. the crystalline form. Meaning from an intellectual property viewpoint compounds can be considered as different even if the chemical constitution is the same. This is because the chemical composition is just one means to describe the material [8].

9.2.3.2 Decision X ZR 58/08 Dated 15 March 15 2011

The claims of the patent in suit were drawn to monoclinic metazachlor (herbicide) having a melting point at 76 °C. Triclinic metazachlor having a melting point of 78–83 °C was known from the prior art, but tended to agglomerate when being provided in the form of concentrated suspensions and thus could not be sprayed. It was an object of the invention to overcome these problems. The Court considered

Table 9.2.2 Hitlist of search for "polymorph" in the decision database of the German Federal Supreme Court (i.e. Bundesgerichtshof) (selected hits being relevant for pharmaceutical solid forms, as of 28 November 2019).

Date of decision	Decision reference	Comment
07 August 2018	X ZR 110/16	Rifaximin α landmark decision
11 November 2014	X ZR 128/09	Repaglinid landmark decision
15 May 2012	X ZR 98/09	Calcipotriol-monohydrate landmark decision
15 March 2011	X ZR 58/08	Monoclinic metazachlor
10 April 2001	X ZR 21/98	Crystallization of chocolate (polymorphic forms of cocoa butter)

Source: Adapted from Bundesgerichtshof [7].

that triclinic metazachlor was metastable and from a theoretical perspective should transform into the more stable monoclinic form. The Court also considered that in dry form, this transition would take a very long time, whereas in suspension it would be accelerated. The Court concluded, however, that this does not inevitably mean that any suspension of triclinic metazachlor always transforms into monoclinic metazachlor. Different crystalline modifications of metazachlor were not known from the prior art and for this reason already, a skilled person was not motivated to overcome the above agglomeration problems by providing another crystalline modification. Thus, inventive step was acknowledged (Bundesgerichtshof, Urteil X ZR 58/08 "Monoklines Metazachlor" [9]).

9.2.3.3 Decision X ZR 98/09 Dated 15 May 2012

The claims of the patent in suit were drawn to Calcipotriol monohydrate, a vitamin D analogue, which was said to exhibit improved grindability under wet conditions and improved stability (reduced decomposition). While Calcipotriol monohydrate was considered novel, it was known from the prior art that other vitamin D analogues structurally closely related to Calcipotriol could be stabilized by providing the monohydrate form. The Court held that this was sufficient motivation for the skilled person to also provide Calcipotriol in the form of its monohydrate. In auxiliary requests, the Calcipotriol monohydrate was further specified by characteristic resonances of the cross-polarization magic angle spinning (CPMAS)-NMR spectrum. The Court, however, was of the opinion that reciting cross-polarization magic angle spinning (CPMAS)-NMR data in the claims would not add anything inventive (Bundesgerichtshof, Urteil X ZR 98/09 "Calcipotriol-Monohydrat," [10]).

9.2.3.4 Decision X ZR 110/16 Dated 7 August 2018

The headnote of this decision in English translation reads:

> "The provision of a crystalline form of a polymorphic substance that a skilled person inevitably obtains when performing a method for the preparation of

said substance, which method is rendered obvious by the prior art, is the result of skilled activity and therefore per se does not involve an inventive step."

The claims of the patent in suit were drawn to purified Rifaximin α, a polymorph of the antibiotic Rifaximin, having a water content below 4.5% and providing 14 characteristic XRPD peaks.

The formation of polymorphic α-, β-, and γ-forms of Rifaximin inter alia depends upon crystallization conditions (especially temperature and water content of solvent). Subsequent drying has particular relevance, as the water content at the end of drying determines the polymorphic form that is finally obtained. In the course of drying, there is no abrupt but continuous transition from the β-form to the α-form. Rifaximin is hygroscopic, and the transition is reversible. When the water content is below 4.5%, the α-form is obtained, and β-form is obtained when the water content is above 4.5%. The transition is visible, as the material changes its color.

The prior art disclosed examples where Rifaximin was obtained in crystalline form. While the prior art did not explicitly mention a subsequent drying step, the Court held that such drying step would be standard procedure. In order to ensure a defined drug content, it would be obvious if not mandatory to the skilled person to perform drying at least until a constant weight of the dried material is obtained, and such procedure would then inevitably yield the α-form of Rifaximin as claimed. This was experimentally confirmed by a series of tests faithfully reproducing the prior art under a variety of different drying conditions. Further, in view of the change of color associated with the transition from the β-form into the α-form, a skilled person would immediately have recognized the phase conversion, i.e. the existence of different polymorphs. Thus, inventive step was denied (Bundesgerichtshof, Urteil X ZR110/16 "Rifaximin alpha," [11]).

9.2.4 Assessing Validity of a Patent or the Chances of Success

When assessing validity of a patent or the chances of success for a new patent application, according to our experience, potential weaknesses can be identified by seeking answers to the following non-exhaustive list of questions:

(a) Do the claims unambiguously define a single polymorph or does the definition cover mixtures of polymorphs?
(b) Is the new polymorph obtainable in a reliable manner when faithfully reproducing the examples that are disclosed in the description?
(c) Do the experimental conditions (e.g. solvent, pressure, temperature, and the like) that make to the new polymorph available have any other relevance for other technical reasons, e.g. with regard to synthesis, further processing, analytical testing, and the like?

(d) Are the above experimental conditions in accordance with routine polymorph screening by means of standard procedures?
(e) Is there any evidence of a prejudice which would have prevented a skilled person from considering the above experimental conditions?
(f) Are other polymorphs of the same drug explicitly disclosed in the prior art?
(g) Which polymorphs are obtained in a reliable manner and thus inherently disclosed when faithfully reproducing the synthesis protocols of the prior art?
(h) Are comparative data available demonstrating technical differences of the new polymorph compared to known polymorphs?
(i) When more than a single polymorph is known from the prior art, do the available comparative data compare the new polymorph against each of the known polymorphs?
(j) Do the demonstrated technical differences seriously translate into technical advantages that are truly relevant, e.g. with respect to processability or pharmacokinetics, or is it necessary to create artificial scenarios in order to argue relevance?
(k) Does the description mention or discuss the technical differences or advantages?
(l) Does the prior art discuss disadvantages of the known polymorphs and if so, what has been suggested in the prior art in order to overcome these disadvantages?

9.2.5 Interaction with Patent Professionals

One of the basic requirements of patentability is that the invention, e.g. a polymorph or solid form, must be novel. Thus, prior to the filing of a patent application, the invention must not be disclosed publicly or to anyone not committed to secrecy.

When considering the filing of a patent application, the help of a patent professional should be sought out early in the process. The approach in drafting a patent application and the language used in the legal field are very different from those of scientific publications and other publications in the technical field. The patent professional will help to avoid serious mistakes that can easily be made early in the process and will also take into account that the patent application will likely be prosecuted in different countries with different legal requirements and jurisdictions.

In order to describe the invention properly in a patent application, the patent professional will need some technical background information about the invention as well as the most relevant experimental findings regarding the novel polymorph or solid forms. The questions under Section 9.2.4 above may serve as a guideline for gathering relevant information.

If you are aware of any problem associated with your invention, e.g. an accidental disclosure in a PhD Thesis or a scientific publication that has been prematurely submitted to a review process, this should also be brought to the attention of the patent professional as early as possible as legal remedies to the problem may be available.

List of Abbreviations

BPatGE	Federal Patent Court of Germany
CPMAS	cross-polarization magic angle spinning
EESR	extended European search report
EPO	European Patent Office
EPC	European Patent Convention
IR	infrared spectroscopy
NMR	nuclear magnetic resonance spectroscopy

References

1 EPO (2011). T 0777/08 (Atorvastatin polymorphs/WARNER-LAMBERT) of 24.5.2011. https://www.epo.org/law-practice/case-law-appeals/recent/t080777ex1.html (accessed 28 November 2019).

2 EPO (2019). Search in the Board of Appeal decisions database. https://www.epo.org/law-practice/case-law-appeals/advanced-search.html (accessed 22 November 2019).

3 EPO (2015). T 1555/12 (Aripiprazole polymorph/OTSUKA) of 29.4.2015. https://www.epo.org/law-practice/case-law-appeals/recent/t121555eu1.html (accessed 22 November 2019).

4 EPO (2016). T 2114/13 (Stable polymorphic form of febuxostat/TEIJIN) of 12.10.2016. https://www.epo.org/law-practice/case-law-appeals/recent/t132114eu1.html (accessed 28 November 2019).

5 EPO (2018). T 2397/12 (Crystalline squalamine dilactate/GENERA) of 12.3.2018. https://www.epo.org/law-practice/case-law-appeals/recent/t122397eu1.html (accessed 22 November 2019).

6 EPO (2018). T 0246/15 (Vardenafil enthaltende Arzneimittel/BAYER) of 13.11.2018. https://www.epo.org/law-practice/case-law-appeals/recent/t150246du1.html (accessed 28 November 2019).

7 Bundesgerichtshof (2019). Aktuelle Entscheidungen des Bundesgerichtshofs. https://juris.bundesgerichtshof.de/cgi-bin/rechtsprechung/list.py?Gericht=bgh&Art=en (accessed 28 November 2019).

8 Bundespatentgricht (1977). *Kristallformen. Entscheidungen des Bundespatentgerichtes BPatGE* 20: 6–9.

9 Bundesgerichtshof (2011). Urteil X ZR 58/08 "Monoklines Metazachlor". http://juris.bundesgerichtshof.de/cgi-bin/rechtsprechung/document.py?Gericht=bgh&Art=en&sid=29830efa12bf94bf8f3b1c154c1d5444&nr=56044&pos=2&anz=7 (accessed 28 November 2019).

10 Bundesgerichtshof (2012). Urteil X ZR 98/09 "Calcipotriol-Monohydrat" http://juris.bundesgerichtshof.de/cgi-bin/rechtsprechung/document.py?Gericht=bgh&Art=en&sid=48e718a0fc08945796e409d60642f587&nr=60632&pos=0&anz=1 (accessed 28 November 2019).

11 Bundesgerichtshof (2018). Urteil X ZR110/16 "Rifaximin alpha". http://juris.bundesgerichtshof.de/cgi-bin/rechtsprechung/document.py?Gericht=bgh&Art=en&sid=05e2864f727502a03035c6c29a4901e3&nr=90515&pos=0&anz=7 (accessed 28 January 2019).

10

Regulatory Frameworks Affecting Solid-State Development

Christoph Saal

Merck KGaA, Frankfurter Strasse 250, 64293 Darmstadt, Germany

10.1 Introduction – The Need for Regulation in Pharmaceutical Industry

Regulatory aspects for pharmaceuticals are described in a plethora of different documents, and it is difficult to get a comprehensive overview. There are documents dealing with many different topics. This is required because pharmaceuticals are complicated products with a high medical and social impact. The goal is always that drugs must be efficient and safe and have an adequate quality. For this purpose, good manufacturing practice (GMP) started to evolve during the first half of the last century originating not only in pharmaceutical industry but also in food and cosmetics industries. The purpose was to set up rules, how industrial processes must be carried out to avoid products with unsuitable quality which have dangerous properties. Examples for such failures have been diphtheria antitoxin that was contaminated with tetanus bacilli causing death of 12 children. This led to the Biologics Control Act in 1902 [1]. Upton Sinclair described in his book "The Jungle" [2] conditions in Chicago's meat industry. Selling rotten or diseased meat was a major problem. Rats fell into the equipment and these might have even led to more severe problems as rats have been poisoned. So, the poison was introduced into human food. Inadequate usage of drugs was another problem: syrups to calm colicky babies contained alcohol, opium, or morphine. Further examples were weight loss drugs that caused death, hair removers that caused baldness, even if not used on the head, lotions and creams that could cause mercury poisoning, hair dyes that cause lead poisoning, and eyelash that blinded women. Diethylene glycol was used as an excipient in an oral elixir of sulfanilamide that led to the death of 107 people, many of them children. This triggered the "Federal Food, Drug and Cosmetic Act" in the United States in 1938. In the 1940s and 1950s, nearly 300 people were killed or injured by sulfathiazole tablets that were tainted with phenobarbital. Due to this, the US Food and Drug Agency (FDA) revised its manufacturing and quality requirements drastically, leading to what was later called GMP. Further examples that require regulation of pharmaceuticals are a case where a company failed to

Solid State Development and Processing of Pharmaceutical Molecules: Salts, Cocrystals, and Polymorphism, First Edition. Edited by Michael Gruss.

inactivate the virus that was used in polio vaccinations resulting in individuals developing polio after the vaccination or side effects from thalidomide in the 1960s, leading to severe disability of newborns. This list of examples demonstrating the need for regulation in the pharmaceutical industry could easily be continued. A good overview of the GMP history is summarized by Immel [3]. Fortunately, nowadays regulations got a more proactive character compared to the evolvement of regulations based on the examples given above where regulations evolved much more in a reactive character to avoid that similar things might happen again.

On the other hand, this resulted in a tremendously increasing number of regulatory requirements and documents dedicated to this purpose. Today, the pharmaceutical industry is certainly the most highly regulated industry. This increase in regulation was driven by a growing need for safer drugs, more efficient drugs, and higher quality requirements. For this purpose, more and more diverse aspects of drugs became relevant. For example, topics such as genotoxic impurities [4], elemental impurities [5], or leachables and extractables [6, 7] have just been addressed in recent years or are just evolving into new regulations. Such topics might be clearly ascribed to be efficacy, safety, or quality topics. But they might also represent a mix of them, e.g. genotoxic impurities have aspects for safety as well as for quality. This increases complexity.

From another perspective, complexity increases as requirements typically change from region to region. This is the case as, for example, different regions have their own pharmacopoeias that describe established pharmaceuticals. Fortunately, there is a certain degree of harmonization that has resulted, for example, in the European Pharmacopoeia (EP), which has mainly replaced national Pharmacopoeias such as the "British Pharmacopoeia" (BP) or the "Deutsche Arzneibuch" (DAB) in Europe. However, there are other areas where this is still not the case. The US Pharmacopoeia (USP) and the EP today are very different, and the same is the case with the Japanese Pharmacopoeia (JP). Harmonization as driven by "Pharmaceutical Discussion Groups" (PDGs) is only slowly moving ahead, and only a few chapters are harmonized in EP und USP.

Beyond the role of legislation for established pharmaceuticals by Pharmacopoeias, there are further needs for regulation for established drugs as well as for drugs in clinical development. These are also addressed locally, e.g. by guidances and other documents issued by national health authorities such as the FDA in the United States, Bundesamt für Arzneimittel und Medizinprodukte (BfArM) in Germany, Agence Nationale de Sécurité du Médicament et des Produits de Santé (ANSM) in France, Medicines and Healthcare Products Regulatory Agency (MHRA) in the United Kingdom, Pharmaceuticals and Medical Devices Agency (PMDA) in Japan, and so on. Also here, we see a certain degree of harmonization that is driven by overarching authorities such as the European Medicines Agency (EMA) in Europe. Establishing still more overarching rules is pushed forward by guidances issued by the "International Council of Harmonization" (ICH), where Europe, the United States, and Japan are involved. Beyond this, ICH guidances play a significant role for countries acting as observers.

Regulatory aspects are addressed by regulatory authorities during drug approval. Like aspects mentioned earlier, these might be national procedures. These might also be already international procedures like drug approvals by the EMA in Europe, which is mandatory for certain pharmaceuticals.

Finally, within industry, consortia such as Pharmaceutical Research and Manufacturers of America (PhRMA) or Biophorum Operations Group (BPOG) exist, which clearly do not have the right and intention to establish regulations, but which allow to establish common sense and standards or certain aspects.

Still another aspect that results in an increasing number of regulatory documents and complexity is driven by increasing complexity of pharmaceuticals by themselves. Pharmaceuticals rely on more and more diverse and complex principles. For example, in the last century, most drugs have been simple small molecules originating from chemical synthesis. Today, we see much more diverse drugs and treatments such as antibodies, even bispecific antibodies, antibody–drug conjugates (ADCs), ribonucleic acid (RNA) drugs, and cell therapies such as chimeric antigen receptor T-cell (CAR-T-cells). Lately, treatments by gene editing such as cluster regularly interspaced short palindromic repeats – CRISPR associated proteins (CRISPR-CAS) are starting to evolve. Beyond these aspects, which are relevant for the active principal of a drug, other aspects come into play, which have more to do with other components of the drug such as excipients. For example, targeted drug delivery gets more and more important nowadays. Other challenges that must be tackled – and where solid-state forms also come into play – are low-soluble active pharmaceutical ingredients (APIs). Low solubility represents a major problem in today's drug discovery, which is driven not only by aspects such as high-throughput screening (HTS) and combinatorial chemistry but also just by a closer understanding of drug targets and their druggable parts and properties.

All these aspects result in regulatory needs. New products and techniques must be assessed for their benefits and their risks. Adequate regulation must be provided to make sure that pharmaceuticals based on new principles are efficient, safe, and have adequate quality. In this context, the role of regulation is to enable scientific progress and bring it to the patient, but at the same time to make sure that this is done in a safe and efficient way. Doing this – if done well – will also result in an economic benefit as this leads to the availability of guidance how certain aspects should be addressed. Considering this is not something which has to be done individually by every sponsor in an inefficient way. At the same time, a certain obligingness is given by regulatory authorities, which makes it clearer for industry what is expected and allows industry to work in a goal-oriented way.

10.2 Solid-State Forms to Be Used for Drugs

Before we embark into the regulatory aspects of solid-state forms to be used in pharmaceutical industry, a brief introduction of the three relevant types of solid-state forms is given, as this represents the basis for the discussion of regulatory aspects.

Pharmaceutical manufacturing comprises two major parts: manufacturing of the drug substance and manufacturing of the drug product. With regard to the drug product, different solid-state forms can be used:

- Pharmaceutical salts are obtained by salt formation between a base and an acid with a salt former. A comprehensive overview of the topic can be found in the book by Stahl and coworkers [8]. The usage of pharmaceutical salts provides a tool to improve properties of the drug substance. For example, dissolution rate typically is higher for pharmaceutical salts compared to the free base or free acid. Melting point, thermal stability, chemical stability, hygroscopicity, and crystal shape are just a few other properties, which can be optimized by using pharmaceutical salts. As a considerable variety of salt formers exist. Salt screening provides a powerful tool to shape a drug substance. An overview of the usage of pharmaceutical salts as used in approved drugs can be found in [9].
- After selection of a pharmaceutical salt or even parallel to it, the polymorph relevant for drug substance is selected. Therefore, typically a polymorph screening is carried out to get to know all relevant polymorphs of the parent drug or the pharmaceutical salt of the API. Polymorphic forms – also called morphic forms – might be true polymorphs that have the same chemical composition. This might also be hydrates or solvates or mixed hydrates–solvates. These might be stoichiometric or nonstoichiometric. Such crystal forms are also named pseudo-polymorphs. In the remaining of this chapter, the term "polymorph" will be used in a way comprising also pseudo-polymorphs. In most cases, the thermodynamic stable polymorph is chosen upon start of clinical development of a drug and for commercial manufacturing. This is the case as metastable polymorphs might convert to the thermodynamically stable polymorph in an uncontrolled way, and this might have severe effects on drug-product performance and manufacturing of drug substance and drug product. The amorphous state still provides another tool, especially to improve bioavailability of a drug. However, drug substances are typically not manufactured for this purpose in the amorphous state but obtained in a crystalline solid-state form as amorphous drug substances might be very difficult to stabilize against crystallization. The amorphous state is typically obtained by a drug-product manufacturing process such as hot-melt extrusion, spray drying, or co-precipitation leading to so-called "bio-enabling formulations". A good overview on polymorphism as relevant for drugs is given in [10].
- Finally, co-crystals provide another means for solid-state forms that can be used in pharmaceutical industry. Compared to pharmaceutical salts and polymorphs, this topic is quite recent and has evolved considerably in the last two decades. Co-crystals have an interesting role as they are close to pseudo-polymorphs as well as to pharmaceutical salts. Where a pseudo-polymorph consists of the active moiety and a solvent or water – which are liquid as a pure substance – forming a common crystal lattice, the same is true for a co-crystal. The only difference for a co-crystal is that in this case the co-former is not a liquid but a solid as a pure substance. The relation between co-crystals and pharmaceutical salts is easily recognized from co-crystals with carboxylic acids as co-formers. Consider the

case where the crystal lattice consists of the active moiety, which might be a weak base and a carboxylic acid. In the case that the acidic proton is transferred from the carboxylic acid to the active moiety, a pharmaceutical salt is the result. In the case that this transfer does not take place, a co-crystal is obtained. From this, it becomes clear that the distinction between a co-crystal and a pharmaceutical salt is not always easy to make. Typically, it requires knowledge of the X-ray crystal structure including the exact position of hydrogen atoms. There might be drugs on the market that have been approved as pharmaceutical salts, but which are in effect co-crystals, as during their development co-crystals have not been a relevant topic nor regulation for them existed.

10.3 General Regulatory Considerations for Pharmaceutical Solid-State Forms

As outlined in the introduction, major goals of regulation in pharmaceutical industry are to make sure that drugs are efficient, safe, and have an adequate quality. Therefore, strict assessment of quality aspects during development and commercial manufacturing are key as well as conducting animal and human studies to assess safety efficacy of drugs. Especially the latter require a huge effort. Consequently, the question arose if the full program of studies is required for different solid-state forms of a drug as the ones required for a totally new chemical entity, or if certain types of studies could be bridged or waved without an increasing risk on efficacy, safety, or quality. Doing this does not just represent a financial benefit with regard to the development of new drugs but also an ethical benefit as the number of studies in animals and humans can be reduced. Still additionally, using such approaches, drug development can be accelerated making drugs earlier on available to patients. This aspect is important for drugs that address unmet medical needs. In recent years, regulatory authorities in the United States, Europe, and Japan have recognized the importance to make drugs addressing such unmet medical needs available to the public and have put in place respective programs such as the "Breakthrough Therapies" in the United States [11], priority medicines (PRIME) in Europe [12], and "Sagigake" in Japan [13]. Finally, generic drugs result from not requiring a full clinical program, but a reduced program based on scientific assessment of the "similarity" of the generic drug to the originator. With regard to the approval process for drugs, this is described as "abbreviated new drug approval" (ANDA) [14] compared to the "new drug approval" (NDA) required for a new chemical entity. Introduction of generic drugs made therapies available at lower prices compared to originators and accordingly had and still has a major financial aspect on our health systems. Development of generic drugs after patent expiry for the originator's drug can be based on a new formulation of the same solid-state form of the drug substance but might also be based on a different solid-state form, a different pharmaceutical salt, a different polymorph, or even a different co-crystal compared to the originator. The specific approach for a certain generic drug will depend on the intellectual property of the originator's drug. For example, there might be situations where the patent for the drug substance has expired, but there is still patent protection for the formulation, for a pharmaceutical salt or polymorph, or for a drug substance or a drug product manufacturing process. Depending on aspects with regard to manufacturing of the generic drug, aspects about already available generic competition and scientific considerations, different approaches might be taken by companies to develop generic drugs. Beyond the abovementioned aspects relevant for generic companies, similar considerations might be also relevant for the originator of the drug himself in the context of lifecycle management of the drug.

10.4 Regulatory Framework for Pharmaceutical Salts

Reducing clinical trials of pharmaceutical salts upon substitution of one pharmaceutical salt by another salt requires assessment of "similarity" of respective pharmaceutical salts within the drug product. Therefore, four concepts play a key role: "pharmaceutical equivalence", "pharmaceutical alternatives", "bioequivalence," and "therapeutic equivalence". Before discussing the regulatory framework for pharmaceutical salts, we will first introduce these four concepts, as this will be an important basis for the regulatory framework.

Additional overviews about the regulatory environment with regard to this topic can also be found in [15, 16].

10.4.1 Pharmaceutical Equivalence and Pharmaceutical Alternatives

The term "pharmaceutical equivalence" is given in the EMA guideline on bioequivalence [17]:

> "Medicinal products are pharmaceutically equivalent if they contain the same amount of the same active substance(s) in the same dosage forms that meet the same or comparable standards.
> Pharmaceutical equivalence does not necessarily imply bioequivalence as differences in the excipients and/or the manufacturing process can lead to faster or slower dissolution and/or absorption."

Unfortunately, the term "same active substance" is not defined in the EMA guideline. This term could refer to different pharmaceutical salts of the substance. However, it is clearly stated that "pharmaceutical equivalence" does not imply bioequivalence. Accordingly, the term "pharmaceutical equivalence" refers much to quality considerations such as amount of active substance and dosage form and less to biopharmaceutical considerations.

In comparison to the European Union (EU) framework, the US FDA provides another definition on "pharmaceutical equivalence" [18]:

> "Pharmaceutical equivalents are drug products in identical dosage forms and route(s) of administration that contain identical amounts of the identical

> active drug ingredient, i.e., the same salt or ester of the same therapeutic moiety, or, in the case of modified-release dosage forms that require a reservoir or overage or such forms as prefilled syringes where the residual volume may vary, that deliver identical amounts of the active drug ingredient over the identical dosing period; do not necessarily contain the same inactive ingredients; and meet the identical compendial or other applicable standard of identity, strength, quality, and purity, including potency and, where applicable, content uniformity, disintegration times, and/or dissolution rates. They may differ in characteristics such as shape, scoring configuration, release mechanisms, packaging, excipients (including colors, flavors, preservatives), expiration date/time, and, within certain limits, labeling."

FDA does not explicitly state that different pharmaceutical salts are within the scope of "pharmaceutical equivalents" but mentions that the active drug ingredient must be the same, e.g. the same salt. Additionally, more information is provided about parameters that must be equivalent (e.g. identity, strength, purity) as well as parameters that might be changed (e.g. inactive ingredients). The parameters that are provided in the list of drug products with therapeutic equivalence evaluations are mainly quality based. However, also parameters are provided that are directly linked to biopharmaceutical considerations, e.g. dissolution rate.

Within the regulatory US framework, a very similar definition is provided in 21 CFR 314.3 [19].

> "Pharmaceutical equivalents are drug products in identical dosage forms and route(s) of administration that contain identical amounts of the identical active drug ingredient, i.e., the same salt or ester of the same therapeutic moiety, or, in the case of modified-release dosage forms that require a reservoir or overage or such forms as prefilled syringes where residual volume may vary, that deliver identical amounts of the active drug ingredient over the identical dosing period; do not necessarily contain the same inactive ingredients; and meet the identical compendial or other applicable standard of identity, strength, quality, and purity, including potency and, where applicable, content uniformity, disintegration times, and/or dissolution rates."

With regard to "pharmaceutical alternatives", definitions for the EU and the United States are given in the EMA guideline on bioequivalence [17] and in 21 CFR 314.3 [19].

> "Pharmaceutical alternatives are medicinal products with different salts, esters, ethers, isomers, mixtures of isomers, complexes or derivatives of an active moiety, or which differ in dosage form or strength.
> Pharmaceutical alternatives are drug products that contain the identical therapeutic moiety, or its precursor, but not necessarily in the same amount or

> dosage form or as the same salt or ester. Each such drug product individually meets either the identical or its own respective compendial or other applicable standard of identity, strength, quality, and purity, including potency and, where applicable, content uniformity, disintegration times, and/or dissolution rates."

Within both frameworks, pharmaceutical salts are explicitly mentioned, and it is said that one way in which "pharmaceutical alternatives" might distinguish is the pharmaceutical salt contained in the drug product. Again, the definition is very much based on parameters relevant for quality, and the US framework also mentions parameters relevant for biopharmaceutical aspects such as dissolution rate.

10.4.2 Bioequivalence

Beyond the concepts of "pharmaceutical equivalents" and "pharmaceutical alternatives", assessment of "bioequivalence" plays a key role with regard to regulation for solid-state forms, especially pharmaceutical salts.

Bioequivalence is described in a guideline by EMA [20] as well as two draft guidances by FDA [21, 22]. One of these two US guidances is focused on bioequivalence studies as used for new drug applications (NDAs) and investigational new drugs (INDs) in clinical trials. The second FDA guidance describes the use of bioequivalence data for abbreviated new drug applications (ANDAs) and is accordingly relevant for the approval of generic drugs including generic drugs containing a different pharmaceutical salt compared to the originator. Another relevant definition of bioequivalence for the United States is given in 21 CFR 320.1. [23].

The EMA guideline provides the following definition of bioequivalence:

> "Two medicinal products containing the same active substance are considered bioequivalent if they are pharmaceutically equivalent or pharmaceutical alternatives and their bioavailabilities (rate and extent) after administration in the same molar dose lie within acceptable predefined limits. These limits are set to ensure comparable in-vivo performance, i.e. similarity in terms of safety and efficacy."

According to this, bioequivalence refers to the drug product and not to the drug substance. The concept cannot be used to compare two pharmaceutical salts with regard to bioequivalence directly, but only to two drug products containing different pharmaceutical salts or the same pharmaceutical salt. In the latter case, the pharmaceutical salt used in the two drug products might differ in properties such as particle size or morphic form. Investigation of bioequivalence as described in the EMA guidance and others is based on a comparison of plasma concentration–time curves to assess the rate and extent of absorption using relevant parameters such as the area under the curve (AUC), the time of maximum plasma concentration after administration (t_{max}), and the concentration of the drug in plasma at that time point (c_{max}) including predefined acceptance criteria. Drug products that are covered by the guidance are immediate release oral and nonoral formulations. For drug products that are parenteral solutions, bioequivalence studies are generally not required as processes such as dissolution of the drug product and absorption through the gut wall are not required for parenteral drugs, and less variability compared to oral drug products is expected.

Within the EMA guidance, a link is given to the Directive 2001/83/EC, Article 10(1), which is describing applications for generic medicinal products. Therein, it is described that the purpose of establishing bioequivalence is to demonstrate equivalence between the generic medicinal product and a reference medicinal product to allow bridging of preclinical tests and of clinical trials associated with the reference medicinal product. In the following Article 10(2)(b), the definition of a generic product is provided as being a medicinal product that has the same qualitative and quantitative composition in active substances and the same pharmaceutical form as the reference medicinal product, and whose bioequivalence with the reference medicinal product has been demonstrated by appropriate bioavailability studies. Different salts, esters, ethers, isomers, mixtures of isomers, complexes, or derivatives of an active substance are considered to be the same active substance, unless they differ significantly in properties with regard to safety and/or efficacy. Additionally, the EMA guidance on bioequivalence states that the concept described there is also applicable to comparative bioavailability studies evaluating different formulations during clinical development of new chemical entities (NCEs). Finally, guidance for a broad range of modified release formulations is provided in a dedicated guidance "Guideline on the pharmacokinetic and clinical evaluation of modified release dosage forms" [24].

According to these documents, within the EU framework a consistent picture is given, which points out the importance to show bioequivalence of drug products but – for good reasons – does not allow to use the concept of bioequivalence on the level of a drug substance: different pharmaceutical salts might or might not exhibit the same solubility and dissolution profiles in relevant media. However, the performance of the respective pharmaceutical salts might be affected by the respective formulation. This is one of the reasons why the concept of bioequivalence has to be used on the drug product level and not on the drug substance level including the pharmaceutical salt.

Within the US framework, the definition of bioequivalence is provided in 21 CFR 320.1 [23] and the FDA guidance relevant for NDAs and INDs [21] in the same way:

> "Bioequivalence means the absence of a significant difference in the rate and extent to which the active ingredient or active moiety in pharmaceutical equivalents or pharmaceutical alternatives become available at the site of drug action when administered at the same molar dose under similar conditions in an appropriately designed study."

In the relevant guidance referring to bioequivalence as used for generic products in ANDAs, the following background is provided:

> "To receive approval for an ANDA, an applicant generally must demonstrate, among other things, that its proposed drug product is bioequivalent to the reference listed drug (RLD, or reference product). The FD&C Act provides that a generic drug is bioequivalent to the listed drug if:
> The rate and extent of absorption of the drug do not show a significant difference from the rate and extent of absorption of the listed drug when administered at the same molar dose of the therapeutic ingredient under similar experimental conditions in either a single dose or multiple doses ....
> For most products, the focus of BE studies is on the release of the drug substance from the drug product into the systemic circulation. During such BE studies, an applicant compares the systemic exposure profile of a test drug product to that of the RLD."

According to these principles with regard to bioequivalence in the EU and US frameworks, similar scenarios are provided in a consistent way in both legislations. Bioequivalence can be established either between pharmaceutical equivalents or between pharmaceutical alternatives.

10.4.3 Therapeutic Equivalence

Within the US concept of ANDAs if pharmaceutical equivalence is given and also bioequivalence is shown, this leads to "therapeutic equivalence".

10.4.4 Biowaivers

Within the description of the regulatory framework for pharmaceutical salts, the concept of bioequivalence was outlined. The usage of this concept to show equivalence of drug products can reduce effort in clinical bioequivalence studies considerably compared to clinical trials to assess efficacy and safety of a drug product. Additionally, the concepts of biowaivers can still further reduce effort especially for the assessment of polymorphs. The concept is described in the FDA guidance "Waiver of In Vivo Bioavailability and Bioequivalence Studies for Immediate-Release Solid Oral Dosage Forms Based on a Biopharmaceutics Classification System" [25]. However, care must be taken about the applicability of this concept. Within the guidance, it is stated that:

> "This approach is useful only when the drug substance belongs to BCS class 1 or 3, and the formulations pre- and post-change are pharmaceutical equivalents (under the definition at 21 CFR 320.1(c)). BCS-based biowaivers are intended only for subsequent in vivo BA or BE studies. They do not apply to food effect BA studies or other PK studies. BCS-based biowaivers may be applicable for pharmaceutical alternatives including other oral dosage forms (e.g., powders), if appropriately justified. The sponsor should contact the appropriate review division in such situations."

Accordingly, for different pharmaceutical salts representing pharmaceutical alternatives, contacting FDA is advised. The situation might be simpler for different polymorphs as discussed later in this chapter.

Biowaivers allow the replacement of bioequivalence studies – at least partially – by in vitro studies. However, this cannot be applied for all kind of drugs, but certain requirements must be fulfilled. Most importantly, the drug must belong to class 1 or 3 of the biopharmaceutical classification scheme (BCS) as introduced by Amidon et al. [26]. Generally, the BCS classifies drugs with regard to their solubility and permeability:

Class 1: High Solubility – High Permeability
Class 2: Low Solubility – High Permeability
Class 3: High Solubility – Low Permeability
Class 4: Low Solubility – Low Permeability

"High Solubility" means that the highest dose of the drug substance is soluble in 250 ml or less of aqueous media in the pH range 1–6.8 at 37 °C. To assess this, thermodynamic solubility of the respective drug substance in the respective solid-state form has to be measured at the extreme pH values of this range (1 and 6.8) and as well for ionizable drugs at the pH of the pK_a, and at pK_a +1 as well as pK_a −1, if these fall into the range 1–6.8.

"High Permeability" means that the systemic bioavailability of the drug substance or the extent of absorption in humans is 85% or more with regard to the administered dose based on mass balance or comparison to intravenous dosing. Permeability can be preferably determined via human pharmacokinetic (PK) studies but also by in vivo intestinal perfusion studies in man. Alternatively, also animal in vivo or in situ intestinal perfusion studies or even in vitro permeability models based on intestinal tissues or cell monolayers such as Caco-2 or Madin-Darby-Canine-Kidney cell line (MDCK) assays might be used. However, especially in the latter case, care must be taken with regard to drug transporters.

The FDA guidance describes that biowaivers can be used for BCS class 1 and 3 drugs. Depending on the BCS class, different requirements must be fulfilled:

For BCS class 1 drugs in addition to being highly soluble and highly permeable, the drug product must be rapidly dissolving. This must be the case for the newly developed drug product and for the drug product that is used for comparison. Finally, the drug product must not contain any excipient that will affect the rate or extent of absorption of the drug. The latter aspect is highlighted as there are excipients that influence uptake of drugs through the gut wall. If such excipients are present, this can influence pharmacokinetics that is beyond effects from the drug substance itself.

For BCS class 3 drugs in addition to being highly soluble, the drug product must be very rapidly dissolving. Again, this must be the case for the newly developed drug product and for the drug product that is used for comparison. Finally, for BCS class 2 drugs, the test drug product must be qualitatively the same and quantitatively very similar.

In this context, "rapidly dissolving" means that 85% of more dissolution is reached after 30 minutes using USP apparatus 1 or 2 in 0.1 N HCl or simulated gastric fluid

(SGF), in pH 4.5 buffer and in simulated intestinal fluid (SIF) with pH 6.8, so over the pH range relevant for the gastrointestinal tract. Further details about the conditions to be used for dissolution testing can be found in the abovementioned FDA guidance. The term "very rapidly dissolving" means that 85% or more dissolution is reached after 15 minutes.

As for BCS class 3 drugs permeation through the gut wall represents a rate limiting step, this must not be influenced by other components such as excipients present in the drug product. In this context, "quantitatively very similar" allows for example changes in the technical grade of an excipient. Allowed differences in concentration of excipients are provided in the FDA guidance for several classes of excipients such as fillers, disintegrants, gliders, etc.

According to the biowaiver concept, comparison of different drug products based mainly on an in vitro assessment can facilitate changes of formulations also containing different polymorphs – as discussed later in this chapter – and possibly – if agreed with the agency – also pharmaceutical salts in an efficient way avoiding the need for a large clinical program.

10.4.5 Regulatory Approval for Pharmaceutical Salts

Neither in the EU nor in the United States, there is a special regulation dealing with approval of pharmaceutical salts. In both legislations, the drug approval process for pharmaceutical salts is described in a general way where certain aspects are relevant for pharmaceutical salts. Therefore, the concepts of "pharmaceutical equivalents", "pharmaceutical alternatives", "bioequivalence", and "biowaivers" represent important tools during the approval process and its preparation starting from establishing relevant in vitro, animal and human studies and putting this into a regulatory dossier in a right and efficient way.

10.4.5.1 Regulatory Approval Pathways in the United States

In the United States, "therapeutic equivalence" is required for a generic drug application. "Therapeutic equivalence" beyond "bioequivalence" requires also "pharmaceutical equivalence". According to the definition of "pharmaceutical equivalence" in the US legislation as discussed above, new pharmaceutical salts do not represent pharmaceutically equivalent entities. Consequently, drug products containing a new pharmaceutical salt must be approved using the NDA. The respective requirements and procedures for such applications are described in Section 505(b)(2) of the Federal Food, Drug and Cosmetic Act [27] and the respective FDA guidance [28]. However, Section 505(b)(2) mentions that the application might be based on safety and efficacy data, which have not been generated by the applicant himself:

> "An application submitted under paragraph (1) for a drug for which the investigations described in clause (A) of such paragraph and relied upon by the applicant for approval of the application were not conducted by or for the applicant and for which the applicant has not obtained a right of reference or use from the person by or for whom the investigations were conducted shall also include – best of his knowledge, with respect to each patent which claims the drug for which such investigations were conducted or which claims a use

for such drug for which the applicant is seeking approval under this subsection and for which information is required to be filed under paragraph (1)..."

Here clause (A) refers to safety and efficacy data:

"(A) full reports of investigations which have been made to show whether or not such drug is safe for use and whether such drug is effective in use"

This approval pathway in the United States is somehow like the "hybrid application" in the EU as discussed in the next paragraph of this chapter. Compared to an ANDA – as discussed in the part of this chapter dealing with regulations for polymorphic forms – this scenario requires more effort. However, as clinical trials beyond bioavailability or bioequivalence studies might be required in this context, also market exclusivity for up to three years can be granted.

Alternatively, the applicant of course can file a full NDA for a new drug product containing a new pharmaceutical salt as described in Section 505(b)(1) of the Federal Food, Drug and Cosmetic Act. Compared to the abovementioned pathway and to an ANDA, this requires considerably more data but provides also the chance for a longer market exclusivity of up to five years.

The regulatory pathway described in Section 505(b)(2) of the Federal Food, Drug and Cosmetic Act has not been developed especially for new drug products containing new pharmaceutical salts but as a general principle for new drug products. However, pharmaceutical salts of course are a topic where this regulation can be very useful. The key principle within this regulation is bridging studies comparing the new drug product and the established drug product. Beyond pharmacokinetic studies, typically such studies will also contain efficacy and safety studies. As successful results from such studies already exist for the established drug product containing the original pharmaceutical salt, the risk of such safety and efficacy studies using a new pharmaceutical salt will be lower compared to the studies carried out with the original formulation and pharmaceutical salt, where the drug has been administered the first time to human beings.

10.4.5.2 Regulatory Approval Pathways in the European Union

In the EU, drug approvals are regulated in Directive 2001/83/EC [29], which leads to a different scenario compared to the United States. The general approach to obtain a marketing authorization within the EU is described in Article 8 of the directive. Article 8 summarizes the information to be submitted by the applicant to the competent authority of the member state.

Beyond this approval pathway, Article 10(1) describes generic applications:

"In derogation of Article 8(3)(i), and without prejudice to the law relating to the protection of industrial and commercial property:

(a) The applicant shall not be required to provide the results of toxicological and pharmacological tests or the results of clinical trials if he can demonstrate:

(i) either that the medicinal product is essentially similar to a medicinal product authorized in the Member State concerned by the application and that the holder of the marketing authorization for the original medicinal product has consented to the toxicological, pharmacological and/or clinical references contained in the file on the original medicinal product being used for the purpose of examining the application in question …"

Generic drug applications are feasible if the generic drug product has the same qualitative and quantitative composition as the reference with regard to the active substance as defined for "pharmaceutical equivalents" and the same pharmaceutical form. Additionally, bioequivalence must be demonstrated as discussed before in this chapter. In this context, different pharmaceutical salts are considered as the same active substance if they do not differ significantly with regard to their efficacy and safety profile. Consequently, the applicant must provide respective data for pharmaceutical salts and their formulation in a generic product in the dossier for a generic drug application.

Particularities for application according to Article 10 are also explicitly described including pharmaceutical salts in a respective notice to applicants by the European Commission Health and Food Safety Directorate [30]:

> "The different salts, esters, ethers, isomers, mixtures of isomers, complexes or derivatives of an active substance must be considered to be the same active substance, unless they differ significantly in properties with regard to safety and/or efficacy. In such cases, additional information providing proof of the safety and/or efficacy of the various salts, esters, ethers, isomers or mixtures thereof or derivatives of an authorised active substance must be supplied by the applicant."

If there are differences with regard to efficacy and safety between the originally used pharmaceutical salt and the pharmaceutical salt in the new formulation, additional data to proof efficacy and safety must be provided. Which data might be necessary for this purpose will be a case-by-case decision. If such data do not show similar efficacy and safety, a generic application within the framework of Article 10(1) is not possible in the EU. In such cases, one scenario of course is that the applicant can file a full drug application according to Article 8.

A third option within the EU is a so-called "hybrid application", which is described in Article 10(3). This can become relevant, if requirements for a generic drug application are not fulfilled, for example, if bioequivalence between the formulation containing the original pharmaceutical salt and a new pharmaceutical salt cannot be demonstrated. For "hybrid applications", additional preclinical and clinical date are required to compare both drug products.

Information about data to be provided for applications under Article 10 are briefly summarized in appendix I and II of the respective notice to applicants by the European Commission Health and Food Safety Directorate [30]:

> "Different salt/ester complex/derivative (with the same therapeutic moiety)
> Additional data usually required:
> Evidence that there is no change in the pharmacokinetics of the moiety, pharmacodynamics and/or in toxicity which could significantly change the safety/efficacy profile (otherwise, to be considered as a new active substance)"

10.4.6 Regulatory Approval for Polymorphs

For polymorphs, the "International Council for Harmonisation of Technical Requirements for Registration of Pharmaceuticals for Human Use" (ICH) provides an important regulatory background. This is done by ICH on a global level with the EU, the United States, and Japan as founding regulatory members and other regulatory members such as Canada, Switzerland, Brazil, and others. Ways to handle polymorphism are described in ICH Q6A "Specifications: Test Procedures and Acceptance Criteria for New Drug Substances and New Drug Products: Chemical Substances" [31] and more lately in ICH Q8 "Pharmaceutical Development" [32]. Again, the focus of such regulation is on the drug product efficacy, safety, and quality. This can also be concluded from the fact that ICH Q11 "Development and Manufacturing of Drug Substances (Chemical Entities and Biotechnological/Biological Entities)" [33] does not mention polymorphism explicitly but only "physical properties of a drug substance" in a more general way. This guidance somehow represents the counterpart to ICH Q8.

Within ICH Q8, the following statement is provided:

> "The physicochemical and biological properties of the drug substance that can influence the performance of the drug product and its manufacturability or were specifically designed into the drug substance (e.g., solid state properties), should be identified and discussed. Examples of physicochemical and biological properties that might need to be examined include solubility, water content, particle size, crystal properties, biological activity, and permeability. These properties could be interrelated and might need to be considered in combination.
> To evaluate the potential effect of drug substance physicochemical properties on the performance of the drug product, studies on drug product might be warranted. For example, the ICH Q6A Specifications: Test Procedures and Acceptance Criteria for New Drug Substances and New Drug Products: Chemical Substances describes some of the circumstances in which drug product studies are recommended (e.g., Decision Tree #3 and #4 (Part 2))."

Therefore, sponsors should justify during the pharmaceutical development of a drug product the selection of the polymorph. There might also be the need to control the morphic form in drug substance as used for manufacturing of the drug product or even the morphic form within the drug product.

Specific guidance on how to assess and handle polymorphism in drug substances and drug product is provided by ICH Q6A, which certainly still today represents the most specific and useful piece of regulatory information.

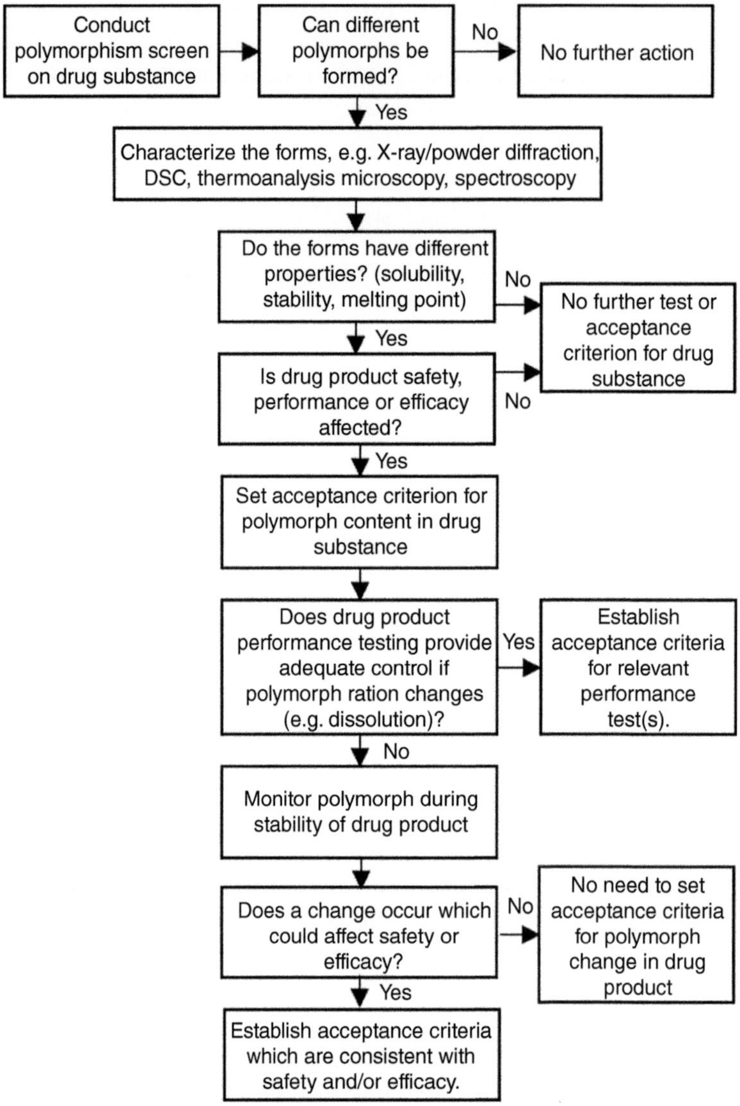

Figure 10.1 Decision trees for polymorphism as given by ICH Q6A. The three decision trees provided in ICH Q6A have been condensed to one in this figure to facilitate reading. Source: ICH Harmonised Tripartite Guideline ICH Q6A [31].

> "Some new drug substances exist in different crystalline forms which differ in their physical properties. Polymorphism may also include solvation or hydration products (also known as pseudo-polymorphs) and amorphous forms. Differences in these forms could, in some cases, affect the quality or performance of the new drug products. In cases where differences exist which have been shown to affect drug product performance, bioavailability or stability, then the appropriate solid state should be specified."

Based on this, three decision trees on further guidance is given (see Figure 10.1).

The decision tree starts with the requirement that a polymorphic screen on the drug substance, e.g. an unionized API, a pharmaceutical salt, or a co-crystal, should be carried out. Today, this can be considered as a routine procedure. There is no guarantee that during the polymorphism screen, all or at least all relevant forms will be discovered and new morphic forms might appear suddenly, as this was the case in the famous ritonavir case [34] and other less prominent cases such as telmisartan [35]. However, conducting an extensive polymorphism screen reduces the risk for unexpected appearance of new polymorphs.

This is followed by the question "*Can different polymorphs be formed?*". If this is not the case, no further actions are required. As most APIs exhibit polymorphism, in most cases further steps are required. The first of them is a systematic characterization of the relevant polymorphs. This should include not only analytical techniques such as powder X-ray diffraction (PXRD), single-crystal X-ray diffraction (SCXRD), spectroscopic techniques (e.g. Raman, infrared [IR] spectroscopy), thermoanalysis (e.g. thermogravimetry [TG], differential scanning calorimetry [DSC]), assessment of particle properties (e.g. by microscopy and laser diffraction), and assessment of hygroscopicity (e.g. by pharmacopoeial methods or better by dynamic vapour sorption [DVS]) but also an assessment of solubility, dissolution, and stability.

The next question is "*Do the forms have different properties? (solubility, stability, melting point)*". To some extent, this question seems to be needless, as by nature and definition different polymorphs will have different properties, at least different crystal structures and these will lead to other different properties such as e.g. different melting points. The question must be translated into "Do the forms have different properties that are relevant?" This means that relevant differences in relevant properties must be followed up. If there are no relevant differences in relevant properties, "*No further test or acceptance criterion for drug substance*" are required. If such differences in relevant properties of polymorphs exist, the next question to be answered is: "*Is drug product safety, performance or efficacy affected?*" From this, it becomes clear that again the drug product is key. If neither drug product safety, nor efficacy, nor its performance is affected, again no specification for the morphic form of the drug substance is required. This might be typically the case for drug substances belonging to BCS class 1 and 3 that exhibit high solubility. However, the fact that a drug substance belongs to these BCS classes does not mean that this provides enough basis to answer this question. A more specific assessment is required for each drug including different pharmaceutical salts or co-crystals of the same moiety.

If such differences exist between different morphic forms of the drug substance, an acceptance criterion for the polymorph content in drug substance must be set by the sponsor. This requires also an analytical method to assess each batch of drug substance for release and to be used for stability testing. Typical methods to be used for this purpose might be PXRD, IR spectroscopy, Raman spectroscopy, or DSC. These might be used in a quantitative way but can also be used as methods for identification of the correct morphic form, provided that the user makes sure that contributions from unwanted polymorphs are recognized.

If a specification of the morphic form becomes relevant, the question applies if the morphic form of the API is altered by the formulation process. There are formulation methods such as hot-melt extrusion or spray drying converting crystalline polymorphs to amorphous formulations. In many cases, the intention of the formulation is to maintain the crystalline morphic form of the drug substance within the drug product. As analysis of the morphic form in drug product can be challenging – especially for low-dose drug products – this is avoided by the last part of the decision tree and represents only the last resort if other approaches fail. Therefore, the question asked is "*Does drug product performance testing provide adequate control if polymorph ratio changes (e.g. dissolution)?*". If this is the case, drug product performance testing should be used for this purpose. This represents a pragmatic approach as for oral formulations dissolution will be tested anyway in most cases, and generally, the most critical aspect for an oral formulation with regard to polymorphism will be its dissolution behavior. However, one must bear in mind that there might be other drug product properties that can be relevant and influenced by the morphic form, e.g. hardness of a tablet or friability.

If drug product performance testing does not provide an adequate way, the guidance requires that the morphic form is monitored during drug product stability studies. At this stage, an analytical method for the content of the morphic form in drug product is required. If the result of such stability studies is that no change with regard to the morphic form in drug product occurs, which affects drug product safety and efficacy, there is no need to establish a specification for the morphic form in drug product. However, if safety or efficacy is affected, there is a need to establish a respective specification.

The abovementioned assessment of morphic forms might also be relevant for the dossier of a submission. The general format – Common Technical Document (CTD) – is described in ICH M4 "The Common Technical Document" [36]. Relevant parts of the CTD, where polymorphism might be addressed, are listed in Table 10.1 [37]. From the list, it becomes clear that there are sections referring to drug substance as well as to drug product. Aspects with regard to selection of the morphic form, manufacturing, and control of the morphic form are listed. The whole list of relevant chapters reflects the procedure as defined in the ICH Q6A decision trees very well.

Compared to ICH Q6A, a slightly different approach is described by the FDA with regard to ANDAs [38] as used for the approval of generic drugs. The focus of the guidance is to facilitate approval of generic drugs by avoiding that for such drugs, a full clinical development or in vivo PK assessment must be carried out. Key elements of this guidance for industry are recommendations on assessing the sameness of polymorphic forms of a drug substance. It also contains decision trees that recommend procedures to monitor and control polymorphs in drug substance and drug product.

Again, the starting point given by FDA for ANDAs is that a sponsor should investigate if for a drug substance different morphic forms exist, e.g. by conducting a polymorphic screen:

Table 10.1 Chapters within the CTD which might be relevant with regard to solid-state forms.

Chapter	Content with regard to selection of morphic form
3.2.S.1.3	If called for, list the polymorphic form(s) present in the proposed active as a characteristic of the drug substance
3.2.S.2.2	Description of manufacturing process and process controls should indicate which polymorphic form is synthesized
3.2.S.3.1	Studies performed to identify the potential polymorphic forms of the drug substance, including study results. Total number of polymorphs should be listed here and those intended to form the active should be summarized in 3.2.S.1.3
3.2.S.4.1	Specification. If a polymorph is to be defined or limited, it should be discussed here.
3.2.S.4.2	Analytical procedures
3.2.S.4.3	Validation of analytical procedures
3.2.S.4.4	Results of batch analyses
3.2.S.4.5	Justification of specification (if appropriate). Reasons why a particular limit on form is appropriate (should also probably refer to 3.2.P.2)
3.2.P.2. 1.1 and 3.2.P.2.2.3	Identifies the influence of polymorphism on the drug substance and dosage form
3.2.P.5.1	Specification. If polymorphs are to be controlled in the drug product, they should appear here.
3.2.P.5.6	Justification of specification (if called for)

Source: ICH Harmonised Tripartite Guideline ICH M4 [37].

> "We recommend that ANDA applicants investigate whether the drug substance in question can exist in polymorphic forms. Polymorphic forms in the context of this guidance refer to crystalline and amorphous forms as well as solvate and hydrate forms…"

The key elements where polymorphic forms might lead to differing drug product performance are solubility, dissolution rate, and pharmacokinetics:

> "Since polymorphic forms differ in their internal solid-state structure, a drug substance that exists in various polymorphic forms can have different aqueous solubilities and dissolution rates. When there are differences in the apparent solubilities of the various polymorphic forms, we recommend that you focus on the potential effect such differences can have on drug product bioavailability (BA) and bioequivalence (BE)."

Without explicitly mentioning the BCS, the guidance makes a similar distinction with regard to drugs and the limiting steps that are influencing absorption of the drug to the blood stream for oral applications:

> "For a drug whose absorption is only limited by its dissolution, large differences in the apparent solubilities of the various polymorphic forms are likely to affect BA/BE. On the other hand, for a drug whose absorption is only limited by its intestinal permeability, differences in the apparent solubilities of the various polymorphic forms are less likely to affect BA/BE. Furthermore, when the apparent solubilities of the polymorphic forms are sufficiently high and drug dissolution is rapid in relation to gastric emptying, differences in the solubilities of the polymorphic forms are unlikely to affect BA/BE."

According to this, within an ANDA, the sponsor must show bioequivalence between the new drug product and the reference listed drug (RLD).

However, beyond the effect of morphic forms on pharmacokinetics, bioavailability, and bioequivalence, the guidance also mentions that the applicant has also to show that the drug product manufacturing process must be robustly delivering drug product with consistent and adequate quality. It is pointed out that different morphic forms might have different physico-chemical properties such as hygroscopicity, particle shape, density, flowability, and compatibility, which might influence drug product manufacturing. This includes also knowledge of potential changes of morphic forms during the drug product manufacturing process.

Additionally, the guidance points out that also stability of the drug product should be assessed, which means effects of the morphic form on drug product stability. This can include different morphic forms for the generic product and the RLD leading to different chemical stability of the drug product as well as changes of the morphic form in drug product over time. The latter becomes especially relevant, if a metastable polymorph is used in the drug product.

With regard to sameness of the drug substance, the guidance links to the CFR and USP and clearly states that different morphic forms might be considered as same:

> "Specifically, 21 CFR 314.92(a)(1) provides that the term "same as" means, among other things, "identical in active ingredient(s)." The drug substance in a generic drug product is considered to be the same as the drug substance in the RLD if it meets the same standards for identity.
> When a United States Pharmacopeia (USP) monograph exists for a particular drug substance, standards for identity generally refer to the definition (e.g. chemical name, empirical formula, molecular structure, description) at the beginning of the monograph. However, FDA may prescribe additional standards that are material to the sameness of a drug substance.
> …. In the context of sameness of active ingredient(s) in the preamble to the 1992 final rule, FDA specifically rejected a proposal that would have required an ANDA applicant to show that the active ingredient in its generic drug product and the active ingredient in the RLD "exhibit the same physical and chemical characteristics, that no additional residues or impurities can result from the different manufacture or synthesis process and that the stereochemistry characteristics and solid state forms of the drug have not been altered." Therefore, differences in drug substance polymorphic forms do not render drug

substances different active ingredients for the purposes of ANDA approvals within the meaning of the Act and FDA regulations."

Several ANDAs have been granted, where generic products contain different morphic forms compared to the RLD, e.g. warfarin sodium, famotidine, or ranitidine. This is not only the case for true polymorphs but also for pseudo-polymorphs, e.g. for terazosin hydrochloride, ampicillin, or cefadroxil.

Decision trees within the ANDA guidance as shown in Figure 10.2 provide information about:

- Decision Tree 1: How to investigate whether a specification with regard to polymorphism is required for drug substance (for solid oral and suspension drug products).
- This is based on the questions if a drug substance exhibits polymorphism and on the BCS classification.
- Decision Tree 2: How to set a specification for a drug substance with regard to polymorphism if this is required based on the previous assessment.
- This is based on the requirements for the drug substance that are given in the USP.
- Decision Tree 3: How to investigate whether a specification with regard to polymorphism is required for drug product (for solid oral and suspension drug products).

  This is based on the "concern" about the requirement of a specification. For example, the guidance states that if the generic drug product and the RLD both use the thermodynamically stable form, no specification with regard to the polymorph in the generic drug product is required. However, the situation changes, if a metastable polymorph is used.

  Finally, like the procedure in ICH Q6A, the ANDA guidance also tries to replace setting a specification for the morphic form in drug product by drug product performance testing, such as dissolution testing that represents a surrogate parameter for the morphic form.

10.4.7 Polymorphism in Pharmacopoeias

Within the European Pharmacopoeia (EP), there is a chapter dedicated to polymorphism [39]. The chapter provides general information about polymorphism including pseudo-polymorphism but does not introduce much with regard to regulatory requirements for polymorphs. Beyond the definition of polymorphism and the introduction of enantiotropic and monotropic systems, the chapter provides an overview of analytical methods suitable for the identification and characterization of morphic forms. As discussed in previous parts of this chapter, polymorphism mainly effects properties related to drug products, drug product performance including bioavailability, bioequivalence, and dissolution. However, monographs for drug products have only be introduced into the EP in recent years and are still few. Accordingly, polymorphism beyond the statement that a substance exhibits polymorphism is rarely mentioned in monographs in a

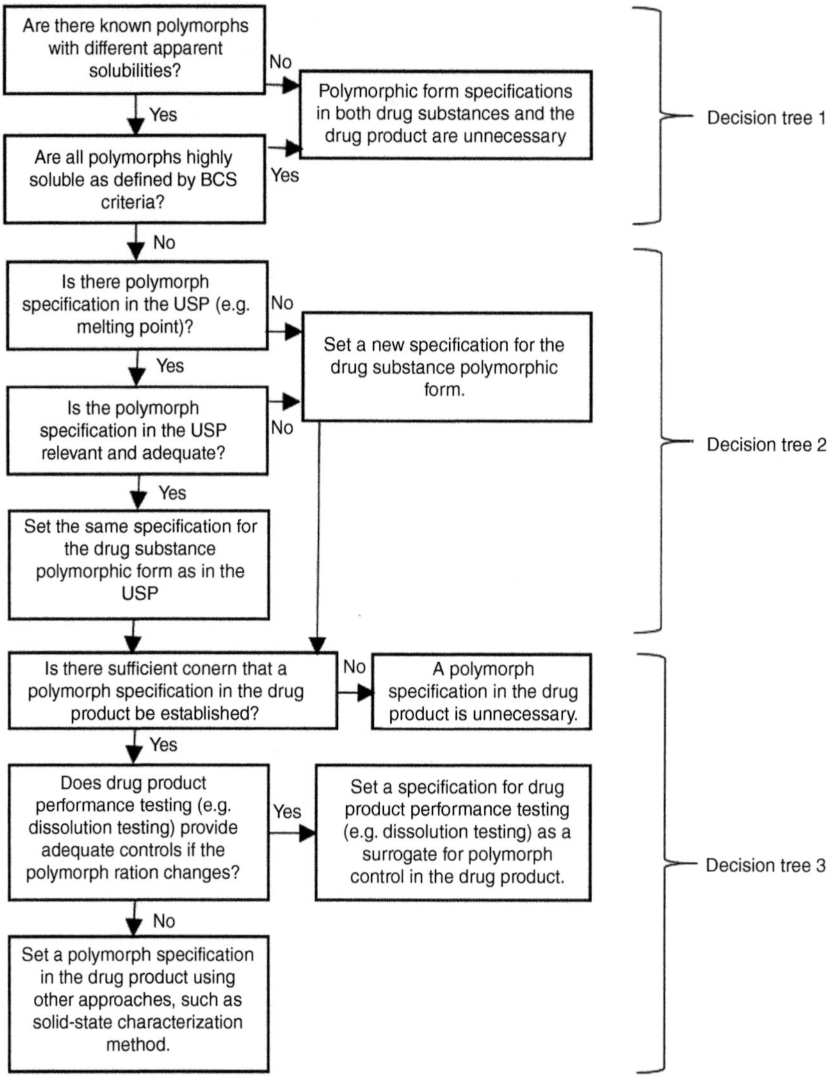

Figure 10.2 Decision trees for polymorphism as given by the FDA "Guidance for Industry – ANDAs: Pharmaceutical Solid Polymorphism" [38]. The three decision trees provided there have been condensed to one in this figure to facilitate reading. Source: FDA (2007) [38].

direct way. However, indirectly morphic forms are covered by identification tests. Frequently IR spectroscopy is used as a first identification test, and determination of the melting point of a substance is described in many monographs as a second identification test. The IR spectrum of a substance as well as its melting point will depend on the respective polymorph. In cases where differences in IR spectra as used for an identification test occur, the description of IR spectroscopy [40] allows to recrystallize the sample to be investigated and chemical reference substance to yield the same morphic form as the reference that is used for comparison.

> "When comparison of the spectra recorded in the solid-state show differences ..., treat the substance to be examined and the reference substance in the same manner so that they recrystallize or are produced in the same crystalline form, or proceed as described in the monograph, then record the spectra again. However, this procedure must only be done for substances where the monograph does not cover a particular form of a substance that exhibits polymorphism."

As mentioned in the last sentence cited above, some monographs including drug substances and excipients contain explicit statements on the morphic form. This is the case as such properties might be relevant, for example, for manufacturing or other tests. This especially holds true for hydrates that might be explicitly mentioned in monographs.

Within the United States Pharmacopeia (USP), the description of polymorphism is much more condensed and can be mainly found in chapter <941> [41] "Characterization of crystalline and partially crystalline solids by X-ray powder diffraction (XRPD)". The chapter is focused on characterization, identification, and quantification of morphic forms.

Within the USP chapter on "Spectroscopic Identification Tests" <197> [42], a similar procedure and applicability of this procedure as described by the EP is provided:

> "Differences between the USP Reference Standard spectrum and sample spectrum that may be observed are sometimes attributable to differences in the solid-state form of the materials, if a solid-state technique is used (e.g., ⟨197A⟩, ⟨197K⟩, or ⟨197M⟩). If a specific crystal form is not specified in the monograph, where spectral differences between the sample and USP Reference Standard are observed, recrystallize both the sample and USP Reference Standard under identical conditions to produce the same solid-state form, unless specific procedures are provided in the individual monographs. Dissolve equal portions of the sample and the USP Reference Standard in equal volumes of a suitable solvent, evaporate the solutions to dryness in similar containers under identical conditions, and repeat the identification test on the residues. Other techniques for recrystallizing the sample and USP Reference Standard based on known scientific principles may be used with appropriate scientific justification."

10.5 Regulatory Framework for Co-crystals

Compared to pharmaceutical salts and polymorphs including solvates and hydrates, co-crystals used in the pharmaceutical industry represent a comparatively new topic. The concept to use co-crystals to improve physico-chemical properties of APIs has become more widespread only about 15 years ago. At this time, no regulation existed with regard to co-crystals, leading to uncertainty about how health authorities might think about these solid-state forms. Such a regulatory uncertainty led to a situation where sponsors have been reluctant to use co-crystals as feedback from authorities was difficult to predict. Accordingly, this is an example where guidance by authorities was required to enable scientific progress.

FDA has been the first agency to come up with a draft guidance on cocrystals in late 2011 "Regulatory Classification of Pharmaceutical Co-crystals" [43]. The guidance provided a definition of co-crystals

> "Solids that are crystalline materials composed of two or more molecules in the same crystal lattice."

Additionally, the role of a co-crystal was defined as a "drug product intermediate".

> "Co-crystals should be classified within the Agency's current regulatory framework as dissociable "API-excipient" molecular complexes. They may then be treated as a "drug product intermediate" rather than as a drug."

Both definitions together yielded a difficult situation:

From a scientific standpoint, it is difficult to draw exact boarder lines between pharmaceutical salts, solvates, hydrates, and co-crystals. All of these represent crystalline two (or more) component systems. The difference between a solvate or hydrate and a co-crystal depends on the question whether the second component is a liquid or solid at room temperature. Additionally, there are solvates and hydrates of co-crystals. In a similar way, solvates and hydrates of pharmaceutical salts are well known. The distinction between a pharmaceutical salt and a co-crystal depends on the fact, whether the components are ionized or not, accordingly on the exact position of a proton, which sometimes can be hard to predict or to locate experimentally. There are examples of co-crystals of pharmaceutical salts. As such an example, fluconazole together with maleic acid forms a co-crystal of a pharmaceutical salt [44]: within the crystal lattice, there are ionized molecules of maleic acid as well as neutral molecules of maleic acid. The first one yield to a classification as pharmaceutical salt and the second one to a classification as co-crystal. Finally, there is also an overlap between pharmaceutical salts, solvates, and hydrates and co-crystals by hydrated or solvated co-crystals of pharmaceutical salts.

As FDA proposed to classify co-crystals as "drug product intermediates", this would have severe implications as neither the classification of a pharmaceutical salt nor the classification of a polymorph including solvates and hydrates leads to this classification. This classification also means that co-crystals do not belong anymore to drug substance manufacturing but to drug product manufacturing. As co-crystals are typically obtained by crystallization, which is a typical drug substance manufacturing step but very exotic as a process in formulation, this definition was not considered as very helpful. The issues raised by the draft guidance were addressed in a publication by authors from industry and academia from the United States and India [45]. Within this publication, the authors suggested to classify co-crystals together with pharmaceutical salts. The authors argued that the distinction of a co-crystal and a pharmaceutical salt just depends on the exact position of a proton, and this might be hard to allocate, especially when the single-crystal structure is not available. There are examples of drugs that have been approved as pharmaceutical salts but represent co-crystals by their crystal structure, e.g. caffeine citrate or escitalopram oxalate. Additionally, co-crystals dissociate upon dissolution in a similar way as pharmaceutical salts and similarly to the common-ion effect that influences the solubility of a pharmaceutical salt, a co-crystal also dissociates in solution, and there is a common-component effect that also has an effect on solubility.

Based on this discussion, FDA came up with a new guidance on co-crystals in February 2018 [46], which is relevant for NDAs as well as ANDAs. The guidance now defines cocrystals as:

> "...crystalline materials composed of two or more different molecules, typically active pharmaceutical ingredient (API) and co-crystal formers ("co-formers"), in the same crystal lattice."

The distinction between co-crystals and pharmaceutical salts is made in the way that a pharmaceutical salt requires ionic interactions, whereas within the crystal lattice of a co-crystal, there is also a defined stoichiometry but non-ionic.

The distinction of co-crystals and polymorphs is made in the way that "true" polymorphs represent single-component solid-state forms that have different arrangements of the molecules in the crystal lattice. According to the FDA guidance, co-crystals "*can be viewed as special case of solvates and hydrates, wherein the second component, the co-former, is not a solvent (including water), and is typically non-volatile.*" Within the guidance this leads also to the important conclusion about the regulatory status of a co-crystal including co-crystals composed of two or more APIs:

> "... has a regulatory classification similar to that of a polymorph of the API. Specifically, it is not regarded as a new API. From a regulatory perspective, drug products that are designed to contain a new co-crystal are considered analogous to a new polymorph of the API. A co-crystal that is composed of two or more APIs (with or without additional inactive co-formers) will be treated as a fixed-dose combination product and not a new single API."

From a scientific standpoint, this definition is meaningful, and from a regulatory standpoint, it facilitates the use of co-crystals much more compared to the definition in the precious version of the guidance.

Beyond a new definition of co-crystals, the guidance provides information to be included in submission documents for co-crystals. Such data should show that the API and co-former build together a crystal lattice and that interactions within the crystal lattice are nonionic. This will be obvious in cases where there are no protolytically active groups in the API and co-former but must be demonstrated by experimental data in cases where the API and co-former can act as acid and base so that a proton transfer can take place. Data used for this purpose might be the difference in pK_a of API and co-former. If these differ by more than one unit, the probability of proton transfer is considered high, and the opposite holds true if the difference is less than one. Additionally, reference is given to the usage of "*spectroscopic tools and other orthogonal approaches*" to elucidate whether a proton transfer has taken place or not within the crystal structure. Of course, X-ray diffraction methods represent powerful tools for this purpose.

Finally, the guidance asks for date showing that the co-crystal dissociates to yield the free pharmacologically active species before it reaches the site of pharmacological activity. This can be shown by in vitro methods such as dissolution or solubility. On the one hand, this requirement is meaningful, as if such dissociation does not take place a different efficacy and safety profile might be the result. On the other hand, it is very unlikely that such dissociation does not take place in the aqueous environment of the human body.

Within the guidance, it is mentioned that the guidance is only relevant for new entities that have not been assessed previously as pharmaceutical salts or other non-co-crystalline solid-state forms. This addresses cases where APIs might represent a co-crystal based on newer data on structural elucidation, but this has not been known at the date of submission and accordingly submissions might have claimed pharmaceutical salts. As in such a case only the nomenclature changes, but not the properties of the API and drug product, it seems reasonable to limit the scope of the guidance only to new co-crystals.

After FDA issued its draft guidance on co-crystals, EMA issued a "reflection paper on the use of co-crystals of active substances in medicinal products" [47]. This reflection paper was adopted by Committee for Medicinal Products for Human Use (CHMP) in May 2015, so after FDA issued its first draft guidance, but before the conceptually revised FDA guidance was available. Even in the introduction of this reflection paper, EMA made clear that the concept as described by EMA was not in line with the concept of the FDA draft guidance.

The definition of a co-crystal as described by EMA is not too different from the one given by FDA:

> "...defined as homogenous (single phase) crystalline structures made up of two or more components in a definite stoichiometric ratio where the arrangement in the crystal lattice is not based on ionic bonds (as with salts). The components of a cocrystal may nevertheless be neutral as well as ionized."

Also here, the nonionic interactions in the crystal lattice are described as a key element of a co-crystal. Nevertheless, the definition by EMA mentions that not necessarily all components that build up the crystal lattice of a cocrystal must be neutral but might be also ionic.

Additionally, the EMA reflection paper mentions that co-crystals need not always be stoichiometric:

> "Homogenous crystalline solids containing variable amounts of co-former (also known as solid solutions) are described in the literature. In such not

> fully stoichiometric solids, the amount of co-former may vary over a given range at a given point in the lattice of a crystal structure."

The reflection paper describes hydrates and solvates as a special case of co-crystals, where water or a solvent takes over the role as a co-former. The definition that the co-former in a co-crystal must be a solid as a pure substance is criticized. However, the terms hydrates and solvates are justified.

With regard to the distinction between pharmaceutical salts and co-crystals, some analogies between the EMA and FDA view exist. For example, attention is paid also by the EMA reflection paper on the difference in pKa between API and co-former. However, EMA acknowledges that there is no strict borderline between a pharmaceutical salt and a co-crystal, as the proton transfer – the position of the proton – might also be incomplete within the crystal structure.

With regard to the regulatory framework as defined by the EMA reflection paper, the key element is the fact that co-crystals upon dissolution dissociate into the pharmacologically active species and the co-former. Therefore, the situation that the solid-state form provides the active species in solution is similar to pharmaceutical salts, hydrates, and solvates – the latter as special cases of co-crystals. As the EMA reflection paper acknowledges this principle, also a similar regulatory approach results where again bioequivalence becomes a key element. Co-crystals – including hydrates and solvates – will become eligible for generic drug applications in a similar way as pharmaceutical salts and as described in directives 2001/83/EC Article 10(2)(b) and 2001/82/EC Article 13(2)(b). Again, no differences in safety or efficacy must occur, e.g. due to the co-former. Co-crystals are not considered as new active substances (NAS) by EMA. However, a differentiation is made between oral administered co-crystals and other application routes. For oral administration, the probability of dissociation of the co-crystal is very high as a first step in the gastrointestinal tract will be dissolution including dissociation. This might not necessarily hold true for other administration routes such as topical or pulmonal application of a co-crystal. For such administration routes, the status about NAS might be different.

Finally, the EMA reflection paper defines the status of a co-crystal in manufacturing similar to a pharmaceutical salt, as typically co-crystals are manufactured by similar processes including crystallization. Accordingly, the EU GMP framework for active substances becomes relevant. However, the reflection paper also considers situations where the formation of a co-crystal takes place during typical drug product manufacturing steps such as wet granulation or hot-melt extrusion. In such cases, the regulatory framework to be used is the one for finished products. This represents a very pragmatic and scientifically sound approach, which does not complicate the regulatory situation of co-crystals in an unnecessary way.

Finally – like the FDA guidance – the EMA reflection paper also mentions co-crystals that contain more than one pharmacologically active moiety. Here, both health authorities come to the same conclusion and treat such co-crystals like fixed-dose combinations.

10.6 Summary

Within this chapter, an overview on the regulatory framework for solid-state forms is provided. Considering the manifold of solid-state forms such as pharmaceutical salts, polymorphs, pseudo-polymorphs, hydrates, solvates, and co-crystals and the fact that frequently – even in a scientific concept – there are overlaps between such concepts, a complicated situation is the result. This complex situation gets even more complex as within different legislations – such as the United States and the European Union – different approaches are taken by regulatory authorities. To enable scientific progress within the pharmaceutical industry and making such progress available to patients an innovative regulatory framework is required. Certain major steps such as the definition of the regulatory landscape for pharmaceutical salts and polymorphs including solvates and hydrates and a lately guidance for co-crystals have been made. However, for different regions still different approaches are valid. Clearly, there is still room to facilitate progress in the pharmaceutical world by harmonization of such approaches that will make pharmaceutical development more efficient.

List of Abbreviations

ADC	antibody drug conjugate
API	active pharmaceutical ingredient
ANDA	abbreviated new drug application
ANSM	Agence Nationale de Sécurité du Médicament et des Produits de Santé
AUC	area under the curve
BA	bioavailability
BCS	Biopharmaceutical Classification Scheme
BE	bioequivalence
BfArM	Bundesamt für Arzneimittel und Medizinprodukte
BP	British Pharmacopoeia
BPOG	Biophorum Operations Group
Caco-2	human colon carcinoma cell line
CAR-T Cell	chimeric antigen receptor T-cell
CHMP	Committee for Medicinal Products for Human Use
CRISPR-CAS	cluster regularly interspaced short palindromic repeats – CRISPR associated proteins
CTD	Common Technical Document
DAB	Deutsches Arzneibuch
DSC	differential scanning calorimetry
DVS	dynamic vapor sorption
EMA	European Medicines Agency
EP	European Pharmacopoeia
FDA	Food and Drug Administration
FD&C	Federal Food, Drug, and Cosmetic (Act)

GMP	Good Manufacturing Practice
HTS	high-throughput screening
ICH	International Council for Harmonisation of Technical Requirements for Registration of Pharmaceuticals for Human Use
IR	infrared
IND	Investigational New Drug
JP	Japanese Pharmacopoeia
MDCK	Madin-Darby-Canine-Kidney cell line
MHRA	Medicines & Healthcare Products Regulatory Agency
NAS	New Active Substance
NCE	New Chemical Entity
NDA	New Drug Application
PDG	Pharmaceutical Discussion Group
PhRMA	Pharmaceutical Research and Manufacturers of America
PK	pharmacokinetic
PMDA	Pharmaceuticals and Medical Devices Agency
RLD	Reference Listed Drug
PXRD	powder X-ray diffraction
RNA	ribonucleic acid
SGF	simulated gastric fluid
SCXRD	single-crystal XRD
SIF	simulated intestinal fluid
TG	thermogravimetry
UK	United Kingdom
US	United States
USP	United States Pharmacopoeia

References

1 Biologics Control Act (1902). https://www.usp.org/sites/default/files/fda-exhibit/legislation/1902.html (accessed 08 July 2019).

2 Sinclair, U. (1985). *The Jungle*. Pinguin Classics.

3 Immel, B.K. (2001). A brief history of the GMPs for pharmaceuticals. *Pharmaceutical Technology* 42: 44–52.

4 ICH Harmonised Tripartite Guideline ICH M4. (2017) Assessment and control of DNA reactive (mutagenic) impurities in pharmaceuticals to limit potential carcinogenic risk. https://database.ich.org/sites/default/files/M7_R1_Guideline.pdf (accessed 08 July 2019).

5 ICH Harmonised Tripartite Guideline ICH Q3D. (2019). Guideline for elemental impurities. https://database.ich.org/sites/default/files/Q3D-R1EWG_Document_Step4_Guideline_2019_0322.pdf (accessed 08 July 2019).

6 USP-38-NF33, <1663>. Assessment of extractables associated with pharmaceutical packaging/delivery systems.

7 USP-38-NF33, <1664>. Assessment of drug product leachables associated with pharmaceutical packaging/delivery systems.

8 Stahl, P.H. and Wermuth, C.G. (2011). *Handbook of Pharmaceutical Salts*. Wiley-VCH.

9 Paulekuhn, S., Dressman, J., and Saal, C. (2007). Trends in active pharmaceutical ingredient salt selection based on analysis of the orange book database. *Journal of Medicinal Chemistry* 50 (26): 6665–6672.

10 Hilfiker, R. (2006). *Polymorphism in the Pharmaceutical Industry*. Wiley-VCH.

11 FDA. (2018). Breakthrough therapy. https://www.fda.gov/patients/fast-track-breakthrough-therapy-accelerated-approval-priority-review/breakthrough-therapy (accessed 7 July 2019).

12 EMA. PRIME: priority medicines. https://www.ema.europa.eu/en/human-regulatory/research-development/prime-priority-medicines (accessed 07 July 2019).

13 PMDA. Strategy of SAKIGAKE by MHLW. https://www.pmda.go.jp/english/review-services/reviews/advanced-efforts/0001.html (accessed 07 July 2019).

14 FDA. (2019). Abbreviated new drug application (ANDA). https://www.fda.gov/drugs/types-applications/abbreviated-new-drug-application-anda (accessed 07 July 2019).

15 B. Schulze, Different Salts of a Drug Substance – Comparison of Regulatory Pathways in the EU and USA, Bonn, Wissenschaftliche Prüfungsarbeit, Zur Erlangung des Titels, "Master of Drug Regulatory Affairs" 2016.

16 A.S. Raw. Regulatory considerations on pharmaceutical solids: polymorphs/salts and co-crystals, FDA-CDER-office of generic drugs. http://www.icdd.com/assets/ppxrd/presentations/11/P24-Andre_Raw-Regulatory_Considerations-Polymorphs_and_CoCrystals.pdf (accessed 21 July 2019).

17 EMA (2010). CPMP/EWP/QWP/1401/98 Rev.1, guideline on the investigation of bioequivalence. https://www.ema.europa.eu/docs/en_GB/document_library/Scientific_guideline/2010/01/WC500070039.pdf (accessed 16 June 2019).

18 FDA (2019). Approved drugs with therapeutic equivalence evaluation. https://www.fda.gov/media/71474/download (accessed 16 June 2019).

19 21 CFR 314.3. (2021). Applications for FDA approval to market a new drug – definitions, subpart A—general provisions. https://www.ecfr.gov/cgi-bin/text-idx?SID=99967e52b6bc64fd27bf728ecfd089b7&mc=true&node=se21.5.314_13&rgn=div8 (accessed 16 June 2019).

20 EMA (2019). CPMP/EWP/QWP/1401/98 Rev. 1/ Corr, guideline on the investigation of bioequivalence. https://www.ema.europa.eu/en/documents/scientific-guideline/guideline-investigation-bioequivalence-rev1_en.pdf (accessed 20 July 2019).

21 FDA (2014). Bioavailability and bioequivalence studies submitted in NDAs or INDs – general considerations, draft guidance. https://www.fda.gov/media/88254/download (accessed 20 June 2019).

22 FDA (2013). Bioequivalence studies with pharmacokinetic endpoints for drugs submitted under an ANDA, draft guidance. https://www.fda.gov/media/87219/download (accessed 20 June 2019).

23 21 CFR 320.1. (2021). Bioavailability and bioequivalence requirements. https://www.ecfr.gov/cgi-bin/text-idx?SID=ae20ae4ee980ed7516223b62b4eaca36&mc=true&tpl=/ecfrbrowse/Title21/21cfr320_main_02.tpl (accessed 20 June 2019).

24 EMA (2014). EMA/CPMP/EWP/280/96 Corr1, guideline on the pharmacokinetic and clinical evaluation of modified. Release dosage forms. https://www.ema.europa.eu/en/documents/scientific-guideline/guideline-pharmacokinetic-clinical-evaluation-modified-release-dosage-forms_en.pdf (accessed 20 June 2019).

25 FDA (2017). Waiver of in vivo bioavailability and bioequivalence studies for immediate-release solid oral dosage forms based on a biopharmaceutics classification system. https://www.fda.gov/media/70963/download (accessed 20 June 2019).

26 Amidon, G.L., Lennernaes, H., Shah, V.P., and Crison, J.R. (1995). A theoretical basis for a biopharmaceutic drug classification: the correlation of in vitro drug product dissolution and in vivo bioavailability. *Pharmaceutical Research* 12 (3): 413–420.

27 Federal Food, Drug and Cosmetic Act, Section 505(b)(2). https://www.fda.gov/regulatory-information/search-fda-guidance-documents/applications-covered-section-505b2. Food and Drug Administration (FDA) https://legcounsel.house.gov/Comps/Federal%20Food,%20Drug,%20And%20Cosmetic%20Act.pdf (accessed 22 June 2019).

28 FDA (1999). Applications covered by section 505(b)(2), draft guidance. https://www.fda.gov/media/72419/download (accessed 22 June 2019).

29 Directive 2001/83/EC of the European Parliament and of the Council of 6 November 2001 on the Community Code relating to Medicinal Products for Human Use. (2001). https://eur-lex.europa.eu/LexUriServ/LexUriServ.do?uri=OJ:L:2001:311:0067:0128:en:PDF (accessed 21 June 2019).

30 European Commission Health and Food Safety Directorate – General Health and Food Systems and Products – Medicinal Products (2019). Notice to applicants, volume 2A procedures for marketing authorization, chapter 1, marketing authorisation, revision 10. https://ec.europa.eu/health/sites/health/files/files/eudralex/vol-2/vol2a_chap1_en.pdf (accessed 21 June 2019).

31 ICH Harmonised Tripartite Guideline ICH Q6A. (1999). Specifications: test procedures and acceptance criteria for new drug substances and new drug products: chemical substances. https://database.ich.org/sites/default/files/Q6A%20Guideline.pdf (accessed 23 June 2019).

32 ICH Harmonised Tripartite Guideline ICH Q8. (2009). Pharmaceutical development. https://www.ich.org/fileadmin/Public_Web_Site/ICH_Products/Guidelines/Quality/Q8_R1/Step4/Q8_R2_Guideline.pdf (accessed 23 June 2019).

33 ICH Harmonised Tripartite Guideline ICH Q8. Development and manufacturing of drug substances (chemical entities and biotechnological/biological entities. https://database.ich.org/sites/default/files/Q11%20Guideline.pdf (accessed 23 June 2019).

34 Bauer, J., Spanton, S., Henry, R. et al. (2001). Ritonavir: an extraordinary example of conformational polymorphism. *Pharmaceutical Research* 18 (6): 859–866.

35 Dinnebier, R.E., Sieger, P., Nar, H. et al. (2000). Structural characterization of three crystalline modifications of telmisartan by single crystal and high-resolution X-ray powder diffraction. *Journal of Pharmaceutical Sciences* 89 (11): 1465–1479.

36 ICH Harmonised Tripartite Guideline ICH M4. (2003). The common technical document. https://www.ich.org/products/ctd.html (accessed 23 June 2019).

37 ICH Harmonised Tripartite Guideline ICH M4 – Questions and Answers. M4Q implementation working group questions and answers (R1). https://www.ich.org/fileadmin/Public_Web_Site/ICH_Products/CTD/M4_R1_Quality/M4_Quality_Questions_Answers_R1.pdf (accessed 23 June 2019).

38 FDA (2007). ANDAs: pharmaceutical solid polymorphism. https://www.fda.gov/regulatory-information/search-fda-guidance-documents/andaspharmaceutical-solid-polymorphism-chemistry-manufacturing-and-controls-information (accessed 29 June 2019).

39 European Pharmacopoeia 10, Polymorphism. European Directorate for the Quality of Medicines & HealthCare (EDQMH).

40 European Pharmacopoeia 10, Absorption spectrophotometry, infrared. European Directorate for the Quality of Medicines & HealthCare (EDQM).

41 US Pharmacopoeia. USP42-NF37, <941>, Characterization of crystalline and partially crystalline solids by X-ray powder diffraction (XRPD).

42 US Pharmacopoeia. USP42-NF37, <197>, Spectroscopic Identification Tests.

43 FDA (2011). *Regulatory Classification of Pharmaceutical Co-crystals, Draft Guidance*. FDA https://www.regulations.gov/document?D=FDA-2011-D-0800-0001.

44 Kastelic, J., Hodnik, I., Sket, P. et al. (2010). Fluconazole cocrystals with dicarboxylic acids. *Crystal Growth and Design* 10: 4943–4953.

45 Aitipamula, S., Banerjee, R., Bansal, A.K. et al. (2012). Polymorphs, salts, and cocrystals: what's in a name? *Crystal Growth and Design* 12: 2147–2152.

46 FDA (2018). Regulatory classification of pharmaceutical co-crystals, guidance for industry, revision 1. https://www.fda.gov/media/81824/download (accessed 05 July 2019).

47 EMA (2015). EMA/CHMP/CVMP/QWP/284008/2015, reflection paper on the use of cocrystals of active substances in medicinal products. https://www.ema.europa.eu/en/documents/scientific-guideline/reflection-paper-use-cocrystals-active-substances-medicinal-products_en.pdf (accessed 06 July 2019).

11

Opportunities and Challenges for Generic Development from a Solid-state Perspective

Judith Aronhime[1] *and Mike Teiler*[2]

[1] *Consultant, Pharmaceutical Polymorphism and Solid State Chemistry. IP Litigations Support, 5a Maor Yosef St., Rehovot, 7621745, Israel*
[2] *Consultant, Portfolio Management and New Product Introduction, 12 Har Kamon St., Kfar Sava 4448418, Israel*

11.1 The Birth of a New Drug and the Generic Siblings that Will Follow – Two Different Mindsets

If you want your project to succeed – work backwards.

While at first this advice might seem a little counterintuitive, anyone who has experience working on projects will tell you that this is true, and the reason isn't that hard to understand. When you set off on a journey, you plan according to your destination.

A family vacation requires very different planning to a business trip, and each requires different planning. In other words, you start from the end, and work backward. Projects are just the same. You start with a clear understanding of the project objectives. In the highly regulated pharma space, this is especially important.

11.1.1 Generics

A generic pharmaceutical project objective differs from a proprietary project objective in many ways. What does the objective of a generic project look like? For one, a generic project most often has a timing objective that is determined by external factors. This may be a patent expiry date or some other legally mandated protection that determines when a dossier can be submitted to the competent authorities or when the product itself may be introduced to the market. The important thing to know is when the drug master file (DMF) or pharmaceutical dossier needs to be submitted or the product should be launched, and work backward from there. Regulatory approval time to meet this date or sometimes data exclusivity will dictate the filing date. Filing date will determine the date by when stability studies must be initiated. From here one can calculate back to formulation development start and active

Solid State Development and Processing of Pharmaceutical Molecules: Salts, Cocrystals, and Polymorphism, First Edition. Edited by Michael Gruss.

pharmaceutical ingredient (API) development start and other interim deliverables. Work backward.

Another objective of a generic project is to ensure that the drug product meets the requirements of "essential similarity" to a known reference product. Besides having the same active ingredient in the same dosage form and the same amount, the formulation must release the active ingredient in such a way that it is absorbed to the body or performs like the reference drug in vivo. The criteria for equivalent performance will differ between dosage forms and products. This is known as bioequivalence (BE). Different dosage forms will have different BE requirements. For a simple solution product, there should be no issue of release from the dosage form. For an oral solid dosage form, anything impacting the absorption of the drug must be considered, particularly solubility and permeability. For each different route of administration and dosage form, and sometimes even drug product, there may exist specific BE requirements that must be considered.

The drug product must be demonstrated to remain chemically and physically stable when stored as directed throughout the labeled shelf-life. In addition, the product must be industrially manufacturable, and that process must be sufficiently robust to repeatedly and reliably give the same quality product batch after batch. These are the objectives of a generic project. Some of these activities may require the generic manufacturer to design around patents designed to protect the proprietary reference product in order to meet the timing objectives of the project.

The patents protecting a drug product will often relate to the drug substance. In addition to patenting the active moiety of the molecule itself, additional protections may include the salt form, cocrystal, ester, hydrate, crystal form, or particle size of the drug substance and may also include the route of synthesis.

11.1.2 Proprietary Products

A project to develop a proprietary product, sometimes called an "innovative" or "ethical" product, is generally much broader in scope and culminates in a new, branded product being launched in the market. While the pharmaceutical objectives may be exactly the same, the safety and clinical objectives are far more complex. An innovative development needs to consider exactly what indications and warnings are going to be approved by the competent authorities to be included in the labeling. This will determine what can be claimed for the new product and how it can be presented to and used by the medical community. A generic, in contrast, will show pharmaceutical and therapeutic equivalence and simply copy other aspects of the label of the proprietary product.

Another key difference from generics is that innovative products generally face no externally generated time limitations, and the development time is the determining factor for commercial product introduction. While the generic team is focused on designing around patents, the innovative development team will be doing their best to erect barriers of protection to ward off other brand and generic companies, generally in the form of patents. As a result of these considerations, the development teams are under immense pressure to complete their tasks in the minimum time possible, while still meeting all quality and project scope requirements.

11.1.3 API and Solid State

11.1.3.1 Generics

The importance of solid-state development of the API for both generic and proprietary drugs cannot be overstated, even as the objectives for each may differ. If a reference product has no patent protection related to the solid state, the generic developer will focus on faithfully duplicating the reference drug substance in every way. However, if valid patents protecting the solid-state attributes of an API could delay the market entry of a generic, these patents need to be designed around, for example, by using an alternative crystalline form of the API. Finding an alternative polymorph or cocrystal may result in changes to the physical properties of the API and need to be compensated for. This may be achieved by modifying the properties of the API itself or during the formulation process. For example, a different polymorph to that found in the reference product may have different solubility characteristics. If solubility of the API is a critical quality attribute impacting upon BE, the different solubility will need to be compensated for which can be done, e.g. by modifying the particle or granule size in the formulation to offset the difference in rate of solubility.

11.1.3.2 Proprietary

For the developer of a new chemical entity (NCE), a critical project objective may be to invent and protect novel solid-state characteristics of the API. The importance of timing may present a dilemma for the development team. For example, it may be very urgent to synthesize even miniscule amounts of API in order to allow preclinical work to begin post-haste; however, the initial route of synthesis devised may not be the most efficient and may need to be improved before the product is finally launched. The initial crystalline form recovered may not be the optimal form for use in the finished pharmaceutical product, and further screening may be required in order to arrive at a more desirable result during development. These routes of synthesis and crystalline forms can serve as the basis for early testing as well as for some of the earliest patents filed to protect the product. There are many examples of changes in crystal form during the development process, and new polymorphs may allow the possibility of issuing new patents, which potentially might extend the innovator's exclusive franchise on a product. At the same time, the earlier attempts and the patents written to protect them might turn out to be just the cracks in the wall that a generic company is looking for to enable introduction of their products to the market earlier than the conventional wisdom might indicate. These are complex issues and the development strategy as well as the budget and timelines must be considered very carefully to achieve the right balance of product protection, project cost, and time to market.

11.2 Portfolio Management – How Is a Portfolio Constructed and Maintained?

Crafting a portfolio for a company can be likened to buying a wardrobe of clothes. You can go to a department store and buy everything off the rack, or you can go to a designer who will tailor the wardrobe to fit your exact desires as well as your dimensions. Just as no two people are built exactly the same, similarly, no two companies are exactly the same with respect to strategy, size, capacity, growth objectives, technological capabilities, and commercial strengths and weaknesses. All of these need to be considered when creating a portfolio strategy and the governance process to support it.

11.2.1 Activities and Timelines

11.2.1.1 Strategy

Portfolio strategy is nothing if not a reflection of company strategy. The first thing for a company to consider when creating a portfolio is the overall financial growth objective of the company. The new product pipeline, whether organically driven or procured from third parties, is the most important engine of growth for any company. How much and how fast does the company want to grow? While the answer to that question may seem obvious, responsible management will know that the obvious answer is probably not the healthiest choice for the company and the right answer will depend upon the answers to many more questions:

How many projects are required to meet the desired level of growth? What is the mix of projects required? The mix of products may be defined as a technological mix, the different market segments strategic to the company, or the level of financial or legal risk the company is willing to assume. Is the company capable of supporting this number of projects? Are there enough research and development (R&D) resources? Are investments in new technologies, equipment of manufacturing infrastructure required? Are there sufficient financial resources to fund them?

Finally, it is important to look at the timing of the projects so that revenues can grow continually and smoothly over the entire horizon of the portfolio, and if gaps are found, they can be filled in early enough to prevent any negative financial impact.

11.2.1.2 Value

How much competition? One of the most important factors impacting the price of a product in the market is the number of players. The substitutable generic market, that is, where substitution is at the prerogative of the dispensing pharmacist or institutional purchasing agent and not dictated by the prescribing physician, is highly competitive. Profit margins tend to be quite narrow and that makes selection of high value products all the more challenging. While there is no secret formula, the classic rule of supply and demand applies. The less competition on a particular product, the more likely that the price of the product in the market will hold up.

How can the solid-state characteristics of a drug substance have an impact on the level of competition in the market and become a competitive advantage? Consider a drug product in which the API in the reference product is protected by a crystal form patent. There are many such examples of blockbuster drugs that were protected this way, from sertraline to clopidogrel to imatinib [1, 2]. This protection, if longer than the protection provided by the patent on the molecule itself, only keeps out those manufacturers who choose to copy the originator's crystalline form and represents an opportunity for a company that is skilled in screening, identifying, and isolating polymorphs. By creating a crystalline, hydrate, solvate, amorphous, or cocrystal form that is not protected by a patent, a company would be creating an opportunity to enter the market earlier than some competitors. If the alternative crystalline form were sufficiently unique and could itself be patented, the company would potentially have an exclusive offering of enhanced value.

A key factor in achieving value in the market is to launch the product first, or at the very least with the first wave of companies, once legitimate patents and exclusivities have expired. Not only should there be less competition, but it is considerably easier to capture market share at the time of market formation and far more difficult to "claw back" share from those who have it once the market has already formed.

Having the right solid-state strategy could be the difference between launching with the first wave and scrambling with a second or later wave of launches.

11.2.1.3 Factors Impacting on Timing – When and How Does a Product Show Up on a Generic Company's Radar Screen?

Like all of our questions, this goes back to company strategy. How aggressive is the company about being first to market? When is "on-time" and where do I get this information? Launching first or at least in the first wave generates a price premium, but it also comes at a significant cost. First to market is not always possible, and sometimes maybe not even be desirable, but for products where a first-to-market strategy is to be pursued, the candidate selection needs to happen early, sometimes 10 or more years before patent expiry. The following sections will explore some of the main reasons why.

First-to-Market Strategies Beginning with the United States. The US market is unique both in the way that patents are addressed and in the way data exclusivity is applied. However, since the United States represents well over 40% of the global pharmaceutical market in value [3], many companies will prioritize this market, and it deserves special discussion. The rules governing generic filings (abbreviated new drug applications, ANDAs) in the United States were originally passed by congress in 1984 as "The Drug Price Competition and Patent Term Restoration Act," more commonly referred to as the "Hatch-Waxman Act." This law has been modified over time and the following information is accurate as of the time of writing of this chapter. It is important to understand a few basic concepts in order to understand the mechanics of a first-to-market strategy.

The **"Orange Book"** (OB), formally "Approved Drug Products with Therapeutic Equivalence Evaluations," is the registry maintained by Food and Drug

Administration (FDA) of all approved new drug applications (NDAs) and ANDAs and excludes biological drugs (BLAs). In addition to the active ingredient, dosage form, strength, brand name, and name of applicant, you will also find information listed about reference listed drug (RLD, i.e. reference product) and therapeutic equivalents, exclusivities granted by the FDA, and patents that may be listed against each drug product. These include patents on the molecule itself, polymorphs, formulations, and methods of use. Route of synthesis patents are not eligible for inclusion in the OB [4].

Data exclusivity is known as "NCE exclusivity" in the United States and is granted to any molecule the first time it is approved as a drug for human use, extending for a period of five years from first approval by the FDA. NCE is the only exclusivity that limits when the FDA may *accept* an ANDA for filing. All other exclusivities, e.g. orphan drug exclusivity (ODE), new dosage form (NDF) are "market exclusivities" meaning that the product may not be put on the market until these exclusivities have expired, but they do not limit when an ANDA for the product can be filed [5, 6].

Patent certification – When submitting an ANDA to the FDA, the applicant must inform FDA how he is going to address any patents listed in the OB relating to that specific RLD. Each patent is addressed separately. There are four certifications. Paragraph I – no patents listed, Paragraph II – listed patents have expired, Paragraph III – the applicant is aware of the patent and is asking FDA to approve the ANDA after expiry of the patent, or Paragraph IV (PIV) – the patent is not relevant to the applicant's product because of invalidity or non-infringement of the patent [7, 8].

PIV – The US congress incentivized generic applicants to find ways to bring products to market as early as possible in order to reduce the cost of pharmaceuticals for their constituents. The first applicant to file an ANDA with a PIV certification is eligible for 180 days of "generic exclusivity" in the market, assuming the applicant is successful with their patent challenge. This means that FDA will not approve another generic until the 180 day period of exclusivity has expired. In case more than one "first" applicant should file on the same day, all applicants filing on the same day would be eligible to share the exclusivity [7].

NCE-1 – In case of a PIV application, the NCE (five years of data exclusivity) is shortened to four years and thus is referred to as "NCE-1," denoting five years minus one year [9]. Since NCE exclusivity is the only limitation to ANDA filing and acceptance by FDA, you will quickly realize that in order to ensure eligibility for the 180 day generic exclusivity as reward for challenging a patent, a generic applicant would have to target the NCE-1 date for filing an ANDA. In recent years, this has resulted in numerous instances of "shared exclusivity."

30 month stay – Once an ANDA submitted to FDA with a PIV certification is accepted by FDA for review, the ANDA applicant must notify the NDA holder. Should the NDA holder sue the ANDA applicant for patent infringement in time, FDA will apply a "30 month stay" during which time the ANDA cannot be granted final approval. Only tentative approval may be granted. The rules surrounding the 30 month stay are complex [7–9] and should be discussed with a patent expert on a case-by-case basis. The 30 month stay is a provision to allow the parties time to complete patent litigation, but may have the added effect of prolonging the protection of the patent holder. Without final approval, the generic product cannot be launched into the US market. Impact – despite the fact that FDA works to a 10 month review timeline, should a 30 month stay be in place, the actual time from submission to final approval is effectively 33–42 months [8, 9]. Just to be crystal clear, an ANDA submission with a PIV certification cannot be granted final approval until at least seven and a half years have passed from the first approval of the RLD [9].

First to Market Can Be an Expensive Proposition for a Number of Reasons **Starting material** – In many cases, in order to be ready to file an ANDA on the NCE-1 date, a generic company will need to start developing their API and finished drug product many years in advance of patent expiry. APIs, advanced intermediates, or key starting materials (KSMs) may not be readily available at such an early date. Those that are available are likely to be expensive both because of lack of competition, as we discussed earlier, or because manufacturers are looking for a quick return on investment (RoI) and don't necessarily want to wait for all protections to run out to start recouping their investment on commercial sales.

Legal expenses – These expenses may include in-depth analysis of the patents and literature to reach a comprehensive understanding of the intellectual property (IP) landscape, legal opinions on the validity of patents, freedom to operate (FTO) opinions – to demonstrate that the intended API and drug product do not infringe anyone else's patents, and litigation costs for PIV litigation in the United States, or simply, any patent litigation needed to clear the way to commercial launch, in other territories. If you have your own IP to protect, there will be costs associated with writing and prosecuting your own patents, as well as submission fees and annuities to maintain each patent in each country.

Risk management – In order to be sure to meet the submission and/or launch dates, it may be necessary to mitigate the risk of delay or failure with a robust development plan that includes back-up plans and redundancies. Any "extra" work performed in order to prevent or minimize delay caused by the failure of the lead strategy will have resource and cost implications. Saving time is often paramount, however, dictating that some level of risk management and mitigation is a must. The level of risk that a company is willing to assume will differ from company to company.

11.2.2 Timing

11.2.2.1 When Is "On-time?"

Remember, we said a successful project is always managed backward, right? So, how long does it take to develop a product? 12–24 months for API? Another 18 months for formulation? Minimum of 10 months for regulatory approval? Do I have to consider any time for litigation or 30 month stay? Project management can calculate these times, add an appropriate project buffer, and report back how long ahead of market formation the product needs to be selected so that the teams can start developing a product. Alternatively, when considering a PIV (patent challenge) submission to the US FDA, calculate back from the NCE-1 date.

There are many good reasons to start a project as late as practical. The later new product development is initiated, and the closer to market entry, various uncertainties decrease. The success or failure of the reference product can be observed and not just estimated, as well as whether or not new alternatives enter the market. More suppliers of KSMs and intermediates will become available and their prices will decrease. Regulatory guidelines may be published as more interest is generated by both companies and governmental and para-governmental organizations, such as pharmacopeia. Net present value (NPV) of the project increases as there are fewer years of early investment to discount. Technology costs may come down as newer technologies become available. For all these reasons a company may legitimately choose to start new product development later, rather than earlier.

Conversely, the longer one waits to start development, the bigger the "head-start" other companies have to develop their own technologies and create new IP to protect them (e.g. route of synthesis, hydrates, true crystal forms). Starting late could mean that the company now has to consider not only innovator patents, but third-party patents, which may have been filed as well. What impact could that have on FTO?

From this example one might conclude that waiting for the latest feasible start date is not really feasible after all, and it is better to start as soon as possible, as soon as one becomes aware that a potential new product is in development, before an NDA is approved or even submitted! Starting too late presented us with risks, particularly of new patents being filed threatening our FTO. Conversely, starting too early presents other risks. If we start before the product is approved, there is always some risk that FDA will reject the NDA, or even earlier, fail in phase III. Do we have the intestinal fortitude to start as early as phase II, where the risk of innovator failure is even higher?

The double-edged sword: In conclusion, though it is left to each company to make its own decisions if an ANDA needs to be submitted to the US FDA on the NCE-1 date, there are exactly four years from the date of approval of the innovator drug to the date of submission of the first generic, and that is not sufficient time to develop both API with polymorph screening and formulation, including performance of BE and pivotal stability studies with any degree of safety.

On the other blade: The earlier a product is selected and development is started, the higher the risk that the innovator product will fail, either commercially, during review, or in phase III clinical trial. Again, when to actually start execution of new project activities comes down to each company's objectives and appetite for risk. Whatever the decision with respect to execution, it would be prudent to identify development candidates, whether NCE-1 or not, before the reference product is approved in the market.

11.2.3 Market-specific Considerations Based on Local Legislation and Administration (OB, PIV, Various Exclusivities – US, EU, JP, etc.)

Section 11.2 so far, while discussing many considerations generally, was presented with a US orientation and modifications for different markets must be addressed. While it is not feasible to discuss all global territories in this chapter, it will be helpful to point out those aspects that apply only to the United States, while providing as much direction as possible for the European Union (EU), Japan (JP), and other highly regulated markets, which represent the lion's share of global value.

11.2.3.1 Patents Through the Eyes of the Regulatory Authorities

US FDA The US FDA is the only competent authority who is tasked by law to behave as a gatekeeper of patents. While FDA claims no expertise as to patent infringement or validity, they are responsible to administer the process. The discussion earlier, with respect to the PIV, process, with its detailed timing requirements as a mechanism to facilitate patent litigation before marketing authorization (MA) is purely a US issue.

EU In the EU, the regulators are not held responsible to be aware of the patent issues that may exist between the innovator and the generic applicant. It is fully the responsibility of the generic applicant to take whatever actions necessary to be able to commercialize their generic product. This is generally the situation in all countries besides the United States.

11.2.3.2 Data Exclusivity (Data Protection)

The purpose of data exclusivity is to incentivize the development of new drugs and allow an NDA holder sufficient time to profit from a new product even in cases where it was considered that adequate patent protection may not be available. For example, consider a naturally occurring substance that has potential therapeutic properties. Because it is naturally occurring, such a molecule could not be patented, per se. Other types of patents affording reasonably broad protection may, of course, be filed and listed against the reference drug, where applicable.

Technically, the regulator is required to hold the safety and efficacy data exclusively for the NDA holder and may not refer to this data in review of a generic application, preventing submission of an ANDA (in EU article 10.1 Marketing Approval Application (MAA)) until the period of exclusivity is expired [10].

Data exclusivity is not applied globally, nor are the rules or exclusivity periods the same across all countries that do apply data exclusivity. In the United States, this is the five-year NCE exclusivity, discussed earlier (see "First to market strategies"). In Europe, the rule is known as "8 + 2 + 1" allowing eight years of data exclusivity and a further two or three years of market exclusivity from the first EU approval for human use [10]. The third year is applied if a significant new indication is approved. Japan applies a 10 year "re-examination period" to the data and will not accept generic applications before this time. New indications in Japan may be granted an additional, shorter, reexamination period, which can extend the data exclusivity [11].

Canada applies eight years of market exclusivity [12], while Taiwan applies five years of market exclusivity [13].

11.2.3.3 Salts and Esters

US FDA considers different salts and esters of the same active moiety to be different products, thus ineligible for submission as generics, while the EU regulators consider them to be the same and thus may be submitted as generics [14]. A good example of how this applies to solid-state development is the case of the antidepressant Seroxat or Paxil from Glaxo SmithKline (GSK) (paroxetine hydrochloride, hemihydrate). In this case the hydrate form was patent protected longer than the active moiety and its salts. In Europe, the first generic product was able to circumvent the crystal form patents by obtaining approval for a different salt of paroxetine once meeting all requirements for essential similarity. It should be noted that if the alternate salt or ester raises safety concerns, some additional testing may be required [14]. Finally, it should be pointed out that such a product could be submitted and registered in the United States, but not as an "AB rated" substitutable generic [15]. There are cases of 505(b)(2) applications using alternate salts or esters being approved by US FDA [16]. The product would not be considered substitutable to the reference and would need to be actively marketed and prescribed by brand name. The marketing effort required represents a costly proposition for a generic company that does not have the infrastructure in place to handle without making significant investments.

11.2.3.4 Think Global, Act Local

For pharmaceutical manufacturers, the idea of having a single, global product seems very appealing. Whether discussing an API or a formulated product, it is easy to see how efficiency and economies of scale could be realized if a product with a global monograph could be manufactured for all countries and customers. Unfortunately, life is not that simple and regional differences by way of patent landscape, legislation, guidelines, and pharmacopeia monographs often dictate different timing, product specifications, and sometimes even different products. Strategically, a company would have to decide whether the additional expenses to develop, manufacture, and maintain a slightly (or completely!) different product are justified, in light of the potential revenue from entering an additional market or launching earlier in that market.

Patents and exclusivities make up the two important legal protections afforded to pharmaceutical manufacturers for their products, and we have already pointed out that the timing of submission of a generic in different countries will often differ because of differences in data exclusivity periods.

Patents, too, are national in nature, and while some aspects of patents may be administered internationally (World Intellectual Property Organization [WO], Patent Cooperation Treaty [PCT]...) [17, 18], ultimately, patents are granted, managed, and maintained on a country-by-country basis. Even when a single international patent application is initially made, ultimately, each patent is reviewed regionally or nationally, which may result in different patents being granted, or whether or how some claims are granted.

Examples of products such as the antidepressant Zoloft from Pfizer (sertraline hydrochloride) or the anticancer drug Gleevec from Novartis (imatinib mesylate) where there was different protection of crystalline forms between territories are important to note. Each of these products was a blockbuster in its own right, and bringing the product to market on time in different geographies was dependent on having the right crystalline form, on time, for the right territory. For high-value products, it is clear that having a country-specific strategy must be considered and the increased revenue weighed against the additional development and operational costs of having two or more products available on time for different markets.

11.2.4 Sources to Evaluate a Project

In the previous sections, we have had a look at portfolio strategy that has focused primarily on questions of "when", "where," and "how much?" This section looks at some of the data sources available to portfolio managers and other commercial and technical managers in the business to support their decision-making processes. Some of the data sources are free and readily available to anyone with access to the internet, while others are proprietary and available commercially through subscription or license.

11.2.4.1 Government and Regulatory Agencies

US FDA **OB** – We have already discussed the OB at some length. The electronic OB is available online to anyone and lists all approved NDA and ANDAs and for each lists the brand name, active ingredient, dosage form, strengths, therapeutic equivalence code, application holder, RLD or reference standard, date of approval, and any unexpired patents or exclusivities and more. In addition, one can find links to regulatory correspondence, supplemental filings, product labeling, and the summary basis of approval (SBA). The SBA, as the name implies, provides an overview of the data that was presented to FDA to gain product approval. This data is a great starting point when developing the quality target product profile (QTPP) and to get a general idea of project timing, though you will not want to draw too many conclusions without expert interpretation.

DMFs – FDA has a database of all DMFs deposited with FDA along with the status, owner, and name of each DMF.

Product-Specific Guidances for Generic Drug Development provides the BE requirements for a generic development. This information facilitates estimating the complexity and cost of generic drug development on a product-by-product basis. Note, this database is updated on an ongoing basis. Some guidance documents are drafts while others are final. Guidance documents under preparation or reevaluation are not available in the database [19].

Purple Book – Both the Center for Drug Evaluation and Research (CDER) and the Center for Biologics Evaluation and Research (CBER), both departments of the US FDA list information on BLAs under their respective responsibility [20].

DailyMed is a very useful and reliable resource for viewing US drug labeling graphics and texts, and **PubMed** (abstracts available for free) is a repository of peer-reviewed scientific articles, both from National Institute of Health (NIH) [21].

European Databases The legislation of the EU is more diverse than that of the Unites States, and this is reflected by the structure of the available databases.

European Medicines Agency (EMA) [22], **European Public Assessment Reports (EPARs)** [23] – All human medicines approved through the centralized procedure including both proprietary and generic MAs. The EPAR is similar in significance and scope to the US FDA SBA, aforementioned. This database also includes summary of product characteristics (SmPC), which provides the full prescribing data for the physician, the product information leaflets (PILs) for the patients as well as administrative data on how and when the product was approved. The latter is helpful when calculating data exclusivity.

MRI Product Index [24] – Abbreviation for "mutual recognition product index" found in the website of the Heads of Medicines Agencies [25] and provides information on all products approved through the decentralized procedure (DCP) or the mutual recognition procedure (MRP). General administrative data can be found including the (lead) reference member state and (other) concerned member states, and in many cases the SmPC and PIL are also included in attachment.

National Databases – Each member state maintains its own database of MAs. A few examples include the UK Electronic Medicines Compendium (eMC) and the Dutch College ter Beoordeling van Geneesmiddelen (CBG), which translates in English to the Medicines Evaluation Board (MEB) drug database [26, 27].

11.2.4.2 Analyst Reports and Company Financial Reports

All publicly traded companies are required to file financial reports with the investment community where the company is listed. In addition to the dry tables of financial accounting data, companies must report any information that can significantly impact revenue. Often, a company will list their leading medicines and report sales data as well as some information on products in the pipeline, their status, and when they may be expected to launch. Investment banks and other industry analysts also create reports about future trends across the industry or by therapeutic segment. These reports or excerpts or summaries from them may sometimes be found online for free, though often are sold individually or as part of a subscription service. A free website with many useful insights is PharmaCompass [28].

11.2.4.3 Pay Data Sources

There are many subscription services and the costs vary, generally in proportion to the amount of unique data they provide. These services range from consolidation and analysis of data in the public domain to own generated collection of data and analyses. Among the best known for market data is IQVIA (International Market Survey [IMS], before they merged with the Contract Research Organization [CRO] Quintiles). IQVIA data is the industry standard for retail and hospital pharmaceutical sales data in the United States and much of the world. The data can be analyzed in many different ways including by mass in kg of API, formulation sales in units or in value and because it is comprehensive can serve as a good tool to filter the pharma universe, country by country. Another well-known commercial database is the Newport database, which provides high-level sales data, patent expiry estimates, and information on current and potential API sources.

11.2.5 Evaluation Tools

Once you have your data sources identified, the next step in the portfolio process is to analyze the data and compare between projects that meet company project selection criteria.

11.2.5.1 Business Case

Creating the business case is critical as no one wants to invest time and resources in projects unless the RoI can be demonstrated. This is a tricky exercise for many reasons. For one, no matter how good your data is, there is always a margin of error. It doesn't matter if the data is collected internally from sources in the public domain, or expensive data purchased for this purpose. All data has some element of error and bias. The second main reason is that the business case is created for a future proposition, by definition. It takes time to develop API and the final formulation, followed by the time to achieve regulatory approval. If you have done well and day 1 launch is your objective, then you probably have to wait for litigation to run its course or for a patent to expire. To paraphrase Yogi Berra, the future is one of the hardest things to predict, hence, even more uncertainty. So, whichever forecasting model you choose, NPV, internal rate of return (IRR), or others to model your RoI, remember that it should not to be taken too literally and should only be considered a tool for comparing between projects.

11.2.5.2 Quality Target Project Profile (QTPP)

The QTPP represents the fundamental information you need to have before initiating a project. The purpose of the QTPP is to set the scope and technical objectives of the project. What is the active ingredient (including what polymorph or cocrystal or other solid-state characteristics), dosage form, content, trade dress (i.e. size, color, and shape), pack size, BE study requirements, batch size and annual forecast (for operational planning), required submission date, required launch date, projected sales (in units and value, if API, mass in kg). This information is key both to guide development and to estimate the full costs, including future investments in a project, and compare this to the sales revenue to calculate RoI. All of these parameters are key to having a successful project and, more importantly, a successful product launch.

11.2.6 Criteria for Identifying Promising Projects

Resources are finite, and so is the number of projects a company can work on. Each project in the portfolio must be capable of generating substantial value to justify the resources dedicated to it, both in absolute terms and relative to other projects in the portfolio. The overall sales value of a project will be the product of the price per unit and the number of units sold. As expected, competition increases, both market share and unit price decrease, impacting on the projected sales. Identifying products that fall into the company's field of expertise will provide a competitive advantage allowing the company to develop higher barrier products with lower risks, both technical and commercial. Being first to offer a product can be both an advantage and a disadvantage. Being first provides the opportunity to protect your own IP, capture higher market share, and recover expenses by supplying R&D quantities at a premium. At the same time, there is more uncertainty around the long-term success of the project as the competitive landscape continues to change. Whatever timing goal you choose, be sure to identify the product early enough to allow for uncertainty in development and regulatory time.

Solid state can be a key part of the answer to this question, by enabling the company to overcome barriers established by the innovator and by setting barriers to other generics through robust and well-protected unique crystalline forms. Using novel crystalline forms may bring formulation challenges, but when these are overcome, still more IP could be generated making it even more difficult for other generics to reach the market with a non-infringing product.

11.2.7 Criteria for Building a Robust Portfolio

In summary, what are those criteria that should be satisfied as we select products for our portfolio? Remember that resources are always finite. Time and other resources expended on one new project are, by definition, at the expense of another project opportunity, and we want to be satisfied that not only the absolute cost is justified, but that the "opportunity cost" is justified as well. By prioritizing the projects, whether using NPV, IRR, or other economic proxies, we can decide which of the projects will bring the highest value. The value of each product should also be sustainable over time. Consider that if the value erodes too quickly, the pipeline will have to supply higher value each year to compensate for value erosion. Fit to the company strategy and objectives should be considered, as even the highest of value in potential has little actual value if we are not equipped to realize that value. Confirm that the value of the portfolio of projects is sufficient to meet desired financial objectives, and if they are not, what steps can be taken to remedy that. Look along the entire time horizon of the portfolio and ensure consistent ongoing growth. It is great to show that you will meet your long-term goal, but if the short- to medium-term results cannot be met, the company will struggle and possibly fail along the way. The level of risk and the portfolio mix must be considered. If all projects are high risk–high value, then missing any one of them may put the company at risk, if some level of delay or failure is not factored in.

Finally, consider the mechanisms in place to approve new projects and to cancel projects should the need arise. It is better to "fail early," before significant resources have been consumed, and the mechanism for periodic review must be in place in order to catch these changes on time. Too often do researchers and other managers "fall-in-love" with their projects and fail to see when a project has lost the potential to bring value. This may be especially true if large efforts have been expended, but remember that monies spent are never coming back, and the focus must be on the remaining cost and the value that the investment will bring to the company.

11.3 Challenges in Developing a Generic Product from the Solid-state Perspective

The complexity of generic drug development stems from the need to ensure that the product to be marketed will have similar therapeutic effect, safety, and efficacy to the RLD. However, drug development is very often further complicated by patents (mainly those held by the innovator), which seek to limit the ability of generic companies to arrive at an equivalent product. In practical terms, a generic product needs to be equivalent to the RLD in terms of absorption kinetics, even if different in composition for design-around reasons.

Here we encounter the high relevance of the physical characteristics of the API and especially the crystal form. The success of achieving a design-around formulation similar in performance to the RLD depends on the availability of an appropriate crystal form of the API. The performance of a solid formulation heavily depends upon the physical characteristics of the API. It is well known that crystal form, morphology, and particle size distribution of the API may have a fundamental impact on the dissolution properties, pharmacokinetic behavior, and bioavailability of the API in the finished formulation. This subject has been extensively discussed in the literature [29–40].

The influence of API physical attributes on the bioavailability is particularly relevant to biopharmaceutics classification system (BCS) class 2 and 4 products [41] where *solubility* and *dissolution* are a limiting factor to drug release and absorption kinetics and thus are critical for bioavailability. In those products, correct solid-state development of a generic drug product will be critical to a successfully satisfying BE requirements [42, 43].

While generic development could seem to be no more than a copy of the innovator's product, with no need for innovation and creativity, the reality is quite different, especially in a scenario in which patent barriers to formulation or crystal forms limit the choice of ingredients. In products where the innovator has patented crystal forms, the generic player that seeks to enter the market at the earliest date possible (e.g. after expiry of the compound patent and all exclusivity protections expire) will want to have a design-around plan, such as an alternative crystal form of the API to formulate the generic product. This plan requires efforts to search for a suitable crystal form, a task whose complexity is not to be underestimated in the crowded arena of polymorph discovery. In practicality, scientists working for the generic industry are faced with the challenge of finding creative ways to produce a design-around product and still be close enough to the RLD to successfully demonstrate BE.

The reality is that the generic industry, in order to design-around patents, frequently needs to develop a formulation with an alternative crystal form with respect to the RLD. And the reality is that in the predominant majority of those cases the alternative crystal form will be thermodynamically less stable than the crystal form in the RLD, which may cause serious challenges in development and BE.

It is the goal of preformulation scientists to investigate and characterize the available crystal forms of a pharmaceutical solid and select a form with the best combination of desired properties to proceed to the formulation stage, considering all the challenges of working with a metastable form.

The search for suitable solid-state properties of an alternative form imposes continuous work with the marketed RLD as a comparison standard for crystal form, morphology, and particle size of the API. Attempts to measure those characteristics in the RLD are an analytical activity employed with variable degrees of certainty, considering that the API is intimately mixed with other, inactive ingredients and the information achieved could be partial or ambiguous.

In summary, the major challenge to the generic R&D formulator working with alternative forms of an API, obtained by reproducing crystal forms known in the literature or by seeking novel crystal forms, is the achievement of dissolution and solubility as close as possible to that of the RLD in the final formulation.

11.3.1 Implications in Developing Formulation with a Metastable API

The RLD formulation contains in the vast majority of, but not all, the most or one of the most stable forms of the API. This thermodynamically stable crystal form could be anhydrous or, as in many cases, a stable hydrate. From the innovator's perspective, working with the most stable forms is the default preference for drug product development (with some noted exceptions) to prevent phase transformations during formulation or storage.

From this it can be understood that in the generic domain, if an alternative crystal form is required, in the vast majority of cases it will be a thermodynamically metastable form. The use of an alternative form that is thermodynamically metastable may present some challenges:

(a) *Control of the crystallization process to produce an API of high polymorphic purity:* Isolation of the API in a metastable crystal form demands the employment of a very selective method of crystallization and conditions that prevent transformation to a thermodynamically more stable form in the process. The drying stage is often the most critical stage.
(b) *Solubility higher than that of the RLD:* Metastable forms are, per definition, more soluble than the thermodynamically stable form. Potential solubility differences between the generic crystal form and the RLD crystal form could be an impediment to demonstrating BE, since a more soluble crystal form might lead to higher bioavailability. From the formulator's point of view, solving this problem by reducing the dissolution rate in a formulation could be an arduous task.
(c) *Need of a stable formulation to preserve the API crystal structure during storage:* Formulations, whether due to a wet processing, or water coming from excipients, combined with the great physical pressure applied during compression create a humid and packed environment in which the metastable form of an API could

potentially undergo a polymorphic transition immediately or over the course of the product shelf-life, although this will not necessarily occur. Awareness of this potential occurrence requires formulation development specific for control of the desired polymorph, as well as monitoring of the crystal structure to demonstrate its preservation throughout the approved shelf-life.

11.3.2 The Stability Question

Before discussing considerations and decision-making in the development of a product, we need to dedicate some attention to the polymorphic stability since this issue is very critical to the generic development. Monitoring stability is a central activity during development, as it is relevant during processing, handling, and storage.

When discussing stability, a clear distinction must be drawn between thermodynamic and kinetic stabilities, which are often misunderstood by the inexperienced person.

Thermodynamic stability is the basic property upon which the entire stability problem stands. It is expressed by the free energy of the crystal structure, which results from a combination of conformational energy of the molecules, physical interactions among the molecules, entropy, and more. Different crystal forms differ in their free energy and may coexist in spite of their different thermodynamic stability. However, at any given temperature, only one form will have the lowest free energy (thermodynamically stable form) and the others (thermodynamically metastable forms) will have higher free energy levels according to an energetic hierarchy. All metastable forms possess a natural predisposition to recrystallize as a more stable form. Since free energy is an intrinsic thermodynamic property of each crystal structure, there is nothing we can do to change it. Furthermore, there is nothing we can do about the natural propensity of any metastable form to recrystallize into the stable form.

Kinetic stability is the resistance of a metastable form toward its tendency to spontaneously transform to a more thermodynamically stable form and is dictated by the energy barrier which stands upon a transformation; the energy barrier depends on the molecules packing modes in the crystals and the structural change needed for rearrangement. Kinetic stability may be managed, in part, by controlling external factors (solvent/humidity, temperature, mechanical stress, seeding) and some physical attributes (particle morphology, particle size, crystallinity, etc.).

11.3.2.1 Polymorphic Stability in Dry Conditions

Metastable forms are not necessarily unstable kinetically under dry conditions, in spite of a natural tendency to transform to more stable forms. Based on real-life experience, it can be asserted that many metastable forms are kinetically stable and processable as long as the dry material is kept away from humidity or solvent vapors, the most predominant factor that facilitates polymorphic transformations. Moreover, in many cases metastable forms are sufficiently stable over time and resistant to processing and storage conditions, allowing the generic industry to safely produce and commercialize drug products containing metastable forms, with acceptable stability throughout the approved shelf-life. The effect of solvent vapor as the most efficient facilitator of solid-state polymorphic transformations in the API should be considered carefully. This effect may be latent during development until it reveals itself during storage in bulk, for example, in large container bags. What may happen during storage is that the powder is tightly packed and the bag is full, leaving a little void space where a small quantity of residual solvent or humidity may produce vapor pressure. This may lead to a vicious cycle of solvent migration, which will facilitate polymorphic transformation and caking. Both water humidity and solvent vapors may expedite transformations. Temperature too is a very dominant factor that accelerates polymorphic transformations. The most devastating effect is, frequently, a synergy of temperature associated with humidity.

Less frequently we see polymorphic transformation triggered by mechanical stress. It should be noted, however, that the presence of solvent could drastically enhance the effect of mechanical stress (e.g. grinding of wet samples before X-ray analysis, drying equipment in which the wet powder is mixed by paddles during drying, etc.).

11.3.2.2 Polymorphic Stability in Wet Conditions (Slurry)

A metastable form in contact with a solvent may readily dissolve and recrystallize into a less soluble form, e.g. a thermodynamically more stable form or a solvate. This occurrence, well known as *solution-mediated transformation*, is usually unwelcome when metastable forms need to be preserved. This is the most recurrent solid-state phenomenon in both chemical processes and formulation processes and might take place with little awareness of the chemist or formulator. Slurry of API at the end of crystallization, seeding, drying, transportation of wet powders, and wet granulation in pharmaceutical processes are some of the critical steps that may induce solution-mediated transformation. Although thermodynamically driven and energetically inevitable, a solution-mediated transformation may be kinetically hindered by controlling one or more of several factors, such as temperature, solvent, time, solid:solvent ratio, stirring regime, purity, morphology, and particle size distribution of the (starting) sample [44, 45].

11.4 Generic Solid-state Development

11.4.1 General

In order to understand how to integrate relevant solid-state activities into the development process, it is important to note the main R&D phases in which solid-state and crystal forms have an impact. First a strategic assessment is needed to decide which crystal form is preferred in the API and what the desired characteristics of such a material should be. Ideally this would be done jointly by the API and formulation development teams, as each has a critical, but potentially different perspective. Second, a crystal form discovery phase may follow based on the agreed strategy. The most appropriate crystal form needs to be selected and then the development of synthetic route and crystallization process for the API can be conducted and scaled-up. Once the API development is complete and the best form is available, formulation development can begin. Formulation development, using a form that differs from that of the API in the RLD, is by no means trivial, as discussed in detail in the previous sections. The teams need to work together so that the formulators have a clear understanding of the characteristics of the API, and the API team can learn from the formulators failures even more than their successes. Risk management may dictate that a second crystal form, whether polymorph, cocrystal, or salt, may be available in case the initial assessments made were incorrect and a fall-back strategy is required. Conclusion of the formulation development complete with stability and a demonstration of BE is the culmination of the development stage, which all started from a fruitful and successful polymorph screening.

Successful drug substance design relies on a number of critical R&D steps involved with major solid-state analysis: review of polymorphic patent landscape, own IP generation of novel crystal forms, analysis of the RLD formulation (specifically polymorph, particle size, morphology of the API), physical form target selection, characterization and control of physical properties of the API batches used for formulation. In the following sections important aspects of some of those activities will be briefly illustrated.

11.4.2 Predevelopment Phase: Solid-state Strategy

An important part of the strategy to be established is the solid-state strategy, meaning what crystal forms to select for formulation, whether a design-around form is needed, or to duplicate the form in the RLD, if a novel form is preferred, should it be protected, or is it sufficient that it does not infringe other patents – commonly known as FTO.

11.4.2.1 Review of the Solid State, Especially the Polymorph Patent Landscape

Development of the solid-state strategy and the course of the solid-state development plan depends on the solid-state patent landscape effective at the planned date and geography of launch. Initially, the only patent barrier on polymorphs comes from the innovator, while additional polymorph patents generated by generic companies follow with time.

The first stage of scrutinizing the patent scenario is to study the information disclosed in the literature, especially patent literature. For best understanding it is recommended that the solid-state chemist work closely with a patent expert because scientists and patent experts will each approach the material from a different perspective and emphasize different aspects. Patent analysis is not only about understanding the patent expiry date, but by analyzing the language and the scientific content with respect to quality of analytical data, procedures from previous publications, reliability of chemical procedures, etc. it is possible to speculate what is covered by the patent and where the weaknesses may lie. During this analysis the innovator's polymorph patent draws the most attention especially when development starts early.

The second stage is monitoring the innovator's polymorph patent with time, and this can be a long process because from the time of PCT publication until allowance of all the divisional applications years may pass and it cannot be known ahead of time exactly which claims will be pursued and granted, and which form, or forms may be protected. The level of uncertainty about how the final claims are granted is even greater as the initial patent analysis and API development are done earlier in time with respect to the PCT publication date.

In addition, at early stages of development, if the innovator's product has not yet, or only recently received regulatory approval, the compound patent expiry date itself is not clear as patent term restoration and extensions and pediatric exclusivity may come later, influencing the earliest market launch date.

Any new information and documents published by either innovator, generic players, or academia, at any time during this long period, could affect the solid-state patent scenario and solid-state strategy of a generic drug product.

As the patent situation evolves with time, the following documents of the innovator should be monitored:

(a) *Publication of the PCT:* A PCT is an international application that does not itself result in the grant of a patent. The PCT includes the body of the patent with all the important scientific information as a basis for the claims. If the application is not followed by national applications within a certain period of time, it will lapse and be regarded as just a publication with no rights of legal protection.
(b) *Publication of the national patent applications in the different geographies of interest:* National patent applications contain the claims requested by the patentee. It is important to follow after the phrasing and the content of the claims. Throughout the examination by the national patent offices, the requested claims may change in phrasing and content, may be added or deleted, and this could bring changes in solid-state patent protection. The examination process could endure several months, during which time there is uncertainty around the subject claims. To go into the specifics, in the solid-state field, changes can be observed in (i) what crystal form is claimed, (ii) number or choice of the X-ray diffraction (XRD) peaks used to characterize the crystal form, (iii) combination of analytical data, (iv) addition or deletion of a characteristic feature, such as "hydrate", "crystalline," "stable," etc.
(c) *Post-granting of a national application:* New divisional patent applications requesting claims for additional crystal forms or claims for additional features need to be monitored, to understand what other forms are sought to be protected.

11.4.2.2 Design-around Considerations

In sections 11.1 and 11.2 of this chapter, we discussed the concept of PIV filings and the timing and competitive advantage a company that is expert in solid-state development could gain in the market by launching in the first wave with a non-infringing product.

From the solid-state point of view, a design around drug product could be accomplished using different approaches.

Alternative Nonproprietary crystal form Selecting a crystal form not protected by a patent and openly described in the literature is one possible strategy to lessen the risks of IP confrontation, though nothing can reduce risk to zero. Following this strategy will facilitate FTO but will not bring any differentiation from other generic companies and will not stop others from following the same strategy. Finding such a scenario is rare, although not impossible.

Alternative Proprietary Crystal Form Having a proprietary (novel and patent protected) alternative crystal form that is stable, processable, and has solid-state properties facilitating formulation of a stable and bioequivalent drug product is a common scenario and is certainly the best choice if your strategy is to differentiate your product from other generic competitors.

While this is an advantageous approach, it may not always be successfully implemented for numerous reasons:

(a) Screening for new (metastable) forms in a polymorph search is a fascinating but also a frustrating activity, because the outcome cannot be predicted. In fact, there is no assurance that any new forms will be isolated.
(b) New forms isolated will likely be metastable forms. The ability to scale up the crystallization process for a metastable form cannot always be predicted from the beginning and unpleasant surprises can occur in advanced stages of development, after significant valuable resources have been expended.
(c) Not all the new forms found eventually in a polymorph search are suitable to be included in the drug product. Also, at the laboratory scale in which the polymorph search and characterization are done, it is not always possible to predict with certainty whether or not a particular crystal form would be suitable for the drug product.
(d) Choosing a novel crystal form discovered in-house through a polymorph screening procedure is usually accompanied by filing a patent for legal protection. Although generic patent protection could be potentially beneficial, there will be periods of uncertainty. It could take several months or a couple of years to discover whether another company has submitted a patent application for the same form before you, and you could find yourself in a situation where you cannot use your own invention. An alternative crystal form disclosed to the public in the academic or patent literature and not protected by a patent would not present risks of this kind, but also would not be patentable if published before you could submit your own patent application.
(e) In cases where generic development starts before the innovator's product is launched, it may not be possible to guess what crystal form is used by the innovator. Consequently it will be harder to select the most appropriate alternative form and particle size to develop without knowing the reference crystal form. This problem becomes even more critical when it is not known whether the API will be a salt or not.
(f) The probability to discover novel proprietary crystal forms diminishes the later the development starts, since more and more generic companies are already engaged in this same activity.

Cocrystals If no alternative crystal form of the API suitable for production and formulation is found, the cocrystal option should be investigated before going to explore the amorphous option.

Cocrystals with pharmaceutically acceptable inactive compounds are the best second-choice option for a design-around crystal form of an API. Regulatory acceptance in the United States of cocrystals as equivalent to polymorphs in 2018 [46] has expanded the horizon of new crystal form options. Typically, but not necessarily, cocrystals are more soluble than the thermodynamically stable form of the API [47, 48].

The main downsides of the cocrystal strategy are:

(a) Finding cocrystals is not an easy task and screening with many coformers may be extensive and time-consuming. Not always can cocrystals be isolated.
(b) Physical instability of cocrystals could lead to the recrystallization of the API and coformer into separate crystals and to a decrease in solubility.
(c) Demonstration that the new API is a cocrystal and not a salt is needed for regulatory acceptance.

Alternative Salts Another design-around strategy commonly used by generic players for specific markets is to look for alternative salts. This strategy is relevant only to specific geographies where alternative salts may be recognized as generics, like in the EU, for instance. US FDA does not accept this approach for ANDA filing.

Alternative salts suitable for formulating a drug product should still need to demonstrate BE to a relevant reference product, and so it would be helpful if they demonstrated the following:

(a) Similar in solubility and dissolution to the innovator's API. This target could be hard to accomplish, considering that on the contrary of polymorphs in which solubility differences are usually around two- to threefold, solubility differences among salts can be much bigger than that.
(b) Stable during storage toward decomposition into the neutral forms of API and salt former.

The alternate salt is a known pharmaceutically acceptable salt. In case of any doubt with regard to safety, additional data may be requested by the authorities. EudraLex – Volume 2b – Pharmaceutical legislation on notice to applicants and regulatory guidelines for medicinal products for human use states "The different salts, esters, ethers, isomers, mixtures of isomers, complexes or derivatives of an active substance shall be considered to be the same active substance, unless they differ significantly in properties with regard to safety and efficacy" [14].

Amorphous If no alternative crystal form suitable for production and formulation is found, the amorphous form will automatically be the next choice, although much less attractive because it is more difficult to develop and produce. The main challenges with amorphous forms of API are found in their potential instability, both physical and chemical, which can express itself both during formulation process and upon storage. Another challenge is establishing a suitable process for API production; amorphous materials are typically isolated from spray-driers, due to the fast precipitation of the solute in the tiny droplets created in this equipment. For APIs with low solubility in specific organic solvents, this procedure may require unmanageably large quantities of organic solvent.

Amorphous Solid Dispersions A widely employed way to physically stabilize noncrystalline API is to produce amorphous premixes or what are commonly called amorphous solid dispersions. Solid dispersions are prepared by coprecipitation of the API and a polymer, and the amorphous state of the API is stabilized because the dispersed polymer inhibits its crystallization. Although the system is thermodynamically unstable and the inhibition is kinetic only, recrystallization may be suppressed for a long time, rendering this option attractive. Attention should be given to the possibility that the quantity of polymer needed to stabilize the amorphous API may exceed the required size of the tablet [49].

11.4.3 Crystal Forms Discovery

11.4.3.1 Importance of the Crystal Forms Discovery Stage

Based on long time experience and on widespread affirmations in the literature, we can state that more than half of drug molecules (not including oligomers and polypeptides) are expected to exhibit polymorphic behavior under normal working conditions, either anhydrates, hydrates, or solvates.

For a solid-state design-around strategy, crystal form discovery is a preliminary research stage performed prior to the API development. It is essential to understand the importance of this stage, performed at very small scale, typically tens of milligrams for each experiment, to a successful drug product development program and BE study.

To develop a design-around formulation bioequivalent to the innovator's drug product may require experimenting with different crystal forms of API to see which one may lead to a successful BE study, which is especially important for BCS class 2 and 4 products. The more crystal forms available for development, the more chances to successfully conclude drug product development. In the likely event that the most stable forms are protected by the innovator, and generic competitors move rapidly to protect all "*left-over*" crystal forms, it is very important for the generic business to discover the largest number possible of forms in a dedicated screening program and protect them and to expand as much as possible FTO in the development of the formulation.

11.4.3.2 New Crystal Forms Unpredictability

In the real world, the most remarkable feature of polymorphs is their unpredictability. It is well accepted that the know-how to accurately predict

(a) what novel crystal forms may exist in the normal working space,
(b) what their physical attributes will be, and
(c) how they can be isolated

is not yet available to science. The only way to come to know novel crystal forms is to enter the laboratory and start looking for them by performing a large variety of crystallization experiments to isolate them.

The unpredictability statement lies in apparent contrast to the fact that novel crystal forms are sought today using systematic procedures generally described in the literature [50–56], so that isolation of new crystal forms looks like it is the result of a predictable, routine procedure with no surprise effect.

In reality, these two apparently contradictory but still widespread opinions may be conciliated by the following consideration: it is only partially true that the procedures to isolate new forms are systematic and described in the literature, because the description is only general. In fact, a skilled chemist knows the general protocol of crystal forms search, which is in essence: identify good and poor solubility solvents, and perform different types of crystallizations (cooling, solvent/anti-solvent, slurry, evaporation, exposure to solvent vapors, etc.). But the degrees of freedom behind this routine agenda are many, and the choice, number, and variety of experimentation are a decision of the chemist during the course of his work. For example, slurry is a standard procedure "known" to induce formation of a different crystal form, less soluble in the specific solvent system. However, the specific conditions such as temperature, time, concentration, solvent or mixture of solvents, etc. need to be explored and those may vary from compound to compound. The same considerations are applicable to cooling crystallization and other procedures.

It should be noted that for the generic player, special efforts are needed to find novel crystal forms suitable for formulation. Undoubtedly, increasing the number of experiments will increase the probability of finding new forms, after the most stable forms have been isolated. In general, the likelihood of finding novel crystal forms besides those disclosed by the innovator relies in part on the API polymorphic tendencies, in part on the extent of the search that was performed by the innovator, and in part on the extent of the search that will be performed by the generic company.

11.4.3.3 Pragmatic Questions About Crystal Forms Search

A generic player investigating novel crystal forms will confront some pragmatic questions such as how to increase experimental efficiency, what is the minimum quantity of material needed for each experiment, what is the minimum chemical purity needed, and how many experiments are required to decide that the polymorph screening is complete?

How to Increase Efficiency? First of all, it is important to understand that a good search for new crystal forms starts with the human factor, i.e. passion and determination to find as many "hits" as possible. A dedicated team of enthusiastic solid-state chemists and solid-state analysts focused on this task is the best recipe for a good search and high efficiency. The other aspect of efficiency relates to the laboratory practices. Parallel experimentation, of course, is one key way to increased efficiency. Several commercial solutions for automation and semi-automation are available; in this way time-consuming manual operations can be replaced by automated equipment and many experiments can be run in parallel. There is a common belief (mainly by people that are not chemists) that by using automated equipment to search for new crystal forms, the routine element of this procedure predominates and renders newly found forms more predictable and obvious. In order to avoid confusion, it is important to realize that automated equipment only replaces the hands of the chemist, but certainly not the mind. The chemist still needs to plan the experiments, observe the outcomes, and learn the results.

How Many Experiments Are Required in a Comprehensive Search? The unpredictability factor in the appearance of novel crystal forms certainly affects the extent of experimentation to be undertaken. Based on the approach that the number of crystal forms found is proportional to the time invested in finding them, the chemistry team will strive for the largest number of experiments in order to maximize the chances to find all possible "hits." On the other hand, conducting a crystal forms search for an open-ended period of time is not realistic in the industrial realm. The dilemma about what is the end point of a crystal forms search could be hard to resolve. As a rough estimation, an exhaustive and comprehensive search may include hundreds of experiments, between 500 and 1000. Even so, there is no guarantee that a new polymorph could have been found by the one experiment that was not preformed, although the chance of missing any "hit" after such broad experimentation is low.

What Is the Minimum Chemical Purity Needed? API chemical purity is an essential requisite for conducting a reliable crystal forms search. For this reason it is recommended that purity of the API used for polymorph search will be maximal, at least 99% [57] and preferably 99.5%.

This can be a real challenge due to the fact that at the stage in which a polymorph search is performed, synthesis of API is usually not optimized. Evidence coming from the literature reveals that specific chemical impurities may interfere with the growth of specific crystal forms [44, 58].

The effect of both the amount of an impurity and its chemical structure may vary from substance to substance. The risk of working with a low-purity API at this early stage is that the search might not be comprehensive and some crystal forms, especially stable ones, might be missed.

What Is the Minimum Quantity of Material Needed for Each Experiment? In the last decades, high-throughput automated technologies for crystal forms screening have made progress. One of the characteristics and declared advantages of the high-throughput equipment were the low quantities of API (up to 10 mg) employed and the use of Raman spectroscopy as a quick tool for crystal form identification. Although the advantage of using only few mg looks quite attractive, experience teaches that crystallization of such small quantities of API for each experiment is not easily reproducible or scalable.

11.4.3.4 Late-appearing Polymorphs

A natural consequence of working with low chemical purity or a chemical purity that does not represent the production lots could be the phenomenon of "*late appearing polymorphs.*" This is a general issue related to the appearance of novel forms at some advanced point in time, in different late stages of development such as pilot scale-up or production, or even after commercial launch. The phenomenon of late-appearing polymorphs encompasses the specific cases of disappearing polymorphs, in which the occurrence of the new crystal form causes difficulty (at least temporarily) in obtaining the previously produced form. A few cases of late-appearing and -disappearing polymorphs in the pharmaceutical industry are documented in the literature including commercial products (Norvir, Rotigotine, etc.) [59, 60].

A late-appearing polymorph is usually thermodynamically more stable than the previously known forms. Since the innovator's general trend is to develop a formulation with the most stable crystal form, the appearance of a new, more stable form, even at late stages of development, will lead to a change in the solid-state objective. As a result, this phenomenon can compromise previously set milestones or even cause recalls from the market and an inability to market the product, at least temporarily, like in the cases of Ritonavir and Rotigotine. In each of those cases only one form was known at the development stage, while a new form appeared in the formulation only after commercial marketing commenced [61–63].

An industrial case in which some crystal forms were found in the preliminary screening phase and other crystal forms were discovered during process development is Axitinib [63, 64].

Possible causes for this undesired phenomenon may be insufficient crystal forms search from the beginning and/or a change in the impurity profile following changes in the synthetic route. The lesson to all those (especially generics) who perform searches for new crystal forms in the laboratory prior to API development is that both incomplete crystal forms screening and low chemical purity of the API may result in missing potential "hits." To increase the probability to find the largest number of "hits" in the discovery stage, it would be a good practice to perform an extensive and thorough crystal form screening with maximal purity material and preferably from different synthesis routes.

11.4.3.5 Irreproducibility of Procedures

Another natural consequence of working with different chemical impurities (coming from synthetic route) could be the irreproducibility problem:

(a) Inability to reproduce crystallization procedures developed by other laboratories around the globe.
(b) Developing procedures that cannot be reproduced by other laboratories around the globe.

This problem may be encountered in some API projects and is not necessarily resolved easily. In summary, the chemical characteristics of the API used for crystal forms screening may affect the crystallization outcome and may depend on the raw material and synthetic route employed. It is important to emphasize, though, that such cases are a minority and do not represent all existing APIs.

11.4.3.6 Analytical Focus

For unequivocal identification of crystal forms, X-ray powder diffraction (XRPD) is the gold standard technique to rely on. Especially when not all crystal forms are yet known, it is not a good idea to rely on spectroscopic techniques. The reason is that sometimes crystal structure differences are difficult to see by spectroscopic techniques due to similarity in molecular packings. In contrast, the XRPD overall pattern of each crystal form must be unique and characteristic. Frequently used Raman spectroscopy may not be selective enough to enable unambiguous identification of new crystal forms. Currently it is accepted that crystal forms searches are carried out on the scale of at least a few tens of mg for each sample, and the crystal form is identified by XRD, which is the most appropriate technique to unequivocally identify novel crystal forms. Attention is required when analyzing and comparing XRD powder patterns, not to confuse artifacts (mainly due to sample preparation) with real differences, since in some cases the differences are subtle. Also it should be understood that it is difficult to recognize polymorphic mixtures; so, as more samples are analyzed, the higher the reliability of the analyses. At the end of the day, capability of distinguishing crystal forms and their mixtures by XRPD relies heavily on the competence of the analyst. Still, in uncertain cases where XRD data are similar but also different and it is not fully clear if polymorphs exist or not, the analyst may find himself doubting about the results. Remember that analysis of polymorphs does not tolerate vague or nonconclusive opinions such as "similar to" or "slightly different from," but is a yes/no test, because at the end of the day the analyst needs to clearly state if the analyzed sample is a distinct polymorph or not. In uncertain cases, this decision, very important for both regulatory and legal reasons, should be made demonstrating significant differences in crystal structure data, such as cell dimensions or space group (significant does not necessarily mean large).

Thermal techniques such as thermogravimetric analysis (TGA) and differential scanning calorimetry (DSC) as well as Karl Fischer titration are complementary tools available to characterize solvation state. Each form isolated should be characterized by additional spectroscopic techniques such as infrared (IR), Raman, solid-state nuclear magnetic resonance (NMR) spectroscopy, and optical microscopy. Final proof of the existence of a single phase of the promising crystal forms might be provided by crystal structure determination.

11.4.4 Target Selection

As already mentioned, the best scenario for generic development, especially in BCS class 2 and 4 products, is to finalize the crystal forms search with more than one alternative form and to choose the one most appropriate for development, preferably close in physical properties to the crystal form found in the API of the RLD. After the crystal forms search, some forms may be ready to scale up, although their physical properties were not yet fully investigated. To avoid redundant scale-up development of unsuitable crystal forms, some basic solid-state characteristics should be analyzed before continuing development. It is very arduous to predict what crystal form will be suitable in formulation using laboratory samples that do not represent the final API product, because many important bulk solid-state attributes (particle size distribution, specific surface area, hygroscopicity, tensile strength, morphology, etc.) can be established only after finalization of production process. However, to mitigate the risks of failure later on in development, there are some basic characteristics of each crystal form that can be analyzed in laboratory samples and compared to the characteristics of the RLD crystal form: solubility, morphology, solid-state stability. The larger the difference in those characteristics between a target form and the RLD form, the larger the risk to fail the BE study in the BCS class 2 and 4 products.

11.4.4.1 Solubility

Solubility is an intrinsic feature of each crystal structure, and it can be measured using material prepared in the laboratory with just a few tens of grams. For drug substances, solubility is the most crucial parameter for bioavailability. The gap in solubilities between two crystal forms, including hydrates, of the same API is usually two- to threefold [41, 65]. Development of a generic drug product with an alternative crystal form of the API two to three times more soluble than the RLD crystal form should not compromise severely the chances to pass the BE study; however, as the solubility gap increases to fourfold and above, the risk that the drug product bioavailability will be too high with respect to the RLD increases. It is not impossible to encounter metastable forms that are 7- or even 10-fold more soluble than the RLD form. Those are usually metastable forms with low kinetic stability and poor morphology. The decision to develop generic drug product with those forms should be weighed carefully.

When discussing solubility analyses for target selection, attention should be given to the procedure for analysis and the solid-state characteristics of the samples analyzed.

Solubility should be measured under the general conditions described in the BCS guidance [66]: slurry in three pH aqueous solutions in the pH range 1–6.8, shake-flask method, 37 °C, and check concentrations. The BCS guidance addresses equilibrium solubility and does not specify the duration of the experiment, even if it may be one of the very important parameters that affect solubility data. A common approach to implement the guidance is to consider that this measurement aims to reflect physiological pH conditions, so practically preferred time points for analysis could be in the span of about –five to seven hours, with one optional analysis after 24 hours in slurry. All the solubility points collected should be recorded and compared.

Measurement bias might deceive if some details pass unnoticed. Although this procedure is meant to represent equilibrium solubility, equilibrium is seldom achieved under measurement, because during the measurement a metastable form may gradually recrystallize. Initial solid-state attributes such as initial particle size distribution or morphology of the sample analyzed may affect the recrystallization kinetics and the results. In addition, it is reported that the amount of excess solids in the slurry may affect the results [67]. Dependence of solubility results on those frequently disregarded parameters is one of the reasons for irreproducibility of solubility data between laboratories. Moreover micronization may (i) increase solubility by reducing the particle size, (ii) introduce disorders in the crystal lattice similar to amorphous regions; those two factors also affect solubility. If the marketed RLD contains micronized API, the solubility results of crystalline samples prepared in the laboratory may not fully represent the actual product.

11.4.4.2 Morphology

Although solubility is a fundamental property to examine, the dissolution rate is a more practical aspect with regard to formulation performance. The morphological similarity or dissimilarity between a generic API and the RLD API may cause a big difference in the dissolution profiles of the generic and RLD formulations and hence compromise the probability of the generic drug product to be bioequivalent. This problem is mainly perceived in drug products in which the API crystals are not micronized and is related to the obvious morphological gap between well-crystallized stable forms and fast precipitated metastable forms. APIs in some of the RLD of BCS class 2 and 4 API products may be found in well-defined large crystals while alternative metastable forms prepared by fast precipitation may appear as very small needles or aggregates of undefined shaped tiny crystals, which cannot be modified. Such a gap in morphologies is a practical obstacle for a generic formulation to attain BE.

11.4.4.3 Solid-state Stability

Preliminary distinction between kinetically more stable and less stable metastable forms may be done by testing their instability tendency under mild stress conditions such as elevated humidity and temperature for a few days or weeks. Testing under those conditions should enable to predict long-term storage stability. Two elements that may decrease solid-state stability should be also tested:

(a) combined elevated humidity and temperature rather than separate conditions
(b) testing the stability of API pellets in addition to powder

Special attention should be devoted to hygroscopicity and possible formation of hydrates due to environmental humidity. Such hydrates can be very stable and their formation will be strongly dependent on storage conditions.

It should be remembered, though, that kinetic stability, which depends primarily on the crystal structure, may also depend upon large-scale physical attributes such as crystallinity, particle size and morphology, micronization, packaging, environment, etc.

11.4.4.4 Additional Factors

Other factors, not connected to solid-state properties, may need to be taken into account as well, such as manufacturability risk assessment. For instance, it is more difficult to control production of a pure metastable crystal form if the process involves drying of a solvate form. IP considerations may also become relevant if the alternative crystal form selected displays analytical characteristics that overlap those mentioned in a patent claim, for instance XRPD peaks.

11.4.5 Process Development in the Laboratory Scale

11.4.5.1 Process Development

Laboratory development aims to establish a crystallization system suitable to crystallize the target crystal form of the API in production. At laboratory scale it is practically impossible to predict all the engineering and scale-up factors that may affect polymorphic composition at large scale, but gathering basic knowledge is highly recommended in order to support later scale-up development and production. The basic input needed to support scale-up is (i) thermodynamic relationships among the forms and (ii) solubility curves.

11.4.5.2 Thermodynamic Stability Relationships

Since understanding a polymorphic system is essential for full control of a crystallization process, the thermodynamic relationships between polymorphs are the most basic information to gather. Thermodynamic relationships between polymorphs are described by their energy–temperature diagrams. Many sources in the literature explain the energy–temperature diagrams of polymorphs and the distinction between monotropic and enantiotropic polymorph systems, a couple are brought here as reference [68–70].

The essence is that it is vital to know for each pair of polymorphs whether they are related monotropically or enantiotropically, and if they are related enantiotropically, what is the transition temperature between the polymorphs. This analysis can be done in either one of the following ways:

(a) measuring solubility curves,
(b) measuring DSC melting curves and evaluating the thermal data,
(c) performing competitive slurries at different temperatures [71].

Each method presents experimental challenges and should be separately discussed. The DSC technique, based on Buerger's rules, is highly recommended, because it is quick, exact, most of the times unequivocal, and requires only a few milligrams of material. However, it is limited to cases in which the DSC melting/transition curves are not concomitant to other thermal events such as decomposition. Various examples in the literature show how DSC melting curves are utilized to generate energy–temperature–energy curves of polymorphs [72, 73].

11.4.5.3 Solubility Curves

In general, solubility curves of polymorphs, if correctly measured, are a mirror image of the thermodynamic energy–temperature diagrams, (regardless in which solvent they are measured) so they contain fundamental information about the polymorphic system that is not altered by engineering factors (see Figure 11.1).

Solubility curves also reflect the temperature–concentration domains where thermodynamic or kinetic factors govern crystallization of polymorphs [74].

For this purpose it is recommended also to measure recrystallization temperatures by cooling back each concentration until turbidity forms, in order to assess the "metastable zone" between solubility and spontaneous crystallization [75–77].

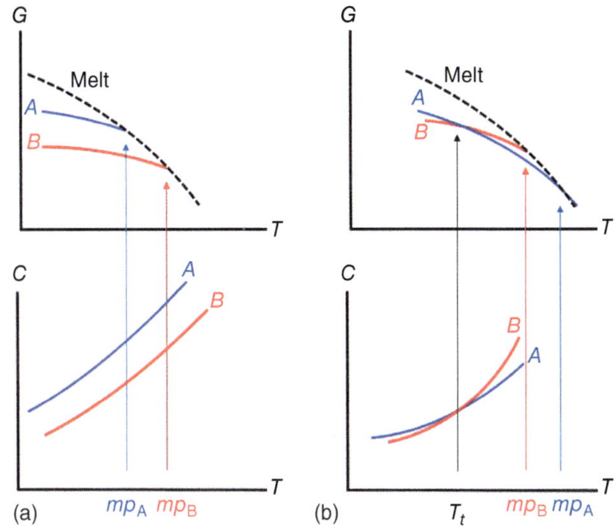

Figure 11.1 Energy–temperature diagrams (top) and solubility curves (bottom) for (a) a monotropic system, (b) an enantiotropic system. In the enantiotropic system the transition temperature is consistent in both energy–temperature diagrams and solubility curves.

Solubility curves generated in the solvent system selected for crystallization deliver information very important to establish at what condition to seed effectively, what temperature-concentration to avoid or to prefer during crystallization of the metastable form, etc. This information may be used to control a selective crystallization process for a metastable form, as initial crystallization temperature and concentration, or rate of crystallization, etc.

From the practical point of view, close attention is needed in measuring solubilities at different temperatures to build the solubility curves. Several experimental factors may affect the precision and accuracy of the solubility measurement. Some factors depend on the experimental procedure, mainly how the solubility point is measured. Other artifacts may derive from undesired polymorphic transformations into less soluble forms during the solubility measurement.

11.4.5.4 API Target

Success in developing a reliable process relies on setting a target specification and achieving it. In most cases, the target is a pure polymorph. However, any process for crystallization of a metastable crystal form encompasses the risk of partially isolating the thermodynamically stable form in varying mixture levels. When discussing polymorphic objectives, the level of polymorphic purity we wish to attain in the API needs to be understood. The specification of polymorphic purity may vary according to the circumstances, based on the crystallization process reliability and risk factors caused by solubility gaps between two polymorphs in a mixture.

For instance, if it is difficult to isolate a pure crystal form, minor quantities of a less soluble crystal form may be tolerated if solubility is not a critical parameter for BE, as in BCS class 1 and 3 products. In contrast, for BCS class 2 and 4 products, minor traces (even 1%) of a less soluble crystal form may severely impact the solubility and dissolution of the API and compromise formulation performance, especially in APIs with very low or no water solubility, so, in those types of products, the final target of the API development should be a pure crystal form. IP considerations may also become relevant with respect to certain crystal forms found in minor quantities.

Mixtures of crystal forms in similar ratios (such as 30%:70% or 50%:50%) are less common as an objectives for final API. Depending on the intended outcome of the crystallization process, it might be very difficult to consistently control polymorph ratios.

11.4.5.5 Analytical Methods for Polymorphic Purity

Having established the specification, a suitable analytical method may be developed. For quantitation of a polymorph mixture, common techniques such as XRD, IR, solid-state NMR spectroscopy may be employed depending on the spectral features of each crystal form. DSC could also be an excellent solution for crystal form quantitation provided that possible artifacts due to sample preparation and heating effects are excluded. For a reliable XRPD analytical method, it is a good practice to acquire a theoretical X-ray powder pattern and the theoretical list of peaks of the target crystal form, both extracted from a single-crystal structure analysis, in order to confirm unequivocally the nature of possible small peaks that may appear from time to time. In reality, the desired metastable form is frequently impossible to achieve as crystals suitable to analysis because often the metastable form can only be produced by fast crystallization that does not allow sufficient crystal growth. In such cases the theoretical X-ray powder pattern and the theoretical list of peak analysis may be extracted from crystal structure data obtained from high-resolution powder diffraction data (synchrotron).

11.4.6 Scale-up Challenges

The two major challenges during scale-up, related to solid-state properties, are

(a) control of crystal form and polymorphic purity and
(b) control of morphology and particle size.

11.4.6.1 Control of Crystal Form

A crystallization process of a pure metastable form may be successful at small scale in the laboratory but may fail when scaled-up.

One of the keys to successfully scale-up a crystallization process is undoubtedly the ability to reproduce the various parameters such as mixing, heat transfer, etc. at large scale. A team of trained chemists and chemical engineers is usually competent to accomplish this task. However, various difficulties related to polymorphic contamination may be encountered in large-scale processes due to local events not directly related to the bulk crystallization itself, but rather to what happens concomitantly or afterward, and may depend on equipment or engineering factors. Some of the occurrences in which contamination at various levels of stable crystal form may occur are:

- uncontrolled cooling after bulk crystallization,
- uncontrolled phase mixing in the reactor (especially relevant in highly polymorphic substances and systems of solvent mixtures),
- material sticking to the wall of the reactor or stirring blade or pipes,
- material remaining on the wall of the reactor or stirring blade or pipes from preceding batches,
- slurry after crystallization,
- precipitation in small amount from the mother liquor due to cooling during filtration,
- storage of the wet cake at an uncontrolled temperature,
- inefficient drying (promoted by the solvent still present in the cake either as liquid or as vapor), and
- stirring during drying.

Some well-known remedies may be considered in order to minimize the risk of these occurrences, although not always relevant or feasible:

- use of small and completely filled reactors,
- controlled cooling of the crystallized bulk based on solubility curve information,
- slurry of the crystallized mass at low temperatures,
- minimized moisture content of the wet cake after filtration,
- replacing the mother liquor solvent with an anti-solvent in the wet cake prior to drying,
- applying a two-step drying procedure,
- increasing drying efficiency.

11.4.6.2 Control of Particle Size and Morphology

Since the scale-up crystallization conditions will be different than laboratory crystallization conditions, and the morphology and particle size highly depend upon those conditions, particle size distribution and other bulk parameters are usually measured during scale-up development but not in early-stage laboratory development.

One of the biggest challenges in controlling particle size and morphology in production scale-up is that those features may be very sensitive to small variations in equipment parameters, so it might happen that as long as production and equipment are not finalized, particle size and morphology might undergo some modifications.

Another challenge occurring at large scale is a possible decrease in crystal quality with respect to crystallinity and morphology. It may happen without notice that production samples show lower stability than laboratory samples because of poor morphology or crystallinity. This phenomenon cannot always be avoided and cannot always can be perceived using conventional analytical techniques. For this reason, it is a good practice to repeat some solid-state characterization analyses, e.g. testing solid-state stability under stress conditions such as elevated humidity and temperature for a few days or weeks.

11.4.6.3 Lot-to-Lot Variability

During API manufacture, slight variations in physical properties of API batches and process parameters are commonplace and mostly pass unnoticed. However, at times such variations lead to the very frustrating problem known as "lot-to-lot variability," meaning that some practically imperceptible differences in physical properties of the API have occurred, which cause serious problems later in formulation. This phenomenon is very easy to miss when analyzing API using conventional tools and usually reveals itself through unexpected and unexplainable catastrophes during formulation manufacture. Most failures related to the API in formulation processes are caused by deviations in physical properties such as flowability, stickiness, cohesivity, etc. These bulk physical properties strongly depend on the particle attributes including surface of the API particles [78], and the effect on the bulk behavior is tremendous. That's where the problem is: these phenomena are not only poorly understood, but also difficult to observe and measure using common analytical methodologies.

A large variety of techniques may be used to characterize bulk properties of an API. Some of them are common and well known, some others are less widespread. Although the "lot-to-lot variability" problem cannot be eliminated completely, a practical suggestion to at least mitigate its occurrence would be continuous monitoring of as many solid-state properties as possible, in the produced batches. Important solid-state attributes that can be fairly easily monitored are specific surface area, amorphous content in micronized crystalline APIs, hygroscopicity, polymorphic and amorphous contamination, crystallinity percent, particle size, morphology, tensile strength, and flowability. Variability among API lots should be monitored to understand the range of values that the formulation may tolerate.

11.4.6.4 Analytical Focus

Characterization of bulk properties, as described earlier, may include a variety of features and be very extensive. Of course, it would not be realistic to perform so many measurements indiscriminately on each lot of API, putting excessive workload on the analytical laboratory. Instead, it is possible to prioritize the tests according to each API's attributes and the needs of the formulation. A few instances are presented as follows:

(a) Highly soluble API powder (nonmicronized) tends to suffer from poor flowability. This problem may be aggravated if a significant amount of fine particles ("fines") are present and the material is exposed to a humid environment. In formulation, this problem would be most pronounced when a dry mix process is employed. In this case, flowability and particle size distribution (with focus on the fines) need special attention and should be monitored. Since flowability can be measured in different ways, several commercial techniques are available depending on the relevant type of flow. The USP <1174> general method recognizes a few types of flow measurement: flow rate through orifice and angle of repose (gravitational), compressibility index, and shear test (for cohesiveness test). The choice of technique should be adapted to the type of flow applied in formulation. In addition, the Hausner ratio and Carr index extracted from the common bulk and tapped density measurements may provide a good sense of differences in flowability.

(b) Low soluble APIs (nonmicronized) may also suffer from poor flowability. This problem will depend on the proportion of large crystals and their morphology. In this case particle size distribution and morphology assessment need special attention.

(c) In case of high load oral dosage forms, API tensile strength analysis should be in focus. In formulations containing API above a certain percent (about 25–30%), the tensile strength of the API may play a dominant role in tabletability properties. Tensile strength may vary according to crystal form, morphology, particle size distribution (especially the span and d_{50}), therefore it may vary among lots of the same crystal form if those parameters are not properly controlled. Analysis of the tensile strength of the API, commonly called hardness, requires preparation of a tablet with API and a lubricant and is measured using a tablet hardness tester commonly found in pharmaceutical laboratories. The main challenge of this analytical technique is consistency in preparation of the API tablet, such as time duration of pressing, thickness, time of relaxation prior to analysis. Several replicate analyses are recommended for each sample.

(d) Amorphous content is frequently encountered in crystalline APIs after micronization, due to the mechanical stress that causes noncrystalline regions on the surface of the crystals (see "the effect of micronization on amorphous content"). Suitable techniques for analysis are microcalorimetry, DVS (dynamic vapor sorption), DSC [79–84]. Microcalorimetry and DVS analysis are based on monitoring recrystallization or vapor absorption in a controlled humidity environment. DSC analysis is based on monitoring recrystallization or glass transition events during heating. One important remark about developing an analytical method: the noncrystalline fraction formed during micronization is usually not a pure amorphous state, but it is rather a "disrupted crystal" [85]. It is important to stress this difference because pure amorphous and "disrupted crystals" are two types of material with different properties, despite that both show no diffraction peaks in the X-ray powder diffractogram. The main difference between a "disrupted crystal" and an amorphous state is that the "disrupted crystal" usually recrystallizes back to the originating crystal structure while the amorphous state not necessarily does this. For this reason the "disrupted crystal" shows a DSC exothermic crystallization peak at a typical temperature, while amorphous material will not necessarily recrystallize in the DSC or at the same temperature. The "disrupted crystal" state standard should be prepared only by milling (a laboratory ball mill is suitable) and not by conventional techniques to isolate amorphous forms.

11.4.7 Pharma Development

Due to the dependence of the formulation on the physical properties of the API, this is a very sensitive stage for the success of the drug development. If the physical properties of an API are not suitable for the desired formulation, failures may occur in formulation processability and performance. The importance of the API bulk physical properties may be underestimated during API scale-up development and problems in API manufacturing often lead to low efficiency, delayed shipment, failed batches, or even the expensive product recall. Three main points needing special attention are mentioned as follows.

11.4.7.1 The Tetrahedron Principle and Consistency Among Lots

The crystal structure and other features such as crystal morphology, aggregation, particle size distribution, surface area, etc. when taken all together form a set of physical properties of the bulk API. But the physical properties of an API alone are quite irrelevant unless the API is combined into a specific formulation. In fact, the process and performance of a drug formulation are intimately dependent on the solid-state properties (and structure) of all the ingredients, not least among them, the API. The concept of the materials science tetrahedron (MST) has been adopted to illustrate the interdependence among the four elements: process, performance, properties, and structure [86]. On top of those four elements, we should not forget to add the environment, i.e. temperature and humidity, on which the four elements also may depend. The practical consequences of the tetrahedron principle are that formulation process and properties strongly rely on the bulk solid-state properties of the API and once the formulation process is finalized, it is imperative to continue to use API similar in physical properties in order not to compromise the performance or processability of the formulation. Although the formulator relies, consciously or unconsciously, on consistency of the lots, the specific solid-state parameter that is most critical for consistency is not always clear, and it could be various parameters. It is advised to use diverse techniques to monitor solid-state properties of samples, as mentioned in the previous paragraph on lot-to-lot variability. Extra special attention should be paid to the API lot used for manufacturing the drug product batch used in the BE study. This lot needs to be characterized by the largest number of techniques possible, because it will become the reference lot for future production (if BE study is successful, of course).

11.4.7.2 The Effect of Micronization on Amorphous Content in Crystalline APIs

One of the most serious problems in micronized crystalline APIs is the effect of micronization on the crystallinity of the particles. Micronization processes are usually quite aggressive and often, but not always, cause destruction of the crystalline order and formation of a phase that looks like amorphous but is actually a "disrupted crystal" as explained in the previous paragraph. The quantity of the disrupted crystal (commonly called "amorphous content") immediately after micronization can reach 15–20% w/w. This disrupted crystal tends to recrystallize back, usually into the original crystal form. The noncrystalline regions may create variability in dissolution, formulation processability, polymorphic contamination, etc. In addition, the inevitable gradual recrystallization with time may generate inconsistencies in "amorphous content" between lots and uncertainty about what is the "amorphous content" in the API used for formulation. The rate of recrystallization is mainly a function of humidity and temperature [82], hence, a function of packaging conditions. Amorphous content should be monitored in micronized APIs from the time of micronization until the time of formulation.

11.4.7.3 Solid-state Stability upon Storage

Recrystallization of the "amorphous" fraction has serious implications during storage. Storage in large bags causes the powder to be compacted under its own weight; and when the bag is filled almost to the end, water or residual solvent vapor begins circulating in the void space creating vapor pressure and migrates between the particles. This vapor migration phenomenon may favor recrystallization of the amorphous part. On top of that, the recrystallization may end up causing a gradual increase in particle size due to deliquescence and efflorescence phenomena of adjacent closely packed particles [87].

Fluctuations in storage conditions and opening and closing of the same bag numerous times might affect the rate of recrystallization. This problem is very common in products for inhalation because they are very heavily micronized. Monitoring the amorphous content or eliminating it by controlled recrystallization could decrease the extent of this problem. For practically insoluble crystalline APIs, uncontrolled amorphous content could compromise BE. It is good practice to measure the amorphous content and particle size caused by recrystallization right before incorporating the API into the formulation. A final note: it should be considered that the APIs in the RLD drug products may also contain unknown levels of amorphous material, which is challenging to measure in the formulated finished drug product.

11.4.8 Impact on Formulation

The challenges to the generic industry when formulating with a metastable form are mainly related to physical instability and risk of failing to demonstrate BE. The risk is most prevalent in BCS class 2 and 4 drug products and increases as the physical properties of the generic API diverge from those of the RLD API.

Differences in solubility and dissolution are the main cause of potential differences in bioavailability of a drug product. We cannot stress enough that the generic metastable form is frequently less thermodynamically stable, hence more soluble than the stable form in the RLD product. The gaps between solubility depend on how large the difference in thermodynamic stability is: the larger the difference in thermodynamic stability, the larger the difference in solubility will be. The bigger the gap, the bigger the risk of failing to demonstrate BE. Another feature that could enlarge this gap is the difference in morphologies. This is relevant in nonmicronized APIs. The gap between the well-formed crystals of an RLD API and the poor morphology crystals (frequently clusters of tiny crystals) found in a metastable API is one of the main responsible factors for hyper-bioavailability with respect to the RLD.

Other challenges in formulation development are related to processability. It is well known that crystal form transformations in formulations can lead to instability in physicochemical, biopharmaceutical, and processing properties of products. Crystal form changes can occur at various stages of the formulation process: wet granulation, drying, milling, compression, coating, and storage. Thus, it is important to determine whether phase transformations occur, and, if so, what factors influence them, (especially in BCS class 2 and 4 products). In recent years, the use of dry formulations in generic industry has become more common. Besides saving precious manufacturing facility time and being more energy efficient, dry processing in the formulation process also reduces the risk of crystal form transformations during manufacture and storage. However, it brings new challenges related to manufacturability, since dry compression formulations are more sensitive to the poor flow of components. Poor morphology and low powder density frequently encountered in metastable forms are two physical parameters that may affect manufacturability, resulting in poor flowability and compaction.

11.4.9 Summary of Timelines for Solid-state Activity

A schematic summary of the solid-state analyses needed for development of an alternative crystal form is exemplified in Figure 11.2.

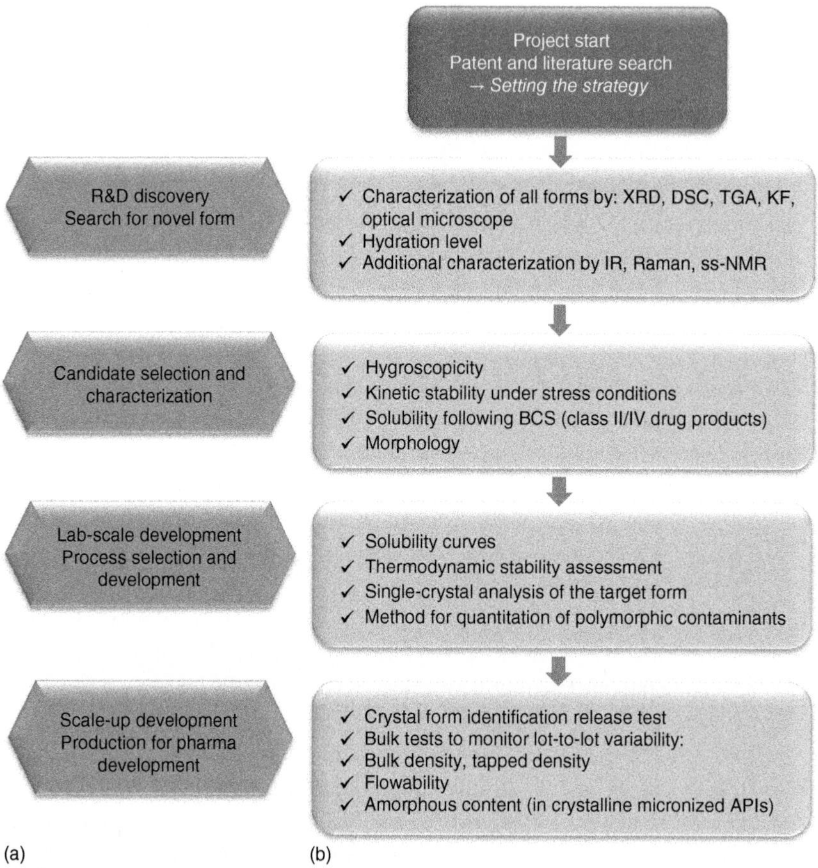

Figure 11.2 Schematic summary of solid-state characterization activities for a development of an alternative crystal form. (a) Phases of development, (b) analytical and characterization activities.

11.4.10 Intellectual Property (IP) Strategies and Activities

Due to the fact that polymorphic form is a distinct characteristic of the API and is easy to identify using the proper analytical techniques, polymorphism of drugs has been used for some decades as a tool to create valuable IP by either innovative companies who seek to delay generic entry to the market and by generic companies who seek to capture increased market share.

For a generic company, the IP activities are performed in two directions:

(1) Activities to ensure FTO
(2) Activities to issue patents on proprietary crystal forms (IP creation)

A generic company that seeks differentiation among the generic players may combine those activities by turning the challenge presented by innovator crystal form patents into an opportunity to create own IP as an obstacle to other generic competitors.

The crystal form patents of the innovators may limit the freedom of generic companies, but not eliminate it, since they will always try to protect the most stable forms and hydrates, but not always cover all possible forms. As a consequence, a generic company may invest efforts in extensive crystal form screening to discover heretofore unknown metastable crystal forms and protect them. This situation presents to an ambitious generic company with strong solid-state capabilities an opportunity to create their own IP and to use it as a tool to create a competitive advantage among the generic API manufacturers and protecting as many alternative crystal forms as possible.

By this, the IP domain becomes a front of generic competition. The IP activity around crystal forms of the generic companies, in fact, has expanded significantly in the last couple of decades and recently has been intensified by the increasing IP activity of companies from the Far East, especially China. The proliferation of generic patents demonstrates that alternative forms can be still discovered alongside those found by the innovator.

The competitive advantage of a generic polymorph patent depends on:

(a) quality,
(b) number of polymorphs covered,
(c) time of filing,
(d) financial capabilities to enforce it.

Quality of the patent refers to the clarity and reliability of the data presented. From reading the generic patents it is not always easy to understand which of the crystal forms disclosed are sufficiently stable, manufacturable, and bioequivalent to the RLD form. As a matter of fact, some of the disclosed forms may not be suitable for pharmaceutical use at all, being mixtures, unstable or irreproducible. Some patents fail to disclose novel forms and seek protection for already known forms. So generic crystal form patents need to be scrutinized with criticism and caution.

The extent of polymorph coverage at the time of patent filing clearly is critical in gaining a competitive edge because the more forms that are protected, the greater the competitive advantage. Considering the heavy competition, and that today several companies invest resources in screening for new, useful crystal forms, more or less simultaneously, the biggest advantage is certainly the priority date.

Some of the more aggressive generic players will try to anticipate patent filing with respect to the innovator's polymorph (and other) patents, understanding that filing time is the key to creating a competitive advantage. When successful, this looks like a brilliant strategy, but remember that this strategy presents some disadvantages as well, particularly related to the high cost and risk undertaken by trying to anticipate development.

11.5 Success Factors

Increasingly, innovator companies are using solid-state development strategies to bolster their IP portfolios, to increase the effective life of their products and to challenge generic manufacturers technically and financially. Therefore, it is not uncommon to see generics turn to unprotected, sometimes novel polymorphs in order to avoid these innovator patents. The regulatory authorities accept that a different crystal form, including cocrystals, is still the same molecule and as long as a generic applicant can demonstrate all aspects of essential similarity, and can routinely and robustly manufacture a commercially stable finished product, meeting all physical and chemical criteria, they are happy to grant MA to the generic.

11.5.1 Successful Biostudy

A different crystalline form may present a formulator with numerous challenges, the most critical often being demonstration of BE. Different forms often exhibit different rates of solubility and even the extent of solubility may differ. Solubility is often rate limiting for the release of an API from the dosage form, which in turn may impact the rate of absorption of the active moiety in vivo. It becomes the formulator's role to design the dosage form in such a way as to overcome these differences and achieve a product that, although employing a different polymorph, is still bioequivalent to the reference product. Various tools are at the formulator's disposal to modify the rate and even extent of solubility of an API. Common efforts would include modifying particle size of the solid API to increase or decrease the rate of solubility. In those relatively few instances where a more soluble material is required, milling, micronization, and nanonization have been successfully employed to decrease particle size and thus increase surface area in order to enhance the rate of solubility of a powder solid. It can be a lot more complicated when decreased solubility is the desired outcome. Crystalizing larger particles may slow the rate of dissolution, but this would require a very well-designed and controlled crystallization and milling process and would not be always feasible with certain metastable forms, which are isolated only by fast crystallization. Different excipients, such as surfactants in low concentrations, may also help to increase both rate and extent of dissolution while certain granulation techniques may serve to slow the rate of release of the API from the formulation.

When in vitro methods of dissolution, be they standard pharmacopeial methods or exotic methodologies invented by the analytical development team, can predict or at least relate to in vivo absorption kinetics, the development team has a powerful tool to accelerate development. While many such in vitro–in vivo correlations (IVIVC) exist, a formulator is often happy to settle for a less predictive in vitro–in vivo relationship (IVIVR) rather than have no tool at all as a guide. Pilot biostudies, that is, quicker, cheaper, but statistically underpowered studies are required to show the mathematical relationship and if none can be found, may be the only tool available to the formulator as a guide to success.

While achieving and demonstrating BE is the most critical challenge faced by formulation scientists when incorporating a different polymorph into a finished dosage form, it is not the only challenge. Many physical characteristics, such as shape of the crystal or its hardness, may impact on the flowability or millability of a powder rendering it difficult to develop a commercially viable high-speed manufacturing process. The new particle size or surface area of an API may render it to be less flowable or compressible, or to garner a static charge, making the commercial manufacture impractical without further formulation, granulation, or processing changes. The need to maintain a particular hydrate or solvate form may limit the ability to use water or other processing solvents in the manufacturing process, even if the liquid is ultimately removed by drying, as a change in form may occur during processing. These limitations on granulation and coating processes can render the formulation process more expensive or totally unfeasible. Even the inherent moisture found in a capsule shell could be sufficient to impact the hydrate form of an API. Different forms of a crystal may be only quasi-stable and over time convert to another polymorph of different solubility to that in the initial formulation [88–90]. This may result in a drug product that was bioequivalent at time zero to slow down or speed up upon storage, limiting the shelf-life of the formulation.

11.5.2 Successful Launch

Launch on time is the final critical success factor of any new project effort. After all of the time, effort and expense to achieve a MA on time, failure to convert this into an on-time launch would be a tragic end to an otherwise successful project. Actual project success can only be measured by realization of revenue from the first day possible. Any delay would allow competitors to gain market share that would be challenging to regain later on and may result in increased price discounts in order to claim lost market share. The launch stage of a project can be very risky. Even projects using the same polymorphic form as the innovator have been known to run into difficulties. For example, are the original suppliers of KSMs and advanced intermediates used during development still in business, still offering the product, and are they able to provide sufficient quantities to satisfy scale-up, validation, and launch quantities? If any of these suppliers, five, seven, or even more years after API R&D was completed, requalifying new suppliers can consume considerable time and put launch on-time at risk. API with special, often metastable characteristics presents even more fundamental risks and commercial-scale manufacture may not succeed

100% the very first time. Lab and pilot-scale parameters may not be easily scalable to full commercial batch size resulting in decreased yield, or, worse, the failure to achieve the desired solid-state characteristics. Even if the scale-up effort is mostly successful, the process may not meet all process validation parameters the first time round. Thus, the lead time to obtain API may be uncertain, even if manufactured internally, with all resources fully under internal management control. Similar concerns exist with formulation batch manufacture.

Allowing extra time for launch activities may be the only way to address these issues. This could take the form of performing partial or full commercial-scale manufacture and validation of API and/or formulation batches as the last stage of development. This may be the lowest risk option, but it may also be the most expensive since those batches may not retain sufficient shelf-life at launch to be commercially acceptable to customers. A slightly higher risk option, but likely more cost-effective is to allow ample time for prelaunch to perform scale-up and validation activities, but not so far in advance as to compromise the useful remaining shelf-life. This is especially true of APIs that often have an effective shelf-life of five years.

Successful launch should not be limited to the first three validation batches. If market demand is high compared to batch size, the goal of a successful launch program should be to be able to supply all market demands at launch and then again to be able to resupply the market once the launch stocks have been depleted. Going out of stock immediately after launch costs more than sales. The cost in reputation as an unreliable supplier can very quickly chase your customers into the waiting arms of your competition and regaining that reputation is an uphill battle. Successful launch includes successful process validation, continued manufacture of successful batches, satisfying market demands, and being able to restock the market as initial launch supplies dwindle.

11.5.3 Generic Commercial Success

The fact is that with all of the innovator polymorph patents that are written and granted, in the final analysis, these patents have not always succeeded in delaying generic product launch dates. For example, generic companies may be able to demonstrate that a patented polymorph was prior art, which could lead to those patents to be found invalid. In other cases, the use of alternative forms, as discussed in depth in this chapter, has been extremely successful in bringing generic versions to market, saving health systems and patients billions in prescription costs. In the war between the brand and generic companies, both sides have the right to protect their products as robustly as possible and solid-state form patents are major tools in the IP portfolio. The race for polymorphs has cost both generic and brand companies enormous amounts of money in research and legal costs and this is not likely to change anytime soon. Consider that even a "medium-sized" product in the US market could easily be worth upward of a million dollars in sales a day. With economics like these, a patent holder would not need to keep the competition at bay for very long to justify developing and defending a very robust stable of patents.

List of Abbreviations

ANDA	abbreviated new drug application (FDA)
API	active pharmaceutical ingredient (used as synonym for DS)
BCS	biopharmaceutics classification system
BE	bioequivalence
BLAs	biologics license applications
CBER	Center for Biologics Evaluation and Research (FDA)
CBG	College ter Beoordeling van Geneesmiddelen (The Netherlands)
CDER	Center for Drug Evaluation and Research (FDA)
DCP	decentralized procedure (EU)
DMF	drug master file
DSC	differential scanning calorimetry
EU	European Union
EMA	European Medicines Agency
eMC	Electronic Medicines Compendium (UK)
EPARs	European Public Assessment Reports
FDA	Food and Drug Administration (US)
FTO	freedom to operate
Gx	generics
GSK	Glaxo SmithKline
IP	intellectual property
IR	infrared spectroscopy
IRR	internal rate of return
IVIVC	in vitro–in vivo correlation
IVIVR	in vitro–in vivo relationship
JP	Japan
kg	kilogram
KSM	key starting material
MA	marketing authorization
MAA	marketing authorization application
MEB	medicines evaluation board (English form of CBG)
MRI	mutual recognition index (EU)
MRP	mutual recognition procedure (EU)
MST	materials science tetrahedron
NCE	new chemical entity
NDA	new drug application (FDA)
NDF	new dosage form
NIH	National Institutes of Health (US)
NMR	nuclear magnetic resonance

NPV	net present value
OB	Orange Book (FDA)
ODE	orphan drug exclusivity
PCT	Patent Cooperation Treaty
PIL	patient information leaflet
PIV	paragraph IV patent certification
QTPP	quality target product profile
R&D	research and development
RLD	reference listed drug (FDA)
RoI	return on investment
SBA	summary basis of approval (FDA)
SmPC	summary of product characteristics (EU)
TGA	thermogravimetric analysis
US	United States of America
WO	World Intellectual Property Organization (abbr of WIPO)
XRD	X-ray diffraction
XRPD	X-ray powder diffraction

References

1 Brown, A. (2014). Novartis settles Gleevec as other patent fights continue. https://www.evaluate.com/vantage/articles/analysis/novartis-settles-gleevec-other-patent-fights-continue (accessed 12 September 2019).

2 Strawbridge, H. (2012). *Wallets Rejoice As Plavix Goes Generic*. Harvard Health Publishing. https://www.health.harvard.edu/blog/wallets-rejoice-as-plavix-goes-generic-201205214727 (accessed 12 September 2019).

3 Mikulic, M. (2019). Global pharmaceutical industry – statistics & facts. *Statista*. https://www.statista.com/topics/1764/global-pharmaceutical-industry/ (accessed 12 September 2019).

4 US Food and Drug Administration (US FDA). (2021) Orange Book: Approved Drug Products with Therapeutic Equivalence Evaluations. https://www.accessdata.fda.gov/scripts/Cder/ob/ (accessed 31 January 2021).

5 US FDA. (2020) Frequently Asked Questions on Patents and Exclusivity. https://www.fda.gov/drugs/development-approval-process-drugs/frequently-asked-questions-patents-and-exclusivity (accessed 31 January 2021).

6 US FDA. (2017) CDER Conversation: Patents and Exclusivities for Generic Drug Products. https://www.fda.gov/drugs/news-events-human-drugs/cder-conversation-patents-and-exclusivities-generic-drug-products (accessed 31 January 2021).

7 US FDA. (2021) Patent Certifications and Suitability Petitions. https://www.fda.gov/drugs/abbreviated-new-drug-application-anda/patent-certifications-and-suitability-petitions (accessed 31 January 2021).

8 U.S. Department of Health and Human Services, Food and Drug Administration, Center for Drug Evaluation and Research (CDER), Center for Biologics

Evaluation and Research (CBER). (2017). Guidance for industry 180-day exclusivity: questions and answers – draft guidance. https://www.fda.gov/media/102650/download (accessed 12 September 2019).

9 US Government Publishing Office (GPO). 21 U.S. Code § 355 – New drugs, https://www.govinfo.gov/content/pkg/USCODE-2010-title21/html/USCODE-2010-title21-chap9-subchapV-partA-sec355.htm (accessed 12 September 2019).

10 European Commission Health and Food Safety Directorate-General. (2019) Notice to applicants, Volume 2A: Procedures for marketing authorization, July 2019, section 6. Data exclusivity and market protection. https://ec.europa.eu/health/sites/health/files/files/eudralex/vol-2/vol2a_chap1_en.pdf (accessed 12 September 2019).

11 SIKs & Co. https://www.siks.jp/contents/case/images/Data_exclusivity.pdf (accessed 12 September 2019).

12 Government of Canada Guidance document: data protection under C.08.004.1 of the food and drug regulations. https://www.canada.ca/en/health-canada/services/drugs-health-products/drug-products/applications-submissions/guidance-documents/guidance-document-data-protection-under-08-004-1-food-drug-regulations.html (Revised 2017) (12 September 2019)

13 Chen, C. (2015). Patents vs. data exclusivity in pharmaceuticals. https://www.lexology.com/library/detail.aspx?g=5fe621cd-4372-49af-933f-714b3e8e7a02 (accessed 12 September 2019).

14 European Commission Health and Food Safety Directorate-General. (2019) Notice to applicants, Volume 2A: Procedures for marketing authorization, July 2019, section 5. Application types. https://ec.europa.eu/health/sites/health/files/files/eudralex/vol-2/vol2a_chap1_en.pdf (accessed 12 September 2019).

15 US Food and Drug Administration. (2021) Orange book preface, Food and Drug Administration, Center for Drug Evaluation and Research, approved drug products with therapeutic equivalence evaluations, preface to the 39th edition. https://www.fda.gov/drugs/development-approval-process-drugs/orange-book-preface (accessed 1 February 2021)

16 US Congress. (1984) Section 505(b)(2) of the Federal Food, Drug, and Cosmetic Act.

17 World Intellectual Property Organization, abbreviated as WIPO and further abbreviated as WO, https://www.wipo.int/pct/en/ (accessed 1 February 2021).

18 Carnes, D. What is a WO patent? https://www.wipo.int/pct/en (Revised 2021), Author Worold Intellectual Property Organization, (accessed 1 February 2021)

19 US FDA. (2020) Product-Specific Guidances for Generic Drug Development. https://www.accessdata.fda.gov/scripts/cder/psg/index.cfm (accessed 31 January 2021).

20 US FDA. (2020) Purple Book: Lists of Licensed Biological Products with Reference Product Exclusivity and Biosimilarity or Interchangeability Evaluations https://www.fda.gov/drugs/therapeutic-biologics-applications-bla/purple-book-lists-licensed-biological-products-reference-product-exclusivity-and-biosimilarity-or (accessed 31 January 2021).

21 US National Institutes of Health, US National Library of Medicine. (2020) Dailymed. https://dailymed.nlm.nih.gov/dailymed/ (accessed 31 January 2021).

22 European Medicines Agency. (2021) Home page. https://www.ema.europa.eu/en (accessed 31 January 2021).

23 European Medicines Agency. (2020) Medicines. https://www.ema.europa.eu/en/medicines (accessed 31 January 2021).

24 Heads of Medicines Agency (EU). (2021) MRI Product Index. https://www.hma.eu/mriproductindex.html (accessed 31 January 2021).

25 Heads of Medicines Agency (EU). (2021). About HMA. https://www.hma.eu/ (accessed 1 February 2021).

26 Datapharm Ltd. (2021) Electronic medicines compendium (UK). https://www.medicines.org.uk/emc/ (accessed 1 February 2021).

27 Medicines Evaluation Board (NL). (2021) Home. https://english.cbg-meb.nl/ (accessed 1 February 2021).

28 Pharmacompass. (2021) Home. https://www.pharmacompass.com/ (accessed 1 February 2021).

29 Hilfiker, R. (ed.) (2006). *Polymorphism in the Pharmaceutical Industry*. Weinheim: Willey-VCH.

30 Saurabh, G. and Kaushal, C. (2011). Pharmaceutical solid polymorphism in abbreviated new drug application (ANDA) – a regulatory perspective. *Journal of Chemical and Pharmaceutical Research* 3 (3): 6–17.

31 Yu, L.X., Furness, M.S., Raw, A. et al. (2003). Scientific considerations of pharmaceutical solid polymorphism in abbreviated new drug applications. *Pharmaceutical Research* 20 (4): 531–536.

32 George, J.Q. (2006). Selection of the drug form in exploratory development. In: *Fundamentals of Early Clinical Drug Development* (eds. A.F. Abdel-Magid and S. Caron). Weinheim: Wiley-VCH.

33 Ende, D.J.A. and Rose, P.R. (2006). Strategies to achieve particle size of active pharmaceutical ingredients. In: *Fundamentals of Early Clinical Drug Development* (eds. A.F. Abdel-Magid and S. Caron). Weinheim: Wiley-VCH.

34 Snider, D.A., Addicks, W., and Owens, W. (2004). Polymorphism in generic drug product development. *Advanced Drug Delivery Reviews* 56: 391–395.

35 Raza, K., Kumar, P., Ratan, S. et al. (2014). Polymorphism: the phenomenon affecting the performance of drugs. *SOJ Pharmacy & Pharmaceutical Sciences* 1 (2): 1–10.

36 Tiwary, A.K. (2001). Modification of crystal habit and its role in dosage form performance drug development and industrial pharmacy. *Drug Development and Industrial Pharmacy* 27 (7): 699–709.

37 Modi, S.R., Dantuluri, A.K.R., Puri, V. et al. (2013). Impact of crystal habit on biopharmaceutical performance of celecoxib. *Crystal Growth and Design* 13 (7): 2824–2832.

38 Longuemard, P., Jbilou, M., Guyot-Hermann, A.-M., and Guyot, J.-C. (1998). Ground and native crystals: comparison of compression capacity and dissolution rate. *International Journal of Pharmaceutics* 170 (1): 50–61.

39 Wei, H., Dalton, C., Di Maso, M. et al. (2008). Physicochemical characterization of five glyburide powders: a BCS based approach to predict oral absorption. *European Journal of pharmaceutics and Biopharmaceutics* 69 (3): 1046–1056.

40 Brittain, H.G. (1999). *Polymorphism in Pharmaceutical Solids*. New York: Marcel Dekker Inc.

41 Censi, R. and Di Martino, P. (2015). Polymorph impact on the bioavailability and stability of poorly soluble drugs. *Molecules* 20 (10): 18759–18776.

42 Kawabata, Y., Wada, K., Nakatani, M. et al. (2011). Formulation design for poorly water-soluble drugs based on biopharmaceutics classification system: basic approaches and practical applications. *International Journal of Pharmaceutics* 420: 1–10.

43 Sun, Z., Yao, D., Yuan, G. et al. (2014). Human pharmacokinetic variations of different nitrendipine crystalline polymorphs. *Latin American Journal of Pharmacy* 33 (7): 1182–1187.

44 Mukuta, T., Lee, A.Y., Kawakami, T., and Myerson, A.S. (2005). Influence of impurities on the solution-mediated phase transformation of an active pharmaceutical ingredient. *Crystal Growth & Design* 5 (4): 1429–1436.

45 Gong, Y., Collman, B.M., Mehrens, S.M. et al. (2008). Stable-form screening: overcoming trace impurities that inhibit solution-mediated phase transformation to the table polymorph of sulfamerazine. *Journal of Pharmaceutical Sciences* 97 (6): 2130–2144.

46 US FDA (CDER) (2018). Regulatory classification of pharmaceutical co-crystals – guidance for industry. https://www.fda.gov/media/81824/download (accessed 6 October 2019).

47 Miroshnyk, I., Mirza, S., and Sandler, N. (2009). Pharmaceutical co-crystals-an opportunity for drug product enhancement. *Expert Opinion on Drug Delivery* 6 (4): 333–341.

48 Weyna, D.R., Cheney, M.L., Shan, N. et al. (2012). *Improving solubility and pharmacokinetics of meloxicam via multiple-component crystal formation. Molecular Pharmaceutics* 9 (7): 2094–2102.

49 Newman, A. (2015). *Pharmaceutical Amorphous Solid Dispersions*. Hoboken, NJ: Wiley.

50 Aaltonen, J., Allesø, M., Mirza, S. et al. (2009). Solid form screening – a review. *European Journal of Pharmaceutics and Biopharmaceutics* 71: 23–37.

51 Morissette, S.L., Almarsson, Ö., Peterson, M.L. et al. (2004). High-throughput crystallization: polymorphs, salts, co-crystals and solvates of pharmaceutical solids. *Advanced Drug Delivery Reviews* 56: 275–300.

52 Alvarez, A.J., Singh, A., and Myerson, A.S. (2009). Polymorph screening: comparing a semi-automated approach with a high throughput method. *Crystal Growth and Design* 9: 4181–4188.

53 Newman, A. (2013). Specialized solid form screening techniques. *Organic Process Research and Development* 17: 457–471.

54 Allesø, M., Tian, F., Cornett, C., and Rantanen, J. (2010). Towards effective solid form screening. *Journal of Pharmaceutical Sciences* 99 (9): 3711–3718.

55 Hilfiker, R., De Paul, S.M., and Szelegiewicz, M. (2006). *Approaches to Polymorph Screening in Polymorphism in the Pharmaceutical Industry*. Weinheim: Wiley-VCH.

56 Ismail, S.Z., Anderton, C.L., Copley, R.C.B. et al. (2013). Evaluating a crystal energy landscape in the context of industrial polymorph screening. *Crystal Growth & Design* 13 (6): 2396–2406.

57 Lee, A.Y., Erdemir, E., and Myerson, A.S. (2011). Crystal polymorphism in chemical process development. *Annual Review of Chemical and Biomolecular Engineering* 2: 259–280.

58 Blagden, N., Davey, R.J., Rowe, R., and Roberts, R. (1998). Disappearing polymorphs and the role of reaction by-products: the case of sulphathiazole. *International Journal of Pharmaceutics* 172: 169–177.

59 Desikan, S., Parsons, R.L. Jr., Davis, W.P. et al. (2005). Process development challenges to accommodate a late-appearing stable polymorph: a case study on the polymorphism and crystallization of a fast-track drug development compound. *Organic Process Research & Development* 9: 933–942.

60 Prashad, M., Sutton, P., Wu, R. et al. (2010). Process research and development of a MTP inhibitor: another case of disappearing polymorphs upon scale-up. *Organic Process Research & Development* 14: 878–882.

61 Chemburkar, S.R., Bauer, J., Deming, K. et al. (2000). Dealing with the impact of ritonavir polymorphs on the late stages of bulk drug process development. *Organic Process Research & Development* 4: 413–417.

62 Bauer, J., Spanton, S., Henry, R. et al. (2001). Ritonavir: an extraordinary example of conformational polymorphism. *Pharmaceutical Research* 18 (6): 859–866.

63 Chekal, B.P., Campeta, A.M., Abramov, Y.A. et al. (2009). The challenges of developing an API crystallization process for a complex polymorphic and highly solvating system. Part I. *Organic Process Research & Development* 13 (6): 1327–1337.

64 Campeta, A.M., Chekal, B.P., Abramov, Y.A. et al. (2010). Development of a targeted polymorph screening approach for a complex polymorphic and highly solvating API. *Journal of Pharmaceutical Sciences* 99 (9): 3874–3886.

65 Pudipeddi, M. and Serajuddin, A.T.M. (2005). Trends in solubility of polymorphs. *Journal of Pharmaceutical Sciences* 94 (5): 929–939.

66 U.S. Department of Health and Human Services, Food and Drug Administration, Center for Drug Evaluation and Research (CDER) (2017). Waiver of in vivo bioavailability and bioequivalence studies for immediate-release solid oral dosage forms based on a biopharmaceutics classification system – guidance for industry.

67 Kawakami, K., Miyoshi, K., and Ida, Y. (2005). Impact on the amount of excess solids on apparent solubility. *Pharmaceutical Research* 22 (9): 1537–1543.

68 Lee, E.H. (2014). A practical guide to pharmaceutical polymorph screening & selection. *Asian Journal of Pharmaceutical Sciences* 9 (4): 163–175.

69 Threlfall, T.L. (2009). Turning DSC charts of polymorphs into phase diagrams: a tutorial paper. *Organic Process Research & Development* 13: 1224–1230.

70 Lohani, S. and Grant, D.J.W. (2006). Thermodynamics of polymorphs. In: *Polymorphism in the Pharmaceutical Industry* (ed. R. Hilfiker). Weinheim: Wiley-VCH.

71 Zhang, T., Wang, L., Bao, Y. et al. (2018). Confirmation of more stable polymorphic form of etoricoxib at room temperature. *Journal of Pharmaceutical Sciences* 107 (7): 1903–1910.

72 Braun, D.E., Gelbrich, T., Kahlenberg, V. et al. (2009). Conformational polymorphism in aripiprazole: preparation, stability and structure of five modifications. *Journal of Pharmaceutical Sciences* 98 (6) p. 2010, 2026.

73 Burger, A., Henck, J.-O., Hetz, S. et al. (2000). Energy/temperature diagram and compression behavior of the polymorphs of D-mannitol. *Journal of Pharmaceutical Sciences* 89 (4): 457–468.

74 Threlfall, T. (2000). Crystallisation of polymorphs: thermodynamic insight into the role of solvent. *Organic Process Research & Development* 4: 384–390.

75 Barrett, P. and Glennon, B. (2002). Characterizing the metastable zone width and solubility curve using lasentec FBRM and PVM. *Chemical Engineering Research and Design* 80 (7): 799–805.

76 Kim, K.-J. and Doherty, M.F. (2015). Crystallization of selective polymorph using relationship between supersaturation and solubility. *AIChE Journal* 61 (4): 1372–1379.

77 Kline, B.J., Saenz, J., Stankovic, N., and Mitchell, M.B. (2006). Polymorph and particle size control of PPAR compounds PF00287586 and AG035029. *Organic Process Research & Development* 10: 203–211.

78 Vippagunta, R.R., Pan, C., Vakil, R. et al. (2009). Application of surface area measurement for identifying the source of batchto-batch variation in processability. *Pharmaceutical Development and Technology* 14 (5): 492–498.

79 Sheokand, S., Modi, S.R., and Bansal, A.K. (2014). Dynamic vapour sorption as a tool for characterization and quantification of amorphous content in predominantly crystalline materials. *Journal of Pharmaceutical Sciences* 103: 3364–3376.

80 Sheokand, S., Modi, S.R., and Bansal, A.K. (2016). Quantification of low levels of amorphous content in crystalline celecoxib using dynamic vapour sorption (DVS). *European Journal of Pharmaceutics and Biopharmaceutics* 102: 77–86.

81 Saini, M.K. (2015). DSC method: determination of amorphous fraction in solid dosage and fragility. *AIP Conference Proceedings* 1665: 070022-1–070012-3.

82 Petzoldt, C., Bley, O., Byard, S.J. et al. (2014). An example of how to handle amorphous fractions in API during early pharmaceutical development: SAR114137 – a successful approach. *European Journal of Pharmaceutics and Biopharmaceutics* 86: 337–350.

83 Forcino, R.G., Williams, G.R., Brum, J., and Vogt, F.G. (2013). The study of micronization induced disorder in active pharmaceutical ingredients. *American Pharmaceutical Review* 16 (6): 40–46.

84 Brodka-Pfeiffer, K., Langguthb, P., Grass, P., and Hausler, H. (2003). Influence of mechanical activation on the physical stability of salbutamol sulphate. *European Journal of Pharmaceutics and Biopharmaceutics* 56 (3): 393–400.

85 Planinsek, O., Zadnik, J., Kunaver, M. et al. (2010). Structural evolution of indomethacin particles upon milling: time-resolved quantification and localization of disordered structure studied by IGC and DSC. *Journal of Pharmaceutical Sciences* 99 (4): 1968–1981.

86 Sun, C.C. (2009). Materials science tetrahedron – a useful tool for pharmaceutical research and development. *Journal of Pharmaceutical Sciences* 98 (5): 1671–1687.

87 Mauer, L.J. and Taylor, L.S. (2010). Deliquescence of pharmaceutical systems. *Pharmaceutical Development and Technology* 15 (6): 582–594.

88 Purohit, H.S., Trasi, N.S., Sun, D.D. et al. (2018). Investigating the impact of drug crystallinity in amorphous tacrolimus capsules on pharmacokinetics and bioequivalence using discriminatory in vitro dissolution testing and physiologically based pharmacokinetic modeling and simulation. *Journal of Pharmaceutical Sciences* 107 (5): 1330–1341.

89 Zhang, X., Wen, H., Fan, J. et al. (2017). Integrating in vitro, modeling, and in vivo approaches to investigate warfarin bioequivalence. *CPT: Pharmacometrics & Systems Pharmacology* 6 (8): 523–531.

90 Rietveld, I.B. and Ceolin, R. (2015). Rotigotine: unexpected polymorphism with predictable overall monotropic behavior. *Journal of Pharmaceutical Sciences* 104 (12): 4117–4122.

Index

Solid State Development and Processing of Pharmaceutical Molecules: Salts, Cocrystals, and Polymorphism, First Edition. Edited by Michael Gruss.

e

f

q